THE GENUS SEPTOBASIDIUM

The University of North Carolina Press, Chapel Hill, N. C.; The Baker and Taylor Company, New York; Oxford University Press, London; Maruzen-Kabushiki-Kaisha, Tokyo; Edward Evans & Sons, Ltd., Shanghai; D. B. Centen's Wetenschappelijke Boekhandel, Amsterdam.

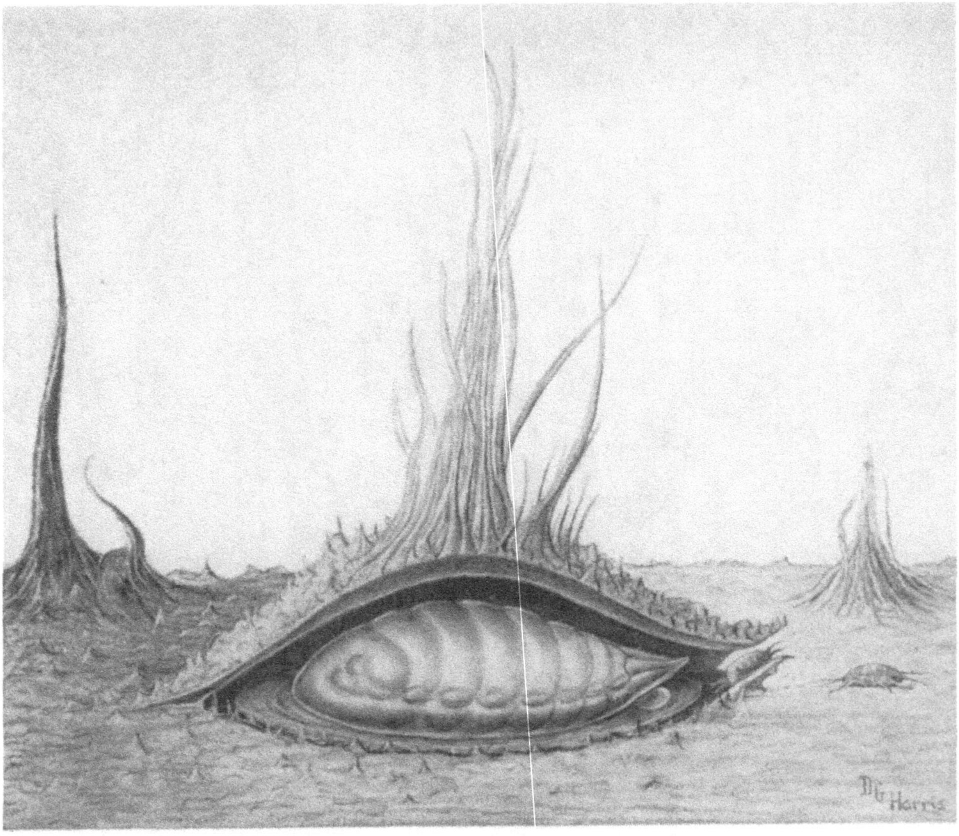

INSECT HOUSE OF *Septobasidium apiculatum* WITH HALF THE ROOF CUT AWAY EXPOSING A SCALE INSECT GIVING BIRTH TO YOUNG, ONE OF WHICH IS CRAWLING OUT THROUGH THE DOOR. × ABOUT 50.

THE GENUS SEPTOBASIDIUM

By

JOHN N. COUCH

Professor of Botany, University of North Carolina

With Frontispiece, 114 Plates,

and 60 Text-figures

CHAPEL HILL

THE UNIVERSITY OF NORTH CAROLINA PRESS

1938

To

WILLIAM CHAMBERS COKER

PREFACE

When this work was started in 1926 about seventy-five species of *Septobasidium* had been named, but few of these had been adequately described and illustrated. It soon became evident that before any real advance could be made in the taxonomy of the genus a careful comparative study of the species already named would have to be made. It is the chief purpose of the present work to assemble for the first time descriptions and illustrations of the type species. The accomplishment of this task would have been impossible without the generous coöperation of the curators of the various collections of type specimens. The following collections have been available. The Patouillard, von Höhnel, and Curtis collections at Harvard; the Berkeley and Curtis, and Berkeley and Broome herbaria and other collections at Kew; the Hennings and von Höhnel and other collections at the Museum of the Botanic Garden at Berlin; the Sydow Herbarium at Berlin; the Raciborski collection at Cracov, Poland; the collections in the New York Botanical Garden; the Burt and Missouri Botanical Garden collections at St. Louis; the Lloyd and other collections, United States Department of Agriculture Herbarium; the Schweinitz Herbarium, Philadelphia; the Coker collections at the University of North Carolina; and several specimens from the Montagne collection in Paris. Duplicate type material of their entire collection from Java has been sent by Boedijn and Steinmann. Mr. T. Petch, King's Lynn, England, has turned over to me duplicates of all his Ceylon material and has given valuable advice from time to time. Large collections of valuable material have been received from J. Rick, Sao Leopoldo, Brazil; A. S. Müller, Vicosa, Brazil; Miss A. M. Bottomley, Pretoria, South Africa; J. H. Simmonds, Brisbane, Australia; G. O. Ocfemia, Los Baños, Philippine Islands; T. Tanaka, Formosa, Japan; John Dearness of London, Ontario, Canada; H. Bourdot, Saint Priest-en-Murat, France; and numerous others, the senders being mentioned under the species.

In the United States Dr. Clair A. Brown, University of Louisiana, Baton Rouge; Professor Erdman West and Dr. George Weber of Gainesville, Florida; Dr. B. B. Higgins of Experiment, Georgia; Dr. George Armstrong of Clemson College, South Carolina; Dr. L. R. Hesler, University of Tennessee, Knoxville, Tennessee; Dr. J. J. Taubenhaus of College Station, Texas; Dr. H. S. Fawcett, University of California, Riverside, California; Dr. D. H. Linder and Dr. W. H. Weston of Harvard University; Dr. C. W. Dodge of Missouri Botanical Garden; Dr. John Stevenson and Dr. C. L. Shear of the United States Department of Agriculture, Washington, D. C.; Dr. Dow V. Baxter of the University of Michigan; Dr. L. O. Overholts of State College, Pennsylvania; Dr. G. W. Martin of Iowa State University; and Dr. F. A. Wolf of Duke University, Durham, N. C., have all contributed valuable material from their own collections. Other American collectors who may have been overlooked here are cited under the species.

One collecting trip of two months was made in Jamaica, B. W. I., and numerous trips have been made throughout the Southeastern States, the expenses of which have been largely met by grants from the General Education Board Research Fund. Another grant was received from the same fund to pay Dr. A. G. Lang for his help in making most of the photographs and still another grant to pay Mr. A. C. Mathews for help with the spraying experiments. For inking in the drawings I am particularly indebted to Miss Alma Holland and Mrs. Else R. Couch. I am also glad to express my gratitude to Miss Holland for her invaluable help in the final preparation of the manuscript and for assuming the major burden of proofreading. Dr. Harold Morrison, Bureau of Entomology, United States Department of Agriculture, has identified the scale insects.

TABLE OF CONTENTS

THE GENUS SEPTOBASIDIUM

I

THE FUNGUS-INSECT RELATIONSHIP

OBSERVATIONS ON SEPTOBASIDIUM BURTII

All the species of *Septobasidium* so far discovered are found on living plants in association with scale insects. Von Höhnel and Litschauer (1907) were the first to announce the presence of scale insects beneath the stroma of species of *Septobasidium*. Their observations were made on dried material but are nevertheless significant and since they have been incorrectly quoted in Engler and Prantl (Basidiomycetes, 2 ed. 1928) and Boedijn and Steinmann (1931) it is well to give their complete statement here:

"Ein Eindringen der Hyphen der Septobasidien in das vegetabilische Substrat findet nicht statt oder doch nur insoweit, als es zur Befestigung des Pilzes an das Substrat nötig ist. Die Septobasidien können daher nicht als eigentliche vegetabilische Schmarotzer betrachtet werden. In der Tat zeigte sich bei sämtlichen oben genannten Arten, daß sie auf Schildläusen leben, die man stets auf jenen Stellen des Substrates findet, welche vom Pilze bedeckt sind. Dies zeigte sich in allen untersuchten Fällen, sowohl bei den europäischen als bei den amerikanischen Exemplaren; ja zum Teil schienen diese Schildläuse einander so ähnlich, daß sie mindestens derselben Gattung angehören müssen. Dieses konstante Vorkommen von Schildläusen unter dem Thallus der *Septobasidium*-Arten, der Umstand ferner, den man bei mehreren Arten leicht konstatieren kann, daß der Pilz auf den Schildläusen wächst und mit ihnen fest verwachsen ist, läßt keinen Zweifel übrig, daß zwischen den Schildläusen und den *Septobasidium*-Arten irgend ein biologisches Verhältnis besteht, sei es, daß Parasitismus oder Saprophytismus vorhanden ist, sei es, daß ein komplizierteres symbiotisches Verhältnis herrscht. Wenn man in der Tat ein gut entwickeltes Exemplar von *Septobasidium crinitum* (Fr.) betrachtet, so kann man sich des Gedankens nicht erwehren, daß der schirmartig entwickelte Pilz für die darunter lebenden Schildläuse ein guter Schutz gegen starke Insolation, tierische Feinde etc. ist.

Der Gedanke, daß die Schildläuse nur zufällig, etwa Schutz suchend, unter den Pilz gelangen, muß zurückgewiesen werden, da wir bei der

Prüfung von mehreren Tausenden von Exemplaren von echten Corticieen niemals Schildläuse unter dem Thallus dieser gefunden haben."

From the above quotation it can be seen that while Von Höhnel and Litschauer recognized a close biological relationship between *Septobasidium* and scale insects they cautiously state that the relationship may be one of parasitism, saprophytism, or that a complicated symbiotic relationship may exist. Since their discovery in 1907, considerable attention has been paid to the fungus-insect relationship (Petch 1911, 1921, Burt 1916, Coker 1920, Snell 1922, Couch 1929, 1931, 1935, Boedijn and Steinmann 1930, 1931). It was not until 1929 and later in 1931 that the writer reported the results of a detailed study of the fungus-insect relationship in *S. Burtii* (formerly treated as *S. retiforme*). Since subsequent studies on several other species have indicated the correctness of the accounts given (1929, 1931) a somewhat detailed report of those researches will be presented here.

Life History and Structure of the Fungus

Septobasidium Burtii is very common in North and South Carolina on several species of oak, *Quercus palustris* (pin oak), *Q. phellos* (willow oak), and *Q. nigra* (water oak), also occurring in certain localities on cultivated pear, apple, and peach trees. It has been reported from several other southern states in Plant Disease Surveys. This species is particularly common around Chapel Hill, N. C., occurring abundantly on several oaks within less than one hundred yards of the Botanical Laboratory and hence very convenient for study. The fungus usually attains its most luxuriant growth on the under surface of the lower branches of trees, very seldom occurring on the main trunk. I have found it on limbs with a diameter of seven centimeters and on twigs only a year old.

The growth is entirely superficial as has been demonstrated by numerous free hand sections of fresh material, and by stained microtome sections of material imbedded in paraffin. Furthermore if a fresh specimen of *S. Burtii* is soaked in water for several minutes, it is possible, if the operation is carried out under water, to separate the fungus from the bark with dissecting needles, save for the connections between the parasitized insects. These insects are firmly imbedded in the fungal tissue with their sucking tubes extending into the bark.

Series of observations over a period of several years on a large number of living specimens have shown that the fungus growth of *S. Burtii* is perennial, the growing period occurring, as one would naturally expect, between April and November, the greater part of the growth taking place in May and June.

This species of *Septobasidium* reproduces normally only by the formation of spores. During the winter months vast numbers of probasidia

TEXT-FIGS. 1–3

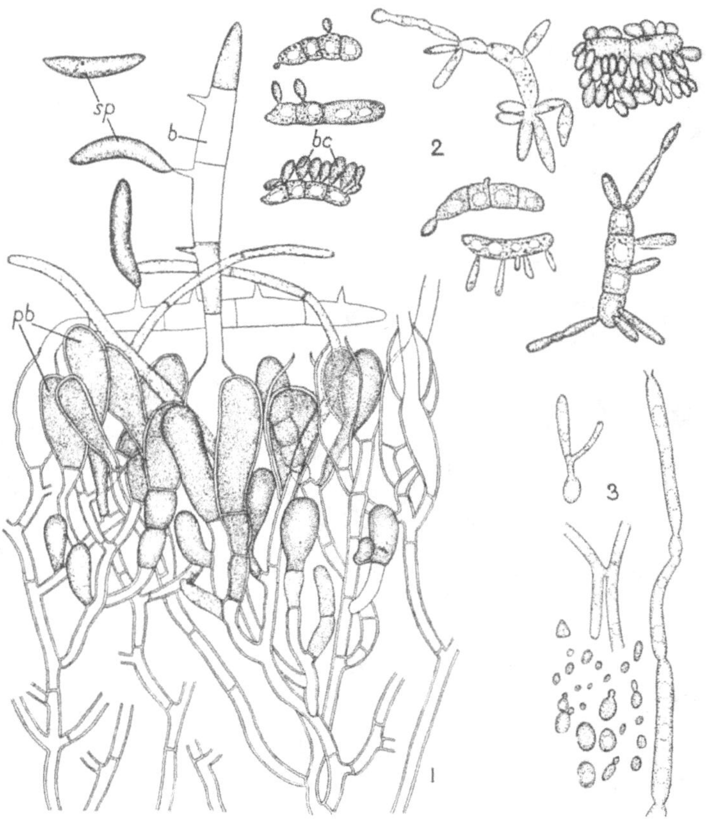

Fig 1.—Section of hymenium or fruiting surface of *S. Burtii*. *pb*, probasidia; *b*, basidium; *sp*. spores. × 640.

Fig. 2.—Budding spores of *S. Burtii* on agar. × 640.

Fig. 3.—Minute bud-cells and hyphae from spores after three weeks on agar. × 640.

are formed near the upper surface of the developing fruit body. When the warm, damp weather of spring sets in, the mature probasidia germinate, forming cylindrical, four-celled basidia, each cell of which usually bears one spore. It is possible at any time between April and August, if specimens are soaked in water and kept in a damp chamber for from 24 to 48 hours, to induce the formation of spores. In spring or early summer the spores may be induced by such artificial treatment to form

TEXT-FIG. 4

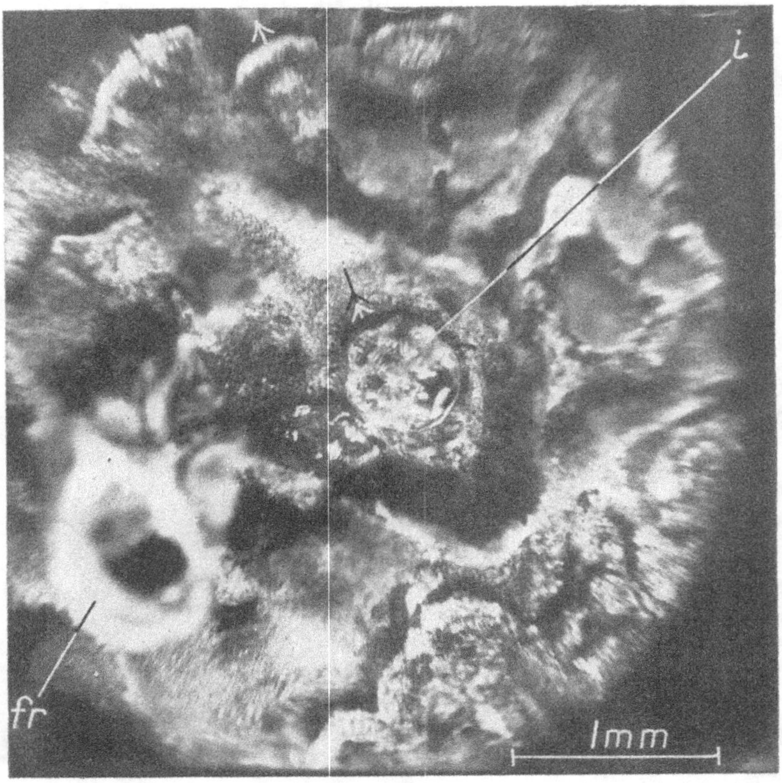

Photograph of living, adult, healthy insect (*i*) in fungal house. Insect lies flat against bark. Fungal roof (*fr*) over insect is shown out of focus. Note tunnel, indicated by arrow, connecting chamber occupied by insect with exterior.

in surprisingly great numbers and for a considerable period of time. In nature, however, the spores are formed only after rains while the fungus body is damp, no spores at all being formed during dry weather.

The fungus body of *S. Burtii* consists of oblong or circular resupinate patches up to several centimeters in diameter. Its structure is very peculiar and complicated. Each patch is composed of a few to many irregular, concentric rings of growth, a new ring being formed each year (Pl. 69). In section, the fungus is composed of top and bottom layers, the top layer supported by ridges. The ridges have mostly a radial direction and branch and anastomose in such a way that between the top and bottom layers there is a labyrinth of chambers and tunnels. Many of the chambers are in direct communication with the outside through natural openings in the top layer. Each opening is partially closed by a flaplike extension of the top layer. The function of this intricate system of chambers and tunnels immediately becomes apparent if the top layer of a fresh specimen is carefully dissected away from the supporting ridges and folded back. Here, during the late fall, winter, and early spring months, regardless of the weather conditions, one sees numerous living, plump, healthy scale insects (Text-fig. 4 and Pls. 1 and 69, fig. 2). Each chamber usually contains only one insect, though a particularly large chamber may contain two or sometimes three. The average chamber with its arched ceiling and concave side walls is well adapted to the shape of the insect's body and is considerably larger. The floor of the chamber is the bark, and since this species does not form a ventral scale, the insect lies directly against the bark, and is not touched anywhere by the fungal threads. The fungus, however, grows over and adheres so firmly to the insect's scale (Pl. 1, sc) that if the top layer is separated from the bottom layer, the scales remain fixed to the under surface of the top, while the living insects remain fastened to the bark.

Life History of the Scale Insect

The scale insect found beneath the fungus has been identified as *Aspidiotus osborni* New. and Ckll. Only once, out of a vast number of specimens of *Septobasidium Burtii* examined, have I found a different species of insect, *Chrysomphalus obscurus*, and I found only one specimen of this.

The changes through which *A. osborni* passes in the course of its development are very similar to those described by Comstock, for the sub-family Diaspinae, to which *Aspidiotus* belongs. The stages in the life-history of *A. osborni* are represented by Text-figs. 5–20. The young are born very slowly and remain motionless near the opening of the vagina a few minutes to an hour or more. Then they break out of their thin pellicle and after crawling for a few minutes, settle down.

Text-fig. 5

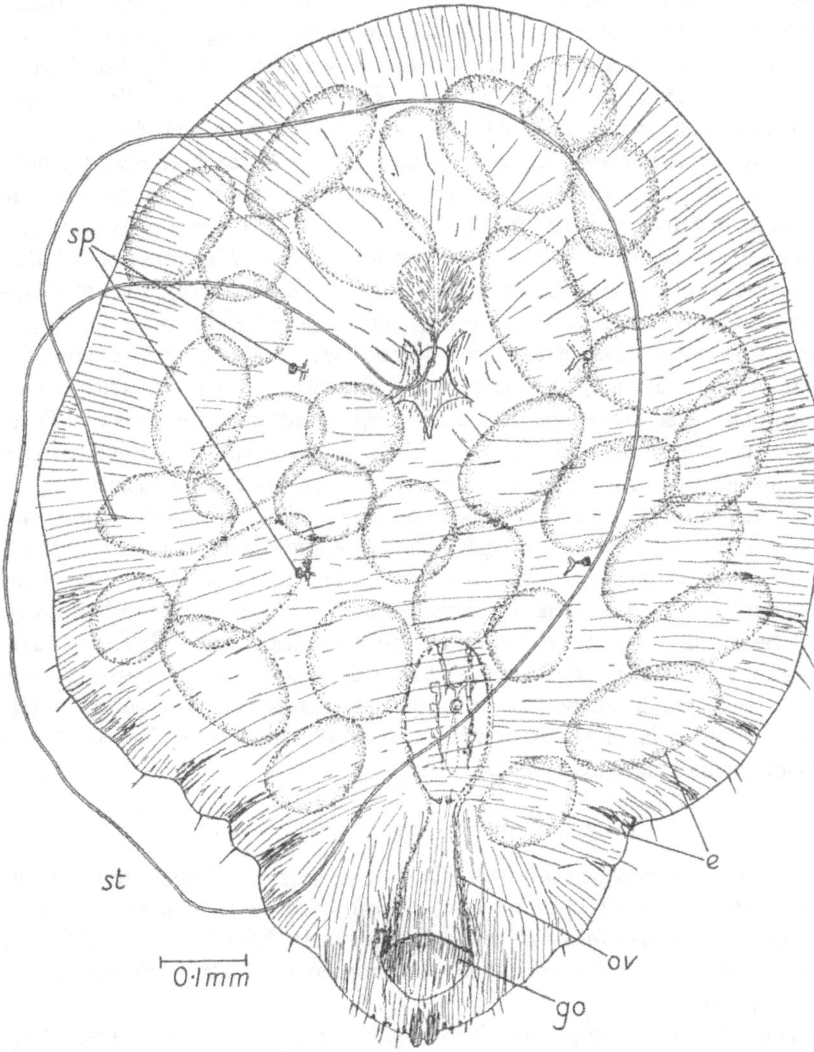

Adult of *A. osborni* giving birth to young. *go*, genital opening; *ov*, oviduct; *e*, eggs; *sp*, spiracles; *st*, sucking tube. × 110.

quickly piercing the bark with their sucking apparatus. Within a few hours the insect begins to excrete fine threads of white wax, which after twenty-four hours has formed a complete covering over the insect's body. This is at first round and whitish, but as development proceeds it becomes slightly eccentric and takes on a dirty grayish colour. About fifteen to twenty days after the larva hatches, it molts, losing legs and antennae. The shed skin sticks to the under side of the thin permanent scale, becoming a part of it. The two sexes are indistinguishable up to this stage, but from now on their development differs.

TEXT-FIGS. 6–13

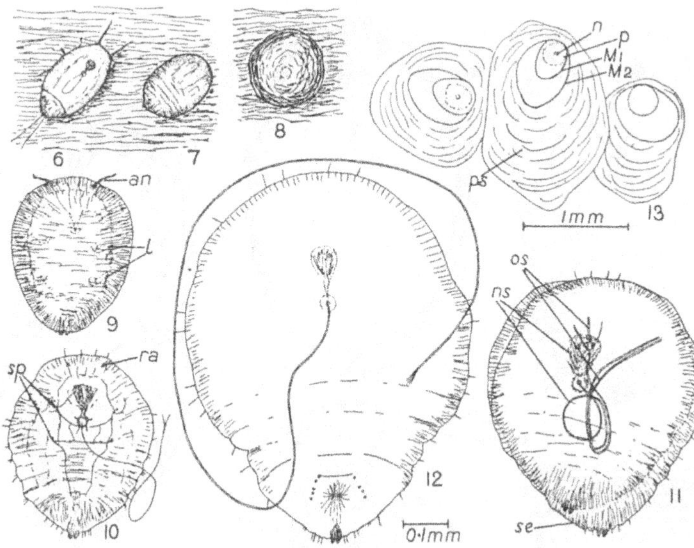

Camera lucida sketches all to same scale except 12 showing development of female.

Fig. 6.—Young insect crawling on bark.

Fig. 7.—Young about two hours after it had settled down.

Fig. 8.—Young about twenty-four hours after settling down, covered with thin white waxy excretion. By this time sucking apparatus has penetrated bark.

Fig. 9.—Shed larval skin, with this are also shed antennae (*an*) and legs (*l*).

Fig. 10.—Insect in second stage, rudiments of antennae persist (*ra*) but legs are completely lost; *sp*, spiracles.

Fig. 11.—Second molt. *se*, second skin; *os*, old sucking apparatus; *ns*, new sucking apparatus.

Fig. 12.—Adult female ready for fertilization.

Fig. 13.—Ventral view of three adult scales. *p*, pellicle with (*n*) nipple-like prominence; m_1, first skin; m_2 second skin; *ps*, permanent scale.

The male can first be recognized by the elongation of its body, and the appearance of two pairs of purplish pigment spots, the eye rudiments, at the anterior end, while the female's body retains its more or less circular outline and lacks the eye rudiments. The stages in its development are shown in Text-figs. 14–20. The mature males emerge about thirty-five days after the larvae hatch. Some of the males possess fully developed wings, while others show a very striking wing abnormality.

TEXT-FIGS. 14–18

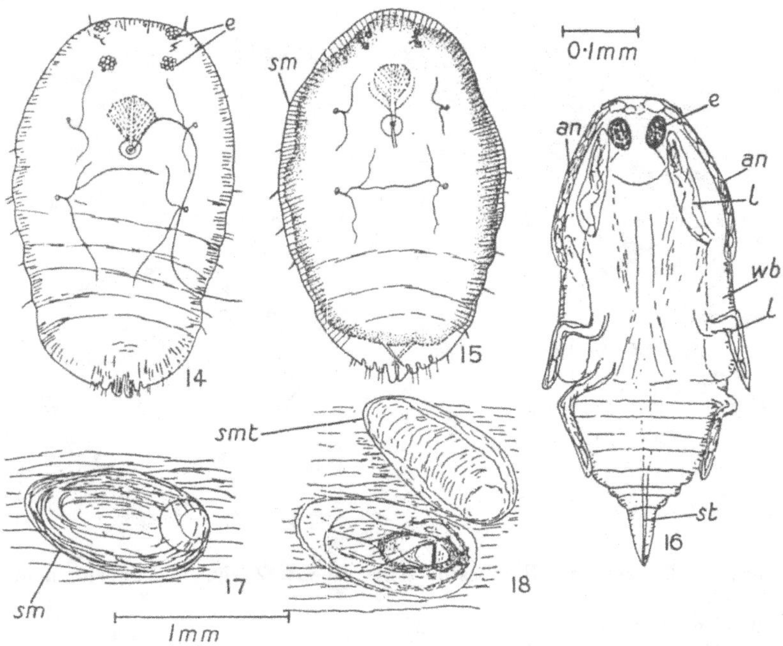

Stages in development of male of *A. osborni* from early stage of differentiation as a male to adult. *e*, eye rudiments; *sm*, second molted skin; *wb*, wing buds; *l*, legs; *an*, antennae; *st*, stylus; *sm*, male scale in position; *smt*, male scale turned back exposing adult male a few minutes before it was ready to emerge.

The second molt of the females takes place shortly before the males emerge. In one species in which I have observed impregnation on several occasions this event occurs shortly after the females have molted the second time. In *A. osborni*, I have been unable, as yet, to observe the act of impregnation. The overwintering females appear, however, to be ready for impregnation about March. At this time a most remarkable event occurs. The females, which until now have been completely hidden by the fungus, project the posterior part of their bodies up through the natural openings of the tunnels in the top layer. It is very likely that impregnation takes place while the females are in this position.

After impregnation the female's body soon becomes greatly distended with developing eggs, and in about two weeks some of the young are ready to be born. Each female may continue giving birth to young over a period of from two to four weeks. In the process of bearing

TEXT-FIGS. 19–20

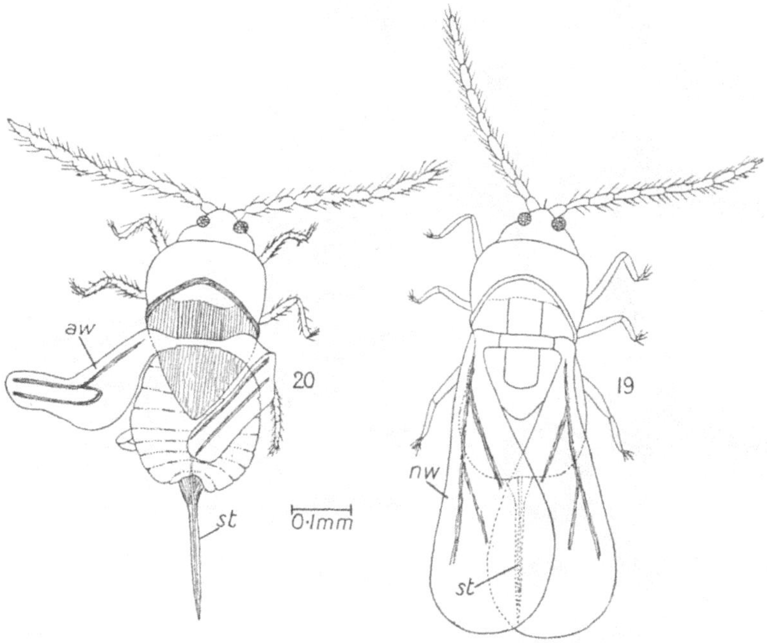

Camera lucida sketches of adult males of *A. osborni*.
Fig. 19.—Form with normal wings.
Fig. 20.—Abnormal wings. Note long pointed fertilizing apparatus, *st*.

young, the female's body is practically entirely used up, so that finally nothing remains but a lifeless, nearly empty cuticle.

The general relation of the scale insects to their host, the structure of their sucking organs, and the general way in which they suck the plant juices, have been studied by numerous biologists. In studying the course of the sucking tube of *A. osborni* in the oak bark, I have been unable to determine which particular tissue it seeks out. Several times in young twigs, I have been able to trace parts of the tube into the cambium; usually, however, it stops before reaching this region. In a species of tropical tree infested with *S. purpureum* and *Chrysomphalus* sp., the tube almost invariably penetrates to the medullary ray-cells in the cambium region (Couch, 1929). In all specimens which I have studied, the tube is surrounded by a sheath which clearly appears to be formed by the plant cells through which the tube passes (Pl. 2).

TEXT-FIG. 21

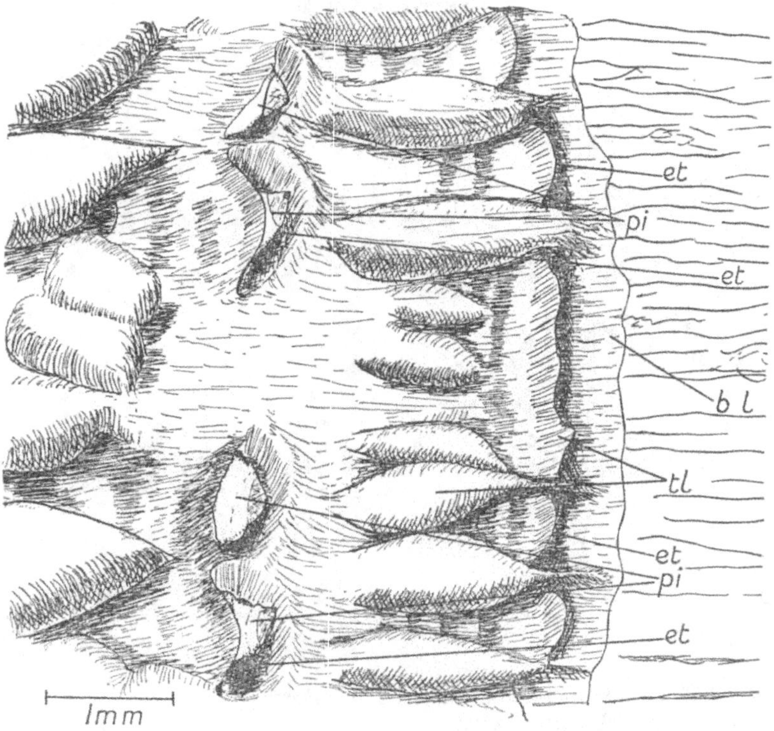

Sketch of part of living colony of *S. Burtii* in March, showing female insects projecting their posterior ends (*pi*) up through openings of tunnels (*et*), perhaps so that they may be impregnated by males.

Observations on the Relationship Between the Fungus and Scale Insects at Different Times of the Year

Dormant Season

In late fall, winter, and early spring, the fungus is in a dormant condition except for the formation of probasidia. During these months the insects too are semi-dormant. The number of overwintering in-

PLATE 1

Semi-diagrammatic drawing of part of a colony of *S. Burtii* on living bark showing condition of fungus and insects in dormant season. Note fungus shows no white margin at this time of year. Top layer (*tl*) of fungus has been partly separated from bottom layer (*bl*) and rolled back showing base of ridges (*r*) which support top layer, and anastomosing tunnels and chambers in which are numerous healthy scale insects (*i*). One of scale insects (*a*) is giving birth to young (*yi*). Note particularly that healthy insects lie flat against bark, fungal floor having been worn away from beneath their bodies by movements of insects in formation of permanent scales. Note also parasitized insects (*fi*), which usually occur imbedded in fungal floor but none are here shown in this position; two mites (*mi*); scales (*sc*) adhering to the top layer; and empty cuticle (*ec*) of scale insect which has exhausted itself giving birth to young. Note tunnel indicated by arrow. Outlined with camera lucida. November 29, 1928. × about 15.

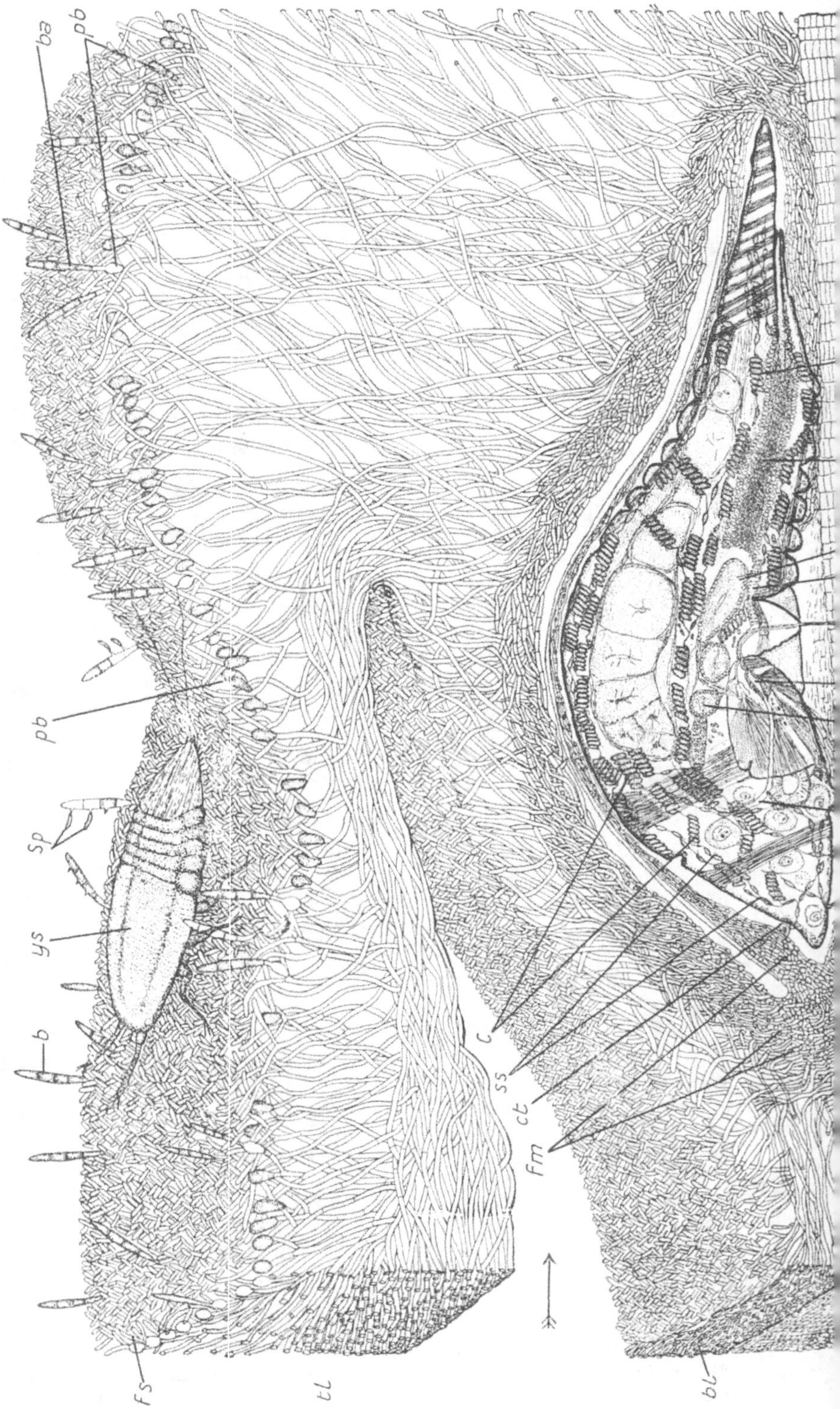

Semi-diagrammatic sectional view of fungus, *S. Burtii*, and bark (*b*), and a sca
tending down through bark into cambium region (*c*). Note fungal coils (*c*) connec
of fine threads pass through dermal pores connecting (*ct*) fungal coils (haustoria) wi
(*fs*) in which are numerous globose or ovoid cells, the probasidia (*pb*); septate basid
surface accidentally picking up spores of fungus and thus becoming infected. Betwe

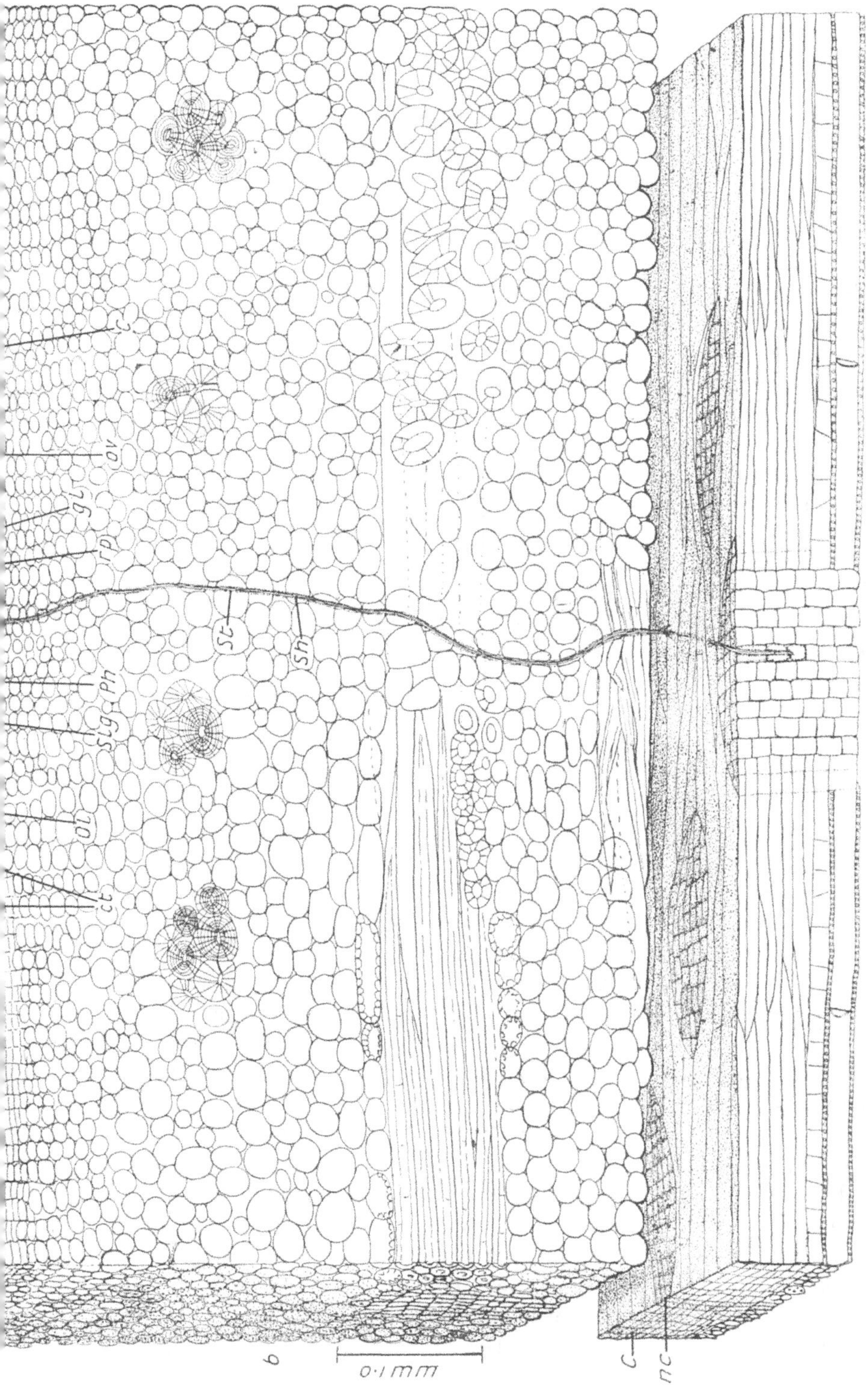

ᴀTE 2

.sect (in the centre of figure) parasitized by fungus. Note sucking tube of insect (*st*) ex-
by very fine hyphae and spindle-shaped cells also connected by fine threads (*ss*). Some
. insect's body with enveloping fungal mat (*fm*). Note also: fruiting surface of fungus
ba), some of which are bearing spores (*sp*); young scale insect (*ys*) crawling over fruiting
op (*tl*) and bottom layers (*bl*) of fungus is part of a chamber or tunnel (indicated by arrow).

sects beneath patches of the fungus varies from one to a hundred or more according to the size of the patch.

Such insects are free from fungal infection, as can be demonstrated by several methods. In the first place, the insects are nowhere in contact with the fungus which covers their bodies. If such insects are examined under the microscope *in toto* or by crushing their bodies or in paraffin sections, they are apparently entirely free from fungal infection. Numerous attempts to get cultures of the fungus by inoculating nutrient agar with the crushed contents of the presumably healthy insects' bodies have been unsuccessful.

A considerable number of parasitized insects is more or less embedded in the lower layer of the fungus. They are more difficult to detect than healthy insects because of their smaller size and position, and can only be found by using dissecting needles under a dissecting microscope. The number of parasitized insects is comparatively much smaller during the colder season than during the warmer months. Such insects are partly covered with a thick, dense mat, covering the dorsal scale and extending downward, surrounding the posterior part of the insect's body. The mid- and anterior ventral regions, the regions of the mouth and spiracles, are free from the enveloping mat. The threads of the mat are, of course, continuous with the threads of the main body of the fungus. The cells of the fungus mat contain conspicuous oil globules and therefore the mat may be looked upon as a food reservoir for the fungus, the food being taken from the insects and stored in the mat.

Although the fungus mat so nearly covers the insect's body, it is nowhere in direct contact with the insect, but the two are connected by means of numerous, two or four times coiled hyphae. This arrangement gives the insect freedom of movement to breathe so that it will not suffocate. These peculiar coiled hyphae are attached to the insect at the dermal pores and setae.

By means of the threads passing through the natural apertures of the body-wall, the haustoria of the fungus within the insect's body are connected with the fungal mat, and thence with the main body of the fungus. The haustoria are found only in the circulatory system or haemocoel of the insect and are exceedingly peculiar in shape, being in the form of coils. The hypha in each coil or haustorium makes from about four to eight symmetrical turns, and thus each haustorium has an appearance suggesting the coil of a hot-water heater. The haustoria are sometimes in clusters of two or three, but more often are connected end to end with one another by very delicate hyphae, which are only a fraction of a micron thick. Such fine threads, in turn, connect the coiled haustoria with the main hyphae within the insect's body. These main hyphae are composed of spindle-shaped, symmetrical segments connected end to end by the same kind of delicate hyphae as described above, so that each hypha resembles a miniature string of sausages.

TEXT-FIG. 22

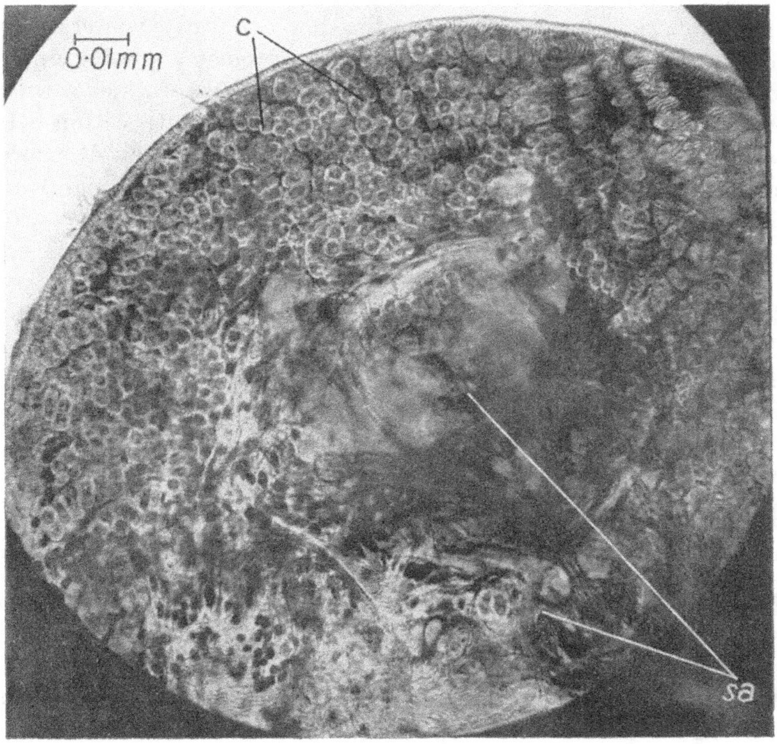

A living parasitized insect from beneath fungus showing haemocoel filled with fungal haustoria or coils (*c*). *sa*, suctorial apparatus. Photograph. × 600.

These hyphae are, in turn, continuous with the threads which pass through the pores in the insect's body-wall. Thus the connexion between the haustoria in the circulatory system of the insect and the main body of the fungus is maintained.

The growth of the fungus within the insect is very slow and such parasitized insects continue to live throughout the dormant season, sucking the juices of the host plant, even after the circulatory system becomes nearly filled with the haustoria of the fungus. The bodies of the insects in which the fungus is so well developed are considerably dwarfed, and they are incapable of reproduction.

Growing Season

With the coming of spring, the fungus starts growth as indicated by the margin changing from a brown colour to white.

The probasidia formed the preceding fall and early winter, stimulated by warmth and moisture, begin to germinate, sending upward above the surface erect four-celled basidia, each cell of which normally bears one spore. The probasidia do not all germinate within a few weeks, but the germination process continues throughout the spring and part of the summer, following rains. During normal spring weather, the spores are formed in greatest abundance in May. From this month on, the number of spores formed after rains diminishes until by the first of August very few or none are formed (Text-fig. 1).

In germinating, each of the spores first becomes divided into four cells, and these by budding may form a number of small, oval bud-cells.

Simultaneously with the renewed development and fruiting of the fungus, the overwintering females begin giving birth to young which pursue one of three courses: (1) some settle down beneath the "maternal roof"; (2) others crawl out and settle down beneath other fungal colonies; (3) and others crawl out and settle down on bark where there is no fungal growth.

There are two ways by which the insects may leave the maternal colony. They may crawl out through the tunnels or over the spore-bearing surface.

It has been shown by direct observations and experimentally that the young are infected by the spores, or rather by the bud-cells, formed by the spores, and are never directly infected by the fungal hyphae.[1] This holds true equally for the young which settle down under the fungal roof on the fungal floor, as well as for the young which settle down on the bark.

The evidence for infection by the bud-cells which come from the spores may be summarized as follows. The possibility of congenital infection has been disproved by the failure to find fungal cells in the eggs or in

[1] For detailed observations and experiments on which this conclusion is based see Couch 1931.

the young immediately after hatching. During dry weather when the spores and bud cells are not forming, only about one percent of the young are infected, whereas during damp weather when the spores and bud cells are forming, about 50% or more of the young which have settled down are infected. It may be concluded that the uninfected young emerge from their maternal chamber through the tunnels and thus do not crawl over the spore surface. About ten percent of the crawling young removed from the surface of the fungus during damp weather show the initial infection stages by the bud cells. Final proof that the bud cells are responsible for infection was obtained by taking the young just ready to crawl from beneath their mothers and placing them on the dampened surface of a specimen of the fungus which was sporulating profusely. Such insects showed abundant infection by the bud cells particularly on the legs and antennae. In several instances parts of legs heavily infected were autotomously thrown off.

Other experiments were carried out whose results tend to show that the young insects must come in contact with the bud cells to become infected. In the first of these the young which had just hatched were transferred to clean uninfected limbs. Of the few that survived to the adult state and to give birth to young, none were infected by the fungus. In another experiment, carried out during a dry season, bud cells from a pure culture were sprayed over the pellicles and scales of a number of young. Of the insects which survived the dry season, none became infected.

Some of the insects which have become infected by crawling over the spore-bearing surface may crawl back beneath the fungus colony under which they were born, others may settle down beneath other fungus colonies, and some settle down on the bark. The first two groups of insects are solely responsible for the survival and continued growth of fungus colonies already established; and the third group which settle down on clean bark are solely responsible for the distribution of the fungus.

A description of the fungus and insects in the later stages of parasitism was given above. The stages in the development of the parasitized insects are essentially the same as for the non-parasitized ones, except that the former usually do not excrete a scale over their bodies, are dwarfed in size, and seldom reach the stage of reproduction. Parasitized insects usually live as long as and sometimes longer than non-parasitized insects which give birth to young.

The infecting fungal cell or cells enter the circulatory system of the

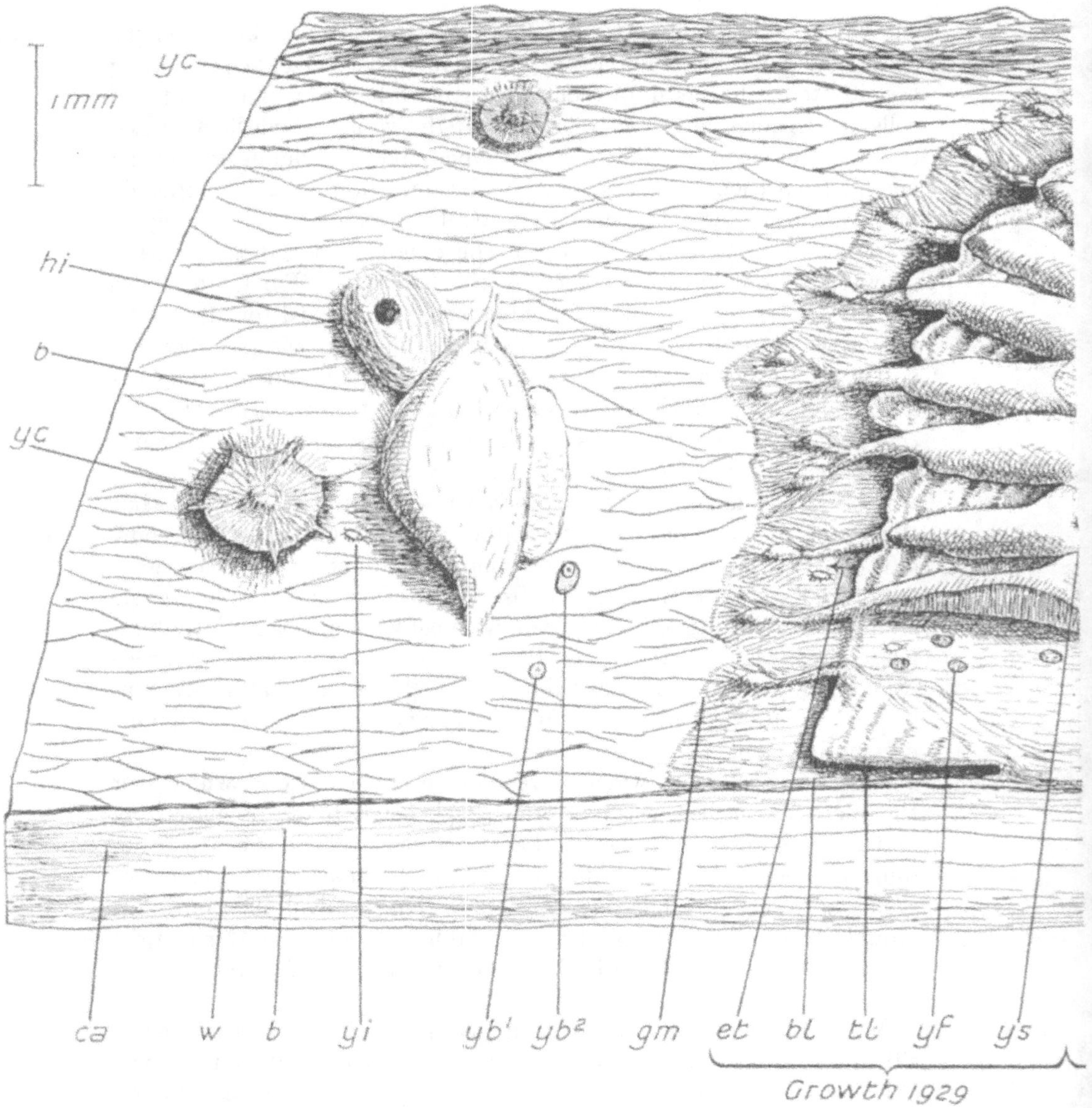

PLA

Semi-diagrammatic drawing of *S. Burtii* showing condition of fungus and insects after beginning c
quarter of patch is shown in drawing. In part shown top layer of first and second years' fungal growth
growth. Note two adults (*a*) giving birth to young (*yi*), some of which are crawling out through tunnel
down on fungal floor (*yf*). Note old cuticles of *Aspidiotus* (*hi*) in which hymenopterous wasps have deve

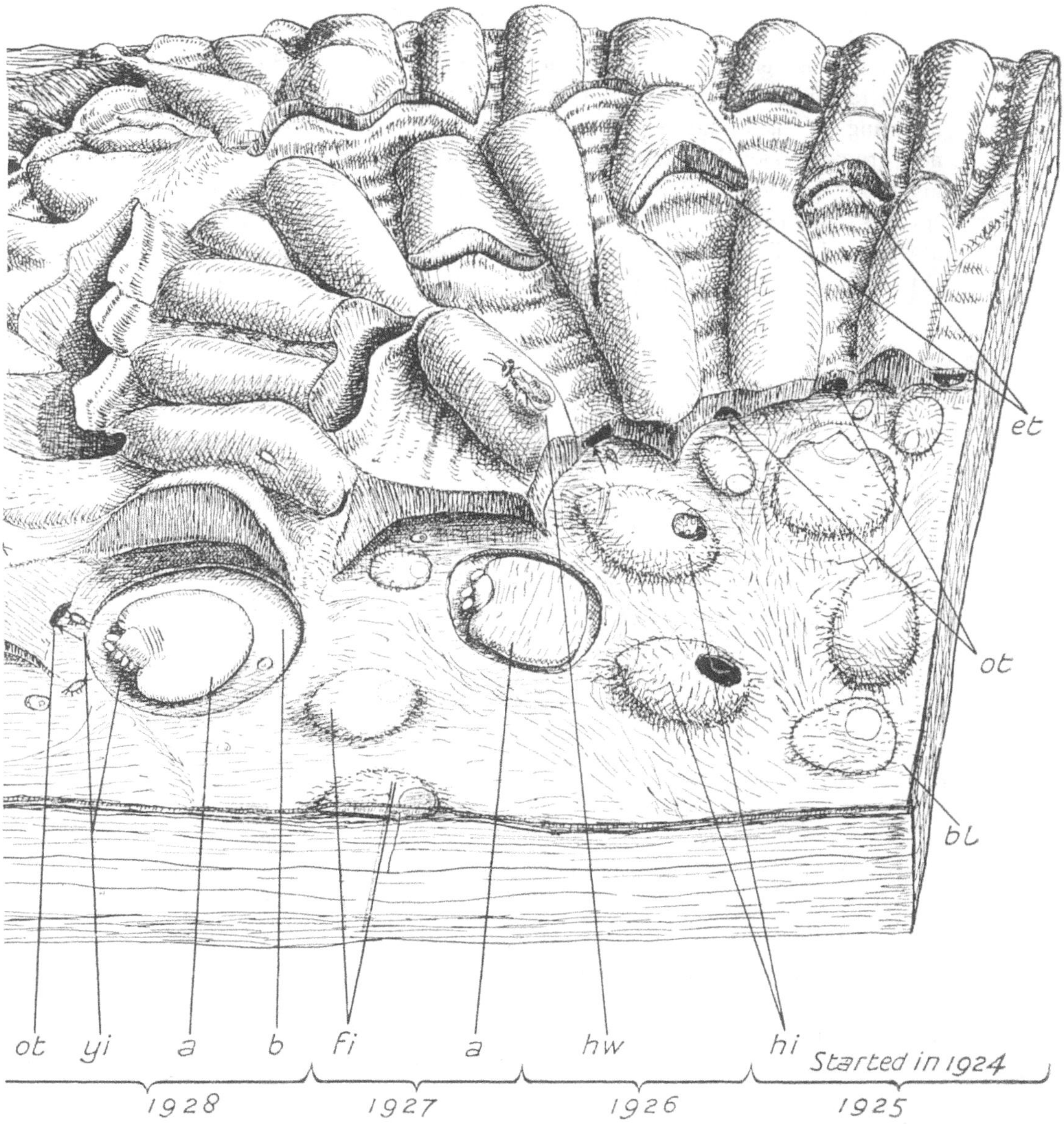

TE 3

growing season, i.e. about last of April. This patch was about five years old. Slightly less than a has been removed (lower right corner) leaving part of top layers of third, fourth, and fifth years' (*ot* and *et*), others are crawling over fruiting surface of fungus (*ys*). A number of young have settled ›ped. Note hymenopterous wasp (*hw*). × about 19.

insect. This cell is usually elliptic, and often may remain with one end attached to the place through which it entered. Sometimes the infecting cell is found free, floating in the insect's blood. This cell now grows into an elongated, more or less irregular thread which usually gives off several bud-cells. Sooner or later each bud-cell or thread sprouts a fine thread which grows for a short distance and then enlarges at the end. This enlargement grows in length and diameter until a club-shaped or spindle-shaped body, pointed at the proximal end and

TEXT-FIGS. 23–29

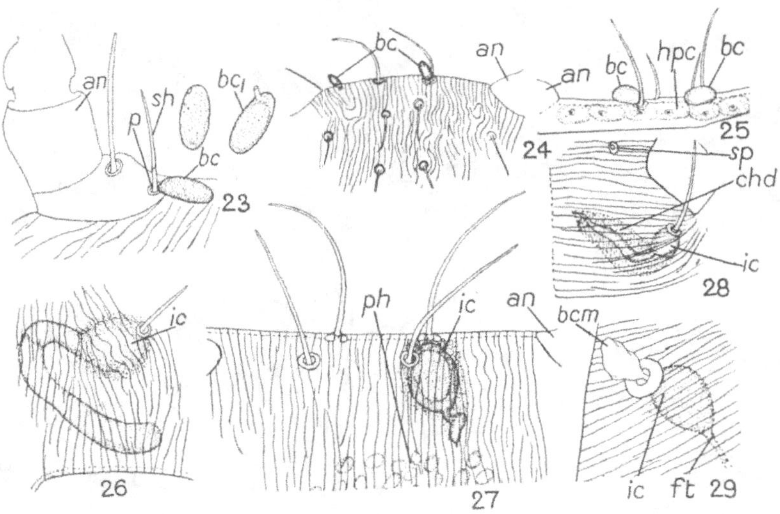

Stages in natural infection of young insect by bud-cells of fungus. *an*, antenna; *bc*, bud-cell; *bcm*, bud-cell membrane; *chd*, chitinous deposit; *ft*, fine thread growing from infecting cell; *hpc*, hypodermal cells; *ic*, infecting cell; *ph*, phagocytes (?); *sp*, spiracular opening; *p*, pore at base of seta; *sh*, seta. Figs. 24, 25 × 700; others × 980.

rounded at the distal end, is formed. This new cell now sends out another fine thread, the end of which enlarges and grows into another spindle-shaped cell, and so the process may continue until chains of spindle-shaped cells connected by fine threads may be seen throughout the entire circulatory system of the insect. These peculiar hyphae, as suggested above, have an appearance similar to a string of sausages. They may often branch by sending out two or sometimes three fine threads from the rounded end of a spindle-shaped cell. Often the infecting cells may give rise to a number of bud-cells which may elongate

TEXT-FIG. 30

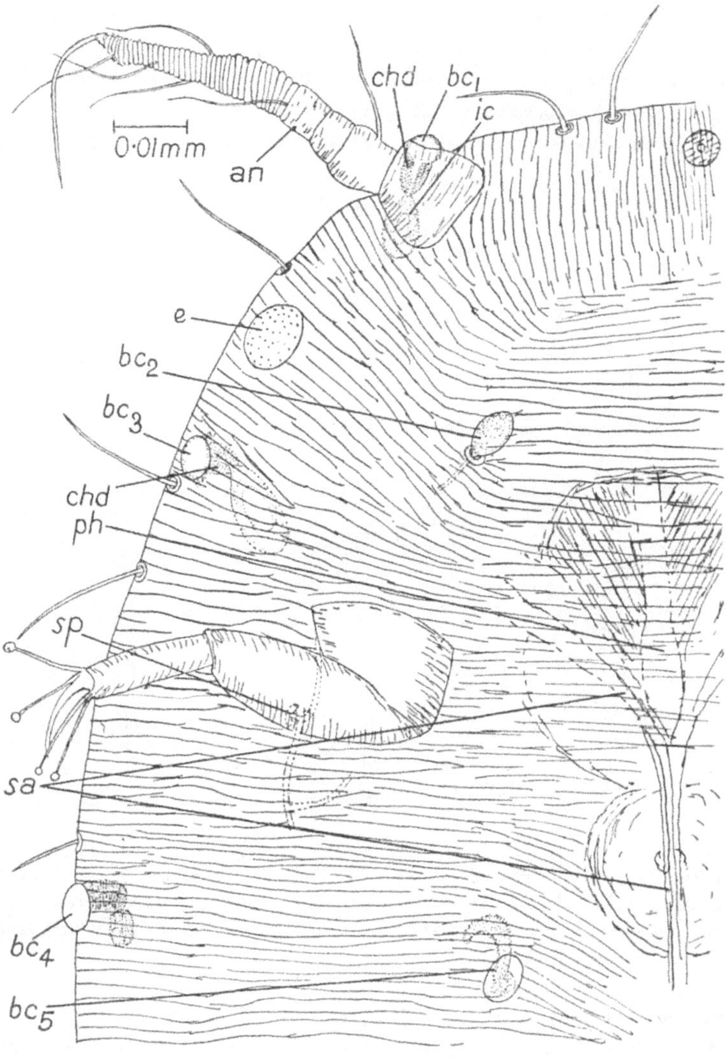

Stages in artificial infection of young; bud-cells infecting body. *an*, antenna; *bc* 1–6, bud-cells; only one of which is penetrating through a natural opening (bc_2), others are apparently penetrating directly through insect's body-wall; *chd*, chitinous deposit; *ic*, infecting cell; *e*, eye; *ph*, pharynx; *sa*, sucking apparatus (not functioning at this time); *sp*, spiracular opening.

and grow into the peculiar hyphae. Usually several weeks after infection and while the insect is in the second stage, the coiled haustoria begin to develop. A coil begins to develop exactly as a spindle-shaped cell starts, i.e., a fine thread sprouts from the rounded end of a spindle-shaped cell, enlarges at the end, but instead of elongating into another straight, spindle-shaped cell it grows around in a symmetrical coil, making from four to eight turns. From the distal end of the coil a fine

TEXT-FIGS. 31–36

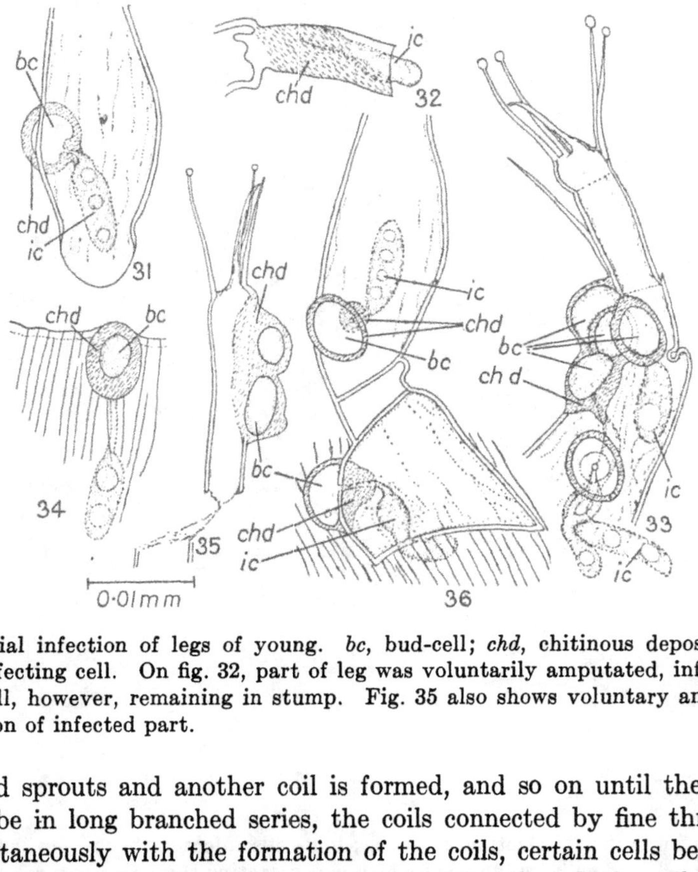

Artificial infection of legs of young. *bc*, bud-cell; *chd*, chitinous deposit; *ic*, infecting cell. On fig. 32, part of leg was voluntarily amputated, infecting cell, however, remaining in stump. Fig. 35 also shows voluntary amputation of infected part.

thread sprouts and another coil is formed, and so on until the coils may be in long branched series, the coils connected by fine threads. Simultaneously with the formation of the coils, certain cells beneath some of the setal and other derm pores send fine threads through these to the outside of the insect. The external end of each fine thread now grows into a comparatively large, spherical cell from which sprouts a twice to several times loosely coiled hypha. These coiled hyphae appear on the outside of the insect only after the second molt. By carefully dissecting out the insects in early June large numbers can be found in this stage.

While the fungus within the infected insect's body has been developing, the hyphal threads making up the fungal floor have more or less overgrown and covered the insect's body. These hyphae, however, never penetrate the insect. As soon as the insect molts the second time, the fungal threads which have grown over its body from the floor grow in

TEXT-FIG. 37

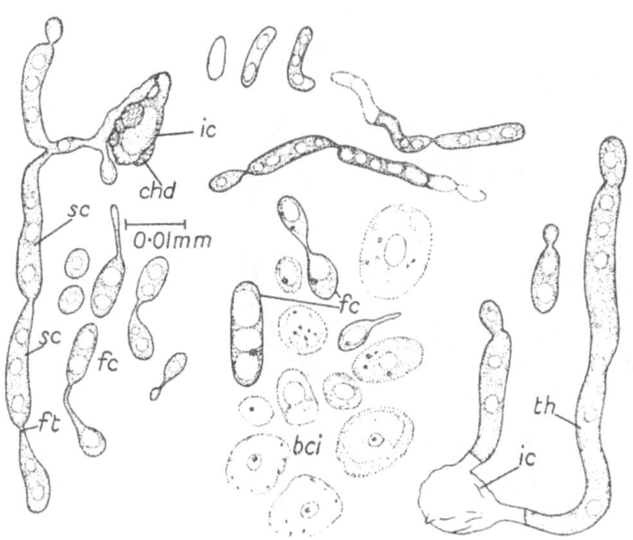

Early stages of development of fungus in insect's body. *bci*, blood-cells; *chd*, chitinous deposit; *fc*, bud-cells of fungus; *sc*, spindle or sausage-shaped cells connected by fine threads (*ft*), the fungal thread. Figures to left and at top × 830, others × 1400.

under the molted skin and more or less cover its lower surface. The coiled hyphae from within the insect's body anastomose with the hyphae which grow over the under surface of the molted skin and thus connexions are made between the fungus within the insect and the fungal threads making up the colony beneath which the insect has settled down. Thus a colony of *Septobasidium* is not a single individual descended from a single spore, but is composed of many strains, and new strains are grafted on the old ones every time the threads from an infected insect anastomose with the threads of the fungus under which the insect has settled.

Nourished by the food absorbed by the coiled haustoria in the circulatory system of the insect, the hyphae covering the insect develop into a mat of threads of considerable thickness. This mat of hyphae is several times thicker on the insects parasitized during the fall and

TEXT-FIGS. 38–40

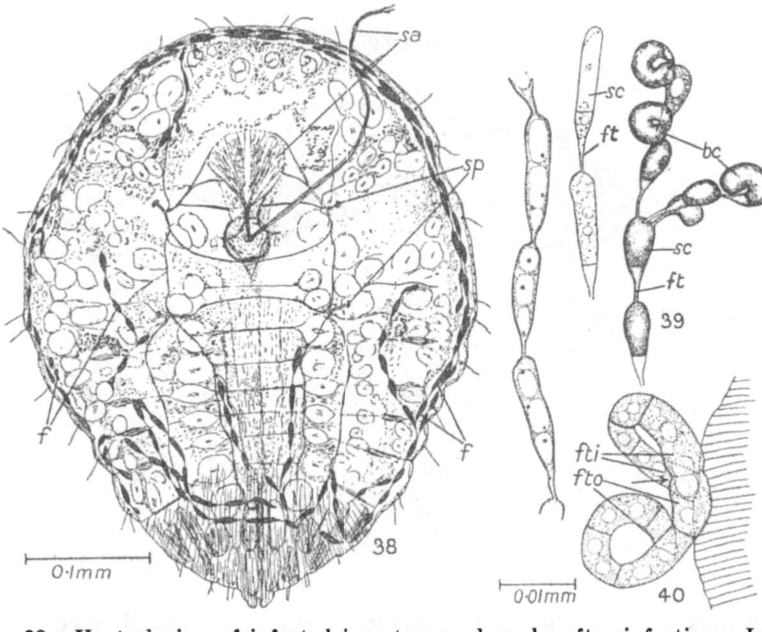

Fig. 38.—Ventral view of infected insect several weeks after infection. Insect in second stage. Fungus (*f*) is mostly in form of spindle-shaped cells held together by delicate threads. A few free fungal cells are shown.

Fig. 39.—Showing spindle-shaped cells held together by delicate threads and beginning of formation of coils (*bc*).

Fig. 40.—Showing fungal hypha from within insect's body fusing with a fungal thread which has grown up over insect from fungal floor.

winter months than on those parasitized during the season while the fungus is growing.

The infected insects which settle down on the bark in new places are responsible for the distribution of the fungus. The young crawl about over the bark perhaps looking for some protecting surface under which they may settle down. They often crawl into small cracks in the bark, or under bits of trash adhering to the bark, or under other patches of fungal growth besides *Septobasidium*; a number, however, settle down on the bare bark, completely exposed. The great number of these latter, whether infected by the *Septobasidium* or not, do not survive. The fungus grows very slowly in the insect's body, appearing on the

TEXT-FIG. 41

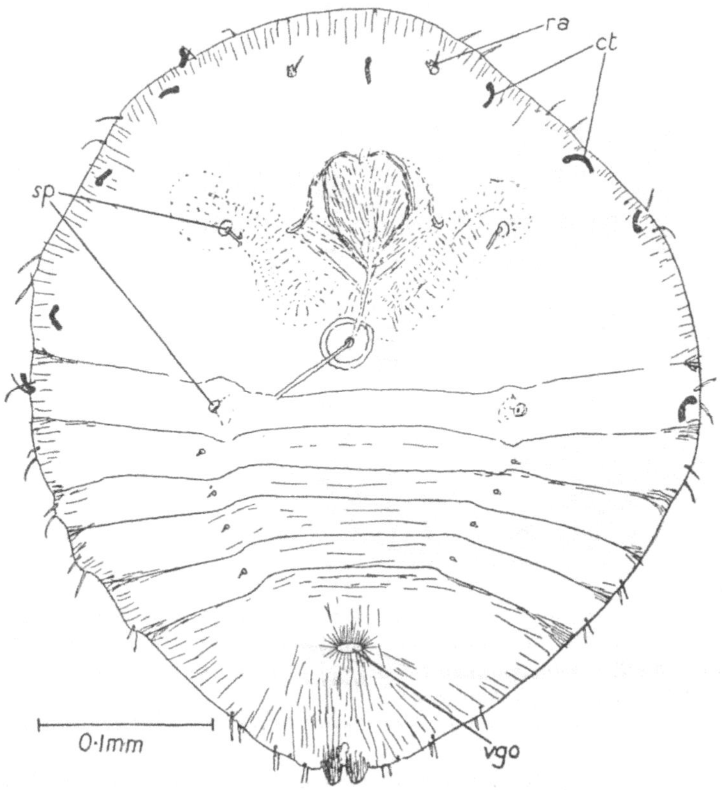

Living parasitized scale insect dissected out from beneath fungal mat showing broken ends of threads (*ct*) connecting fungus within insect's body with fungal mat. *ra*, rudiment of antenna; *sp*, spiracles.

outside only after the second molting, and the insects which settle down in plain view on the bark usually do not survive to reach this stage. Usually, therefore, the young which survive until the fungus appears outside their bodies are partly or entirely hidden beneath a bit of trash of some kind and one can recognize the very minute whitish patch of fungal growth extending, over the bark, out from beneath the trash. If the trash is removed with needles under the dissecting microscope, the minute insect may be discovered more or less covered by the whitish threads. Only in a very few cases, however, have I been able to recognize the patches early enough to be certain that the fungal growth came from within the insect's body. I have examined at least a hundred such patches and each one invariably more or less covered the body of one living insect. The fungal hyphae and haus-

TEXT-FIGS. 42–43

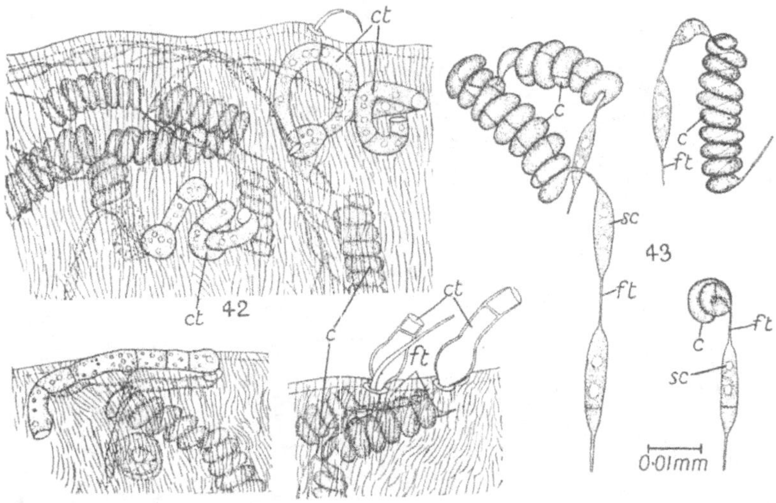

Fig. 42.—Parts of infected insects after fungus has developed considerably in their bodies showing: coiled fungal haustoria (*c*) within insect's body; delicate threads (*ft*) passing through the dermal pores connecting coils with threads (*ct*) which pass into fungal mat.

Fig. 43.—Coils from body of an infected crushed insect. (Letters as in Fig. 40.)

toria within the bodies of such insects were the same as were found in the parasitized insects beneath a large patch of *Septobasidium*. These minute patches of whitish growth appear early in June. By the middle of July the insect's body is entirely hidden by the fungal growth which is now about a millimeter thick though hardly more than 1 or 2 mm. wide. The fungus even at this early stage shows a differentiation into a top and bottom layer, the top layer supported by an exceedingly thick column. The parasitized insect may be found at the base of the column, surrounded, except for the mouth, spiracular, and anal regions,

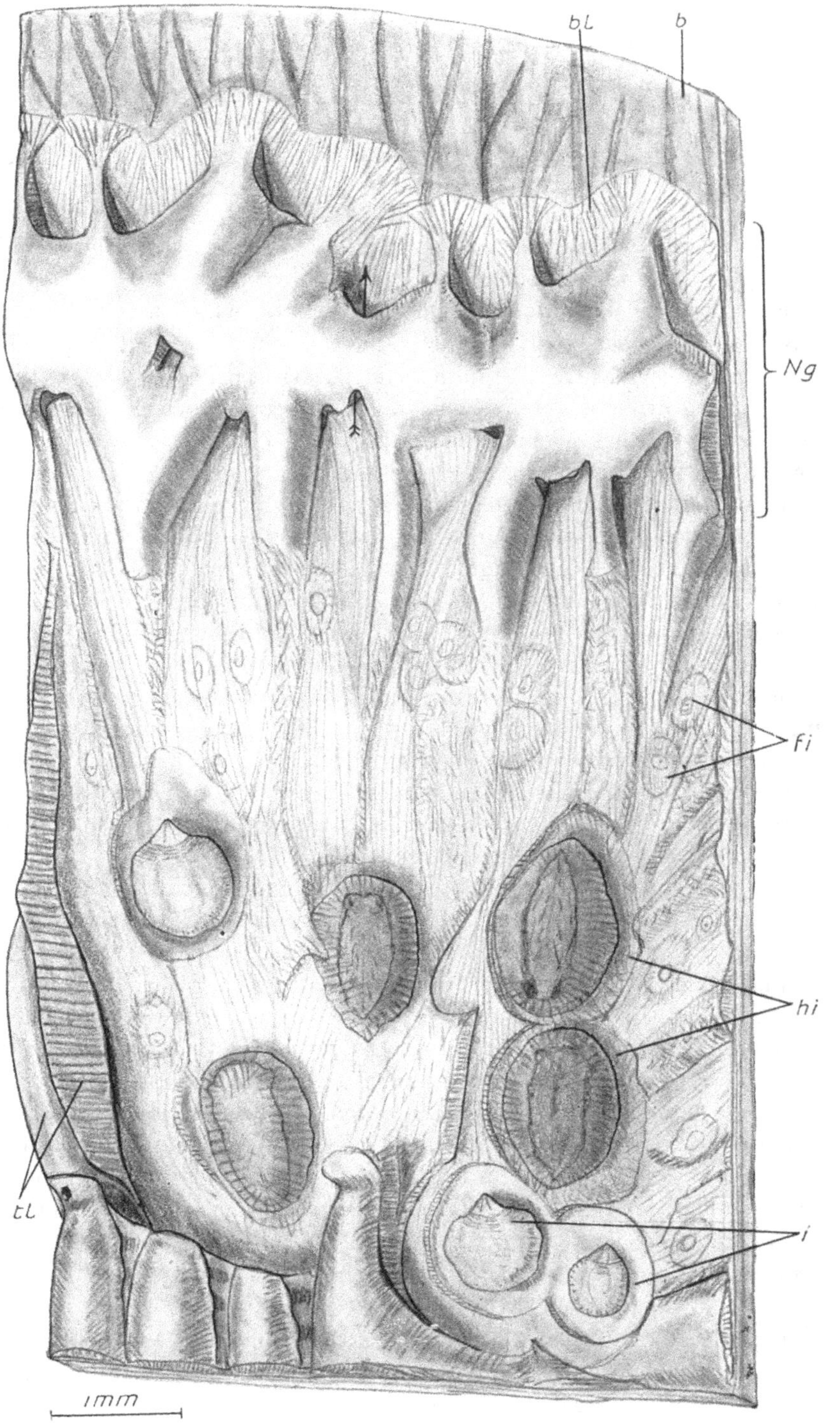

PLATE 4

Semi-diagrammatic drawing of part of colony of *S. Burtii* near end of growing season. Top layer (*tl*) of fungus is almost entirely removed except in region of new growth (*ng*). Note conspicuous pathways leading out through tunnels, one of latter being indicated by arrow. Note four large, dark insects parasitized by hymenopterous wasps (*hi*) and three healthy insects (*i*) on bark; also numerous smaller oval outlines (*fi*). These are scale insects of same generation as others, but parasitized by fungus. July 15, 1930. × about 18.

by the mat of hyphae. Thus the new growth of fungus is established. It now remains for other parasitized insects to settle down beneath this colony in order for the fungus to continue growth.

By the middle of June the infected insects, which settle down in April and May beneath the fungus, have reached the adult stage and are easily recognized. The number of parasitized insects at this time beneath a growing colony of *Septobasidium* is comparatively very large, and the regular arrangement of the parasitized insects between or just behind the new region of growth presents a very striking picture if the top layer of the fungus is removed. It can be shown by simple observations that the amount of fungal growth depends upon the number of parasitized insects. The number of parasitized insects beneath a colony 5 cm. in diameter was estimated at 248 by counting the parasitized ones in a quarter of the colony; another colony 1 cm. in diameter had 44 parasitized insects beneath it; another 0.5 cm., 14 parasitized; another 1 cm., 28. The four fungal colonies showed new growth varying from 1.5 to 3 mm. in width (July 16, 1930). Six colonies were then examined which showed no new growth except in certain localized regions. Where there was no new fungal growth, no parasitized insects were present; but in the small localized regions of new growth there were one or several parasitized insects.

The young, which crawl out through the tunnels even when the bud-cells are forming, as well as those which crawl out over the top surface during dry weather, are free from infection. As with the infected young, some settle down on the bark and others settle down beneath the fungus on the fungal floor. Of those which settle down on the bark, a surprisingly small number survive to reproduce; the vast majority perish; some dry up; others are parasitized by hymenopterous wasps, and some simply disappear, perhaps eaten by birds or insects.

The non-parasitized young which make their home beneath the fungus crawl in through the tunnels and settle down on the floor of the chambers and tunnels. The non-parasitized adults, however, as described above, are seated directly on the bark and are nowhere in contact with the fungus. Moreover, the fungal floor which was beneath the young insect partly covers the adult. Just how this change of the insect's position in relation to the fungal floor was brought about puzzled me greatly for a long time.[2] Daily observations have shown that the fungal floor is worn away beneath the insect by its

[2] Finding the insects beneath the fungus on the bark would naturally lead to the erroneous conclusion that the fungus overgrows colonies of scale insects.

movements in the formation of the permanent scale. The fine waxy threads of which this scale is made are excreted through pores on the

TEXT-FIG. 44

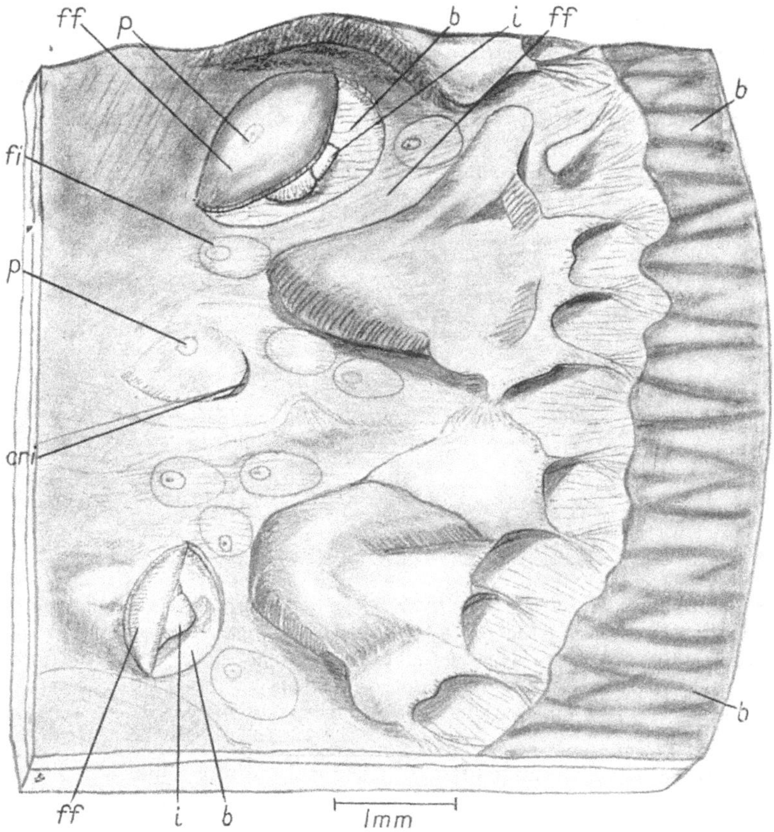

Top layer of fungus removed except in region of new growth: showing healthy insects (*i*) which settled down on top of fungal floor (*ff*) (as indicated now by pellicle (*p*) being on top of floor) now beneath fungal floor flat against bark (*b*). Note crack (*cr*) forming in floor in middle of figure; larger crack below; while floor over insect at top of figure has separated almost entirely from floor not over insect. Outlined with camera lucida, July 7, 1930.

anal segment. The scale covers the entire body, but is not fastened to the insect at any place and hence the insect is free to move about beneath the scale. That the insects do move a great deal is shown by the fact that old insects often have the spines on their anal segments very nearly worn away. By the time of the second molting the floor beneath the insect has been almost completely worn away and fungal threads from the floor have overgrown the insect's body, so that its position can only be recognized by the pellicle which is left on top of the fungal floor, and by a slight elevation of the floor over the insect. As the insect's body increases in size this elevation in the floor is pushed up, and soon a large part of the floor is broken off and becomes firmly fixed to the ceiling of the chamber or tunnel. The permanent scale formed of the two molted skins and the waxy excretion adheres to the

Text-fig. 45

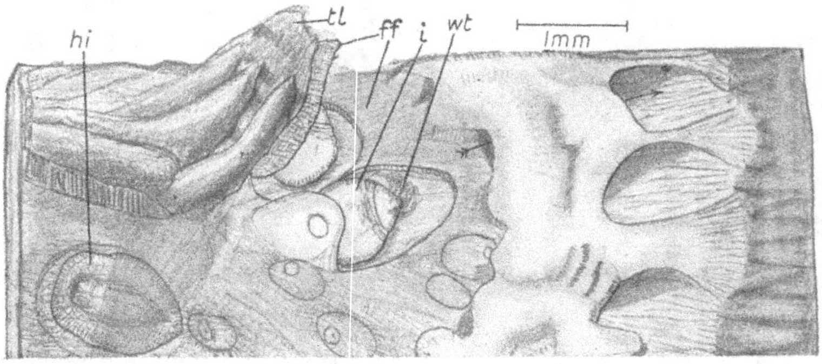

Showing how part of fungal floor over insect after separating from floor not over insect becomes attached to top layer of fungus (see also Pl. 2 showing scales attached to top layer of fungus). *ff*, fungal floor; *i*, healthy insect; *hi*, insect parasitized by hymenopterous wasp; *wt*, wax threads excreted by insect. Arrow indicates passage-way through tunnel.

under surface of the ceiling of the insect's chamber; while the floor of the chamber is the bark thinly covered with a whitish waxy excretion.

The insects which are born in April and May and do not become parasitized by the fungus, are fertilized in the latter part of May and June and begin giving birth to young during the latter part of July. The males develop as a rule on the bark, and only rarely beneath the fungus. I have never found a male parasitized by the hymenopterous wasp and have seen only one parasitized by the fungus. In the latter a considerable number of sausage-shaped cells were formed but no coils.

During the actively growing season of the fungus the number of parasitized insects outnumbers several times the number of non-parasitized. But even with the fungus growing and forming abundant spores the insects are able to hold their own, since some of them do not become infected. The proportionate number of parasitized insects to non-parasitized is reversed as we pass from the actively growing back again to the dormant season.

Transition between Growing and Dormant Seasons

By the end of July, the fungus has practically ceased the production of spores. During July the second generation of insects for the season is born. If the season is early, the second generation may appear as early as the latter part of June; but if late, this generation may be delayed until the latter part of July. A considerable number of the insects born during July, provided the weather conditions favour the production of bud-cells, become infected. I have found young crawling on the bark, July 29, which had just been infected. It is possible that the bud-cells which remain on the hymenial surface of the fungus may retain their vitality and infect the young insects a considerable time after spore production has ceased. After growth has ceased the bud-cells retain their vitality in artificial culture for several weeks; but of course the damp surface of an agar plate is not comparable to the much drier surface of the *Septobasidium*.

About the end of September the third generation of young is born. Exceedingly few of these become parasitized, but just how these few become infected, I am unable to say. If the fall season is late, some of the males may reach maturity, impregnate some of the females, and a fourth generation may be born in one season. The great majority of the young born in September, however, overwinter in the adult unimpregnated condition.

Several times during late summer and fall I have observed a peculiar condition, which, although not of common enough occurrence to play an important part in the life-history of the insect or fungus, is nevertheless interesting in itself. At this time the crushed bodies of certain insects may show many of the coils surrounded by large bodies which I take to be giant phagocytes. In some of the phagocytes the coils were distinct in outline, while in others the walls of the coils were hydrolyzed and very indistinct. In such insects the connexions between the fungal mat on the outside of the insect's body, and the coils and fungal hyphae within the insect's body are broken, and the fine hyphae which passed through the body-wall of the insect are surrounded by an amorphous chitinous deposit.

Text-figs. 46–50

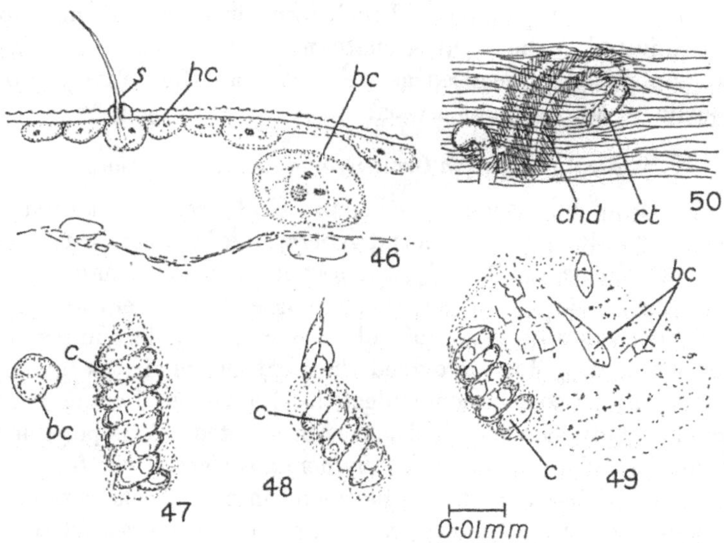

Fig. 46.—Section of healthy insect showing blood-cell (*bc*) probably phagocyte. *s*, seta; *hc*, hypodermal cells.

Figs. 47–9.—Stages in digestion of fungal coils by phagocytes. *c*, coil; *bc*, small part of coil and bits of fungal cells.

Fig. 50.—Surface view of body-wall of insect which has digested fungal cells. Note openings through which fungus penetrated are closed with chitinous (*chd*) material.

Advantages of the Fungus-Insect Association to the Fungus

I have never examined a living healthy patch of *S. Burtii* without finding living insects beneath the fungus. Colonies of the fungus with only dead insects underneath are exceedingly rare, being outnumbered one hundred to one by colonies harboring insects. If a patch of fungus is found without living parasitized insects underneath it, one may be sure that the fungus is either dead or in a dormant condition. Indeed, as has been shown above, the amount of fungal growth is directly dependent upon the number of parasitized insects. If there is no new growth, there are no parasitized insects; if there is much new growth, then there are many parasitized insects beneath the fungus.

Several experiments support the conclusion drawn from the foregoing observations. I have made numerous efforts to establish cultures of *S. Burtii* on the limbs of trees free from insects, but without success. A much more satisfactory means of demonstrating that the insects are necessary for the growth of the fungus in nature is to remove all traces of insects, both living and dead, from actively growing colonies of *Septobasidium*, leaving only a few small patches of the white, growing margin of the fungus. Then, by wrapping the limbs above and below each patch with a thick wide mat of absorbent cotton, young insects may be kept from crawling to the isolated fungal patches. Eight colonies of fungal growth were treated in this manner just after growth had started, and in all cases the remaining white patches of fungus stopped growing, dried up, and eventually disappeared from the bark. In the control experiment all traces of insects were removed leaving a few small patches of the white growing margin with enough of the top layer to offer the migratory young a protected home under which to settle. In this experiment the limb was not wrapped with cotton so that the young might crawl under the small patches from near-by colonies. By the middle of June considerable new fungal growth was apparent on three of the five patches left; and upon examination with a lens, one or more parasitized insects were found beneath each patch which showed new growth, whereas the two patches which showed no new growth had no insects beneath them.

Not only is the fungus dependent upon the insects for food, but it also depends upon the young insects for distribution. This fact brings up an interesting question. How is the fungus distributed from tree to tree? Of course there is a possibility of the bud-cells being blown from an infected to an uninfected tree and lighting on the body of a young, crawling, scale insect. A much more likely means of distribution would seem to be by birds picking up infected young on their feet and carrying them from tree to tree. However the distribution of the infected young may occur, it seems to be solely through them that new colonies of the fungus are established.

Although the fungus depends upon the parasitized insects for its food in nature, it may be separated from the insects and from the tree and cultivated on a variety of nutrient agars and in liquid media. Pure cultures have been made from single spores and from several spores, from single coils removed from the parasitized insect's body, and from slices of the plant. The fungus grows well in a great variety of media but so far no probasidia or spores have been seen in any of the artificial cultures. The failure to fruit in artificial culture may be due to the lack of the proper nourishment, furnished in nature by the scale insects; or it may be due to a complicated heterothallic condition.

Advantages of the Association to the Insects

Whereas the fungus could not exist in nature without the insects, the insects do live without the fungus. Such insects are comparatively rare and their life exceedingly precarious. We do not know enough yet about the resistance of scale insects to extremes of heat, dryness, and cold; but it is certain that insects beneath the thick stroma of *S. Burtii*

Text-figs. 51–59

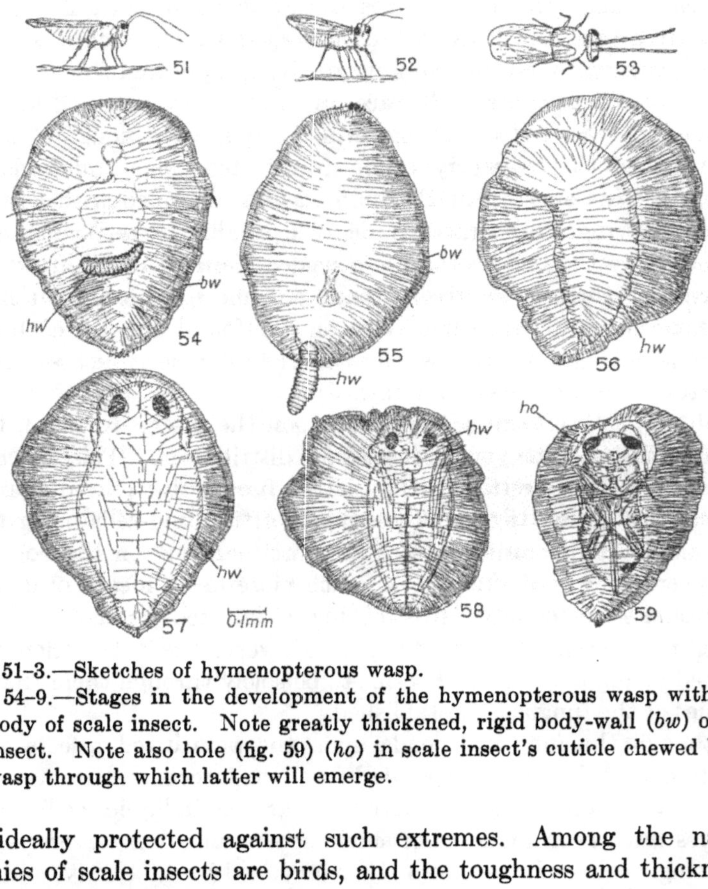

Figs. 51–3.—Sketches of hymenopterous wasp.

Figs. 54–9.—Stages in the development of the hymenopterous wasp within the body of scale insect. Note greatly thickened, rigid body-wall (*bw*) of scale insect. Note also hole (fig. 59) (*ho*) in scale insect's cuticle chewed by the wasp through which latter will emerge.

are ideally protected against such extremes. Among the natural enemies of scale insects are birds, and the toughness and thickness of the fungal covering would protect the insects to a certain extent, at least, from being eaten by them. I have noticed the more delicate fungal covering of *S. pseudopedicellatum* torn away as though by birds and the bodies of the insects removed. Perhaps the most deadly of all the scale insect's enemies are certain hymenopterous wasps. The female of this group punctures the body of the female scale insect and there deposits one egg. The egg may be deposited while the scale insect is in the second, or last stage of development. Curiously enough, the development of the parasitic egg takes precedence over either the development of the fungus, in case the insect is parasitized, or the development of the eggs of the scale insect, if it has been impregnated. The egg develops into a grub which grows at the expense of the scale insect. While the grub is developing, the scale insect undergoes a striking metamorphosis. The formerly thin, elastic, partly-segmented body-wall becomes greatly thickened and rigid, loses its segmentation, and changes in colour from yellow to brick red. The grub grows until it finally occupies most of the scale insect's body. It then undergoes a slow metamorphosis into the adult, winged wasp. This now chews a hole through the cuticle of the former scale insect's body and emerges. It takes just about as long for the hymenopterous wasps to go through a complete generation as for the scale insects. During the period while the young scale insects are emerging the hymenopterous wasps may be found crawling about here and there over the surface of the fungus, or rarely over the surface of the bark, searching for scale insects in which to deposit eggs. They may also be seen between the periods when the young are born. The wasps, of course, cannot see the scale insects through the thick fungal covering, but nevertheless they are able to detect the position of their prey with uncanny accuracy. The wasp pushes her ovipositor into the fungal covering to deposit an egg in the scale insect's body. The number of scale insects parasitized by the hymenopterous wasps varies greatly for those beneath the fungus, but may often be as many as 50 per cent.

A few simple observations and measurements will show that the fungus often affords a very distinct protection to the scale insects against their most formidable enemies. The ovipositor of the wasps varies from 200 to 300 microns in length while the thickness of the fungal roof covering the scale insects varies from 150 microns in the thinner regions to over 1,000 microns in the thicker. A scale insect beneath the thinner region can easily be reached by the ovipositor of the wasp; while those beneath the thicker cannot. If we examine a colony beneath which there are scale insects parasitized by the wasps and others not parasitized by them, we find the hymenopterous parasites only in the bodies of the scales over which the fungal roof is thin, whereas the non-parasitized scale insects are usually in the regions under the thick roof.

Nature of the Association between Fungus and Scale Insect

In trying to find an appropriate term to apply to the relationship between *S. Burtii* and the scale insects, it must be borne in mind that a large number of the scale insects are parasitized by the fungus, but it is clear that the term parasitism is inadequate to describe completely the relationship found here. Parasitism implies the living together

Text-fig. 60

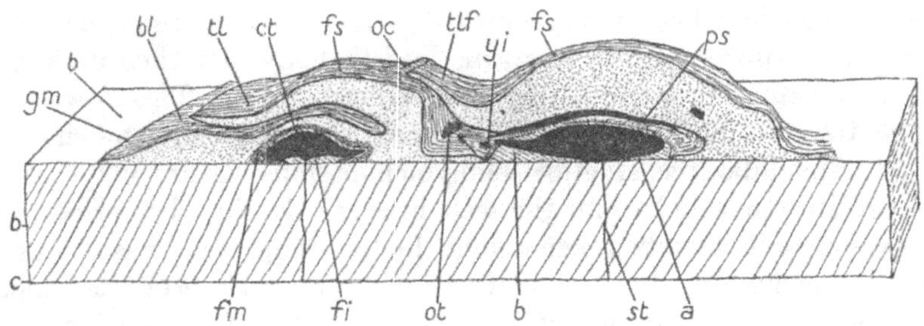

Combined surface and section diagram to show: relationship of fungus to parasitized and non-parasitized insects and relationship of fungus and insects to wood. Surface of fungus indicated by lines; section of fungus by stippling. Insects are black and bark is represented by diagonal lines. Note the suctorial tube of insects (*st*) through which insect draws its food from tree. Insect (*fi*) is parasitized by fungus and through connecting hyphae (*ct*) the fungus draws its food from insect. Insect (*a*) covered by fungus and protected by it is not parasitized and is giving birth to young (*yi*). The young may crawl out through tunnel (*ot*) and avoid infection or they may crawl out through opening of tunnel (*oc*) over surface of fungus (*fs*), and may become infected if fungus is forming spores. *b*, bark; *bl*, bottom layer of fungus; *c*, cambium; *gm*, growing margin of fungus; *ps*, permanent scale of insect sticking to lower surface of top layer of fungus; *tlf*, flap over opening to tunnel.

of two organisms to the benefit of one and the detriment or even death of the other. The relationship found here between the fungus and insects and between both of these and the tree is represented diagrammatically in text-fig. 60. The fungus and insects live together interdependently; the fungus furnishes a home and protection for the insects, while, in return, the insects furnish food and a means of distribution for the fungus. Some of the insects are sacrificed, but to the advantage of the insect colony as a whole; for, from the juices taken from the parasitized insects, the fungus grows and forms more houses for the non-parasitized as well as the parasitized insects. As a result of this association an entirely new type of organism is produced. The relationship here is, therefore, obviously one of symbiosis; both fungus and insects are benefited by the association at the expense of the tree.

OBSERVATIONS ON OTHER SPECIES

In several species in addition to *S. Burtii* a similar symbiotic relationship has been found. In *S. apiculatum*, which occurs on *Cornus* and other woody plants and is associated with *Aspidiotus ancylus* and sometimes with other species of scale insects, there is no differentiation into layers. The fungus is whitish and the surface, except in the marginal region, is more or less covered by irregular mounds, each of which is crowned by one or more upright tufts of fungal hyphae. Each of these mounds is or was the home of a healthy or parasitized scale insect. If the mounds are overturned in a fresh piece of fungus collected in midwinter, numerous healthy overwintering females are exposed. There is also a considerable number of parasitized insects beneath the fungus but such insects are much less abundant during the dormant than during the growing season. A photograph (Pl. 67, fig. 2) shows a healthy, living insect, the roof of whose house has been lifted off.

In the spring the fungus begins to grow and fruit and the overwintered female insects, which have been impregnated some weeks earlier, begin giving birth to young. Just before the young begin to hatch, a crack lined with wax excreted from the anal segment of the insect appears in the fungal tissue at the lower edge of the mound. As the eggs are laid and the young hatch, they emerge to the open through this crack (Frontispiece).

The fungus appears to have a chemotactic attraction for the insects for though they are free to settle down where they will, by far the greater number settle down on the surface of the fungus "body" rather than on the bark. This is very easy to demonstrate throughout the seasons while the insects are hatching and is shown in the accompanying photograph (Pl. 67, fig. 1). On a small patch of fungal growth about 2 sq. cm. in area, fifty-six young had settled down on the surface of the fungus while out on the bark for a distance of one centimeter from the edge only one insect had settled. Many counts similar to this have been made.

A few days after the young settle down, their bodies become completely covered by the white pellicle (white in *Aspidiotus ancylus*) excreted by the insect. In a few more days the fungus begins to grow up around the insect on all sides, forming a sort of circular pen around it. The insect continues to increase in size. After a week to ten days the fungus has almost completely disappeared beneath the insect, perhaps digested by "enzymes" secreted by the insect or worn away by the movement of the insect's body and thus the insect comes to lie directly on the bark. Meanwhile the fungus continues to grow upward over the insect's scale, forming finally a mound-like covering on the top of which are one or more long vertical peaks. This peculiar reaction of the fungus to the presence of the insect either healthy or parasitized is very specific. The fungus may overgrow pieces of trash, the living larvae of other kinds of insects, which are not infrequently found on the surface of the fungus, or even the dead bodies of the scale insects which have been killed and placed on its surface, but the upright pillars are never formed except when the living scale insect settles down on the surface of the fungus.

The adult healthy female lives in a remarkably beautiful house which because of its whiteness and shape suggests the snow houses of the Eskimos. The ceiling of the fungal house is formed by the insect's scale which is so thoroughly overgrown by the fungus that the two adhere firmly together, and the floor of the house is the bark. The healthy insects, though covered by the fungus are nowhere in contact with it. This is shown in the frontispiece in which half of one of the fungal houses has been cut away exposing a healthy, pregnant female laying eggs. The young which have hatched are shown crawling, one in the act of emerging through the crack or "door" of the house.

In another somewhat similar species, *S. rugulosum*, the fungus forms a thin, rugulose, brownish incrustation on the bark (Pl. 77, figs. 5, 6). This species occurs commonly on *Liquidambar* and *Magnolia virginiana* and is associated with *Chionaspis gleditsiae* (?). In material collected in March the adult insects were giving birth to young and the fungus was bearing abundant spores. Consequently a splendid opportunity was afforded for some observations on the fungus-insect relationship. It is impossible to detect the presence of the insects beneath the fungus except by a careful examination. If one examines the surface of the fungus with a hand lens or binocular microscope, one may see numerous, small slits, shaped like a new moon. These slits are the openings or doorways to the insects' chambers. By inserting a needle in one of these cracks and lifting up the inner edge of the "half moon" the adult insects are exposed, each one in a sort of chamber, the roof and walls of which are formed of fungus material and the floor of the bark (Pl. 68, fig. 2). At the time the material was first examined the adults had already given birth to large numbers of young, the empty egg shells being shown. The young emerge through the slits, one of which is seen in the photograph (Pl. 68, fig. 1), crawl about over the top of the fungus for a while, and then settle down on the surface, as a rule picking out the depressions between the ridges. As in *S. apiculatum* the fungus here seems to exert a chemotactic influence on the insects, for although they may crawl away and settle on the surface of the bark by far the greater number settle on the surface of the fungus (Pl. 68, fig. 3). The insects give off something which causes a profound change in the fungus in their immediate vicinity. After a few days, but before the insect has excreted its first scale, the fungus becomes lighter in color beneath and around the young insect. A few days later the insect has excreted its first scale and the fungus around the insect has become almost black (Pl. 68, fig. 4). The insects are so abundant that the surface of the fungus at this time is literally peppered with these dark dots (Pl. 77, figs. 5, 6). After about two weeks the fungus beneath the insects has completely disappeared so that the insects are against the bark. As the insects sink down into the fungus, new growth begins at the surface, the threads extending over the insect so that it eventually becomes completely covered by the fungus.

Some of the living insects examined after the first molt showed the fungal haustoria within their bodies. I suspect that infection here takes place much as in *S. Burtii* and that the spores furnish the source of infection. I have been unable, however, to demonstrate this. I have examined ten young which were crawling about over the sporulating surface and although spores could be seen on the surfaces of the insects' bodies, I was unable to find any infecting cells within them.

Septobasidium fuscum occurs in Louisiana and Mississippi, forming a clove brown or fuscous, thin but extensive growth on the bark of ash (*Fraxinus*) and oak (*Quercus*). Here and there on the surface are numerous elevated tufts of loosely entangled hyphae. These hyphae contrast strikingly with those of the surface of the fungus, being much darker colored, almost black, and over twice as thick. Beneath each of these tufts is a parasitized or healthy scale insect. Such a covering is rather crude in comparison with some of the types of fungal houses found in *Septobasidium* but doubtless this loosely thatched covering affords considerable protection to the scale insects (Pl. 24, fig. 1).

In such forms as *S. pseudopedicellatum,* where the fungus is differentiated into top and bottom layers, the top layer supported by pillars, the young insects settle down on the fungal floor out toward the margin. Both parasitized and non-parasitized insects become surrounded by a mound of fungal growth. In these species the insects are doubly protected, being covered by both the top and bottom layers.

In the above species the insect houses are not preformed but are built around the insect after it has settled down. In many species "insect houses" or "traps" are formed on the margin of the fungus before the young hatch. There are so many species of this type that only a few examples of these studied in detail will be given.

In *Septobasidium Sydowii,* which occurs in Texas on *Ulmus crassifolia* and in the Philippine Islands on *Pterocarpus,* the brownish surface is cracked in such a way as to have the appearance of a miniature bit of baked, finely cracked clay (Pl. 73, figs. 6–8). I have not studied living material but the fungus-insect relationship is obviously a very interesting one. If the surface of the fungus is examined under a high power binocular dissecting microscope, numerous dried but uninfected adult female insects can be seen partly buried in the cracks, the posterior part of the ventral and dorsal scales projecting up through the fissures, the scales reminding one of the bivalved shell of an oyster (Pl. 67, fig. 3). These adults contained many young and judging by the widely opened condition of the genital pore, were in the act of giving birth to young. In the tunnels near the margin many dried, but non-parasitized young were found and farther back in the tunnels there were half-grown insects. A considerable number of insects were infected, and such insects were always completely covered by the fungus. It is quite apparent that the cracks in the surface of the fungus are in some way caused by the insects, though the insects do not actually make the cracks. The evidence for this is found in the very small cracks which are just beginning to be formed. Into each of these one finds the anal segment of a healthy insect protruding.

In *S. Curtisii,* which is one of the most widely spread and injurious of all the species of *Septobasidium,* the fungus forms slate-colored to coal black incrustations on the limbs of a large variety of trees. In this species, during the growing season, one or several rows of irregularly shaped insect houses are formed along the margin of the fungus. Under a powerful hand lens or a dissecting binocular, this marginal region looks like an aerial view of an Indian village. Some of these enclosures are at first V-shaped or U-shaped or semicircular in outline or sometimes like an open umbrella. The fungal hyphae which make the enclosures continue to grow upward and finally the upper margins of the enclosures bend over and anastomose so that a minute "house" is formed. Often the tops of these structures are bordered by radiating fascicles of hyphae. As a rule the roofs grow out horizontally anastomosing with each other and with the older part of the fungus and in this way the top layer is formed. The houses are formed successively throughout the growing season: the younger houses are on the outer edge of the growing margin while the older ones are on the inner edge.

These peculiar structures on the margin appear quite meaningless until one follows the development of the insects. As the young insects hatch beneath the older parts of the fungus, some of the young crawl out and settle down in the fungal houses (Pl. 68, figs. 5, 6). There are many more houses than there are young insects to occupy them and consequently it is difficult to find the insects within these houses unless a rather careful search is made. The photographs made in early August show several houses of different shapes and in different stages, two of which contain a young insect. The further fate of the young insects is essentially the same here as in *S. Burtii*. Some of the young are parasitized by the fungus, others remain free from infection and though covered by the fungus are unharmed by it.

The house of the mature adult female in *S. Curtisii* is very similar in structure to that described above in *S. rugulosum*. Indeed the same new moon-shaped crack appears in the surface of the fungus at the edge of the house and the young which emerge make their exit through this crack.

Just how infection of the young occurs in this species I am still unable to say.

The insect houses or traps in *S. Curtisii* are not arranged with any striking symmetry, nor are they of uniform shape, and the opening to these traps may be toward the margin or away from it or on any side of the house.

In another species, *S. Schweinitzii*, which has two horizontal layers as in *S. pseudopedicellatum*, the insect houses are of uniform design and are placed on the margin in a strikingly symmetrical fashion (Pl. 67, fig. 4). The houses are formed of two plates of hyphae which grow upward from the subiculum forming a structure which viewed from above with a powerful hand lens is somewhat V-shaped or U-shaped, the bottom of the V or U always pointing toward the outer margin. The hyphae forming the sides of the V or U continue to grow upward, the two sides slanting toward each other and eventually uniting, the hyphae anastomosing to form a beautiful little house usually shaped like an army pup-tent. The adult healthy female insects are beneath the main body of the fungus some distance back from the margin. When the young insects hatch, some of them crawl out toward the margin and settle down in these preformed houses.

It is exceedingly interesting to observe the movements of the young as they crawl out from between the top and bottom layers. The specimen from which the enlarged photograph was made was brought into the laboratory August 11 and examined immediately under the binocular. It was a medium sized specimen completely encircling a limb 1.8 cm. thick and extending for a length of 6.5 cm. The houses were very abundant, with about two to four irregular rows on each margin, each row with about two hundred houses. At this time I could count only three insects in the houses; two days later there were fifteen, and with a little patient searching it was possible to find several insects crawling. The movements of the insects appear to be more or less aimless as they crawl about over the surface of the fungus, over pillars, over houses, between the houses, first in one direction and then in another. However, several times I have seen an insect crawl out from between the two layers and go immediately to one of the houses to settle down. Frequently an insect may enter one of the houses and turn around and come out if the house is large enough. However, I have never seen an insect back out of a house. Most of the houses are just slightly wider than the width of the insect's body. Moreover, the inside shape of the house conforms nicely to the shape of the insect's body, and hence it might be that the insects would have difficulty in turning around in such close quarters. The insect's body is about 120 x 204μ while the houses are usually about 140–170μ wide on the inside. These houses may frequently serve as "traps" since an insect might crawl into one which was so narrow that it would have difficulty in extricating itself.

In *S. Mariani*, a brownish species which occurs on the apple (*P. malus*) and pear (*P. communis*) and other woody plants, the fungus consists of three horizontal layers, the adult insects, if parasitized, inhabiting the space between the middle and bottom layers and, if not parasitized, between the lower layer and the bark. During the growing season a series of "houses" or traps are formed between the middle and lower layers and an impenetrable (to the young scale insects) tangle of hyphae is formed over the middle layer. This tangle of hyphae would make it impossible for the young to crawl away from the fungus without crawling over the margin of the lower layer where the "traps" are located. This arrangement would doubtless increase the number of insects which settle in the traps.

In *S. castaneum* and *S. myrsinae* from Brazil, the structure of the fungus in relation to the insects seems to be quite similar to that described in *S. Mariani*.

In some tropical species which I have not been able to study in the fresh condition, the preformed insect houses or traps may be truly fantastic in structure. In *S. lacunosum* from Costa Rica, the fungus forms an irregular, resupinate, grayish growth about 2 mm. thick over the bark. To the unaided eye the surface of the fungus has a finely retiform appearance, in fact the fungus was first called *S. retiforme* by Patouillard, but if the surface is examined with a binocular microscope, one discovers that the depressions are in reality little sunken chambers each one partly covered by a flap of fungal tissue. In spite of the fact that the material was in the dried condition when I studied it, I could still find young, dried, but not parasitized scale insects within several of the chambers. I also found the shrunken shell of two large females whose bodies had apparently been exhausted by giving birth to young. These adult females were not in the sunken chambers but were beneath the thicker part of the fungal tissue in cavities which communicated with the sunken chambers of the tunnels. This remarkable type of insect trap is perhaps the most highly developed of any in the species of *Septobasidium* which I have seen (Pl. 112, fig. 4).

I have described briefly a few of the gross structural adaptations found in certain species of *Septobasidium* which enable these species to live symbiotically with scale insects. In trying to answer the question "*Why does the fungus construct such peculiar houses?*" it is of interest to recall that the scale insects are closely related to those insects which cause the formation of such fantastic galls on leaves and the other parts of plants. Indeed, one of the scale insects, *Cryptophyllaspis liquidambaris*, which is reported to cause the gall formation on certain leaves also lives with certain species of *Septobasidium*. It is highly probable that the large upright spines formed over the insects in *S. apiculatum*, the upright pillars in *S. pseudopedicellatum*, and other species, the chambers and tunnels in *S. Burtii* and the peculiar sunken "traps" in the species from Costa Rica, are the result of the same general type of response as is found in the insect galls on leaves. Another line of evidence to support the assumption that the form of the fungus is due to its response to a stimulus from the insects is found in the fact that when the fungus is grown on artificial culture media it never shows the pillars, upright spines, chambers and tunnels so characteristic of the growth of certain species in nature. Moreover it takes the living insect, parasitized or not parasitized, to cause the peculiar response. A piece of trash, a dead scale insect or the larvae of other species, or a piece of nutrient agar placed on the fungus when growing on the living tree will not cause the formation of pillars or spines, etc.

The "gall development" seems to reach an extreme degree in *S. Clelandii* collected by Dr. Cleland in Australia (Pl. 67, fig. 5, and Pl. 92, figs. 3–5). The insect here is related to the phylloxera gall insects and like those insects the larvae form galls in the woody parts of plants, in this case in the small branches. The larvae become infected with the *Septobasidium* which produces typical haustoria within the insects' bodies but the fungus, instead of forming a typically resupinate growth, produces a dense cluster of club-like structures, each about a centimeter long. Nothing further is known about the fungus-insect relationship in this species.

The above account will suffice to show some of the remarkable adaptations between the scale insects and certain species of *Septobasidium*. For other examples which differ considerably from any of these, see under *S. Rickii* (Pl. 48, fig. 10), *S. Petchii* (Pl. 63, fig. 12), *S. verrucosum* (Pl. 63, fig. 8), and *S. Mülleri* (Pl. 99, fig. 5).

II

PATHOLOGICAL CONSIDERATIONS AND CONTROL METHODS

All species of *Septobasidium* which I have studied cause damage to the trees on which they grow; the damage is due not to the fungus directly but to the combination of fungus and scale insects. In such species as *S. apiculatum* where the growth covers a small area the damage is slight, but in the greater number of species the growth may be quite extensive, in which case the damage is considerable. I have seen numerous small trees of *Fraxinus* and *Nyssa* killed outright by *S. pseudopedicellatum* and *S. Curtisii*. Large trees of both these genera may be heavily infected, in which case they are very unhealthy, showing many dead limbs. Types of wood injury caused by these two species are shown in several of the photographs. In a few species (e.g. *S. makilingianum* and *S. areolatum*) the fungal hyphae may penetrate slightly into the living tissue of the leaf or bark. These species are, however, also associated with scale insects.

The commonest type of injury is manifested as a cracking of the bark, sometimes associated with an abnormal swelling of the underlying tissue. On *Carpinus caroliniana* the type of injury caused by *S. pseudopedicellatum* is rather unusual. The fungus occurs on the main stem and branches, frequently girdling the stems of small trees. Such stems are actually thinner in the infected region than above or below and break easily here if bent.

In addition to *S. pseudopedicellatum* and *S. Curtisii* there are several other very destructive species in the southeastern United States. *Septobasidium castaneum* is abundant on *Quercus phellos*, *Q. nigra* and *Ilex americana*, causing considerable damage in certain regions. Dr. G. M. Armstrong of Clemson College, South Carolina, reports in a letter that *S. castaneum* greatly injures cultivated azaleas in parts of South Carolina. *Septobasidium fumigatum* is very abundant on *Acer* sp. and very injurious, as is *S. Langloisii* on *Magnolia virginiana*.

Septobasidium Burtii though less abundant and less conspicuous than any of the above species causes slow but considerable damage. E.g., the lower limbs on several specimens of *Quercus palustris* in the Coker Arboretum have been killed off during the last ten years.

Usually the smaller species do not cause marked injury. A striking exception to this is *S. leprosum* which is very destructive to *Ilex decidua*.

Another common type of injury caused by the fungus-scale insect combination is the witches' broom. This abnormality is most commonly observed on *Fraxinus americana* or *Nyssa sylvatica*, caused either by *S. pseudopedicellatum* or *S. Curtisii*, and also on *Ilex americana* caused by *S. castaneum*.

In Jamaica the writer has noticed considerable damage done to *Solanum punctulatum* by *S. jamaicaense* (for details see under that species).

Other observers have also noted injury to trees and shrubs due to *Septobasidium*. Galloway (1890) and G. W. Wilson (1912) report *Septobasidium* harm to pear, apple, and certain forest trees in the United States. Rick reports from Brazil that *Septobasidium* causes serious damage to *Citrus* and various other trees.

In Java, Gäumann (1927) reports that he found in the "Berg garten

Tjibodas," cultivated hybrid roses brought from Europe suffering so severely from *S. bogoriense* that they were nearly all killed off.

Of the numerous species of *Septobasidium* reported by Petch (1923) from Ceylon on dadap, tea, *Citrus*, etc., none are reported by him as causing any serious injury though he does report considerable damage from one species in Northern India. Indeed he (1923) states, "These fungi are beneficial, in that, when a bush is badly infested by scale insects, they grow over the insects and destroy them. . . . Unfortunately, however, these fungi in many cases fail to maintain their (beneficial) reputation. After the colony of scale insects has been destroyed, they may attack the stems of the bush and kill them. This occurs in India in the case of Velvet Blight, the hyphae of which penetrate through the bark and into the wood, chiefly along the medullary rays. Similarly, the Ceylon species (*S. rimulosum*) killed the tea stems after it had finished off the insects." Petch has very kindly sent me some dried material of this latter species and I find that even in the oldest part of the specimens there are scale insects which are unparasitized, even several adult females which had been giving birth to young. From this simple observation it seems that we have the same symbiotic relationship between the fungus and scale insects here as in *S. Burtii* and that the damage done is due to the combined action of fungus and insect and is not direct parasitism of the fungus on the tree.

Septobasidium acaciae has been reported from Formosa by T. Tanaka (1918) as sometimes killing *Acacia* trees. The author gives no details, merely saying, "The affection is closely related to the attack of scale insects, and in many cases the dead insects were found embedded in the mycelial strands. The fungus sometimes kills *Acacia* trees as was discovered by a forest inspector, so immediate treatment is desirable." I have examined dried material sent by Professor Tanaka and find non-parasitized insects in all stages of development as well as parasitized insects. It seems probable therefore that the damage to the *Acacia* trees is due to the fungus-insect combination. It may be in some cases that the fungus first kills the scale insects then attacks the tree but no one has yet given sufficient evidence to prove such a statement.

Control Methods

Gäumann (1922) reports unsatisfactory results from spraying experiments to control *S. bogoriense* in Java.

Boedijn and Steinmann (1930) give much information concerning sprays but state that more emphasis should be placed on getting rid of the scale insects by pruning the infected branches.

In 1927 and 1928 we carried out some experiments to see if *Septobasidium* could be destroyed by sprays. *Septobasidium Curtisii*, on ash, *S. Burtii*, on oak, *S. Cokeri*, on oak, and *S. apiculatum*, on dogwood, were all abundant in the arboretum and hence convenient for spraying and observation. Four different sprays were tried: Volck, Bordeaux mixture, kerosene emulsion, and lime sulphur. Kerosene emulsion was also used as a paste. We found that even two heavy applications of the spray at intervals of about a month gave unsatisfactory results. Even if the superficial parts of the fungus were killed, the lower parts and particularly the haustoria in the insects survived and the old patches came back in a year or so. Of all methods tried the only effective one was smearing the kerosene emulsion paste on the patches of *Septobasidium* with a rag. This was tried with *S. Curtisii* on ash.

With our new knowledge of the fungus-insect relationship, it would seem that pruning the infected branches during the dormant season would indeed be the most effective means of control.

In summary, it appears at present from our limited knowledge that there is no immediate cause for alarm over *Septobasidium*. Though the fungus is wide-spread and abundant in certain localities and damaging to certain trees, the trees most seriously injured, as *Nyssa sylvatica* and *Fraxinus americana*, are not important timber trees. On cultivated plants I have only rarely found *Septobasidium* on well kept apple, citrus, or pecan trees. The fungus, however, may be very abundant on neglected fruit and nut trees. Generally the same conditions favoring the growth of scale insects alone also favor the growth of the *Septobasidium*-scale insect partnership. In this connection Comstock's remarks (1916) are of interest, "The general testimony of fruit growers is that sickly trees are much more liable to be attacked by scale insects than those which are healthy."

Heretofore several phytopathologists have considered *Septobasidium* as harmless, even as beneficial, since they were under the impression that the fungus overgrew and killed off the scale insects. Since this conception has been shown to be erroneous, that the fungus protects rather than kills off the insects, it is imperative that we look upon the *Septobasidium*-scale insect combination as distinctly harmful.

III

GEOGRAPHICAL DISTRIBUTION, HOST TREES, AND HOST INSECTS

Except for the Southeastern United States, no region in the world has been carefully combed for *Septobasidium*, and even there it is doubtless true that many species remain to be discovered. With very few exceptions the species from this region have been on native, uncultivated host plants. In the tropics many of the species have been collected on cultivated plants. For example, seven of the eighteen species described by Boedijn and Steinmann (1931) from Java occur on tea, and four of these have been reported as occurring only on tea. In short, it appears that nowhere in the tropics has anyone gone out into the native jungles to collect *Septobasidium*. Until this is done on a large scale in a number of regions, no discussion of geographical distribution will be of much value.

NUMBER OF SPECIES COLLECTED FROM VARIOUS COUNTRIES

North America
- Canada 1
- U. S. A. 36
- Mexico 4

Central America
- Nicaragua 2
- Costa Rica 1
- Panama 2
- Guatemala 2

West Indies
- Cuba 8
- Jamaica 10
- Guadeloupe 5
- San Domingo 1
- Puerto Rico 1
- St. Lucia 1

South America
- Trinidad 1
- Guiana 1
- Brazil 26
- Ecuador 4
- Chile 2
- Venezuela 1

Europe
- Portugal 1
- France 4
- Italy 3
- Poland 2
- Sardinia 1

Asia
- China 9
- Japan 4
- Formosa 4
- Burma 1
- Ceylon 18

East Indies
- Pelew (Palau) Is. 1
- Java 20
- New Guinea 2
- Madura 1

Philippine Islands 18

Australia 7

Madeira Islands 1

Africa
East Africa 1
S. Africa 6
Liberia 1
Uganda Protectorate 2
Tunis 1

Unlike many other fungi, it seems that *Septobasidium* is not disseminated by wind blown spores. There seem to be two means of distribution. In *S. Burtii* it has been shown that the fungus is spread over the host tree by the infected young which may crawl out, settle down on a fresh spot and start a new colony. Although the mode of infection is not certainly known in other species, it is known that the young patches of growth start from one young infected insect. The spread of the fungus from tree to tree and over wide areas seems to depend not upon the dissemination of wind blown spores but upon the dissemination of the infected young. This can be successfully done only during a very short period, i.e., while the young are in the crawling state, for once the sucking tube has pierced the bark the insect is anchored for life. The young have no wings and could hardly crawl from tree to tree unless the limbs of two trees were touching. It is known that several species of birds eat scale insects (Newstead, 1901) and it is quite possible that while feeding on the adults the young, both healthy and infected, might crawl on the birds' feet and thus be carried from tree to tree and even for considerable distances. The young might also be transported by other insects and possibly by the wind. The other means of distribution is by the transplanting of infected plants. This is certainly the most important and possibly the only way by which *Septobasidium* is transported for any great distance across natural barriers.

These two restrictive means of dissemination would seem to account for some of the peculiarities in geographical distribution. The species in Florida and other southern states are entirely different from those in Cuba, Jamaica, or other islands of the West Indies. Also the species in Jamaica appear to be distinct even from those of Cuba, except for *S. rhabarbarinum*. A few of the West Indian species do occur in South and Central America and one from Guadeloupe is also found in Mexico. However, the forms from the islands which also occur on the mainland are not exactly the same; they differ enough to justify the establishment of varieties and in some instances distinct species. *Septobasidium septobasidioides* from Brazil and Jamaica is an excellent illustration.

The form from Brazil has a more distinctly winged margin, a thicker hymenium and longer and thinner spores than has the Jamaican form and yet the two are so strikingly alike in other characters that they can hardly be made distinct species.

Some species have a very wide distribution, doubtless having been transferred from their native home to far distant lands on infected stock. *Septobasidium pseudopedicellatum*, which occurs abundantly in the United States on native as well as cultivated trees and shrubs, occurs in Brazil on *Citrus*. *Septobasidium Curtisii*, also very common here, has been found in Java on tea. *Septobasidium bogoriense*, perhaps the commonest species in the East Indies, has been found once on *Citrus* in the West Indies. These introduced forms are apparently indistinguishable from the parent stock. *Septobasidium rhabarbarinum* and its varieties occur throughout the tropics: Central America, Africa, and the Orient. There is no way to account for its very wide distribution, since it does not occur on cultivated plants, except that it became widely dispersed before the land bridges connecting South America, Africa, and the Orient were broken. The brilliant color of this species makes it easily recognized and this may account for the widely dispersed collections. Other less conspicuous species when carefully searched for may be found to have an equally wide distribution.

Another important factor in the distribution of *Septobasidium* is that the tropical scale insect hosts do not become acclimatized to the colder weather of the temperate zones, nor do the temperate zone scale insects thrive in the tropics (Newstead, 1901). This probably accounts for the fact that very few of the tropical species occur in temperate zones and vice versa.

Since the southeastern United States is the only region in the world in which any considerable collection data have been accumulated a brief summary is given here.

The distribution of the species of *Septobasidium* by states is shown in the accompanying table 1. The most extensive collecting trips have been made in North Carolina, South Carolina, Florida, and Louisiana and this fact might well explain a greater number of species from these states than from Georgia, Alabama, Mississippi, and Texas. It is quite likely that as more collections are made in these latter the number of species from them will be greatly increased. Only five species are known to occur in the northern states, but no one has made an extensive search for the genus in that region.

In the United States the commonest of all species appears to be *S.*

TABLE 1

Distribution of Species of Septobasidium by States as Indicated by the Number of Collections

	Va.	N.C.	Tenn	S.C.	Ga.	Ala.	Miss.	Fla.	La.	Tex.	Ark.	Calif.	Pa.	N.J.	N.Y.	Canada	Total Collections.
S. alni		several		9	1												15
var. squamosum				6													6
S. apiculatum	1	27		3				2	1		1						35
S. canescens												5					5
S. Carestianum																7	7
S. castaneum		5		8	1	1	3	2	7								27
S. Cokeri	1	18						2						1			22
S. cremeum								2									2
S. Curtisii	2	60		14	6	2	2	6			3						95
S. filiforme	1	21															22
S. fumigatum				4	2			9	2								17
S. fuscum							1		1								2
S. grandisporum				1													1
S. Hesleri			2														2
S. Langloisii							6	13	2								21
S. lepidosaphis								2									2
S. Leprieurii				3			5	4									12
S. leprosum	1	6		2					1								10
S. lilacinoalbum		6		1													7
S. Mariani		14	2	1					4		1						22
S. Patouillardii		9		1				3	5								18
S. Peckii															1		1
S. pilosum								1	1								2
S. pinicola		1											several		several		1+
S. pseudopedicellatum	4	30	2	4		1		14	12								67
S. rugulosum				2				3	1								6
S. sabalis									4								4
S. sabal-minor								1									1
S. Schweinitzii		10		1													11
S. sinuosum		3		14			3	11	8								39
S. Sydowii										1							1
S. taxodii									1								1
S. tenue				1			5	3									9
S. Burtii		40		2	1	1			1								45
Total species so far collected in each state	6	15	3	18	5	4	7	16	15	1	3	1	1	1	2	1	
	Va.	N.C.	Tenn	S.C.	Ga.	Ala.	Miss.	Fla.	La.	Tex.	Ark.	Calif.	Pa.	N.J.	N.Y.	Canada	

TAB

HOST TREES OF SPECII

Species of Septobasidium / Host Trees	Acer floridanum	Acer negundo	Acer rubrum	Acer saccharinum	Albizzia	Alnus incana	Alnus rugosa	Aristolochia	Azalea	Betula	Carpinus caroliniana	Ceanothus	Cercis canadensis	Citrus aurantifolia	Citrus grandis	Citrus sinensis	Cornus Amomum	Cornus florida	Cornus stolonifera	Crataegus	Fagus grandifolia	Fraxinus americana	Gleditsia triacanthos	Hicoria alba	Hicoria aquatica	Hicoria glabra	Hicoria pecan	Ilex decidua	Ilex glabra	Ilex opaca	Ilex verti-	Jun-
S. alni																1								1			3				1	
var. squamosum																								1								
S. apiculatum											1		2				18	9							1	1						
S. Burtii				2																												
S. canescens																																
S. Carestianum																			7													
S. castaneum					1				1									2			1			1						3		
S. Cokeri		2																2			1				1							
S. cremeum																													1			
S. Curtisii											6						3	2		2		35	2	2		2		2		1	1	1
S. filiforme		1					1				4	1	1				3	5		1				1								
S. fumigatum	1		11														2	1														
S. fuscum																																
S. grandisporum																	1															
S. Hesleri											2																					
S. Langloisii											10									2		2										
S. lepidosaphis														1	1	3																
S. Leprieurii					1																											
S. leprosum																	1	2		1		1		1				4				
S. lilacinoalbum																								1	5							
S. Mariani			2														2	3		2								1				
S. Patouillardii																						13	1					2				
S. Peckii						1																										
S. pilosum																																
S. pinicola																																
S. pseudopedicellatum			1					1		1	12			2	2	2	3	5		3		15	3		1			4				
S. rugulosum																																
S. sabalis																																
S. sabal-minor																																
S. Schweinitzii																						10										
S. sinuosum										1	1							3												1		
S. Sydowii																																
S. taxodii																																
S. tenue																																
Total species on hosts	1	2	3	1	2	1	1	1	1	2	7	1	2	2	2	3	8	10	1	6	2	6	3	7	4	2	1	5	1	4	1	1

LE 2

…S OF SEPTOBASIDIUM

(…elata, …ans nigra)

Ligustrum sinense	Liquidambar styraciflua	Magnolia grandiflora	Magnolia virginiana	Mangifera indica	Myrica cerifera	Nyssa sylvatica	Nyssa sp.	Pinus strobus	Prunus	Pyrus angustifolia	Pyrus communis	Pyrus japonica	Pyrus malus	Quercus agrifolia	Quercus alba	Quercus laurifolia	Quercus marilandica	Quercus montana	Quercus nigra	Quercus palustris	Quercus phellos	Quercus rubra	Quercus tinctoria	Quercus virginiana	Sabal Deeringiana	Sabal minor	Staphylea trifolia	Syringa vulgaris	Taxodium distichum	Thea	Tilia sp.	Ulmus americana	Ulmus alata	Ulmus crassifolia	Viburnum	Unidentified	Total hosts
	2		1															1		2																	8
	5																																				2
	4		1									1							1	3																	11
5									1		1					1	1	2	33	3				1													10
														5																							1
																																					1
	2		2			1					1							3		7	1			1													14
																	1		2	3	8	1															9
	1																																				2
	1	1	2		3	28	10				1	1			1			1	3	3	2	1								1					2	4	29
1	1											2																1									13
																															1					1	6
																		1																		1	2
																																					1
																																					1
			12																																	3	5
				1																																	5
			9																																		2
																																					6
																																					2
	2									1	2	8	1					5	1										1								13
																																					3
																																					1
			1	2																										1							3
								12																													1
	6		6			1						1						2									1		1				5	1			24
	3		2															1																			3
																									3												1
																										1											1
																																					1
	7		3		3										1	1		9		4																4	12
																																			1		1
																													1								1
			5															4																			2
2	11	1	11	2	2	3	1	1	1	1	4	5	1	1	2	2	1	1	9	5	7	3	1	2	3	1	1	1	3	2	1	1	1	1	1		

Curtisii which occurs on twenty-eight different species of trees, its most frequent hosts being *Nyssa sylvatica* and *Fraxinus americana*. The next most abundant is *S. pseudopedicellatum* which occurs on twenty-three different species of trees, its most frequent hosts being *Fraxinus americana* and *Carpinus caroliniana*. *Septobasidium sinuosum* has been found on fifteen different trees, but is most frequent on *Quercus nigra* and *Liquidambar styraciflua*. *Septobasidium castaneum* occurs on fourteen different trees, but is usually found on *Quercus phellos*. Other species show a more limited range in the choice of tree and several have been found on only one species as, for example, *S. canescens* on *Quercus agrifolia*, *S. Carestianum* on *Cornus stolonifera* (in this country), *S. grandisporum* on *Cornus florida*, *S. Schweinitzii* on *Fraxinus americana*, and *S. sabalis* on *Sabal Deeringiana*, etc.

In my collections seventy-six different species of trees have been found to be subject to attack by *Septobasidium*. Of these, some are very susceptible to the fungal-insect combination. Eleven different species of *Septobasidium* have been found on *Magnolia virginiana*, and *Liquidambar styraciflua*, ten on *Cornus florida*, nine on *Quercus nigra*, eight on *Cornus stolonifera*, seven on *Quercus phellos* and *Hicoria alba*. The trees most subject to attack, however, are not always the ones which harbor the greatest number of different species. The species most susceptible are *Fraxinus americana*, *Nyssa sylvatica*, *Magnolia virginiana*, *Quercus phellos*, and *Q. nigra*. Many species of trees and shrubs appear to be free from attack by *Septobasidium* and others may be practically free, as *Fagus grandifolia*, *Tilia*, *Juglans*, and *Azalea* sp., on each of which only one single collection has been reported (Table 2).

Nineteen different species of scale insects have been found associated with *Septobasidium* in this region. Some species as *S. sabalis*, *S. Schweinitzii*, etc., which occur on only one host plant are associated with only one species of scale insect. Other species which are found on a wide variety of host plants may be associated with several species of scale insects, e. g. *S. alni* and its variety *squamosum* with seven different species. In some cases, as in *S. apiculatum* on *Cornus amomum*, as many as three different species of scale insects have been found under the same specimen, *Aspidiotus ancylus*, *Chrysomphalus obscurus*, and *Parlatoria proteus*, but only individuals of the first were parasitized. *Septobasidium Curtisii*, which usually occurs on ash and black gum with *Chionaspis sylvatica*, may also parasitize species of *Aspidiotus*, as for example *Aspidiotus perniciosus*. From the results so far obtained it appears that the greater number of species of *Septobasidium* are not limited to an association with one species of scale insect but may be associated with several.

The scale insects associated with *Septobasidium* in the United States are shown in the list below. The figures after each species of insect indicate the number of species of *Septobasidium* with which the insect has been found. Identifications of the scale insects have been made by Dr. Harold Morrison, Bureau of Entomology, U. S. D. A., Washington, D. C.

Aspidiotus ancylus 5
Aspidiotus diffinis 3
Aspidiotus forbesi 4
" *juglans-regiae* 10
" *lataniae* (not typical) 1
" *liquidambaris* 3
" *osborni* 8
" *perniciosus* 2 (?)
" (?) *sabalis* 1
" *uvae* 1
" sp. 1
Cerecoccus ehrhorni 1
Chermes sp. apparently 1
Chionaspis corni 1
" *gleditsiae* 6
" *sylvatica* 2
Chrysomphalus obscurus 2
" *tenebricosus* 1
Lepidosaphes beckii 1

In most of the foreign material the scale insects have not been identified. The species which have are given in the lists below.

Scale insects with Jamaican species of *Septobasidium* (identified by Dr. Harold Morrison).

Aspidiotus fabernii 3
Chrysomphalus odontoglossi 3
Chrysomphalus sp. 4
Diaspis sp. ? 1
Pseudoparlatoria ostreata 1

Scale insects with other foreign species of *Septobasidium* (authority for identification unknown).

Aspidiotus rapax 1
Aspidiotus sp. 13
Aulacaspis pentagona 1
Chionaspis biclavis 2

Chionaspis sp. 5
Diaspis sp. 1
Lepidosaphes sp. ? 1
Mytilaspis sp. 1
Odanaspis schizostachyi 1
Parlatoria proteus 2
" *proteus pergandi* 1
Parlatoria sp. 1
Phylloxera sp. 1
Saskiaspis pentagona 2

IV

STRUCTURAL FEATURES OF TAXONOMIC IMPORTANCE

The size, structure, and color characters in *Septobasidium* are of considerable diagnostic importance. The different species vary greatly in size; some are characteristically small and inconspicuous, as, for example, *S. minutulum*, *S. apiculatum*, *S. Hesleri*, *S. pilosum*, etc. Other species, though forming small patches of growth may, because of the large number of patches formed, be quite conspicuous, e.g. *S. Burtii*. Other species form large and conspicuous patches of growth, e. g. *S. pseudopedicellatum*, *S. jamaicaense*, *S. castaneum*, *S. sinuosum*, *S. Langloisii*, etc. Some species show a characteristic definiteness of shape and have a sharply delimited margin, while in other species the margin may be indeterminate and the outline of growth quite irregular. I am using the term "determinate margin" when the margin is definite and distinct and its limits can easily be seen. Such a margin may be smooth or irregular, fimbriate or fibrillose. The "indeterminate margin," on the other hand, is indefinite and so vague that its limitations cannot be determined. The thickness of the growth is also characteristic for the different species, some producing patches the top surfaces of which stand up from the bark as much as a centimeter, as, for example, *S. jamaicaense;* while other species may be so thin as to appear as mere discolorations on the surface of the bark, as, for example, *S. tenue*.

Just as the different species vary in size, they also vary strikingly in the structure of the context. About half of the species described below show three distinct structural regions or layers. First the subiculum which is next to the bark and from which arises the second layer composed of vertical pillars or columns of hyphae. The hyphae of the pillars branch out to form a second horizontal layer which extends as a flat roof over the tops of the pillars. Some of the species with pillars have three horizontal layers: the subiculum, the top layer, and then another, incompletely formed, layer parallel to these two and slightly raised above the subiculum. *Septobasidium pseudopedicellatum* and *S. Cokeri* are excellent examples of forms in which there are two horizontal layers and in which the pillars are very conspicuous. *Septobasidium Mariani* and *S. castaneum* illustrate types with three horizontal layers. The size and structure of the pillars vary greatly in the different species and are therefore of considerable diagnostic value.

The species in which pillars do not occur show such enormous variations in vegetative structure that space does not permit a detailed discussion.

Another structural feature of great taxonomic importance is the type of insect trap in which the young settle and the kind of house in which the adults live. Since these characters are discussed in detail elsewhere they will be omitted here.

Some species of *Septobasidium* are brilliantly colored; for example *S. rhabarbarinum* is golden yellow, *S. rubiginosum* reddish purple, *S. Langloisii* grayish blue, *S. Peckii* tawny. A number of species are almost black, while others are nearly pure white when fresh. By far the commonest color is some shade of brown.

In some species a distinct, compact, hymenial layer is formed. This layer may be composed of entangled threads and probasidia as in *S. Burtii*, *S. pseudopedicellatum*, *S. bogoriense*, etc., or of upright sterile threads and basidia in various stages as in *S. albidum*, *S. jamaicaense*, etc. In other species there is no distinct hymenial layer; the probasidia and basidia are loosely scattered over the upper surface.

In reproductive structures the various species show wide variation. Indeed at least nine distinctly different types of probasidial and basidial structures have been recognized as indicated in the list below. The probasidial cell may be present or absent. If present, it may persist as an empty cell at the base of the basidium or it may elongate and become septate to form the basidium. The basidia may be one, two, three, or four-celled; straight, slightly bent, or coiled. By far the two commonest types are those in which the basidium is four-celled and straight or four-celled and coiled, each with a persisting probasidial cell.

	Species
(1) Basidia 1-celled, with persisting probasidial cell	2
(2) Basidia 2-celled, without persisting probasidial cell	7
(3) Basidia 2-celled, with persisting probasidial cell	13
(4) Basidia 3-celled, without persisting probasidial cell	1
(5) Basidia 3-celled, with persisting probasidial cell	1

	Species
(6) Basidia 4-celled, without persisting probasidial cell, basidia coiled	15
(7) Basidia 4-celled, with persisting probasidial cell, basidia straight.	49
(8) Basidia 4-celled, with persisting probasidial cell, basidia coiled.	41
(9) Basidia 4-celled, without persisting probasidial cell, basidia straight or slightly bent	10
No. of species in which basidia have not been found	31

At first glance it might seem advisable to divide the genus up into nine subgenera on the basis of the above characters. Such a division would, in the writer's opinion, retard rather than advance constructive research in the taxonomy of this group, for these nine different groups show a remarkable overlapping in practically all other structural characters. Nor does it seem advisable to recognize the subdivisions "Typicae and Gausapia" proposed by Patouillard (1892), nor the new subgenus *Noackia* proposed by the same author (1907), nor the three genera *Ordonia, Mohortia,* and *Septobasidium* proposed by Raciborski (1909) (see Couch 1929, and Boedijn and Steinmann 1931), nor the new subgenus *Unisporum* proposed by Couch (1929).

Some species produce conidia as well as basidia and basidiospores and in *S. Westoni, S. Curtisii, S. paulense,* and *S. lanosum,* etc., these two types of spores appear to be formed on different mycelia. In still other species only the conidial stages have been found.

In several species, e. g. *S. cirratum, S. cervicolor, S. lanatum, S. tuberculatum,* etc., a peculiar thin-walled, bladder-shaped cell is formed, about 10–15μ thick, first observed and illustrated by Boedijn and Steinmann (1931, figs. 4c, and 11) for *S. theae* and *S. tuberculatum.* These cells have thin protoplasm and a columella-like structure at the base. They may easily be mistaken for probasidia, for which, however, there is no justification since they always occur in the context and usually near the scale insects. The function of these cells is uncertain. Another peculiar type of cell has been observed in *S. apiculatum, S. theae, S. paulense, S. lanosum,* etc. These cells arise from the hyphae and have an ovoid or cylindrical base with a long narrow neck. Frequently on the ends of these long tubes is formed a globose body with glistening contents. In *S. apiculatum* these bottle- or jug-shaped cells are formed out on the margin where the young insects settle down and it is quite possible that they have something to do with infecting the young.

All species which I have studied in the living condition (*S. Burtii, S. pseudopedicellatum, S. Cokeri, S. apiculatum, S. Schweinitzii, S.*

Patouillardii, S. filiforme, S. Curtisii, and *S. lilacinoalbum*) are perennial and the evidence indicates that those species studied in the dried condition are also perennial. Such species undergo a regular cycle of development. In this region the probasidia are formed during late fall and winter, germinating to form the basidia throughout the spring and early summer. Also with the coming of spring the fungus starts new growth, continuing to grow during the summer and early fall. In the greater number of species a new zone of growth appears around the margin. In some species in addition to a new marginal region of growth, an entirely new top layer of growth may be formed over the old layer. Growth is characteristically slow in all the above species, amounting to only a few millimeters each season.

I have found in *S. Burtii* and *S. Curtisii*, that the spores when mature are shot away from the sterigmata in the same manner as described by Buller (Researches, Vol. 2) for various Basidiomycetes. A small droplet of liquid suddenly appears at the spore-hilum. This droplet increases in size until it becomes about equal to one third the volume of the spore and then suddenly the spore and droplet disappear. The spores are shot away so quickly that one cannot follow the course of one spore but can see a single spore as it suddenly flicks across the field of view.

The haustoria within the parasitized scale insects are very highly specialized as to structure; they are perhaps the most highly specialized haustoria in all the fungi. The type of haustorium has been found to be constant in any one particular species, even in the species of *Septobasidium* which are associated with several different species of scale insects. Since haustoria are always present, while the reproductive structures are not, the structure of the haustorium is of great diagnostic value in separating species, and sometimes even in determining if a particular fungus belongs in the genus. Six types of haustoria have been recognized: (1) glomerulus type, e. g., *S. pinicola, S. Clelandii, S. accumbens,* etc.; (2) irregular, long-coil type, e.g. *S. bogoriense, S. pseudopedicellatum, S. filiforme, S. lichenicolum,* etc.—commonest type; (3) regular coils connected by fine threads and spindle-shaped segments, e.g. *S. Burtii, S. cinereum, S. fuscum*; (4) tufts of hyphae composed of spindle-shaped segments connected by fine threads, e.g. *S. apiculatum, S. rhabarbarinum, S. tuberculatum,*—a common type (the haustoria of *S. arboreum* representing a variation of this); (5) threads composed of spindle-shaped segments which give rise to branches of the same structure but which coil, e.g. *S. Langloisii, S. fumigatum, S.*

pilosum, S. theae; (6) threads of spindle-shaped segments which give rise to irregular coils, e.g. *S. alveolatum.*

The importance of the haustoria from a taxonomic standpoint as well as from the standpoint of understanding the fungus-insect relationship should be strongly emphasized. Although it seems that none of the other students of *Septobasidium* have observed the haustoria, they are not really difficult to find. The infected insects are smaller than the healthy ones of the same age and are more difficult to see since they are covered by a thick, compact mound of fungal tissue. However, they can be dissected out with needles under a binocular dissecting microscope and mounted in water if the material is fresh, or in seven percent potassium hydroxide solution if the material is dry. Sometimes in very old material it is necessary to boil the insects in a weak potash solution to cause the haustoria to swell. Usually the haustoria can be seen through the body-wall of the insect, but not infrequently it is necessary to tear the insect's body open with needles or crush it by gently pressing the cover slip, before these structures can be clearly seen. By careful manipulation it is possible to remove the parasitized insects from the fungal covering and to mount them while they are still alive, in which case one can observe not only the haustoria but also the muscles of the pumping apparatus at work.

It has also been found that the mode of connection between the haustoria within the insects and the external fungus is characteristic for certain groups of species. Five different types of connections have been found: (1) connections passing through dermal pores, e.g. *S. Burtii, S. apiculatum;* (2) through skin around mouth, e.g. *S. sabalis;* (3) through mouth, passing in by sucking tube and entering retreat pouch for sucking tube, e.g. *S. prunophilum, S. albidum, S. Henningsii;* (4) through vaginal pore, e.g. *S. Curtisii;* (5) through anus, and sometimes through both anus and vaginal pore, e.g. *S. pseudopedicellatum.* Just exactly how and when the connection between the fungus within the newly infected insects and the external fungus is made, has been determined only in *S. Burtii.*

It has been observed that the species in which the haustoria are connected by fine threads, as *S. Burtii, S. apiculatum,* etc., never penetrate the insect through the anus, vaginal pore, or mouth region, but enter directly through the minute pores in the insect's derm. On the other hand, those species in which the haustoria are in the form of irregular coils connected by hyphae of the same thickness, as in *S. bogoriense, S. pseudopedicellatum,* etc., never penetrate the insect through the minute derm pores but enter through the anus, vaginal pore, or mouth region. This appears to be a remarkable adaptation. Those species which are able to form fine threads can penetrate through the minute pores in the insects' skin, while those species which cannot or do not form fine threads, must rely upon entering the insect through the larger body apertures as the anus or vaginal pore.

V

HYBRIDIZATION

It frequently happens that one finds several species of *Septobasidium* growing on the same tree and even on the same limb. Numerous times I have found *S. pseudopedicellatum*, *S. Schweinitzii*, *S. Curtisii*, and *S. Patouillardii* on the same limb. The patches may come in contact along their margins and not infrequently one patch may appear within the old circle of growth of another. Such mixed patches doubtless originate if the infected insects from one species of *Septobasidium* crawl up under another species. As a rule such intermingling of colonies of fungal growth has no apparent influence on the type of growth or reproductive bodies formed, i.e. even though the two may appear to be grafted along their contiguous margins both retain their vegetative and reproductive characteristics. One good example of hybridization, however, has been found (Couch, 1935) to occur between *S. pseudopedicellatum* and *S. Schweinitzii*. The former is characterized by distinct pillars on the margin, probasidia, and straight basidia, with short sterigmata; while the latter has tents on the margin, no probasidia and coiled basidia, with very long sterigmata. The hybrid combined to a remarkable degree the vegetative and reproductive characters of both species, having both pillars and tents on the margin, probasidia, slightly curved basidia with sterigmata of intermediate length. (For further details and figures see p. 206). This is the only unmistakable example of hybridization yet reported in the genus, though there are several others which have all the appearance of being hybrids. Rick reports from Brazil that he has seen several apparent examples of hybridization.

On several occasions I have observed that when two species are growing in intimate contact, one or both species may show a marked increase in basidial and spore production along the margin where the two are in contact. Such an increase in fertility was noted in *S. filiforme* growing in contact with *S. Mariani* and in *S. crustaceum* growing with *S. aligerum*.

VI

CYTOLOGY

Some preliminary work has been done on the cytology of *S. jamaicaense*, *S. pseudopedicellatum* and *S. grandisporum*.

In *S. jamaicaense* sporulating material was killed in formol-acetic-alcohol. Several months later, after embedding in paraffin, sections were cut 5μ thick and stained with Haidenhain's iron alum haematoxylin.

The cells of the hyphae contain one to several nuclei. The young basidium originates as a lateral or terminal enlargement and even in the earliest stages is uninucleate. The nucleus and basidium enlarge simultaneously until the latter has almost reached mature size, upon which a longitudinal or diagonal spindle appears and the nucleus divides mitotically. Then a cross wall separates the basidium into two uninucleate cells, each of which gives rise to a uninucleate spore. The same nuclear behavior is found in several rusts (Jackson, 1935).

Septobasidium pseudopedicellatum and *S. grandisporum* were killed in Flemming's weak solution and 5μ sections stained in Gram's gentian violet. Some slides of the latter were also stained in Flemming's triple stain and some in Haidenhain's iron alum haematoxylin. In *S. pseudopedicellatum* the cells of the subiculum, pillars, top layer, were all binucleate, as were also the young probasidia. As the latter mature, nuclear fusion occurs so that the mature probasidium contains a single large nucleus. Division stages were not observed but the mature basidium is composed of four uninucleate cells each of which gives rise to a uninucleate spore. In *S. grandisporum* the cells of the context and hymenial region are usually binucleate though occasionally uninucleate. The young probasidia are binucleate, the mature ones have a single large nucleus. The probasidium germinates and the basidium nearly reaches its full length before the nucleus moves out into it. Two nuclear divisions occur so that the mature basidium contains four longitudinally arranged nuclei. These, with the cytoplasm, pass out through one long sterigma to form a single large spore. Shortly thereafter the spore is cut up into four uninucleate cells and then by other divisions into numerous, small, irregular, uni-, bi-, or trinucleate cells.

VII
RELATIONSHIP

Patouillard (1900) considered the Septobasidiaceae as a saprophytic family under the order Auriculariales next to the rusts. Coker (1920) treated *Septobasidium* with the Auriculariaceae but recognized its doubtful position here because of its parasitism on scale insects and its fibrous, leathery texture. Gäumann (1926) divides the Heterobasidiomycetes with transversely septate basidia into three orders, the Ustilaginales, Uredinales, and Auriculariales. The latter is then divided into four families, the Septobasidiaceae, Cystobasidiaceae, Auriculariaceae and Phleogenaceae. Gäumann includes only the genus *Septobasidium* in the family Septobasidiaceae. In view of the unique symbiotic association of *Septobasidium* with scale insects it seems to me that it is well to go a step further than Patouillard or Gäumann and establish *Septobasidium* as a separate order on a par with the Auriculariales, Uredinales, and Ustilaginales.

Heterobasidiomycetes

(Only the forms with transversely septate basidia are included here.)

1. Symbiotic with scale insects. Culturable as a saprophyte. Texture dry, fibrous, leathery, or flocculent. With or without probasidia. With basidiospores, bud cells, and sometimes conidia. *Septobasidiales*
2. Mostly saprophytic on dead wood, rarely parasitic. Texture cartilaginous, gelatinous, or waxy. With or without probasidia. With basidiospores, bud cells, and sometimes conidia.... *Auriculariales*
3. Obligate parasites on vascular plants. Mycelium greatly reduced; endophytic except in *Goplana*. With or without probasidia. Usually with several spore forms.......................... *Uredinales*
4. Parasites on vascular plants, completely or partially culturable as a saprophyte. Mycelium endophytic. With probasidia, basidiospores, and bud cells............................... *Ustilaginales*

Septobasidium shows several characteristics which indicate an unmistakable relationship to the rusts. There can be no doubt that the teliospores or probasidia in the two groups are homologous. In *Septobasidium* the teliospores show very little diversity in form, are usually unicellular, colorless and spherical, and lack a germ pore, while in the rusts these organs show a great diversity in structure. In *Septobasidium* the teleutospore is embedded in the fungal tissue and thus protected throughout its development, while in the rusts the teleutospores are usually subjected to adverse environment, some overwintering on dead plant parts while others may overwinter in the soil. It is quite possible that the uniform conditions with which the teleutospores of *Septobasidium* are faced, and the adverse and variable conditions encountered by the teleutospores of the rusts, may be correlated with the structure of these organs. The basidia of *Septobasidium* on the other hand show much greater diversity than in the rusts. In the latter the basidia, with only a few exceptions, are four-celled. In spite of the diversity of form there can be no doubt of the close relationship of the basidia of *Septobasidium* and the rusts.

There are two genera which show a striking connection between the rusts and *Septobasidium.* One of these, *Goplana,* represented by three species from Java, produces an extramatrical sorus on the underside of leaves which macroscopically resembles a simple species of *Septobasidium.* There is no distinct probasidium, the basidium is four-celled and straight as in *S. cirratum* or *Coleosporium.* The basidiospores are the only known spore forms. This genus is so close to *Septobasidium* that if it were associated with scale insects it would have to be transferred to this genus. The other intermediate genus is *Uredinella* represented by one species in the southeastern United States. This fungus, parasitic on scale insects, with haustoria like *Septobasidium,* has rust-like teleutospores with brown walls and an apical germ pore (Couch, 1937).

The genus *Herpobasidium* considered by Jackson (1935) as belonging in the rusts is parasitic on ferns. It has a basidium somewhat similar to *S. jamaicaense* but the character of great interest here is the coiled haustorium which strikingly resembles the ones found in certain species of *Septobasidium.* The relationship between the two genera is doubtless rather remote.

Septobasidium also shows a close relationship to the Auriculariaceae (sense of Killermann, 1928). That family however is characteristically gelatinous and saprophytic, though some of the genera included are parasitic and not gelatinous. The saprophytic, gelatinous genera as *Auricularia, Hirneola, Saccoblastia, Platygloea,* are certainly only remotedly related to *Septobasidium. Helicobasidium* resembles *Septobasidium* in the resupinate habit, fibrous texture, and transversely septate basidium. Aside from the septate characteristic, the basidia of *Heliocobasidium* are entirely unlike any found in *Septobasidium.* Furthermore, the association of the latter with scale insects emphasizes the rather remote relationship of these two genera. Other genera in this family which perhaps show a closer relationship to *Septobasidium* are *Cystobasidium*, and *Patouillardina*. The species *Septobasidium album* Burt should perhaps be transferred to the Auriculariaceae as a new genus close to *Patouillardina*. Both of these species are saprophytic on dead wood and *S. album* has clamp connections and ovoid spores, all of which are characters sharply distinct from *Septobasidium*. Both genera have the dry, fibrous texture and the septate basidia as in *Septobasidium*.

VIII

GENERIC DESCRIPTION

Septobasidium Pat. Jour. de Bot. **6**: 61. 1892.

Glenospora Berk. & Desm. Jour. R. Hort. Soc. **4**: 243. 1849.
Ordonia Rac. Bull. Acad. Sci. Cracovie **3**: 360. 1909.
Mohortia Rac. Bull. Acad. Sci. Cracovie **3**: 361. 1909.

Fungus body usually resupinate but very variable, dry, crustaceous or spongy. In the commoner species composed of three regions: the subiculum which grows over the bark, the middle region composed of upright slender or thick pillars or mounds or ridges of hyphae which support the top layer in which the hymenium is formed. In some species the context is more or less compact throughout. Threads septate without clamp connections. Probasidial cell undergoing a long period of rest before germinating or elongating immediately into the basidium; wall typically hyaline but rarely colored; probasidial cell sometimes lacking. Basidium usually transversely septate into two, three, or four cells, rarely one-celled, straight or curved. Basidiospores elliptic or bent-elliptic, colorless, becoming divided into two to many cells soon after formation and budding into numerous sporidia if kept moist. Some species with conidia.

Inhabiting the stems, branches, leaves, and sometimes the fruits of living woody plants and symbiotically associated with scale insects, some of which are parasitized, while others are left to reproduce. Widely distributed in the tropics and the warmer regions of the temperate zone.

According to priority the name *Glenospora* (see Couch 1933) should be used and the name *Septobasidium* thrown into synonymy. Such a change, however, would lead to endless confusion, for the name *Septobasidium* is well established in the literature and is known throughout the world, and furthermore the name itself has some significance. It can truly be said that the first time any member of the genus was described with a fair degree of understanding was when Patouillard (1892) described it and applied the name *Septobasidium*. We therefore follow Boedijn and Steinmann (1931) and retain *Septobasidium* as a *nomen conservandum*.

Although the genus *Septobasidium* was not established until 1892, certain species have been known for a long time. These have been distributed among the Thelephoraceae and other fungi and the lichens, e.g., *Septobasidium Michelianum* was first illustrated 1729 by Micheli as a lichen and *S. Schweinitzii* was first described by Schweinitz in 1822 as *Thelephora pedicellata*. Others have been described as *Corticium*, *Hypochnus*, *Daedalia*, *Hydnum*, *Glenospora*, etc. Doubtless there are still specimens in herbaria among these and other genera which should be transferred to *Septobasidium*.

In 1892 Patouillard founded the genus, describing the species *Septobasidium velutinum*, and since then has made numerous contributions to the taxonomy of this group. Other contributions have been made by Raciborski, Bresadola, Sydow, Petch, Burt, Coker, Couch, Boedijn and Steinmann, and others. (See Literature List.) All of these papers, however, were either brief descriptions of one or two species or a description of the species in a single geographical area. However, a distinct advance in the application of modern taxonomic methods to this genus was made by Boedijn and Steinmann (1931) with their full descriptions and excellent illustrations.

IX

KEY TO THE SPECIES

PAGE

I. Basidia 1-celled, with persisting probasidial cell.
   1. Sterile parts blackish brown, spongy, with distinct pillars, hymenium pale smoky gray; in section up to 2.5 mm. thick; margin with distinct rhizomorphs. South Carolina........*S. grandisporum* 79
   2. Sterile parts purplish brown, context soft, without pillars; in section 0.3–0.7 mm. thick. Jamaica..................*S. purpureum* 80

II. Basidia 2-celled, without persisting probasidial cell.
   A. Context with distinct pillars.
      1. Fungus body large, 3-layered, pillars very abundant, slender, up to 2 mm. tall. Jamaica..................*S. jamaicaense* 81
      2. As in (1) but much thinner and with less distinct and fewer pillars and smaller basidia and spores. Brazil. *S. fusco-cinereum* 83
      3. Fungus body small, surface gray with nodules and cracks, with a thick basal stromatic tissue. Brazil.......*S. sclerotioides* 84

PAGE

4. Fungus body small, surface blackish brown with purplish tint, sometimes lighter to snuff brown, velvety, pillars slender, straight, subiculum tan. Southeastern U. S. A. *S. Patouillardii* 85

B. Context without pillars.

1. Fungus body small, consisting of compact subiculum and upright pteruloid processes. Australia................ *S. Clelandii* 87
2. Surface brown, areolate with numerous rounded holes. Jamaica.. *S. areolatum* 88
3. Surface brown, reticulate with numerous elongated holes. Cuba.. *S. retiforme* 90

III. Basidia 2-celled, with persisting probasidial cell.

A. Context with distinct pillars.

1. Context 2-3 mm. thick, blackish, surface gray with numerous cracks, texture firm and hard throughout. South Africa. *S. protractum* 91
2. Context up to 2 mm. thick, surface brown, broken into sinuous zones, texture soft throughout. Ceylon......... *S. dictyodes* 92
3. Context up to 1 mm. thick, surface brown, smooth, texture soft throughout. Louisiana; on leaves of *Sabal*..... *S. sabalis* 93
4. Context up to 1.5 mm. thick, surface nodulose, rich, velvety, Brussels brown, with three horizontal layers. Chile. *S. cinnamomeum* 95
5. Context 1–1.5 mm. thick, stratose by the formation of several hymenial layers; on *Hevea*. Brazil............. *S. stratosum* 96
6. Context up to 1.5 mm. thick, surface mummy brown, velutinous, subiculum and top layer thin, pillars long and slender; on *Citrus*. Ecuador.............................. *S. velutinum* 97
7. Context 1–1.3 mm. thick, surface dotted with minute bumps, sometimes cracked, pale violet-gray, pillars blackish purple; on *Citrus*. West Indies and Brazil............... *S. griseum* 98

B. Context without pillars.

1. Surface divided into sinuous, anastomosing ridges; basidia 6–7 x 21–29μ. Southeastern U. S. A............ *S. Langloisii* 98
2. Surface much as above but ridges more rounded; basidia 3.8–4.4 x 20–32μ. North Carolina to Fla. and La...... *S. sinuosum* 100
3. Surface with sinuous ridges and dark dots in marginal region, elsewhere with vast number of cracks; basidia 5–5.4 x 21–27μ; on *Thea*. Ceylon............................ *S. rimulosum* 102
4. Surface smooth and velvety or with irregular cracks and holes; basidia 8.4–9.8 x 36–47μ. Brazil.............. *S. Bresadolae* 103
5. Surface smooth or minutely fibrillose, basidia 2–4 celled; on *Pinus rigida*. Massachusetts.................... *S. Linderi* 104
6. Surface smooth and glabrous in places; color bluish gray. South Africa........................... *S. griseopurpureum* 105

IV. Basidia 3-celled, without persisting probasidial cell.

A. Context without pillars, surface with upright spines.

1. Color pure white. Southeastern U. S. A......... *S. apiculatum* 106

PAGE

V. Basidia 3-celled, with persisting probasidial cell.
  B. Context without pillars, surface with bumps and nodulations.
    1. Color white; on *Quercus*. California.............. *S. canescens* 107

VI. Basidia 4-celled, curved, without persisting probasidial cell.
  A. Context with distinct pillars.
    a. With three horizontal layers, the top layer frequently extending out to form lateral margins.
      1. Hymenial surface brownish with numerous large rounded elevations. Philippines....................... *S. alatum* 108
      2. Hymenial surface light drab or purple gray. Ceylon and Australia.................................... *S. aligerum* 110
    b. With two horizontal layers, subiculum and top layer, without a winged margin.
      1. Surface smooth or minutely warted, velvety, color near benzo brown, usually with purplish tint, margin with tent-like structures. North and South Carolina. *S. Schweinitzii* 112
      2. Surface as in (1); summer color white with pale vinaceous tint, winter color as in (1); margin with pillars; on *Hicoria*. North Carolina.......................... *S. lilacinoalbum* 114
      3. Surface flaky, due to incompletely formed top layer, margin with pillars or irregular tents. Tennessee...... *S. Hesleri* 117
      4. Surface smooth at first, becoming slightly wrinkled or cracked, clay color to drab. South Africa.... *S. natalense* 118
      5. Surface pale drab gray, finely powdery when fresh, context frequently spongy, up to 2 mm. thick; on *Citrus*. Philippine Islands................................ *S. Reinkingii* 119
      6. Occurring in small round sometimes confluent patches; on *Ananas magdalenae*. Panama................. *S. Westoni* 120
  B. Context without pillars.
    a. Usually more than 0.5 mm. thick.
      1. Lateral margin winged, surface grayish purple, minutely granulose. New Guinea................... *S. granulosum* 121
      2. Lateral margin not winged, surface yellowish to old gold, margin frequently reticulate. Tropical America, Africa, and East Indies......................... *S. rhabarbarinum* 122
    b. Usually less than 0.5 mm. thick.
      1. Surface smooth except for flattened mounds about 2–4 mm. wide, tawny olive; on *Schizostachyum*. Philippine Islands. *S. subolivaceum* 126
      2. Surface whitish with large and small flattened bumps; hyphae consisting of T, or L or + shaped segments; on leaves. Ceylon.. *S. Petchii* 126
      3. Surface mouse gray, with numerous small dark mounds. Southeastern U. S. A............................ *S. tenue* 128
      4. Surface near sepia, minutely retiform. Southeastern U. S. A. *S. rugulosum* 129
      5. Infertile part blackish brown, myceloid, hymenium light

PAGE

VII. Basidia 4-celled, straight with persisting probasidial cell.

A. Context with pillars.

a. Pillars usually tall and distinct.

1. Surface smooth, glabrous, buff colored, pillars dark brown, unbranched, subiculum whitish. U. S. A. *S. pseudopedicellatum* 132
2. Surface much as in 1 except darker; pillars branching and forming a dense tangle; with three horizontal layers. U. S. A. and Europe .... *S. Mariani* 134
3. Much as in 2 except pillars and rhizomorphs are less distinct and hyphae in hymenium extend in vertical direction. Japan .... *S. Mariani* var. *japonicum* 137
4. As in 1 except pure white throughout and basidia and spores smaller. U. S. A. .... *S. Cokeri* 138
5. Occurring in small orbicular or irregular patches; deep brown; on bark. Java .... *S. obscurum* 140
6. As in 5 but wood to buffy brown; on leaves. Java .. *S. triviale* 141
7. Surface velvety, snuff brown, context very thick, 3–5 mm.; on *Dendrocalamus*. Burma .... *S. pachydermum* 141
8. Surface in form of numerous orbicular patches, Verona brown. Ceylon .... *S. hakgalanum* 142
9. Surface smooth, fuscous to nearly black, margin of subiculum with tent-like structures. Guadeloupe .... *S. atratum* 143
10. Surface near cinnamon-brown, pillars very dense, margin of subiculum with conspicuous (with lens) rounded insect houses. Ceylon .... *S. suffultum* 144

b. Pillars only fairly distinct, usually entangled.

1. Pillars distinct at first, anastomosing to give honeycomb appearance, usually without distinct top layer, rich brown in color, basidia formed mostly over the marginal subiculum. Florida, Brazil, Cuba .... *S. lepidosaphis* 146
2. Surface smooth, shiny, chocolate brown to nearly black, with 3 horizontal layers in marginal region. Southeastern U. S. A. .... *S. castaneum* 147
3. Surface blackish brown to black, with cracks and holes; with large, branched, black rhizomorphs, pillars densely tangled. Brazil .... *S. heveae* 149

c. Pillars short and stubby.

1. Surface smooth, papery, cinnamon to Prout's brown; probasidia irregular, new probasidia often formed in old probasidial cells; basidia 5–5.4 x 50–60μ. Portugal and U. S. A. *S. alni* 150
2. Surface with a few conical spines, wood brown, margin finely toothed. Brazil .... *S. alni* var. *brasiliense* 153
3. As in 1 except surface flaky. U. S. A. .. *S. alni* var. *squamosum* 153
4. Surface with mottled brownish and whitish areas, with large conical spines, otherwise smooth; probasidia irregular. South Africa .... *S. grandispinosum* 154

PAGE

5. Surface smooth except for pin holes and cracks due to the incomplete formation of the top layer, cream color to cinnamon brown; probasidia pyriform, borne in clusters; basidia 4–5 x 33–47μ. Europe, Canada.... *S. Carestianum* 155
6. Surface less compact than in 5; basidia and spores larger. South Africa................ *S. Carestianum* var. *natalense* 157
7. Surface smooth, pale buff to dark brown, lateral margins frequently spongy. Brazil............. *S. alveomarginatum* 158
8. Surface rich brown, rarely lighter, very irregular, usually with only a subiculum and minute spongy mounds, sometimes 3-layered. France...................... *S. Galzini* 160
9. Surface cinnamon to clove brown, smooth or bumpy, occurring in small circular thin patches. Africa... *S. ugandae* 161
10. Surface broken into small (1 mm. dia.) patches, pure white throughout. Southeastern U. S. A............ *S. leprosum* 162
11. Surface nearly smooth or with holes, cracks, spines or mound-like insect houses, purplish black throughout. Widely distributed.......................... *S. Curtisii* 164
12. Surface clove brown with lighter margin, smooth at first, becoming wrinkled; on *Ficus*. Philippine Islands. *S. ficicolum* 167

B. Context without pillars.
  a. In section more than 500μ thick.
    (I) Texture tough.
      1. Surface formed of branched, more or less radial ridges, brownish, 1–2 mm. thick. U. S. A........... *S. Burtii* 168
      2. Surface nodulose, 0.6 mm. thick. North Carolina. *S. Burtii* var. *acerinum* 171
      3. Surface with minute bumps and ridges and with half-moon-shaped openings. Brazil............. *S. Mülleri* 171
      4. Surface smooth, brownish, context with labyrinthiform tunnels. Southern Europe................. *S. Cavarae* 173
      5. Surface smooth, glabrous but with minute holes and sinuous cracks, Rood's brown; basidia 4.2–4.6 x 28–35. Guiana, Guadeloupe, U. S. A.......... *S. Leprieurii* 174
      6. Surface inherently smooth but with numerous small bumps and ridges; pale smoke gray with a faint lilac tint. Trinidad.......................... *S. lilacinum* 176
      7. Surface with broad mounds, snuff brown or lighter, forming small patches; on bamboo. China. *S. nodulosum* 177
      8. Surface very irregular, brownish, forming small patches; on *Pinus strobus*. Mountains, North Carolina to New England............................ *S. pinicola* 178
      9. Surface bister to warm sepia in younger parts, hymenium grayish, young parts and context with distinct honeycomb structure. Cuba.................... *S. siparium* 179
    (II) Texture fragile, i.e., easily torn with needle.

PAGE

1. Surface light to dark gray, context mummy brown to sepia, forming very extensive growth, over 1 mm. thick; usually on *Acer*. Southeastern U. S. A. *S. fumigatum* 180
2. Surface fawn colored, less than 1 mm. thick. Brazil. *S. cervicolor* 182
3. Surface light gray to mouse gray, irregular due to bumps or cracks; on *Citrus*. Brazil............*S. cinereum* 183
4. Surface light brownish drab to a deep rich chocolate, hymenium very thin, papery, and easily separable from context; usually on *Citrus*. Brazil..*S. separans* 185
5. Surface cottony to compact and finely granulose, ochraceous buff, context deep brownish with purplish tint. Puerto Rico and Brazil.................*S. Stevensoni* 186
6. Surface smooth or pulverulent, near cinnamon brown; context alveolate, hair brown, spongy; basidia 5.8–7 x 25–30μ. Cuba..........................*S. spongium* 187
7. Brownish, cottony or spongy; on fruits of *Sabal minor*. Florida..............................*S. sabal-minor* 188

b. In section less than 500μ thick.

1. Surface firm, smooth, glabrous in places, Sayal brown, dotted with black slightly elevated dots, the color of the dots being due to the insect's black scale; on *Acacia*. Formosa.................................*S. acaciae* 189
2. Surface firm, pulverulent, glabrous in places with dots much as in 1, but dark dots here due to darkened fungal tissue forming roof over insect's house. Jamaica...............................*S. atropunctum* 190
3. Surface soft not tough, smooth except for small elevated patches of loosely entangled hyphae, clove brown to fuscous. Mississippi and Louisiana........*S. fuscum* 191
4. Surface smooth at first, irregularly bumpy in the older parts, minutely cracked, dirty dark brown to clove brown. Philippine Islands..........*S. schizostachyi* 192
5. Surface roughened by cracks in bark and sponge-like over insects, dirty brown; on *Taxodium*. Louisiana. *S. taxodii* 193
6. Surface near margin divided into sinuous ridges, older surface with cracks, light brown with purplish tint. France and Italy.....................*S. Michelianum* 194
7. Surface soft, finely alveolate, margin indeterminate, bister to mummy brown. Jamaica..........*S. fragile* 196
8. Surface soft, with a flocculent grayish sheen, context fuscous black. Ceylon............*S. subcarbonaceum* 197

VIII. Basidia 4-celled, curved, with persisting probasidial cell.

A. Context with pillars; 3- or 4-layered.

a. Pillars tall, distinct, mostly unbranched.

1. Occurring in small patches, usually on leaves. Java. *S. Boedijnii* 198

PAGE

2. As 1 but on *Ficus* stem; hymenium stratose. Java. *S. stratiferum* 199
3. In section 250–420μ thick; on leaves. Philippines. *S. minutulum* 200
4. In section up to 2 mm. thick, with 3 horizontal layers. Philippines .... *S. Merrillii* 202
5. Pure white throughout. Ceylon, China .... *S. leucostemum* 203
6. Surface near bister, nodulose, subiculum with cave-like insect houses. Philippines .... *S. laxum* 203
7. Usually in small patches, with a wide whitish margin and brownish central part. Virginia and North Carolina. *S. filiforme* 205
8. Margin with tents and pillars. South Carolina. *Septobasidium* hybrid 206
9. Elastic in texture, pillars at first in longitudinal rows, with 3 horizontal strata. East Africa .... *S. ussanguense* 208
10. Pillars not in longitudinal rows, much entangled below, free above. Mexico .... *S. tropicale* 209
11. White at first, becoming brownish, surface glabrous with numerous small bumps. Java .... *S. cinchonae* 211
12. Hymenial surface in patches, subiculum with insect houses which are half covered over. Philippines .... *S. clethrae* 212
13. Surface some shade of gray, usually pale smoke gray, smooth, pillars stubby or tall. Java, Ceylon, China, Japan, Australia, West Indies .... *S. bogoriense* 213

b. Pillars anastomosing particularly in the older parts.
1. Forming small patches, pillars blackish, hymenium stratose; probasidia large, 16–25μ. Cuba .... *S. pedicellatum* 215
2. Hymenium in small patches, context alveolate .. *S. alveolatum* 216
3. Growth extensive, purplish, pillars distinct in younger parts, not evident in older parts. Java .... *S. rubiginosum* 218
4. Margin with close-set, elongate insect houses, each surrounded by spines, surface bumpy. Africa .... *S. irregulare* 219

c. Pillars short, unbranched.

(I) In section more than 500μ thick.
1. Surface smooth, light cream. Formosa ..... *S. citricolum* 220
2. Surface nodulose, brownish, forming small patches, 600–800μ thick, with conidia; on *Citrus*. China. *S. sinense* 221
3. Surface broken into sinuous anastomosing parts, densely warted in marginal region, hymenial surface smooth, not shining. Ceylon .... *S. Thwaitesii* 223

(II) In section less than 500μ thick.
1. Surface reticulate. Poland .... *S. fusco-violaceum* 224
2. Surface flocculent, pure white. U. S. A. .... *S. cupressi* 225
3. Surface smooth, occurring in very small patches. Java. *S. piperis* 226
4. Surface smooth except for cracks, top layer very compact. U. S. A. .... *S. cremeum* 227

PAGE

5. Surface smooth in older parts, speckled at first due to vast number of minute umbrella-like pillars; in section less than 200μ thick; hypophyllous. Java. *S. humile* 228
6. Occurring in irregular small patches, surface firm, granulose, becoming cracked when mature, the cracks due to incomplete formation of top layer, pillars not present throughout. Ecuador................. *S. Lagerheimii* 229
7. Surface at first smooth, later cracked, pillars at first umbrella-like, occurring throughout. Australia. *S. macadamiae* 230
8. Surface smooth, not shining, Sayal brown near margin, grayish brown in older parts; superficially like *S. alni*. Mexico................................ *S. mexicanum* 231

B. Context with pillars but without a third top layer.
  1. Pillars abundant in small clusters surrounding insect houses, deep brown to blackish. Java................ *S. proliferum* 232
  2. Pillars few, surface irregularly alveolate, deep brown to blackish. Java..................................... *S. neglectum* 233

C. Context without distinct pillars, alveolate; probasidia arising cymosely.
  1. Context regularly alveolate in marginal region, alveolations indistinct in older parts, remaining soft and spongy. China. *S. gossypinum* 234
  2. Context usually alveolate throughout, top layer becoming firm and hard. Mexico and Guadeloupe........ *S. pannosum* 235

D. Context without pillars, otherwise not as in C.
  a. In section 500μ or thicker.
    1. Very compact and hard, surface with numerous deep fissures, brownish. Texas and Philippines.............. *S. Sydowii* 236
    2. Context compact, texture brittle but not hard, basidia S-curved, surface grayish lavender. Ceylon.... *S. murinum* 237
    3. Context of loosely arranged hyphae; basidia usually making one half turn; surface about hair brown. Ceylon. *S. arachnoideum* 238
  b. In section less than 500μ thick.
    1. Composed of indefinite subiculum and elevated circular patches of hyphae, brownish. Formosa.... *S. formosense* 239
    2. Surface retiform with slightly elevated dark dots, pinkish buff, context compact. Philippines......... *S. punctatum* 240
    3. Context very thin, 50–100μ thick, arachnoid at first, later compact, easily separable, brownish. Australia. *S. crustaceum* 241
    4. Forming small anastomosing patches, intermingled with mosses and liverworts, fruiting surface smooth with upturned margins, brownish, rhizomorphs extending between cracks of bark. France.................. *S. Bagl, iettoanum* 241

E. Not as above.
  1. With a resupinate basal part from which extend branched

PAGE

IX. Basidia 4-celled, straight or slightly curved, usually without a probasidial cell.
- A. Context with distinct pillars.
  - a. With closely packed parallel upright threads in hymenium.
    1. Context 1–2 mm. thick, pillars tall, abundant, surface alveolate where hymenium has not formed, context rich brown, hymenium conspicuously cracked. Java.... *S. Henningsii* 243
    2. Context 270–370μ, rarely up to 850μ thick, pillars short, spreading out above to form top layer; hymenium conspicuously cracked; spores recurved at distal end. Ecuador and China.............................. *S. albidum* 245
    3. Context 400–700μ thick; hymenium conspicuously cracked; spores not recurved at distal end. Philippines and South Africa.................................... *S. philippinense* 247
    4. Context 270–750μ thick, margin flocculent, nearly white, hymenium without conspicuous cracks, pillars short; basidia longer and thinner than any above. Java. *S. flavo-brunneum* 248
    5. Context 300–850μ thick, pillars short, inconspicuous, surface with bumps and ridges. Ceylon.......... *S. lichenicolum* 250
    6. Context up to 5 mm. thick, hymenium often with winged margin, hymenial threads curved at tips, pure white throughout when fresh. Brazil................ *S. crinitum* 252
  - b. Hymenial hyphae irregularly arranged.
    1. Context up to 500μ thick, pillars short, surface wrinkled, not cracked, tilleul buff; basidia 7–8 x 29–34μ. Brazil.... *S. pallidum* 253
    2. Context about 1 mm. thick when mature, pillars tall, slender, hymenium with winged margin, surface smooth, near tilleul buff, hymenial hyphae irregularly upright, branched, with recurved tips. Brazil, Jamaica.... *S. septobasidioides* 254
- B. Context with indistinct pillars.
  - a. Hymenial hyphae 1.5–2μ thick, coiled.
    1. Context 1–1.5 mm. thick, surface of hymenium smooth except for cracks, light cinnamon drab. Cuba.. *S. cirratum* 256
- C. Context without pillars.
  - a. Composed of a subiculum and numerous hard rounded warts, about 1 x 1 mm.; on *Quercus*. California.... *S. verrucosum* 257

X. Conidial forms, basidia not yet observed.
- A. Context not divided into distinct subiculum, pillars, and top layer.
  - a. Context more than 1 mm. thick.
    1. Consisting of confluent, cushion-like patches, color about Prout's brown, conidia in rounded compact clusters up to 50μ thick. Ceylon, China................ *S. accumbens* 258
    2. Deep brown, sooty-like; conidia in *Nostoc*-like chains. Java. *S. theae* 259
    3. Consisting of usually isolated cushion-like patches, surface

PAGE

b. Context less than 1 mm. thick.
   1. Consisting of continuous patches of blackish brown, compact, sooty-like growth; conidia hyaline, abundant, minute, 1.5 x 4μ. Florida................. *S. conidiophorum* 262
   2. Consisting of rounded or irregular cushion-like patches (0.5–2 mm. wide), dirty dark brown in color; conidia 3-celled, brown, 8 x 24μ. Philippine Islands... *S. elastostemae* 263
   3. Surface blackish brown or dull slate color; probasidia spherical, 10–14μ; conidiophores brush-like, on distinct hyphae from probasidia, conidia 3 x 6μ, hyaline. China. *S. lanosum* 264
   4. Surface smooth and velvety, buffy brown or drab; conidial and probasidial hyphae distinct, much as in 3. Brazil. *S. paulense* 265
   5. Consisting of fluffy, russet colored hyphae; conidia in clusters; probasidia in groups, with warted walls. Philippines. *S. molle* 266
   6. Surface powdery, benzo brown, woolly, matted, usually 175–300μ up to 600μ thick; conidia in branched, slightly twisted chains. Australia................... *S. tomentosum* 267

B. Context with subiculum, pillars, and top layer reddish brown to blackish.
   a. Surface tan, broken into narrow, irregular anastomosing segments, context stratose. Java.................. *S. tigrinum* 268

XI. Forms in which basidia have not been observed.
   A. Context with long distinct spines but usually without a top horizontal layer.
      1. Consisting of a tuft of upright brownish spines 1.3–2 cm. tall, with black, pointed tips, subiculum indistinct. Brazil. *S. scopiforme* 269
      2. Spines arising evenly from distinct subiculum, 6–14μ tall, margin of subiculum with distinct insect houses. East Indies. *S. pteruloides* 270
      3. With blackish brown, veined top layer through which spines extend obliquely from surface, spines 0.2–0.3 x 1–5 mm., with pointed tips. Chile.......................... *S. spiniferum* 271
   B. Context with distinct pillars, 3- or 4-layered.
      a. Usually more than 500μ thick.
         1. Surface smooth at first, later cracked, 1.5–4 mm. thick, nearly black throughout; on *Schizostachyum*. Philippines. *S. Bakeri* 272
         2. Surface smooth, whitish to buff, in section 1–2 mm. thick with three distinct horizontal layers in marginal region, pillars abundant, dark brown. Brazil........ *S. myrsinae* 273
         3. Top surface incomplete, composed of sinuous plates, tawny to russet, hymenium with upright parallel hyphae. New York.. *S. Peckii* 274
         4. Pillars arising from a thick spongy context, pinkish buff to

PAGE

5. Surface at first finely alveolate due to entangled pillars, snuff brown, later smooth except for wide cracks and minute bumps, nearly black, up to 2.5 mm. thick. Java. *S. robustum* 278
6. Forming small, cushiony patches, surface smooth, like kid, dry, pillars short, whitish throughout; probasidia spherical to pyriform. Australia................*S. Simmondsii* 279

b. Usually 500μ or less thick.
1. Forming small rounded or irregular patches 3–10 mm. wide, slate to dark neutral gray, smooth. Nicaragua. *S. sublilacinum* 279
2. Surface smooth and velvety, rich brown (Natal or warm sepia), margin lighter, hymenium with branched upright hyphae. Japan..........................*S. prunophilum* 280
3. Surface smooth, bluish black, margin minutely fimbriate. Japan..................................*S. indigophorum* 282
4. Surface smooth or minutely ridged, snuff brown, margin indeterminate with fine dots, the young pillars; on leaves. Philippines..............................*S. molliusculum* 283
5. Surface smooth or cracked in older parts, near bister throughout, usually 210–315μ thick. Louisiana. *S.* sp. hybrid on *Liquidambar* 284

C. Context without pillars or spines.

a. Less than 500μ thick in thickest part.
1. Surface smooth and shiny except for a few pin holes; probasidia 12.6–14.8μ. Ceylon..................*S. capnodes* 286
2. Surface composed of brownish, small (0.5–1 sq. mm.) flakes, scattered or gregarious, sometimes in concentric circles; on leaves. Ceylon..........................*S. scabiosum* 286
3. Surface whitish, smooth except for numerous rounded elevations, with branched circinate hyphae in context; on leaves. Brazil...............................*S. abnorme* 287
4. Surface mummy brown, smooth except for mounds over insects; on leaves. Philippines..........*S. makilingianum* 288

b. More than 500μ thick in thickest part.
1. Surface hair brown, pulverulent, with cracks and holes, haustoria sausage-shaped; on *Citrus*. China. *S. carbonaceum* 289
2. Surface at first wood brown, punctured with minute (0.1–0.3 mm. wide) holes which sometimes are in broken concentric rows. Brazil.......................*S. perforatum* 290
3. Surface gray to deep brownish, with numerous circular, or oblong depressions, each partially covered by a flap of fungal tissue. Costa Rica.................*S. lacunosum* 291
4. Surface ashy gray, sometimes with purplish brown tint, with rounded elevations 2 mm. high by 2.5–7 mm. wide. Java...................................*S. tuberculatum* 291

X

DESCRIPTION OF SPECIES

Septobasidium grandisporum Couch. Jour. Elisha Mitchell Sci. Soc. **51:** 72. 1935.

Plates 43 and 73

Resupinate, forming extensive patches on the living bark of dogwood, often girdling the trunk for a height of several decimeters. Hymenium usually pale smoke gray (Ridgway) when well developed, vegetative parts brownish black. Surface spongy except where the hymenium is well formed, then with numerous cracks and holes; even the smoothest places are distinctly bumpy under a lens. Margin usually determinate, often with distinct rhizomorphs. In section up to 2.5 mm. thick, composed, in mature specimens, of the hymenial region and the context but with no distinct subiculum. Context composed of rhizomorphs which extend about over the bark, from which a vast tangle of fascicles of hyphae arise, giving the context a spongy texture. Hyphae of context much branched, septate, without clamp-connections, 3 x 4.2μ in diameter. Hymenial region 150–250μ thick, composed of more or less upright, much branched, hyaline and brownish hyphae, 3–3.5μ thick, and probasidia and basidia in various stages of development. Probasidium usually slightly subspherical, 12–17μ thick, sprouting and emptying its entire contents to form the basidium. Basidium long-cylindrical, often curved, thickest at the distal end, 8.4–9.6 x 46–56μ, giving rise to an apical sterigma, 8–22μ long, on the end of which is borne one spore. Spores elliptic, 12.6–17 x 29–32μ, most about 16.8 x 29.5μ, becoming irregularly septate into a number of cells, hyaline at first but becoming brownish if kept on slide for several days.

Symbiotically associated with *Chrysomphalus obscurus* (Comst.), some of which are parasitized by hyphae which are in spindle-shaped segments (as in *S. fumigatum*) and others of which are free from parasitism. Both parasitized and healthy insects, the latter in all stages of development, are very abundant beneath the fungus.

This species is apparently very rare, having been collected but once, and then on two trees of *Cornus florida* in woods by highway about 30 miles north of Charleston, S. C., March 19, 1930, Else R. Couch, collector. The fungus was very abundant on the infested trees and both trees were very unhealthy, there being a number of dead limbs and other living ones growing in the form of witches' brooms. As in other species the fungus does not penetrate into the wood, the injury being done by the insects.

This species may be confused with *S. fumigatum* Burt, which occurs in the same locality and sometimes on dogwood. They look considerably alike to the naked eye, both having a grayish hymenium and spongy context. In *S. fumigatum* the context is brownish, the hymenium mouse gray, and the plant has a smooth, velvety feel. In the present species the context is brownish black, the hymenium smoke gray, and the surface of the plant has a distinctly rough or coarse feel. The haustoria in the two plants are alike but the probasidia, basidia, and spores are distinct.

Specimen examined:

South Carolina: About 30 miles north of Charleston, on two trees of *Cornus florida*, March 19, 1930, Else R. Couch, coll., U. N. C. Herb., 8465, type.

Septobasidium purpureum Couch. Jour. Elisha Mitchell Sci. Soc. **44:** 255. 1929.

Plates 5, 11, 14, 15

Resupinate, often covering an area of one to several square decimeters on the bark of living trees; brownish with a purplish tint except where the hymenium is developed and then dull, neutral gray. Surface in older regions cracked by irregular fissures. Margin indeterminate. In section 300–700μ thick, not distinctly differentiated into layers, pillars absent. Threads of context much branched and entangled, septate without clamp connections, about 3.8μ thick, deep brown under microscope. Hymenium indeterminate, the basidia apparently being formed perennially, the old, empty, probasidial cells scattered throughout the context, the younger probasidial cells and basidia near the surface. Probasidial cells spherical, 8.5–13.8μ thick, usually about 10μ thick, remarkable in having a wall which is much thickened in one half and very thin in the other; sprouting into a cylindrical basidium which immediately gives rise to a long sterigma on the end of which is borne a single spore. Basidium 6.6–8.4 x 14.7–25μ, usually thickest in the middle or at the distal end; without septa, but separated from the empty probasidial cell by a wall, sterigma usually apical, rarely lateral, about half as long as the basidium. Spores 7.4–10.4 x 15.5–19.6μ, elliptic, slightly flattened on one side, pointed or rounded at the distal end, usually becoming divided into eight cells by transverse and longitudinal septa.

Symbiotically associated with scale insects, many of which are parasitized by peculiar, spindle-shaped segments of hyphae, the segments being connected end to end by fine threads, thus resembling strings of sausages. Other insects, however, are not parasitized. This species may be recognized in the field by the striking color, thin context, and absence of pillars and by the non-septate basidium which bears a single spore.

A very striking observation, in the species of *Septobasidium* studied by the writer, is that in the species in which the basidium is divided into four cells, as in *S. alveolatum*, the spores are normally only once septate and thus two-celled; and in the species in which the basidium is two-celled, as in *S. jamaicaense* Burt, the spores are normally four-celled; while in the present species, the basidium is one-celled and the spore is eight-celled. The regular septation in the spores occurs shortly after they are discharged if kept in a moist chamber; if, however, the spores are kept in water the septations are quite irregular and if kept in a nutrient solution or on agar they form many sporidia by budding.

Specimens examined:

Jamaica: Cinchona, on living branches of deciduous tree, July 1926, type, 8506 in U. N. C. Herb., also 8507 and 8508.

Septobasidium jamaicaense Burt. Ann. Mo. Bot. Gard. **3**: 333. 1916.

Plates 5, 6, 13, 14, 54

Resupinate, sometimes stratose, often covering an area of several square decimeters on the bark of living trees; up to 1 cm. thick in large fresh specimens, drying down to 2–4 mm. thick. Plants usually consist of two distinct regions: the very thin, byssoid, marginal, vegetative region and the thick, fruiting region. Marginal region up to 3 cm. wide, composed of numerous blackish brown rhizomorphs with a purplish tint, from which branch out delicate mycelial threads; rhizomorphs and threads closely adherent to the bark, but separable from it when wet. Fruiting region smoke gray or pale neutral gray over a bister or bone brown context. In section composed of three layers: (1) the subiculum, which runs horizontally over the bark, and is composed of numerous very conspicuous rhizomorphs and mycelial threads, 100–300μ thick and from which the pillars arise in vast numbers; (2) pillars nearly straight and perpendicular in the growing region of young plants but becoming considerably entangled, passing obliquely from the subiculum to the hymenium, sometimes branched, very variable in diameter, sometimes composed of as few as three threads and then about 6.2μ thick, often up to 60μ thick, rarely up to as much as 180μ, most 20–30μ thick; single pillars very long, often 2 mm.; threads of pillars about 3.7μ thick, rarely branched, septate without clamp connections; pillars branching and then the branches breaking up into single hyphae as the pillars approach the hymenium; (3) hymenium compact in certain areas, hypochnoid in others, 70–100μ thick, composed of basidia borne laterally in various stages of development; varying in size from the youngest initial subglobose bud up to the fully mature basidium, 8–14 x 40–62μ, straight, once septate, a transverse partition thus giving the "basidium" the appearance of a teleutospore of *Gymnosporangium macropus*, each cell sprouting a long sterigma on the end of which is borne a long spore which is more or less pointed at both ends and distinctly recurved at the distal end; spores white, 6.8–11 x 23–39μ, including the bent end, most about 8 x 27μ; usually becoming thrice septate and not rarely seven times septate; often forming, after the septation process is complete, numerous little spores by a budding process.

Associated with colonies of scale insects, some of which are parasitized by irregular coils of hyphae, others of which are not parasitized. Scale insects very abundant and evidently the primary direct source of injury to the trees.

This species has apparently been collected only once before by Dr. L. M. Underwood. It was first described by Dr. E. A. Burt ('16) from Underwood's material and was very appropriately named *S. jamaicaense* since I have later found this species to be by far the commonest in Jamaica. My material has been seen by Dr. Burt and has been compared with his type material and agrees. Dr. Burt gives the spore size as 13 x 5½μ, which is much too small. The shape of the spore in his figure is also incorrect since the bent end of the spore should be opposite the mucro end.

This species can be distinguished by its very thick and extensive growth, by its habitat on *Solanum punctulatum*, by the dark brown mycelium with a faint purplish tint when fresh (when dry the purplish tint almost or entirely disappears), and the grayish slate colored hymenial surface in which the probasidia are so often borne in minute clusters. In actively growing specimens the marginal region with its extensive rhizomorphs is very distinctive.

Field observations on this species showed that the fungus-scale insect combination did considerable damage to the host, *Solanum punctulatum*. The fungus often covers the bark on the leeward side of trees for areas of several square decimeters and in a few specimens was seen to extend from near the base of trees as much as thirty feet high almost to the top. Rarely in deeply shaded gaps or ravines small trees were seen girdled by the fungus. Such heavily infected trees were usually quite unhealthy, showing a large number of dead limbs. When the fungus was found on dead limbs or dead trees, it also was dead. The following illustration of this was brought to my attention. On the north side of Portland Gap in the Blue Mountains of Jamaica there occurred a large land slide several years ago. Many trees of *Solanum punctulatum* were uprooted by the slide, and most of them are now dead. The fungus was abundant on many of the dead trees and was invariably dead itself. Another striking illustration is afforded when the main trunk and a small branch are infected at the junction of the two. If, as is often the case, the small branch is girdled by the fungus and dead, then the part of the fungus on the small branch is also dead while the part of the fungus on the main trunk which is still living, is alive.

Specimens examined:

Jamaica: Jim Crow Peak, altitude 5500 feet, April, L. M. Underwood, coll., type in Mo. Bot. Gard. Herb.; also twenty-five collections, mostly on *Solanum punctulatum*, in Blue Mountains, June, July 1926, J. N. Couch, coll., in U. N. C. Herb.

Septobasidium fusco-cinereum Bres. (Herbarium name)

Plate 54, figs. 6–9, and pl. 92, figs. 1, 2

Resupinate, covering extensive areas (the largest piece seen by me was 15 cm. long but was only a small part of the original); the subiculum and marginal region nearly blackish brown, the old washed out hymenial surface near hair brown, the new hymenial surface grayish brown. Hymenial surface nearly smooth to the unaided eye but not at all shiny, sometimes minutely cracked and areolate, mottled with lighter grayish areas, near drab (Ridgway) where the new hymenium is forming, pulverulent under a lens. Margin of subiculum determinate, whitish at the very periphery, composed largely of closely appressed innate rhizomorphs. In section 300–700μ thick in the younger parts where there is only one stratum, becoming much thicker in the older parts by the formation of successive strata, up to 2.5 mm. thick. The primary stratum composed of the subiculum, a middle region of upright hyphae which are mostly united into pillars in the older parts, and the top layer. Subiculum 60–80μ thick, much thicker where the rhizomorphs are abundant or where there are scale insects. Hyphae of subiculum compactly arranged, dark colored, 3.8–4μ thick. From the subiculum arises the middle layer. In the younger parts the middle region is composed of upright hyphae which mostly arise separately from the subiculum, middle region in younger parts about 130–160μ thick, hyphae branched, mostly upright, dark colored, 3.8–4.2μ thick. In the older parts, the hyphae of the middle region are united into pillars, which are 20–160 x 400–500μ. The pillars are abundant in the older parts and branch and anastomose freely, forming a dense spongy tangle. Top layer of variable thickness, 80–170μ thick, composed of much branched, tufted, upright, wavy, hyaline or pale brownish hyphae and basidia in various stages of development. Apparently there is no resting probasidial stage. Basidia long-cylindrical, 8–11.7 x 37–50μ, once septate and thus two-celled, with a distinct but small, sterile, thicker-walled portion at the base, sterigmata up to 46μ long. Spores about 6.7 x 25μ, broadest at the proximal end and sharply recurved at the distal end.

Associated with scale insects some of which are parasitized by irregular coils.

The species is close to *S. jamaicaense* Burt but may be separated from that species by being much thinner in section, by lacking the exceedingly abundant, thin, entangled pillars and by having smaller basidia and spores. Furthermore *S. jamaicaense* always shows a distinct purplish tint both in the hymenial and subhymenial regions while in the present species that shade of color is vague or lacking. There is also a slight resemblance between this species and *S. spiniferum* Burt but the two are readily distinguished. Patouillard suggested on the label that this species was a form of *S. velutinum* Pat. The relationship however is rather remote.

The species is characterized by the comparatively thin structure in section, by the hair brown color and by the conspicuous adnate rhizomorphs.

Specimens examined:

Brazil: Sao Leopoldo, Rick, coll., Oct. 1905, in Pat. Herbarium under herbarium name *S. fusco-cinereum* Bres.; also on deciduous tree, Rick, coll., October 1935, U. N. C. Herb. 10042, and Dec. 1936, Rick, no. 23, U. N. C. Herb. 10396.

Septobasidium sclerotioides n. sp.

Plate 54, figs. 4, 5

Resupinate, covering small areas on the stem of a leguminous plant. Drab gray to light drab. Texture firm but not tough. Surface irregular due to the incomplete formation of the top layer, there being numerous isolated rounded bumps (0.1–0.2 mm. wide) or anastomosing ridges between which are irregular holes and channels; also minute cracks may appear in the top surface after it has been formed. Margin determinate. In section quite variable in thickness, 200–800μ thick, usually about 400μ, consisting in the marginal area and in certain other places of only two regions, a subiculum and the hymenial layer; in many places, however, the hymenial layer is supported by pillars or tufts of hyphae. Subiculum very variable in thickness, 60–580μ thick, consisting of dark brown, densely packed, irregularly arranged hyphae, 3–5.4μ thick, with rather thick walls. The threads of the subiculum are so densely packed as to form a stromatic pad of hard, blackish tissue in which scale insects are imbedded. The hymenial layer may arise directly from this layer or the latter may give rise to pillars or upright tufts of hyphae; pillars 30–60 x 200–250μ, composed of dark brown, thick-walled hyphae which are most about 4.8μ thick. The pillars branch out above to form a thin frequently inconspicuous, horizontal layer of thick-walled, dark brown hyphae which contrast strikingly with the lighter colored hymenial threads; hyphae about 4.8μ thick. This horizontal layer gives rise to the hymenium. The isolated pillars branch out above to form small mounds or flattened areas of hymenial tissue and this gives the irregular bumpy surface. Hymenium 84–130μ thick, composed of branched, upright, wavy threads, the basal parts of which are brownish while the distal parts are hyaline and the region just back of the tip is encrusted with minute crystals; hyphae about 3.8μ thick near base, about 2.8μ thick near tips. Probasidia not formed. Basidia 9.6–12 x 25–30μ, 2-celled, frequently slightly constricted in the middle, usually with rounded ends, the top cell usually slightly larger than the basal one; with a small empty cell at the base. Sterigmata sometimes as long as the basidia. Spores sharply recurved at the distal end, 7–8.4 x 23–25μ.

Associated with scale insects, some of which are parasitized by irregularly coiled haustoria. Penetration through vaginal pore.

This species is readily distinguished by its small size and by the gray surface with nodules, holes, and cracks. The exceedingly thick, basal, dark-colored stromatic tissue from which pillars arise to support the hymenium or from which the hymenium arises directly is also characteristic. In color and surface characters this species suggests an aberrant form of *S. Langloisii* but the latter is coarser and has a distinct probasidial cell which remains as an empty bladder after the basidium has been formed. Also the spores are very distinct. The present species belongs in the *S. jamaicaense* group, as indicated by basidial and spore characters.

Specimens examined:

Brazil: Rio Jurá, on branches of a legume, Nov. 1900. Type, no. 2733, in Berlin Museum under *S. velutinum*, also a fragment in U. N. C. Herb.

Septobasidium Patouillardii Burt. Ann. Mo. Bot. Gard. **3**: 332. 1916.

Plate 42 and pl. 83, figs. 7–11

Resupinate, forming irregular patches which are usually rather small but which may be up to 20 cm. in length; snuff brown or sepia to mummy brown or clove brown to blackish brown with a velvety appearance and a purplish tint, the latter combination of colors being the most characteristic. Surface smooth in the younger regions, cracked in the older parts. Margin irregular in outline, sharply determinate, the lower layer usually not extending beyond the top layer. In section 300–460μ thick, composed of three regions: (1) the subiculum which extends over the bark and is very thin (about 20μ), composed of interwoven, septate hyphae, about tan colored and contrasting strikingly with the darker top layer; hyphae about 3μ thick; (2) a region of very slender, upright pillars which are 20–54 x 140μ, hyphae of pillars about 4μ thick, branching out toward the top to form (3) the top layer which bears the hymenium. Top layer 100–210μ thick, sometimes stratose, the lower part of top layer composed of loosely interwoven, much branched threads which are about 3.5–4μ thick; hymenium 28–38μ thick, sharply distinct from the rest of the top layer by its much darker color, the darker color being due to the paraphyses; paraphyses vertical, thick-walled, about 3.2μ thick. Probasidia absent unless the small, thin-walled, elongated, transitory cell (see figure) which elongates and enlarges to form the basidia may be considered as such. Basidia elongated, 2-celled, 5.2–5.6 x 18–23μ, each cell giving rise to a sterigma and a spore, sterigmata 20–25μ long. Spores 4.2–5.8 x 10.9–16.6μ, rarely up to 21μ long, bent-elliptic, hyaline, becoming thrice septate.

Symbiotically associated with *Chionaspis gleditsiae* some of which are parasitized by irregularly coiled hyphae, others of which are free from fungal infection. Perennial, the spores forming during spring and early summer.

This species often occurs along with *S. Schweinitzii*, *S. pseudopedicellatum*, *S. Curtisii*, *S. Mariani*, and *S. castaneum*, but may easily be distinguished from any of those on the basis of color and structure. The blackish velvety top surface, contrasting with the lighter (usually tan) colored subiculum, the thinness of the plant body, the minute slender pillars, and the distinct, abruptly determinate margin are macroscopic characters by which the present fungus may be distinguished. The vertical paraphyses in the hymenium and the two-celled basidium without a persistent probasidial cell are the most striking microscopic characters. It is very easy, indeed, to recognize this plant from Burt's description. Burt says, "This species may be recognized by its thin fructification resembling a piece of black velvet, slightly raised from the substratum on such short and slender pillars as to be barely visible without the aid of a lens." However, Burt describes the plant as having probasidia. He states that the material with probasidia was "near a specimen of another species." I have examined a large amount of sectioned material collected in February and March, both just before and during the formation of basidia, and have been unable to find any probasidia as described and figured by Burt. It is my opinion that the probasidia belonged to another species.

Specimens examined:

North Carolina: Near Pee Dee, on live *Fraxinus*, March 24, 1926; near Wilmington, on *Fraxinus*, with *S. Schweinitzii*, Dec. 28, 1921, J. N. Couch, coll., 5876; near Winston-Salem, on *Fraxinus*, with *S. Schweinitzii*, Oct. 28, 1928, Kenneth B. Raper, coll., 8370, also on *Fraxinus*, with *S. Schweinitzii*, Dec. 27, 1933, John R. Raper, coll.; Chapel Hill, rather common in swampy regions, on *Fraxinus*, Feb. 5, 1929, 8386; also March 8, 1929, 8393, and March 14, 1929, 8396, the last two collections fertile; also same locality, Oct. 28, 1928, 8373 and 8374.

South Carolina: Lynch's river bottom near McBee, on *Fraxinus*, June 30, 1928, A. B. Couch and J. N. Couch, colls., 8319.

Florida: Daytona, R. Thaxter, no. 75b, in Farlow Herb. and Mo. Bot. Gard. Herb., 43895; near Kissimmee, on *Fraxinus*, March 13, 1932, J. N. Couch, coll., 9298, fertile, sepia to clove brown, the fertile specimens slate colored; near Lisbon, on *Fraxinus* (?), Jan. 24, 1891, L. M. Underwood, coll., 2908. Determined by J. R. Weir as *S. fumigatum* Burt? in U. S. D. A. Herb., also fragment in U. N. C. Herb.

Louisiana: St. Martinsville, A. B. Langlois, 3005, type, in Burt Herb.; near Baton Rouge, very abundant on *Ilex decidua*, 9168, and 9170, on *Fraxinus*, Dec. 31, 1931, J. N. Couch, coll., 9169; on *Gleditsia triacanthos*, with *S. pseudopedicellatum*, Dec. 31, 1931, J. N. Couch, coll., 9161.

Septobasidium Clelandii n. sp. See *S. pteruloides* in Lloyd, Myc. Notes **7:** 1195, figs. 2429, 2430. 1923.

Plate 54, figs. 1–3 and pl. 92, figs. 3–5

Resupinate on branches of *Leptospermum* sp.; composed of a basal pad of tissue from the central region of which arises a dense cluster of upright spines or hydnoid processes. The entire patch of fungal growth is oval in outline and about 3–4 x 7–15 mm. in area, though frequently several of the patches may anastomose. The central region of the pad from which the spines arise may be as much as 1 to 2 mm. thick, the outer region, about 2–4 mm. wide, is much thinner, about 100–200μ thick, and sometimes fibrillose. Blackish brown in color in the dried condition. Upright spines blackish brown, 3–4 mm. tall in one collection, about 7 mm. tall in another, and about 75–550μ thick, simple or sometimes forked, usually thicker in the distal half and with pointed tips. From each pad of growth twenty-five to fifty hydnoid processes arise. These are closely crowded at their bases but spread out above and are composed of a central region of compact, longitudinally arranged, hyaline or pale brownish hyphae, 3.8–5.4μ thick, septate, without clamps. The hymenial region covers the spines and is composed of probasidia and basidia in various stages of development and dark colored hyphae about 3.4–3.8μ thick. These hyphae contrast strikingly with the pale colored ones from the center of the spines. Probasidia spherical to ovoid, about 15μ thick, thin-walled, elongating to become the basidium. Basidia 8.4 x 32–40μ, straight, either 2-celled or 4-celled; sterigmata about 5–10μ long. Spores abundant, bent-elliptic, 5.4–7.5 x 18.9–23.1μ, hyaline.

As noted by Cleland, each patch of fungal growth is associated with a gall or enlarged part of the twig and in each of these galls may be found a cavity lined with fungal tissue, and in the cavity one or two larvae of an insect. The insect larvae appear to be related to the *Phylloxera* gall insects. Each of the insect larvae which I examined was parasitized by the fungus, the insect's body being partly filled by haustoria of the glomerulus type. I did not find any unparasitized larvae but I think that if more material were available one could find such larvae. The fungal insect relationship in this species is evidently a very interesting one. It appears that the gall-forming substance not only has affected the tree but has also stimulated the fungus to form an exceedingly peculiar type of growth. It is possible that this tough, spiny, fungal tissue may have some protective value to the insect larvae.

Patouillard (1925) considered this fungus the same as *S. pteruloides* (Mont.) Pat., having seen material sent by Lloyd (1923) and collected by Dr. J. B. Cleland in Australia. The fungus, however, is undoubtedly distinct from *S. pteruloides*, as I have determined by an examination of the original type material (part of which was lent me by Professor Roger Heim from the Montagne Herbarium in Paris) and the original description and figures of Montagne. For a comparison of the two see under *S. pteruloides*.

Specimens examined:

South Australia: Encounter Bay, on *Leptospermum scoparium*, Dr. J. B. Cleland, coll., Jan. 1935, in U. N. C. Herb. and Iowa State Univ. Herb.; also a collection in Lloyd Herb. from same locality collected by Cleland, June 1922.

Septobasidium areolatum Couch. Jour. Elisha Mitchell Sci. Soc. **44:** 248. 1929.

Plates 5, 6, 13, 54, 88

Resupinate, growing extensively over the bark of living trees and often forming an area of as much as 60 sq. cm. Deep rich brown over the vegetative portion, the hymenium grayish. Entire surface marked with numerous holes which are 0.2–0.6 mm. in diameter and 1 to 3 mm. apart. Margin indistinct, except with a lens, often indeterminate, irregular, thin and byssoid. In section up to 2 mm. thick when fresh, about 1 mm. thick when dry, often stratified by the formation of several hymenia over the same area. Composed of three regions: (1) subiculum, about 20–45μ thick, which grows over the surface of the bark and very noticeably penetrates into the phellogen cells destroying their contents and walls; (2) the middle region, 0.8–1.5 mm. thick, composed of brown, loosely packed, branched threads which extend in all directions, threads 3.5–5μ thick, septate, without clamp-connections; (3) the hymenial region, which is about 40μ thick, and composed of probasidia and basidia in various stages of development and of delicate branched hyaline threads, the extensions of the coarser brown threads; basidia at first spherical or subspherical, with a slightly thickened wall, 10–12.6μ thick, elongating into an irregularly cylindrical, once septate basidium, each cell of which sprouts a long sterigma and bears on its end a spore. Basidium 8–10 x 24–33μ, sterigmata 12–25μ long or sometimes longer; basidium as in *S. jamaicaense* except smaller. Spores bent-elliptic, bent like a fish hook, the bend being opposite the mucro, 5.5–6.8 x 25–30μ, usually becoming thrice septate.

Symbiotically associated with scale insects. Many insects which were not parasitized were dissected out from beneath the fungus, others, however, were parasitized by very irregular coils of hyphae.

The basidia of this species closely resemble those of *S. jamaicaense* but the two plants can readily be separated on vegetative characters.

The small and numerous areolations, the rich brown color contrasting with the grayish hymenium when present, and the absence of pillars characterize the plant. The areolations are formed in an interesting way. In the absence of upright pillars the middle layer is formed by the development of ridges of hyphae which arise from the subiculum. These ridges branch and anastomose with each other thus forming a network of ridges. The hyphae at the top of the ridges grow out horizontally parallel with the subiculum. This growth always ceases before the space between the ridges is completely bridged and thus the little holes (areolations) in the upper layer are formed.

The hyphae of this species penetrate deeply into the periderm of the bark and develop extensively there. Penetration is most extensive underneath the old, dead remains of scale insects.

Specimens examined:

Jamaica: Mossman's Peak, Blue Mountains, July, 1926, on bark of living deciduous tree, J. N. Couch, coll., type 8537, cotype 8538, in U. N. C. Herb.

Septobasidium retiforme (Berk. & Curtis) Pat. Bull. Soc. Myc. Fr. **16**: 55. 1900. (Not *S. retiforme* of American authors.)

Plate 54, figs. 11, 12, and pl. 92, figs. 6, 7

Resupinate, forming irregular, soft, reticulate patches on the bark. Color of sterile parts about snuff brown (Ridgway), the fertile parts about drab-gray. The surface of young sterile parts under lens flocculent, with a grayish, pulverulent sheen. The top surface of the infertile regions is incompletely formed, being marked by numerous, rounded, oblong or sinuous holes which are usually less than 1 mm. in their greatest dimension, the top surface of this region is thus cut up into an irregular reticulum; surface of fertile parts complete, not showing the reticulated condition. The branches of the reticulum are 0.3–3 mm. wide, the surface of the branches flattened with abrupt margins, the margin having the appearance of being chewed. Margin determinate or more usually indeterminate. In section 250–1000μ thick, composed of a subiculum made up of a very thin, loosely packed layer of hyphae which extends over the bark and from which arise the anastomosing ridges which spread out horizontally above to form the reticulated top surface. Hyphae of context 2.1–4.2μ thick, usually about 3.8μ, brownish, septate, considerably branched, sometimes circinately coiled, but more frequently nearly straight, very loosely packed, arising at an angle of about 45°. Basidia at first ovoid or pyriform, apparently elongating and enlarging without a rest period to form the mature basidium. Basidium elongated, straight or slightly bent, apparently 2-celled as in *S. jamaicaense* and *S. areolatum*, with a distinct stalk, 8.4–10.5 x 20–30μ. Sterigmata and spores not seen.

Associated with scale insects, some of which are parasitized by hyphae in the form of irregular coils.

This species is easily distinguished from *S. Burtii* (= *S. retiforme* of American authors). For a comparison see under that species. Burt's description of *S. retiforme*, though he had type material available, appears to have been drawn up from the American material and not from the type material from Cuba. And since Patouillard himself did not give a description, the only description previously made is the one very significant line by Berkeley and Curtis, Fungi Cubenses, p. 330, "Subiculo dense tomentoso umbrino; hymenio reticulato pulverulento griseo."

This plant is very close to *S. areolatum* Couch, described from Jamaica. Both are brownish, both have the surface marked with minute holes and the hyphae, probasidia, and basidia in each are quite similar. The two differ in that *S. areolatum* is considerably larger in all respects than *S. retiforme*. *Septobasidium retiforme* is also lighter colored than *S. areolatum*. A very striking difference is in the size and shape of the holes. In *S. areolatum* the holes are larger and rounded while in *S. retiforme* the holes are smaller and elongated. Thus the surface of the former is areolate while the latter is reticulate.

This species somewhat resembles *S. Langloisii* Pat. but may easily be distinguished from that species. It also resembles *S. fusco-violaceum* Bres. but is quite distinct.

Specimens examined:

Cuba: Part of cotype in Curtis Herbarium, labelled *T. retiformis* (B. & C.), no. 384, C. C. Wright, coll. Also a slide of no. 288 prepared by Dr. Burt and kindly lent me by Dr. Linder of the Farlow Museum. Also in New York Bot. Gard. Herb. part of no. 384, C. C. Wright, coll. Probasidia and young basidia present. Type collections nos. 288 and 214 from Kew Herbarium were loaned by Dr. Arthur W. Hill. One of these, no. 214, proved to be not a *Septobasidium* but either a sterile lichen or a blue-green alga.

Septobasidium protractum Sydow. Ann. Mycol. **10:** 33. 1912.

Plate 51, figs. 7–9, and pl. 93, fig. 2

Resupinate, covering extensive areas with a thick firm growth. Surface gray (about smoke gray) over a blackish context; inherently smooth but cut into numerous irregular areas (1–10 mm. wide) by deep, sinuous cracks which may be so narrow as to be seen only with a lens, varying from such microscopic dimensions up to a mm. wide. The wider cracks expose the blackish, underlying context. Margin determinate, covered with dense tufts of coarse, branched pillars. In section usually about 2 mm. thick, composed of three distinct regions: (1) the subiculum which is 200–500μ thick and composed of exceedingly compactly arranged hyphae; from the subiculum arises (2) the region of pillars which are very abundant and closely set; pillars 33–300 x 1500–2000μ, arising more or less vertically and anastomosing freely with each other, pillars very dark brown, almost blackish in color; hyphae of pillars interlaced, forming a rather regular pattern, 4–5μ thick. Towards the top the threads of the pillars branch profusely, the branches extending vertically, uniting with the branches of other pillars to form an exceedingly compact layer of tissue; (3) the top layer, 250–360μ thick. The hymenial layer is formed over the top layer, the hyphae of both regions being indistinguishable except that the tips of the hymenial hyphae are hyaline and about 3μ thick; hymenial region about 80–90μ thick. Probasidia arising laterally from the vertical hyphae, spherical or ovoid, hyaline or pale brownish, 9–10μ thick when spherical, or 9–10 x 12.6μ when ovoid; germinating and becoming entirely empty to form an elongated, once septate basidium,

7–8 x 38–46μ, thickest at the distal end, sterigmata 8–14μ long or longer on the basal cell; spores bent-elliptic, 5–6.3 x 17.6–21μ.

Associated with scale insects (*Parlatoria* sp.) some of which are parasitized by haustoria in the form of irregular coils; other insects are free from parasitism. The thick, tough fungal covering with the dense region of pillars forms an ideal habitation for scale insects as is indicated by the vast number of insects beneath the fungal stroma. A number of insects were dissected out from beneath the stroma but none were found parasitized by the hymenopterous wasps. Evidently the fungal stroma and the dense pillars furnish an ideal protective covering.

This is one of the most distinct of all species of *Septobasidium.* The thick, firm texture, the irregularly cracked grayish surface, with the coarse, dense, blackish, stiff tangle of pillars; the hymenial layer of upright, parallel threads with the two-celled basidia are the characters by which one may recognize this species. It was considered the same as *S. atratum* Pat. by Neuhoff (note on label in Berlin Museum) but is quite distinct. The species from South Africa mentioned by Lloyd (Myc. Notes **7:** 1336, figs. 3088, 3089. 1925) is also undoubtedly the present fungus.

Specimens examined:

Southeast Africa: Transvaal, District of Pretoria, Plat River, on living trunk and branches of *Acacia nigrescens* var. *pallens*, type, P. J. Pienaar, Jan. 31, 1912. Parts of original specimens distributed by Sydow in N. Y. Bot. Gard. Herb., Herb. Bot. Gard. Ber., and Herb. U. N. C.

Septobasidium dictyodes (B. and Br.) Pat. Bull. Soc. Myc. Fr. **24:** 2. 1908.

Thelephora dictyodes B. and Br. Fungi of Ceylon, Jour. Linn. Soc. Bot. **14:** 63. 1873.

Plate 52, fig. 5, and pl. 93, fig. 3

Resupinate, on the bark of living trees, covering an area of several square centimeters. Older part where hymenium is well formed about Mars brown. Surface net-like in the younger parts due to the incomplete formation of the top layer. Margin with pillars which grow upward for a short distance and then bend over anastomosing with other pillars to form a dense tangle. In section up to 2 mm. thick, composed of a loose subiculum from which arise scattered pillars. Pillars 100–200μ thick, up to one mm. tall, branched, and sometimes anastomosing with each other, composed of loosely packed, pale brownish hyphae which are about 3–4.2μ thick. From the pillars arises the top layer which is up to 1,000μ thick and composed of loosely packed, obliquely ascending, branched hyphae. Toward the top surface the hyphae become arranged in loose, open, branched coils. Scattered through this region are numerous spherical, partially empty cells about 7–9μ in diameter. Hymenial layer about 40–50μ thick, composed of hyaline, coiled, rather closely packed threads about 2.8–3μ wide. Probasidia spherical, 9.6–11.3μ, becoming empty upon the formation of the basidium; wall of empty probasidial cell eccentrically thickened, brownish. Basidia cylindrical, once septate and hence 2-celled, sometimes constricted in two places and then apparently 3-celled, 5.8–6.3 x 26–30μ. Spores not seen.

Associated with *Aspidiotus* sp. some of which are parasitized by spindle-shaped hyphae (haustoria) held together by fine threads; other insects are free from parasitism. Haustoria as in *S. Langloisii* Pat. This species is easily recognized by the brown color, the fine netlike structure, and the densely arranged, recumbent, anastomosing pillars.

The probasidia, basidia, and haustoria indicate a relationship to the *S. Langloisii* group, but the characters of the context will serve to separate this fungus from that group.

Specimen examined:

Ceylon: Central Province, Thwaites, no. 1027, Dec. 1868, type in Kew Herbarium.

Septobasidium sabalis Couch. Jour. Elisha Mitchell Sci. Soc. **51**: 69. 1935.

Plate 40 and pl. 82, fig. 1

Resupinate, spreading about over the surface of the petiole and blade of palmetto leaves, covering an area up to 100 sq. cm. or perhaps more; vinaceous buff to wood brown (Ridgway). Surface smooth, almost shiny. Margin usually determinate but irregular in outline. In section up to 1 mm. thick, composed of a very thin subiculum about 50μ thick from which arise short pillars, 80–125μ tall, which support another, thicker horizontal layer, 80–125μ thick. This second layer is often incompletely developed. Between these two layers the scale insects live. From the bottom layer and sometimes from the second arise the pillars. The pillars pass up through the second layer branching out at their tops to form the third or top layer. Pillars quite inconspicuous, about 15–20 to a square mm., 20–70 x 275–375μ, branching out at the top to form a thick region of loosely interwoven hyphae, about 4.2μ thick, septate without clamps, this region about 500–600μ thick. Hymenium borne on surface of top layer, about 120μ thick, composed of much branched, entangled hyphae which are more densely packed than in the underlying context; and probasidia and basidia in various stages of development. Probasidium subspherical or pyriform, 9.6–12.6 x 15–22μ, sprouting and emptying its contents to form a cylindrical basidium. Basidium 8.4–9.6 x 42–50μ, usually thickest in the distal part, sometimes thickest in the middle, once septate by a transverse wall when mature. Each cell of basidium giving rise to a long sterigma up to 30μ long on the end of which is borne a spore. Spores 5.4–8.4 x 19–27μ, most 6.7 x 25μ, recurved at the distal end, broadest at the mucro end, becoming divided into a number of cells by cross and longitudinal partitions.

Symbiotically associated with *Aspidiotus* (?) *sabalis* Comst., some of which are parasitized by coiled hyphae while vast numbers, in all stages of development, are free from parasitism. The species of scale insect agrees well with Comstock's description of *Aspidiotus* (?) *sabalis*. The scale is snow white and nearly circular. The leaf blades and petioles are remarkably free from the insects except beneath the fungus. In this species the subiculum penetrates the stomatal openings, developing tufts of hyphae within the openings. These hyphae apparently do not enter the cells.

It may be distinguished from other species by its habit on the palmetto, by having a subiculum and then a second layer a short distance above the subiculum in addition to the usual horizontal top layer. In the possession of three horizontal layers this species resembles *S. Mariani* Bres. The two bottom layers may grow together in old specimens so that only one layer, the subiculum is evident. The narrow pillars and the very thick top layer composed of very loosely woven hyphae are other distinguishing characters. Microscopically, the large two-celled basidium and the irregularly septate spore, recurved at the distal end, may help in identification.

Specimens examined:

Louisiana: Near Baton Rouge, on leaves and petioles of *Sabal Deeringiana*, Dec. 26, 1931, C. A. Brown, coll., 317, also U. N. C., 9151, type; near Denham Springs, on leaves and petioles, *Sabal Deeringiana*, Jan. 10, 1932, C. A. Brown, coll., 325; also U. N. C., 9210; near New Orleans, on leaves of *Sabal Deeringiana*, Oct. 26, 1933, M. L. Bomhard and W. T. Swingle, colls., U. S. D. A. Herb. and U. N. C. Herb., 9959.

Septobasidium cinnamomeum Burt. Ann. Mo. Bot. Gard. **13**: 332. 1926.

Plate 52, figs. 8, 9, and pl. 93, figs. 5, 6

Resupinate, extending over the bark for several square centimeters (one piece was 5 x 7.5 cm.) and frequently overrunning lichens and liverworts; color a rich velvety Brussels brown. The surface is roughened in the younger parts by a dense tangle of upright pillars which in the material examined by me are mostly tufted at the top and in the older part where the hymenium is developed the surface is roughened by numerous, rounded elevations which are 0.5–1.5 mm. wide. Margin determinate. In section 0.8–1.5 mm. thick, composed of four distinct layers or regions: (1) the basal layer or subiculum 50–210μ thick, composed of densely compacted, pale brownish hyphae 3–4.2μ thick; the basal layer gives rise to short thick tufts of hyphae which branch out to form (2) another horizontal layer parallel to the subiculum. This second layer is considerably thinner than the subiculum; the hyphae of the two layers are similar. Between the subiculum and the second layer there is usually a space 25–125μ tall, thus affording ample space for the movements of the young scale insects. The tufts of hyphae which have given rise to the second horizontal layer continue to grow upwards, each tuft giving rise to (3) a group of several pillars. Frequently however the hyphae arise from this second layer without being united into pillars and not infrequently the pillars may arise from part of the second layer which is not over one of the thick tufts. Pillars abundant, 21–70 x 300–580μ, hyphae of pillars 3.5–4μ thick, septate, brownish. The hyphae of the pillars branch out along the sides and particularly at the top to give the pillars a brush-like appearance. By continued branching and growing of the hyphae near the tops of the pillars (4) the top layer is formed; 270–450μ thick, frequently stratose by the formation of successive hymenial layers; hyphae of top layer 3–4μ thick. In mature specimens the supporting pillars are very few and inconspicuous, due to the growth of hyphae between the second and top layers. Hymenium 65–84μ thick, formed of upright, much branched threads with hyaline or pale brownish tips. Probasidia arising laterally from the vertical threads, rather sparingly formed, in my material spherical, about 10.5μ thick; germinating and becoming empty to form a club-shaped, 2-celled upright basidium, 5.4 x 29–34.4μ. Spores bent-elliptic, almost crescent-shaped, 4.2–5 x 16.8–21μ.

Associated with scale insects some of which are parasitized by haustoria in the form of irregular coils, while others are free from parasitism.

This species is readily distinguished by the bright Brussels brown color; the abundant, tufted pillars near the margin; the irregular, bumpy hymenial surface; the three horizontal layers; the upright hyphae in the hymenium; and the small two-celled basidia.

Specimen examined:
Chile: Corral, Thaxter, coll., Dec. 1903, type in Farlow Herb. and Mo.

Septobasidium stratosum n. sp.

Plate 51, figs. 5, 6, and pl. 93, fig. 7

Resupinate, covering extensive areas on the bark of *Hevea*. Texture rather friable. Color of subiculum a rich dark brown, near bister; the first hymenial layer is near Sayal brown, while the older hymenial layers are grayish brown. The surface of the plant as a whole is very irregular due chiefly to the formation of the hymenium in patches. These patches are about 2–5 sq. cm. in area and frequently anastomose with each other. The surface of the subiculum, much of which is exposed, is roughened by large rhizomorphs, and a few blackish, minute, mostly unbranched pillars. Margin determinate. In section 1–1.5 mm. thick, distinctly and strikingly stratose by the formation of several hymenial layers one over the other, each hymenial layer being separated from the one above and below by a thin layer of scattered upright threads. The fungus consists of three main regions. (1) The subiculum, which is rather indefinite in extent, the entangled hyphae running in between the cracks of the bark, up to about 700μ thick. From the subiculum arise (2) the pillars which are thin and up to 600μ tall. Over the primary pillars arise (3) the stratified hymenial zones. The pillars do not branch out above to form the primary hymenial layer but this layer appears to arise more or less directly from the subiculum or from short tufts of hyphae. The primary top layer including the loosely arranged, vertical, supporting hyphae is 160–200μ thick; the other layers are 126–138μ thick, thus varying very little in thickness. Hyphae of subiculum and context 3.8–4.2μ thick, deep brownish in color. Hymenial layers uniformly about 30–40μ thick, composed of compactly arranged probasidia and much branched hyphae the distal parts of which are thin-walled and hyaline; with one to seven strata. Hyphae about 2μ thick. Probasidia globose to subglobose, hyaline at first, becoming brownish when old, 9.2–11.7 x 12–17μ. Germinating and elongating somewhat and becoming empty to form a straight or slightly bent, cylindrical, two-celled basidium, 6–7 x 29–36μ. The basidia usually separate from the probasidia and stick to the surface of the hymenium. Sterigmata not seen. Spores bent-elliptic, 5–5.4 x 14.7–21μ; many seen adhering to hymenial surface.

Associated with scale insects some of which are parasitized by irregular coils.

This species is easily recognized by the stratose hymenial layers, which occur in small sayal or grayish brown patches, and by the bister colored subiculum with its innate rhizomorphs. The thin hymenium with the delicate, branched, wavy threads and the two-celled fragile basidia are distinctive. The species is very distinct from *S. stratiferum* Boed. & Stein.

Specimens examined:
Brazil: Cochocira, Grande Manaos, August 23, 1923, James R. Weir,

coll., on young *Hevea brasiliensis*, covering the main stem, in Herb. U. S. D. A. no. 62139.

Septobasidium velutinum Pat. Jour. Bot. **6**: 62. 1892.

Plate 52, figs. 6, 7, and pl. 94, figs. 5–7

Resupinate, covering several square centimeters. Color mummy brown (Ridgway) throughout except for the lighter margin. Surface velutinous to the unaided eye, puberulent under a lens. Margin determinate (?). In section up to 1.5 mm. thick, composed of three regions: the subiculum which is usually thin; the middle region composed of rather widely separated slender pillars (15–50μ thick) which extend upward in an oblique direction; the top layer which varies from 60–300μ thick, usually about 140μ thick, the lower part of the top layer composed of much entangled, branched hyphae between which there is considerable anastomosing; hyphae 3.8–5.4μ, usually about 4.2μ thick. Hymenium borne on upper surface of top layer, composed of upright branched hyphae, the bases of which are brown while the tips are hyaline. The tips of the hyphae stand upright, a condition which gives the surface its characteristic, velvety appearance. Probasidia on the sides of these upright hyphae. The probasidial cell forms the basidium by elongating and after a certain length is reached the basal part of this tube becomes empty of its protoplasm and is separated off by a cross wall from the basidium. Sometimes hardly any basal cell is left. Empty probasidial cell, subspherical, 10.5–14.7μ thick. Basidia 10–11.5 x 31–38μ, becoming 2-celled by a transverse septum. Spores not seen.

Associated with scale insects some of which are parasitized by irregularly coiled hyphae. In the Farlow Herbarium, Harvard University, there is a plant collected by Dr. Thaxter on *Baccharis* from Concepcion, Chile, which is labelled *S. velutinum* Pat. This fungus, however, may easily be distinguished from the type. In the Berlin Museum there are three collections labelled *S. velutinum* Pat. (1) From tropical S. America, Cubatao Cai Santos; leg. F. Noack VI, 78. This agrees with the type in microscopic characters and in the slender pillars; the hymenium, however, is light gray, due to the presence of vast numbers of hyaline hyphae. (2) No. 2733. On branches of a species of legume, Rio Jurá, Bom Fim., Nov. 1900. This is quite distinct from the type in color, surface characters, the absence of pillars, and in the hymenial characters. (3) A third collection labelled *S. velutinum* Pat. collected by Santos is apparently a species of *Hymenochaete*, as indicated by Neuhoff on the label.

This species is distinguished by the surface characters, the delicate oblique pillars, and above all by the hyaline upright threads in the hymenium. It is related to *S. Patouillardii* and to *S. jamaicaense* from both of which it may be easily distinguished.

Specimens examined:

Ecuador: Chimbo, on *Citrus*, August 1891, de Lagerheim, coll., type in Pat. Herb., Harvard University.

Septobasidium griseum n. sp.

Resupinate over extensive areas, the top surface forming a more or less continuous growth and not veined or retiform, dotted with numerous minute whitish bumps which give the surface a finely granular or powdered appearance, nearly smooth in some collections, sometimes cracked in places; near pale Quaker drab or pale mouse gray when dry. Margin of subiculum indeterminate, of fruiting surface irregular, fibrillose. In section 1–1.3 mm. thick, composed of three distinct regions: (1) the subiculum made up of a layer of darkly colored threads with a few black rhizomorphs, over which extends an arachnoid growth of fine hyphae, thus giving the subiculum a cottony appearance; hyphae of subiculum 2–4.2μ thick; (2) the middle region composed of numerous blackish purple, upright or branched and anastomosing pillars which give rise to (3) the top layer 80–170μ thick, hyphae brownish toward the base, much branched and with hyaline upright crystal-encrusted tips. Probasidia 10–12.6μ thick, spherical, remaining as an empty cell after the formation of the basidia. Basidia club-shaped, thickest in the distal half, 2-celled, about 8 x 25μ.

Associated with scale insects. Haustoria as in *S. Langloisii.*

This species is very close to *S. Langloisii*, differing in having pillars, a lighter surface, and in lacking a reticulate structure.

Specimens examined:

San Domingo: Sabana Grande, August 16, 1918, on trunk of *Citrus sinensis*, John A. Stevenson, coll., in Mo. Bot. Gard. Herb. no. 55655, as *S. lilacinum* Burt.

Cuba: La Prenda, on *Citrus* sp., December 1921, sent me by C. L. Shear from U. S. D. A. Herb.

Costa Rica: Limon Province, Finca Castilla, July 25, 1936, C. W. Dodge and V. F. Goerger, colls. Type in Dodge Herb., no. 9351, and U. N. C. Herb.

Brazil: Province de Rio de Janeiro, Glazion, coll., with *S. Langloisii*, nos. 2169 and 1979, in Pat. Herb.

Septobasidium Langloisii Pat. Bull. Soc. Myc. Fr. **16**: 54. 1900.

Plate 41, pl. 76, fig. 4, and pl. 94, fig. 8

Resupinate, covering very extensive areas, sometimes up to several of trees. Color pale violet plumbeous (Ridgway), or about grayish blue. Surface usually very irregular, retiform or with conspicuous cracks in the older areas, rarely nearly smooth or with minute holes. Margin usually determinate and with conspicuous rhizomorphs. In section 400–500μ thick, compact except for tunnels which are abundant in the younger parts. Context 330–420μ thick, composed of rather loosely packed, much branched, entangled, brown, septate threads, without clamps, 2.8–3.8μ thick, often swollen. Hymenium 60–80μ thick, composed of brownish threads which branch out to form very thin (about 1.5μ thick) hyaline upright threads and probasidia and basidia in various stages of development. Probasidium 8.4–12.6μ thick, spherical or slightly subspherical, germinating to form a somewhat ovoid basidium, and remaining as an empty cell at the base of the basidium. Basidia 2-celled, 6.7–12 x 21–29μ, most about 8 x 23μ, usually thickest in the middle part, sometimes thickest in the distal half; sterigmata 10–25μ long, borne on both cells of the basidium. Spores elliptic, flattened on one side, sometimes pointed at each end, 6–8.4 x 13.8–21μ.

Symbiotically associated with scale insects (*Aspidiotus* sp.) some of which are parasitized by hyphae composed of spindle-shaped segments connected by very fine threads, some of these threads being more or less straight while others are irregularly coiled. Other insects are free from parasitism and are in all stages of development.

This fungus may be distinguished by its color, the absence of pillars, and by the peculiar marginal and surface characters. Some of the specimens are remarkably beautiful in color and symmetry. The fungus has doubtless been passed over as a crustose lichen.

Septobasidium sinuosum has been confused with this species. Both species occur abundantly in the Gulf States but while *S. Langloisii* occurs only in the Gulf States, *S. sinuosum* extends as far north as southeastern North Carolina. Though the two species may easily be confused, they are quite distinct. *Septobasidium sinuosum* is usually lighter in color being a pallid mouse gray or when darker, Quaker drab to purplish gray, while *S. Langloisii* is usually pale violet plumbeous. The two may also be distinguished by the surface characters, *S. sinuosum* having a distinctly bumpy and sinuously ridged surface, the ridges being rounded; *S. Langloisii* also has a surface with ridges but the ridges are not so distinct, and their surfaces are more flattened and their margins are not rounded but appear to have been eaten away. The margin of *S. sinuosum* is usually without conspicuous rhizomorphs, while in *S. Langloisii* the dark colored rhizomorphs are quite distinct. Microscopically the two are distinct. In *S. sinuosum* the probasidia are 5.8–7.2μ thick, the basidia 3.8–4.4 x 20–32μ, the spores 4–5 x 11.5–16.8μ; in *S. Langloisii* the probasidia are 8.4–12.6μ thick, the basidia

6.7–12 x 21–29μ, spores 6–8.4 x 13.8–21μ. Thus in *S. Langloisii* the probasidia, basidia, and spores are distinctly larger than in *S. sinuosum.*

This species is common on the Gulf Coast and in Florida, apparently being limited in its American distribution to this region. Usually found on *Magnolia virginiana,* but also occurring on *Fraxinus, Carpinus, Nyssa,* and a few other trees. Often found on the main trunk near the base as well as on the limbs.

Specimens examined:

Louisiana: Type, near St. Martinsville, on *Crataegus arborescens,* May 11, 1899, Langlois, 2995, in Patouillard Herb., Harvard, fertile; near Baton Rouge, on *Crataegus,* Dec. 31, 1931, J. N. Couch, 9172.

Mississippi: Pascagoula, very abundant on *Magnolia virginiana,* Jan. 1, 1932, J. N. Couch, 9186, 9194, 9195, 9196, 9198; also near Ocean Springs, June 1896, in N. Y. Bot. Gard. Herb.

Florida: Daytona, Jan. 1898, Thaxter, Farlow Herbarium; Brooksville, on *Carpinus,* very abundant, almost every tree infected, March 17, 1932, J. N. Couch, 9281; Sebring, on *Fraxinus,* March 14, 1932, 9268, also on unidentified trees, 9271, 9272, 9273; near Kissimmee, on *Fraxinus,* March 13, 1932, J. N. Couch, 9247, also on unidentified trees, 9237, 9249; near Tallahassee, on (?), March 17, 1932, J. N. Couch, 9286; near Orlando, on *Magnolia virginiana,* very common, March 12, 1932, 9956; Ichtucknee Springs, near Fort White, on *Carpinus,* Feb. 1, 1935, Dow V. Baxter, coll., 9967, surface nearly smooth; same locality, Feb. 3, L. E. Arnold and E. West, colls., Univ. Fla. Herb., 9939.

Septobasidium sinuosum Couch. Jour. Elisha Mitchell Sci. Soc. **51:** 65. 1935.

Plate 29, figs. 1–3, and pl. 76, figs. 5–7

Resupinate, often covering an area 20 cm. long by two or more centimeters wide. Surface repeatedly divided by irregular, sinuous, anastomosing ridges, especially toward the margin and often throughout; sometimes nearly smooth in the center with a few minute holes; old areas often becoming cracked; pallid mouse gray or pallid purplish gray to Quaker drab or purplish gray, sometimes with dark dots here and there. Margin thin, whitish, myceloid. In section 400–600μ thick, composed of three rather indistinct regions or layers: (1) the subiculum, a thin inconspicuous layer (about 20μ thick) which grows horizontally over the surface of the bark, threads of subiculum about 2.2μ thick, hyaline in regions or deep mouse gray to sordid purplish brown; (2) middle layer concolorous with the subiculum, usually 300–400μ thick, composed of rather loosely packed, nearly dichotomously branched, entangled hyphae; hyphae 2–2.4μ thick, hyaline or brownish under the microscope, often with dark colored hyphae intermingled with hyaline threads, without clamp connections, peculiar in that under an oil immersion lens the walls are distinctly warted, punctate or spiny; (3) hymenial layer, which is usually hyaline, 50–120μ thick, composed of very densely packed, twisted and much entangled hyphae, the ends of which are slightly coiled, probasidia and basidia in various stages of development. Probasidia spherical or slightly subspherical, 5.8–7.2μ thick, germinating to form a cylindrical basidium, the probasidium becoming entirely empty during the process. Basidium 3.8–4.4 x 20–32μ, becoming once septate by a transverse wall, each cell of the basidium sprouting a long sterigma which grows to the outer surface of the hymenium and there bears a spore. Sterigmata 18–25μ long. Spores smooth, sausage-shaped, 4–5 x 11.5–16.8μ, most 4.2 x 12.6μ.

This species shows a great variation in color, sometimes being very light colored, nearly white, and at other times a very deep, dull purplish brown. It can readily be recognized by the peculiar surface which is divided up by anastomosing, sinuous ridges as in some crustaceous lichens. This species has been confused with *S. Langloisii* (for a comparison of the two see under *S. Langloisii*).

Associated with scale insects, some of which are parasitized by peculiar sausage-shaped hyphae while others are free from parasitism.

Very common (represented by 41 collections) from southeastern North Carolina southward along the coastal region and westward to New Orleans and perhaps farther.

Specimens examined:

North Carolina: Near Wilmington, on *Carpinus caroliniana*, Dec. 30, 1921, J. N. Couch and F. A. Grant, colls., 4955; near Wilmington, on *Liquidambar styraciflua*, Dec. 28, 1922, Couch and Grant, colls., 5934; near Magnolia, on *Cornus florida*, in fruit, Jan. 1, 1928, A. C. Mathews, coll., 8301, type.

South Carolina: Lynch's river bottom, near McBee, on *Liquidambar styraciflua*, June 30, 1928, A. B. Couch, coll., 8316; also on *Ilex opaca*, J. N. Couch, coll., 8321; near Georgetown on *Myrica carolinensis*, July 2, 1928, J. N. Couch, coll., 8322; near Georgetown, on *Quercus phellos*, July 3, 1928, 8323 A; Myrtle Beach on *Quercus nigra*, August 29, 1931, associated with *Aspidiotus osborni*, J. N. Couch, coll., 9099.

Florida: Very abundant as far south as Sebring. Near Jacksonville, Kissimmee, Brooksville, and Tallahassee, J. N. Couch, coll.; on *Quercus laurifolia*, Gainesville, G. F. Weber, coll., 9410, 9964; also

Gainesville, H. E. Stevens, coll., Mo. Bot. Gard. Herb., 44211, under *S. Langloisii* Pat.

Mississippi: Pascagoula, common on several species of *Quercus*, *Cornus florida*, *Liquidambar*, also a variety on *Magnolia virginiana*, with a nearly smooth not sinuous (or only slightly sinuous in places) surface, somewhat cracked in older parts. Color dark vinaceous gray. Probasidia, basidia, and spores smaller than in type. Spores 3–4 x 10–13.8μ. Jan. 1, 1932, J. N. Couch, coll.

Louisiana: New Orleans, on *Liquidambar styraciflua*, Dec. 31, 1932, Donald Rogers, 9935; Baton Rouge, on *Liquidambar styraciflua*, Jan. 8, 1932, Clair A. Brown, 9705; also 9706 from same locality on *Quercus nigra*; and 9178 on vine (*Aristolochia*?) Dec. 31, 1931, J. N. Couch.

Septobasidium rimulosum Petch and Couch n. sp.

Septobasidium sp. Petch, Diseases of the Tea Bush, Plate 2. MacMillan & Co. London, 1923.

Plate 51, fig. 4, and pl. 93, fig. 1

Resupinate, clothing the stems of tea bushes for a length of several feet, when fresh "slate colored toward the advancing margin, the older parts purplish-grey" (Petch), becoming tilleul-buff to avellaneous in the herbarium. In the marginal region the surface is dotted with dark dots less than a mm. wide and with sinuous ridges. The surface abruptly changes from the darker dotted region to the lighter colored part, the surface of which is conspicuously and regularly cracked into a vast number of polygonal areas which are mostly less than a mm. wide. The surface thus has the appearance of finely cracked clay. The margin is determinate in places but quite indefinite in other places. In section 200–600μ, usually about 350μ thick. Consisting of a subiculum, 30–60μ thick, composed of closely packed, brownish threads 3.6–4μ thick, from the subiculum arising the context which does not form a continuous horizontal layer but is broken by the frequent cracks; context 100–200μ thick; threads of context fairly compact, entangled, sometimes dichotomously branched and circinately curved, brownish in color, about 3.4–3.8μ thick, sometimes with minutely rough walls due apparently to the deposit of fine granules of material. From the context arises the hymenium which is 60–200μ thick, and composed of very densely compacted hyphae which extend mostly vertically but are sometimes dichotomously branched and circinate. The hyphae of hymenium brownish toward the base, hyaline toward the top. Probasidia spherical or subspherical, 6.7–8.4μ thick, sprouting and becoming empty to form an elongated two-celled basidium, 5–5.4 x 21–27μ. Sterigmata long. Spores 4 x 9.6–11.5μ, bent-elliptic.

Associated with scale insects some of which are parasitized by haustoria of the spindle-shaped type as in *S. sinuosum*; others are free from parasitism. Several females were found even in the older parts of the fungus which were in the act of giving birth to young.

This species is closest to *S. sinuosum*, resembling it in probasidial, basidial, and spore characters. The threads of the context and subiculum in the two are much alike and the haustoria in both are of the same structure. The two differ in color and in surface characters.

Specimens examined:

Ceylon: Pambagama, May 1918, on tea, Petch, coll., no. 5759, type in Petch Herb. and U. N. C. Herb.

Septobasidium Bresadolae Pat. Bull. Soc. Myc. Fr. **16:** 181. 1900.
Described as *S. velutinum* by Bresadola in Hedwigia **35:** 290. 1896.

Plate 51, figs. 1–3, and pl. 94, figs. 1–4

Resupinate, extending for 30–40 cm. over the bark. Fertile parts avellaneous to wood brown or buffy brown, sometimes with a slight purplish tint; older parts sordid dark brown. Surface smooth and velvety in the fertile parts, roughened by many irregular cracks and holes in the older parts. Margin very irregular, apparently indeterminate. In section up to 1.3 mm. thick from surface of bark to outer surface of fungus, often stratified by the formation of several hymenial layers. The subiculum is not distinct as a layer or region but the fungal threads extend through the irregular cracks of the bark and twine about the vast numbers of mostly empty scales of the insects. Context 500–1100μ thick, composed of loosely woven branched hyphae, 2.8–4.3μ thick; between the hyphae there is considerable anastomosing. Hymenial region 60–80μ thick, composed of hyaline sub-dichotomously branched, recurved hyphae, probasidia and basidia. Probasidia spherical, 10.9–13.4μ thick, frequently with minutely rough walls when mature, germinating to form the basidia, which arise either from the side or top of the probasidial cell. Probasidial cell remaining as an empty cell at the base of the basidium. Basidium club-shaped, typically broadest at the distal end, 8.4–9.8 x 36–47μ, becoming transversely septate to form 2 cells. Rarely the basidia may be constricted in the middle and may consist of three or four cells, one or two of which are small and non-functional. Such atypical basidia produce sterigmata and apparently normal spores. Sterigmata 8–12μ long. Spores short-elliptic, flattened on one side, 6.3–10.5 x 12.6–21μ, most about 8.4 x 18μ.

Associated with scale insects (*Aspidiotus* ? sp. with black, circular, dorsal and ventral scales), many of which were parasitized by haustoria of the *S. fumigatum* type while others were free from fungal infection.

This species shows a striking resemblance to *S. fumigatum* Burt and is apparently closely related to that species, the two differing mainly in the present species having a two-celled instead of a four-celled basidium. Indeed the occurrence of four-celled and three-celled basidia with only two large sized functional cells suggests that the present species might have originated from *S. fumigatum* Burt by the suppression of two of the basidial cells.

Specimens examined:

Brazil: Sao Leopoldo, Rick, coll., in Pat. Herb., probably type; also from Rick, collected near Sao Leopoldo, Sept. 1933, nos. 9424, 9425, 9426, in U. N. C. Herb., all in fruit; also from Rick, same location, no. 10380 and no. 10381. In Lloyd Collection no. 15957, U. S. D. A. Herb.

Septobasidium Linderi n. sp.

Plate 50, figs. 1–11

Resupinate, forming thin, closely adherent, small areas (up to 4 cm. wide) of growth on the small branches of *Pinus rigida.* Color whitish to wood brown. Surface smooth except for irregularities in the bark, sometimes minutely fibrillose. Margin indeterminate. In section very thin, usually hardly more than 100μ thick, not differentiated into layers or pillars. Hyphae fairly compactly arranged, entangled, hyaline to pale brownish, septate, branched, 3–5μ thick. Probasidia and basidia arising from the horizontal threads without being arranged in a definite hymenial layer. Probasidia globose to pyriform with a smooth or faintly rough wall, 11–14.5μ thick; arising singly or in clusters of two or three, usually with one or two stalk cells; germinating and becoming empty to form a long cylindrical basidium which is usually two-celled but may frequently be three- or four-celled. Basidia usually thickest in the distal part, 5.8–6.5 x 34–42μ when two-celled, 5–5.8 x 52–63μ when 4-celled, each cell as a rule giving rise to a short conical sterigma on the end of which is formed the spore. Not infrequently each cell of a two-celled basidium may give rise to two sterigmata or to one sterigma which branches to form two. Such cells which give rise to two sterigmata appear to form only one spore. Spores 5–6.7 x 18.4–21.4μ, the larger spores being formed on the two-celled basidia, the smaller on the three- or four-celled. Some of the spores were once septate.

Associated with scale insects which are very difficult to find, being hidden in the cracks of the bark. Some of the insects are parasitized by haustoria of the glomerulus type.

This species is characterized by its inconspicuousness; its whitish to wood brown color; its exceedingly thin, dryish, papery texture, and indeterminate margin, and by the probasidia usually borne in clusters and the two-, three- or four-celled basidia. It is easily distinguished from *S. pinicola* Snell.

This is the only species found so far which shows such remarkable variation in basidial structure and for this reason would be an interesting form for cytological study.

It is of interest to note that at the same time this species was collected a collection of *S. pinicola* was made by Linder. The latter however was not in fruit while the former was.

Specimens examined:

Massachusetts: Near Pembroke, on *Pinus rigida*, April 11, 1937, D. H. Linder, coll., type in Farlow Herb. and U. N. C. Herb.

Septobasidium griseopurpureum n. sp.

Plate 52, figs. 1–4, and pl. 93, fig. 4.

Resupinate, covering extensive areas (extending for 8 cm.) on the bark. Color dark purplish gray, frequently the surface may be partly covered with lichen soredia thus giving a greenish tint. Surface broken by numerous cracks due either to the splitting of the surface of the fungus or to the incomplete formation of the upper layer; except for the cracks the surface is smooth and in places glabrous. Margin determinate, irregular, due to the fact that the fungus in this region consists of sinuous, anastomosing plates between which the bark is visible, sometimes the marginal region may be roughened by the formation of tufts of hyphae or plates of hyphae which rise from the subiculum to form minute tent or umbrella-like enclosures beneath some of which young insects may be found. In section 200–550μ thick, composed of three regions which in most parts are not sharply differentiated from each other. (1) The subiculum, 20–65μ thick, composed of dark brownish, thick-walled, frequently septate hyphae, 3–6.3μ thick. From the subiculum arise the loosely arranged hyphae of (2) the context. This region is usually not more than 85μ thick; hyphae 3.5–6.3μ thick, rather sparingly branched. The hyphae of context branch out above to form a dense tangle, (3) the top layer, which varies much in thickness, 60–170μ thick, and is composed of profusely branched hyphae, on which are borne numerous probasidia. Probasidia apical or lateral, usually formed in clusters, spherical, 9.5–11.7μ thick, hyaline at first but becoming dark colored upon maturity, wall usually minutely roughened, germinating and becoming entirely empty to form a two-celled, club-shaped basidium, which usually is attached to the probasidium by means of a short stalk. After the formation of the basidia the probasidia may partially collapse leaving about half of the old probasidial wall. Basidia characteristically straight but sometimes curved, 6–6.3 x 30–33μ, thickest in the distal part. Sterigmata about 2.8 x 8.4μ. Spores hyaline, bent-elliptic, 4.2–5 x 15.9–18μ.

Associated with scale insects (near *Lepidosaphes*) some of which are parasitized by irregularly coiled haustoria, while others are free from infection. In the parasitized insects the fungus enters the insect in the mouth region.

This species slightly resembles *S. Curtisii* in color and structure. Also the probasidia are borne as in *S. Curtisii* and have a roughened and dark colored wall at maturity. In the present species the probasidia and basidia are much smaller than in *S. Curtisii* and in the former species the basidium is two-celled while in the latter it is four-celled.

It may readily be distinguished by the dark, purplish gray color, the thin structure, the cracked surface, the absence of pillars, the sinuous structure in the marginal region, the minutely rough, dark colored wall of the probasidia, and the two-celled basidia.

Specimens examined:
South Africa: Coll., P. van der Bijl, in Lloyd Collection 39226.

Septobasidium apiculatum Couch. Jour. Elisha Mitchell Sci. Soc. **51:** 62. 1935.

Frontispiece and plate 39, figs. 1–10, pl. 67, figs. 1, 2, and pl. 80, figs. 6–10

Resupinate, forming rather inconspicuous, small, more or less circular or irregular patches on living bark, up to 5 x 10 cm. in area; typically sordid white in color, when fresh, sometimes with a fleshy or buffy tint. Surface inherently finely granulose but with numerous bumps and irregularities partly due to the roughness of the bark, also with numerous mounds on the top of each of which is a single spine about 1.5 mm. high, mounds up to 1.5 mm. wide. In section 250–550μ thick in well developed specimens, usually about 275μ thick, compact, often stratose due to the formation of successive hymenial layers. Threads of context 4–5μ thick, septate, constricted at joints, without clamp connections. Probasidia thin-walled, hyaline, ovoid, elongating and enlarging to form the basidia. Basidia elongated, club-shaped, 6.8–8.4 x 23–32μ, 3-celled, each cell giving rise to a long sterigma on the end of which a spore is formed. Spores 4.6–6.3 x 13–25μ, most about 5.4 x 17.8μ, bent-elliptic. Haustoria composed of elliptic cells held together by very fine threads.

Symbiotically associated with several species of scale insects, the species depending upon the host plant. In such an association some of the insects are parasitized by spindle-shaped hyphae while others are left free to reproduce. On *Cornus amomum*, associated with *Aspidio-*

tus ancylus which are parasitized through the body wall by the mouth, also sometimes associated with *A. juglans-regiae* and *A. forbesi*. Large numbers of the two species *Chrysomphalus tenebricosus* Comst. and *Parlatoria proteus* are present beneath the fungus on *Cornus amomum*, but none are parasitized. On sweet bay (*Magnolia virginiana*), associated with *Aspidiotus diffinis* Newst. and on pin oak (*Quercus palustris*), associated with *A. osborni* Newst. & Ckll.

Easily recognized by the thin white growth with the upright fascicles of hyphae.

Specimens examined:

Virginia: Mecklenburg Co., on *Cornus florida*, with *S. filiforme*, 8179.

North Carolina: Chapel Hill, on *Cornus florida* and *C. amomum*, common at all times of the year, twenty-seven collections; on *Hicoria* sp., 9323; on *Quercus phellos*, March 11, 1929, 8390, type; on *Quercus palustris*, 8492; on *Liquidambar styraciflua*, 9334; on *Pyrus japonica*, 9078; on *Carpinus caroliniana*, 9391; on *Cercis canadensis*, 9386, 9387; near Raleigh, on *Cornus florida*, May 15, 1929, B. B. Fulton, coll., 8410; Pasquotank County, on *Cornus florida*, Willard Hewitt, coll., 9366.

South Carolina: Near Charleston, on *Quercus*, March 19, 1930, J. N. Couch, coll., 8459; St. Helena's Island, on *Liquidambar styraciflua*, March 21, 1930, 8486; on *Magnolia virginiana*, March 21, 1930, 8484.

Florida: Near Brooksville (host ?), March 17, 1932, J. N. Couch, coll., 9280; near Tallahassee, on *Liquidambar styraciflua*, March 17, 1932, J. N. Couch, coll., 9281.

Louisiana: Near Baton Rouge, on *Liquidambar styraciflua*, Nov. 21, 1932, C. A. Brown, coll., 567, and U. N. C., 9952.

Arkansas: Near Little Rock, on *Quercus phellos*, April 12, 1933, Dow V. Baxter, U. N. C., 9377.

Septobasidium canescens Burt. Ann. Mo. Bot. Gard. **3:** 342, fig. 13. 1916.

Crocynia moxleyi Plitt. The Bryologist **26:** 16–19, figs. 1 & 2. 1923. See The Bryologist **39:** 42. 1936.

Plate 38 and pl. 75, figs. 1–7

Resupinate, covering several square centimeters on the bark of *Quercus*. Pale olive buff, the olive color due to the presence of algae mixed among the hyphae. Surface irregular, due to numerous bumps and cracks, cottony in places. Margin thin but determinate. In section 450–1,000μ thick, not differentiated into layers. Hyphae in the subiculum rather closely packed and extending more or less horizontally from the subiculum; the hyphae arising obliquely to form the loosely woven, branched threads of the context; hyphae 3–5μ thick, hyaline, septate, without clamps. In the older parts the hyphae are thick-walled, much septate, and constricted at the joints. Hymenial region irregularly arranged, not forming a distinct layer, composed of probasidia and basidia in various stages of development. Probasidia spherical, 10.5–19μ, usually 15–16μ thick, basidia arising from the probasidia which are left as empty cells. Basidia 9.2–10.5 x 50–63μ, thickest at the distal end, divided at maturity into 2, 3, or 4 cells by transverse walls, usually three-celled, the distal cell being much larger than the proximal two cells. The basidia are typically very irregular in the material examined by me and very few have any sterigmata. Usually the 2- or 3-celled basidia have thickened walls, an adaptation perhaps to development in dry climate. Spores 6.3–9.6 x 18–21.8μ, hyaline, bent-elliptic.

Associated with scale insects (*Cerecoccus ehrhorni* Cockerell), some of which are parasitized by irregular coils of the glomerulus type while others are free from infection. Material collected August 27, 1921, showed large numbers of young settled down beneath the fungus. The reddish bodies of the young are very conspicuous against the white fungus. The parasitized insects instead of occurring singly under one mound of fungal tissue occur in considerable numbers under large mounds. The tissue of these mounds is very compactly arranged and solid and each parasitized insect is housed in a separate chamber hollowed out of the tissue of the mound.

This fungus may be recognized by the whitish, irregular, cottony surface (under lens), the lack of pillars, the spherical, hyaline probasidia, and the irregular basidia. The mycelium and the irregular basidia suggest a relationship to *Septobasidium apiculatum.*

Specimens examined:

California: In Farlow Herbarium, 1915; Pasadena, on *Quercus,* A. G. Smith, June 15, 1914; same location, March 13, 1915; Tustin, March 1921, on *Quercus agrifolia,* H. S. Fawcett, coll.; Pasadena, Dec. 22, 1933, on *Quercus agrifolia.* Also six collections under the name of *Crocynia moxleyi* Plitt on *Quercus agrifolia* and *Q. chrysolepis,* in Herb. of C. A. Plitt in U. S. D. A. Herb.

Septobasidium alatum Lloyd. Myc. Notes **6:** 888, fig. 1538, pl. 131, 1919.

Plate 58, figs. 8–10, and pl. 114, fig. 3

Resupinate, extending laterally to form winged margins which may be up to a cm. wide. Color dark brown or snuff brown. Surface with irregular, rounded elevations which are 0.5–1.5 mm. wide and up to 0.5 mm. high. Margin of winged part with distinct (under lens) fibrils, some of which are appressed to the surface while others are free. Margin of subiculum determinate, with conspicuous black, shiny pillars which may arise singly or in groups. In section 1–1.5 mm. thick, composed of four distinct regions. (1) The subiculum, 40–270μ thick, made up of very compactly arranged hyphae which appear somewhat pseudoparenchymatous in cross section; the hyphae in lower part of subiculum dark colored, in upper part hyaline. (2) The pillars which arise abundantly from the subiculum in the marginal region but are much less conspicuous in the older parts after the top layer has been well formed. (3) The second horizontal layer which arises from the pillars about 40–50μ above the subiculum. This layer originates from each pillar as a very compact, thin, dark colored circle of hyphae. By the continued horizontal growth of the hyphae, adjacent circles touch each other, their margins anastomosing to form this thin (about 40μ thick) horizontal layer, the pillars arising from the subiculum and extending up through this horizontal layer; pillars single or several pillars may anastomose near their bases, jet black in color, 30–85μ thick by 0.8–3 mm. tall. (4) The top layer which may arise from the tops or sides of the pillars. The lateral wings consist of the top layer supported by pillars which extend out as rhizomorphs. Not infrequently the pillars in the wing branch out to form a palm-leaf-fan-like structure. Top layer 300–500μ thick, characteristically irregular in outline, due to the bumps and elevations, and sometimes minute granulations as in *S. granulosum* Sydow. Hyphae of pillars and lower part of top layer 4–5μ thick, pale brown, of hymenial region 2–3μ thick, hyaline, branched, curved. Probasidia lacking, basidia curved, 4.8–5.4 x 36–46μ, 4-celled, hymenium with many empty collapsed basidia; spores bent-elliptic (only a few seen).

Associated with scale insects, some of which are parasitized by irregular coils while others are free from infection.

This species is very close to *S. granulosum* Sydow and to *S. aligerum* Petch. I have studied type material of all three and though closely related they appear to be distinct. All three have winged margins and the same type of basidia. Sydow (1922) was of the opinion that the present species is the same as *S. granulosum* Sydow. The two differ, however, in the following ways: *S. alatum* is a considerably coarser plant and has conspicuous branched pillars while pillars are lacking in *S. granulosum*; the hyphae are thicker in *S. alatum* and the large fibrils give a distinctly coarser appearance under the lens; *S. alatum* is a lighter brown color in the context than *S. granulosum*, which is a deep chocolate brown. The granulated surface of the hymenium also serves to distinguish *S. granulosum* from *S. alatum*.

This species is close to *S. aligerum* Petch in having a winged margin, three horizontal layers, and in having the same type of basidia. In *S. aligerum* the hymenial surface is light drab or purplish gray, while in *S. alatum* the surface is near snuff brown even when the basidia are abundant. In *S. alatum* the subiculum is light colored and compact and the pillars frequently anastomose and branch, while in *S. aligerum* the subiculum is dark brown, less compact, and the pillars arise singly and as a rule are much shorter than in *S. alatum.*

The present species may be recognized by the conspicuous marginal pillars, the branched fibrils in the wings, the three horizontal layers, the coarsely bumpy surface, and by the whitish or buff colored subiculum.

Specimens examined:

Philippine Islands: Luzon, subprovince Ifugao, Mt. Polis, February 1913, R. C. McGregor, coll., in Herb. Bur. Sci. 20385. Parts of this collection are in the Lloyd Herb., Farlow Herb., N. Y. B. G. Herb., the last mentioned being the only material showing a good margin and conspicuous pillars.

Septobasidium aligerum Petch. Trans. Brit. Myc. Soc. **12**: 276, pls. 18, figs. 1, 2, and 19, figs. 1, 2. 1927.

Hymenochaete ramealis Berk. Jour. Linn. Soc. Bot. **14**: 68. 1873.

Stereum rameale (Berk.) Massee. Jour. Linn. Soc. Bot. **27**: 187. 1890.

Septobasidium rameale Bres. in Herb. Kew.

Septobasidium rameale (Berk.) Bres. in Petch. Trans. Brit. Myc. Soc. **7**: 34. 1921.

Plate 58, figs. 4–7, and pl. 96, figs. 8, 9

Resupinate on the stems and leaves of various shrubs. On orange trees the fungus may cover all the branches for almost their entire length, also extending over the leaves (Petch). Upper surface light drab, or purple gray (Petch). The original type material is deep brownish. Surface with numerous minute bumps or conical elevations where the pillars push up into the top layer. Marginal region usually determinate, except on leaves where it may spread out indeterminately (i. e., margin of the subiculum). In the marginal region and also in the older parts there are a few rhizomorphs which arise from the subiculum and extend diagonally upward ending usually in a fine point. In section 1200–1500μ thick, composed of four layers. (1) The very thin subiculum, 20–40μ thick. The hyphae of the subiculum are brownish, compact and extend horizontally, except for numerous upright short, branched, hyaline hyphae with coiled or recurved tips. These upright hyphae give to the subiculum a characteristic grayish, powdery appearance under the lens. (2) From the subiculum arise numerous, slender, straight black pillars, 21–115 x 1000–1500μ; threads of pillars 3–4.2μ thick, rather thin-walled. (3) From the bases of the pillars there grow out circular tufts of hyaline branched hyphae quite similar to the hyphae which arise from the subiculum. And in some cases these hyphae extend outward until they meet hyphae from adjacent pillars, thus forming a second layer parallel with the subiculum. The young scale insects usually settle down beneath that region of the fungus where the middle horizontal layer has been formed. The bristles or pillars branch out above to form (4) the top layer, 100–200μ thick, over which is developed the hymenium. The top layer often extends outward forming lateral wing-like margins. Hymenium composed of basidia in various stages of development and branched, curved, sometimes circinately curved, hyaline threads 2–2.5μ thick; there are also present in the hymenium seta-like threads which project 30–45μ beyond the hymenial surface. These structures are hyaline except for a little brownish color at the base and arise from thick-walled hyphae that are about 5.4μ thick and frequently brownish. They may be single and taper to the tip as the true setae in *Hymenochaete* or they may branch forming a small tuft of threads. Basidia coiled, frequently forming a complete turn, 5–5.4 x 30–40μ, 4-celled, each cell giving rise to a sterigma on which is borne one spore. Spores short, bent-elliptic, 4.2–5.4 x 11–14μ.

Associated with scale insects some of which are parasitized by irregularly coiled hyphae while others are free from parasitism.

This fungus was first described as *Hymenochaete ramealis* Berk., later as *Stereum rameale* (Berk.) Massee. Bresadola transferred it to *Septobasidium* as *S. rameale* (Berk.) Bres. Due to the fact that this species name had been used for another fungus (*Lachnocladium rameale* B. & Br.) which has been found to be a species of *Septobasidium*, Petch (1927) renamed the fungus *S. aligerum* Petch to avoid the confusion.

The above description has been made from material collected by Petch in Ceylon and material from Australia since this material is in excellent condition. The type material (collected 1847) is brownish, showing no purple, the pillars lack the tufts of hyphae at the base, and the hyphae of the pillars are thicker, being 4.2–5.4μ thick with rather thick walls. The hyphae of the hymenium are thin, about 2.1μ thick, somewhat circinately curved, branched, thin-walled and hyaline and the basidia though mostly collapsed can be seen with a water immersion lens and agree with those in Petch's material.

This species is very close to *S. granulosum* Sydow, agreeing with that species in having a free projecting margin and in having coiled basidia which arise directly from the hyphae without a probasidial stage. They differ in that *S. granulosum* lacks pillars and the hymenial layer is formed by separate tufts of hyphae which give the upper layer its granulated appearance. The basidia and spores in *S. granulosum* Sydow are considerably larger. The present fungus is also related to *S. bogoriense* Pat. but the two can easily be distinguished. The irregular surface, the distinct pillars, the finely granulose appearance of the subiculum, the lack of a persistent probasidial cell are some of the characters by which this species may be distinguished from *S. bogoriense.* For a comparison with *S. alatum* see under that species.

Petch (1927) has well illustrated this fungus except for the basidia and spores. He reports it as being a common "up country" species in Ceylon occurring on shrubs in the jungle, overrunning stems and leaves, also on orange trees, at elevations of 4000 feet or more.

Specimens examined:

Ceylon: Glenugie, on orange leaf and stems, T. Petch, coll., Feb. 20, 1919; also August 1919, Herb. Kew. Part of the same collection under *Hymenochaete ramealis* (Berk.) in Pat. Herb., Harvard University; also Nuwara Eliya, Gardner, coll., January 1847, as *Hymenochaete ramealis* Berk., type.

Australia: North Queensland, on lemon, occurring with *S. bogoriense* Pat., July 17, 1936, no. 4673, in Herb. Plant Pathology, Brisbane. Spores 4–4.2 x 12–14μ. Basidia and spores abundant.

Septobasidium Schweinitzii Burt. Ann. Mo. Bot. Gard. **3**: 324. 1916.

Plates 33, 79, fig. 1, and pl. 83, figs. 1–6

Perennial, resupinate, forming isolated or confluent patches growing over the surface of the bark, covering areas of a few to 15 or more sq. cm. Color quite variable, usually brownish with a purplish tint (benzo brown, or Van Dyke brown, or light cinnamon brown). Surface velvety, not glabrous, often minutely warted, sometimes considerably cracked in old specimens. Margin determinate, whitish, usually finely fimbriate (upper layer), margin of lower layer often more or less covered with v- or u-shaped plates of upright hyphae from which the pillars arise. In section 0.6–1 mm. thick, composed of three distinct regions: (1) the subiculum which extends over the surface of the bark and is whitish with a tint of purple in the young, fresh condition; usually extending only about 2 mm. beyond the margin of the upper layer; 60–90μ thick; from the basal layer numerous, short (60–70μ) upright, white, usually unbranched hyphae arise, bent at the upper end like a shepherd's crook; (2) the middle region composed of upright pillars which arise from the subiculum; pillars numerous, about 10–12 to a square mm., 20–65μ thick, most 30–40μ thick by 0.5–0.6 mm. high; (3) upper layer 125–270μ thick, arising from the pillars and composed of much entangled, branched hyphae, hymenial region 60–100μ thick, composed of upright branched threads between which there is considerable anastomosing, threads wavy and with slightly thickened walls toward the ends, about 3μ thick. Basidia formed directly from the hyphal threads. Basidia 5 x 35–40μ, coiled one to one and a half times, becoming divided by transverse walls into 5 cells, the basal cell being functionless. Basidia formed during the late fall and germinating during the spring and summer; each cell, except the basal one, germinating into a long tube which becomes swollen toward the distal end. The tubes are arranged more or less regularly side by side. From the distal end of each tube a long narrow (2 x 10–15μ) sterigma sprouts, which in turn bears a single spore on its end. Spores very irregular in size, 4.2–6.2 x 12.6–23.1μ, bent-elliptic, white in print, rarely becoming once or thrice septate.

Common on ash, often occurring on the same limbs with *S. pseudo-pedicellatum* Burt, *S. Patouillardii* Burt, and *S. Mariani* Bres., and not rarely confluent with these species. Symbiotically associated with scale insects (*Chionaspis gleditsiae* Sanders), some of which are parasitized by irregularly coiled hyphae, others remaining free from fungal infection.

This species may be recognized with the unaided eye by the striking color (benzo brown, etc.) and the peculiar marginal characters. Through the courtesy of Dr. Pennell, I have been able to study the type material of this species which is in the Schweinitz Herbarium at the Philadelphia Academy of Natural Science. The type material, consisting of a piece about half the size of one's smallest finger, was collected by Schweinitz near Winston-Salem, N. C., over one hundred years ago. Dr. K. B. Raper has collected this species for me near Welcome, N. C., only a few miles from Winston-Salem. The microscopic structure of the type material is identically the same as that of my specimens. The color of the type has faded to a drab color, but the colored picture of the type as illustrated by Schweinitz in "Synopsis fungorum Carolinae Superioris," Tab. 2, fig. 3, agrees remarkably well with the color of my material. In my study of the type material, I have been unable to find the "erect probasidia" figured by Dr. Burt (Fig. 1a); nor did I find the "spore bearing organ" attached as shown in the same figure.

This species is fairly common around Chapel Hill and I have therefore been able to keep it under observation during the past ten years and to work out some of the difficult points in its life history. No persistent probasidia are formed in this species, the basidium arising directly from the hyphal threads. This structure starts as a lateral tuberous cell which elongates and becomes coiled or irregularly twisted and finally becomes divided into five cells by transverse septa. Of these five cells the basal one is apparently functionless and may indicate the remains of a once normal probasidial cell, the other four cells are the "basidium." This structure formed during the early winter remains quiescent during the winter months. In the early spring each cell of the basidium sprouts a long tube which grows up to the surface of the hymenium. From the end of this tube a slender, long sterigma grows and on the end of this a spore is borne. During the formation of the spore the cell of the basidium, the long tube, and the sterigma become empty, the protoplasm, of course, passing into the spores.

This species is fairly abundant in certain localities, but does not seem to have a wide geographic range, being known with certainty only from North and South Carolina. It is represented by eleven collections in our herbarium. Usually occurring with *S. pseudopedicellatum* and *S. Patouillardii.*

Specimens examined:

North Carolina: Near Winston-Salem, Schweinitz, coll., in Schweinitz Herbarium, Philadelphia; same loc., on *Fraxinus*, with *S. Patouillardii* and *S. pseudopedicellatum*, Oct. 28, 1928, K. B. Raper, coll., 8370; same location, Dec. 27, 1933, J. R. Raper, coll., 9341; near Chapel Hill, on *Fraxinus americana*, J. N. Couch, coll., nos. 8373, 8374, 8379, 8397, 8398, 9989; near Wilmington, on *Fraxinus americana*, Dec. 28, 1921, J. N. Couch, coll., 5876.

South Carolina: Near McBee, on *Fraxinus americana*, June 30, 1929, A. B. Couch, coll., 8319. In this collection hybridization between *S. Schweinitzii* and *S. pseudopedicellatum* was taking place.

Septobasidium lilacinoalbum Couch. Jour. Elisha Mitchell Sci. Soc. **51**: 56. 1935.

Plate 34 and pl. 84, figs. 9–13

Perennial, resupinate, occurring on the younger limbs, forming rather extensive confluent patches which often girdle the limbs and extend for a distance of 40 cm. The summer color of young fresh specimens is almost pure white but with a faint pallid-vinaceous tint, older regions ecru-drab, oldest regions dark, almost grayish black; the winter color is benzo brown or mummy or clove brown with a whitish mottled marginal region much as in typical *S. Schweinitzii.* Surface inherently smooth or sometimes minutely granulose with a lens in the young regions, often with circular elevations (the surface above each pillar being slightly elevated); in older regions becoming cracked into numerous irregular patches. Margin determinate, pure white in actively growing material. In section 360–430μ thick, composed of three distinct layers:

(1) subiculum rather thin, 60–80μ thick, composed of a mat of brownish threads which extend over the bark and from which arise a weft of much thinner, hyaline, branched, curved threads which are partly encrusted with crystals, margin of subiculum extending about 1–2 mm. beyond margin of top layer; (2) middle layer composed of upright pillars which arise from the subiculum, pillars about 8–12 to a sq. mm.; very variable in thickness, 35–150μ thick by 210–250μ tall, single or in groups; hyphae of pillars 3–4.2μ thick, septate, often slightly constricted at the septa, without clamp connections; (3) top layer arising from and supported by the pillars (the threads at the top of the pillar growing outward horizontally, each pillar forming a structure like an open umbrella; threads from the tops of adjacent pillars growing together finally forming a continuous top layer), at first very thin and composed only of horizontal hyphae; hymenium arising from and finally forming the larger part of the top layer, 60–100μ thick, thin in the young regions, becoming thicker with age, composed of closely packed, more or less upright, much branched and twisted hyaline threads between which there is much anastomosing. Probasidia not formed. Basidia arising from the upright threads as small, elongated cells which directly grow into the coiled basidia. Basidium 4-celled with a fifth, basal, functionless cell; each cell of basidium giving rise to a long, thick sterigma on the end of which a spore is borne. Basidia coiled, 4.2–5 x 25–35μ, usually not more than 4.2μ thick. The basidia in old regions are often very irregular in shape. Entire top layer 92–262μ thick, the thickness depending upon age, hymenium 60–100μ thick, hyphae of top layer 2.2–3.5μ thick. Spores white, smooth, bent-elliptic, 3–4 x 10–17μ, becoming one to three times septate.

Symbiotically associated with scale insects (*Chionaspis gleditsiae* Sanders?), some of which are parasitized by irregularly coiled haustoria while others are not parasitized.

This species is closely related to *S. Schweinitzii* Burt. The two plants are associated with the same scale insect (*Chionaspis gleditsiae* Sanders?), but on different host plants, and the hymenial, basidial, and spore characters are much the same except for the smaller size of the basidia and spores in *S. lilacinoalbum*. The two plants may be easily distinguished. The color of *S. lilacinoalbum* is very distinct,—white with a lilac tint when fresh. The present species, moreover, is thinner than *S. Schweinitzii*. Finally the marginal characters of the two are quite different. In *S. Schweinitzii* the margin of the subiculum shows the characteristic "tent-like" elevations, while in *S. lilacinoalbum* the corresponding region is characterized by the pillars, the tops of which spread out like umbrella tops.

This species is almost certainly a hybrid formed by the crossing of *S. Schweinitzii* and *S. Cokeri*. It was first looked upon as a distinct species closely related to *S. Schweinitzii* but with the discovery of abundant material in Chapel Hill it has been possible to follow this fungus for several years during which time its hybrid nature has become quite evident. On the same tree there are a few small patches of *S. Cokeri* but no *S. Schweinitzii*. I have found several pieces (nos. 9327 and 9384) which were colored exactly like typical *S. Schweinitzii*, benzo brown with the same mottled effect toward the margin. In these specimens which otherwise looked so much like *S. Schweinitzii* there were pillars on the margin instead of the "tent-like" structures. Usually, however, the whitish color predominates over the benzo brown. The pillars predominate over the "tent-like" marginal structures though in some places very rudimentary tent-like structures are formed. In *S. Schweinitzii* the subiculum is densely covered with upright hyphae which are curved above like a shepherd's crook; while in *S. Cokeri* the threads of the subiculum are nearly all horizontal but at the bases of the pillars there are tufts of threads covered with crystals. In the hybrid these two characters are mingled, the subiculum being covered with more or less upright encrusted threads. In *S. Schweinitzii* the sterile threads in the hymenium are vertical and slightly wavy; in *S. Cokeri* the sterile threads are recurved and neither vertical nor wavy. In the hybrid the sterile threads are vertical but much more wavy than in *S. Schweinitzii*, and are frequently coiled. In both *S. Schweinitzii* and the hybrid the sterile threads of the hymenium anastomose freely but not in *S. Cokeri*. In reproductive structures the coiled basidial characters of *S. Schweinitzii* dominate over the probasidial and basidial characters of *S. Cokeri*, i.e. in the hybrid species the probasidium is usually lacking and the basidium is coiled as in *S. Schweinitzii*. The "probasidial cell" character is latent however, since quite frequently a type of probasidial cell about the same size and shape as in *S. Cokeri* may be formed from any one of the basidial cells. The spores are shaped like those of *S. Schweinitzii* but are considerably smaller in the material from Highlands, N. C.

Specimens examined:

North Carolina: Type near Highlands, N. C., on *Hicoria alba*, August 6, 1931, Else R. Couch, coll., 9064; Chapel Hill, near Laurel Hill on *Hicoria glabra*, October 23, 1932, nos. 9322 and 9327; May 7, 1933, no. 9384, August 24, 1934, no. 9843, and Nov. 26, 1934, no. 9899. Very abundant on this one tree and causing great damage to the tree, having killed over half the limbs. Occurring on the same tree: *S. Mariani*, *S. Curtisii*, *S. Cokeri*, and *S. apiculatum*. By marking certain specimens on this tree and following their development through the seasons it has been possible to detect that this fungus passes through remarkable color changes from season to season. During late summer and early fall the color is a beautiful white with a pale lilac tint. Specimens collected in October are pinkish buff. This color becomes darker, passing through wood brown to mummy brown by the last of November. In the spring as new growth begins, the specimens become lighter in color and pure white with the characteristic lilac tint by mid-summer.

South Carolina: Near Jacksonboro, on *Hicoria* sp., March 20, 1930.

Septobasidium Hesleri Couch. Jour. Elisha Mitchell Sci. Soc. **51:** 54. 1935.

Plate 18, figs. 14–17, and pl. 81, figs. 1–4

Resupinate, rather inconspicuous, the patches of growth being small, up to about 10 cm. long by 2–3 cm. wide. Pallid, purplish gray in the regions of new growth, becoming a dull Sayal or snuff brown in the older parts. Surface characteristically flaky due to the incomplete formation of the top layer; sometimes, however, the top layer may be complete. Margin determinate or indeterminate, dotted here and there with pillars or irregular "tents" or minute flakes. In section 250–400μ thick, composed of three layers: (1) subiculum 12–20μ thick, which is covered by upright, rarely branched, encrusted, hyaline threads, about 40μ tall; (2) pillars 30–140μ thick x 100–190μ tall, usually rather few; (3) top layer 40–100μ thick. Hyphae of subhymenial region 3–4.2μ thick, dark-colored, branched; hymenial region 30–40μ thick, composed of nearly hyaline, frequently dichotomously branched, coiled hyphae, 2.8–3.2μ thick, between which there is much anastomosing, and basidia which are one to one and a half times coiled, about 4.5–5.8 x 32–38μ, divided into four cells with an additional functionless basal cell. Basidial germination as in *S. Schweinitzii*. Each cell, except the basal one, gives rise to a long tube on the end of which a narrow sterigma is formed and on the end of which, in turn, a spore is formed. Spores 5–6.3 x 13–18μ.

Symbiotically associated with scale insects (*Chionaspis gleditsiae*), some of which are parasitized by irregularly coiled haustoria, while others of the insects are free from parasitism and are in all stages of development.

The two collections of this species were in fruit but rather sparingly except on one piece of wood where it was growing in contact with a much more luxuriant growth of *S. pseudopedicellatum*. Along this line of contact in a zone about 0.3 cm. wide *S. Hesleri* was producing an abundance of basidia and spores. The *S. pseudopedicellatum* was not visibly affected by the contact. I have noticed other cases of greatly increased vigor in fruiting on the part of one species where two were in contact (e.g., *S. filiforme* and *S. Mariani*).

This species is closely related to *S. Schweinitzii* and *S. lilacinoalbum*, being perhaps a hybrid between the former and some unknown species.

The present species may be distinguished by its small size, thin, usually flaky structure, dull brownish color, and the inconspicuous "tents," "Eskimo huts," and pillars on the margin. The basidia and spores are as in *S. Schweinitzii* but in that species the hymenium is thicker and the sterile threads are wavy but upright while here the sterile threads are dichotomously branched and recurved, frequently coiled.

I am naming this species in honor of Dr. L. R. Hesler, who collected it.

Specimens examined:

Tennessee: Near Gatlinburg, on *Carpinus caroliniana*, June 20, 1931, U. N. C., 9906, and July 19, 1934, U. N. C., 9907, type, L. R. Hesler, coll.

Septobasidium natalense n. sp.

Plate 95, fig. 5

Resupinate, forming small (0.5–3 cm. diam.), easily separable, thin, papery patches. Clay color to drab in the older parts, sometimes about auburn. Surface nearly smooth or sometimes slightly wrinkled, becoming minutely cracked in the older parts as the hymenium ages. Margin determinate, regular in outline or minutely fimbriate; margin of subiculum whitish, sometimes with a few inconspicuous tufts of upright hyphae. In section 250–500μ thick, composed of three layers or regions, subiculum, pillars, and top layer. The subiculum is very thin, somewhat membranous or papery and composed of whitish hyphae. The pillars are few, 40–50μ tall, and usually 100–200μ thick. Top layer 100–300μ thick, at first composed of loosely entangled hyphae but as the hymenium is developed the hyphae become densely arranged. Hyphae 3–4μ thick. In old specimens the top layer is covered over by a layer of hyaline amorphous material up to 80μ thick. Probasidia lacking, basidia arising laterally from the hyphae, becoming irregularly twisted or s-shaped and four-celled when mature (only a few good basidia seen). Spores not seen.

Associated with scale insects some of which are parasitized by haustoria of the irregularly coiled type.

This species is close to *S. Carestianum* Bres.; in fact one collection (no. 21011) has been identified at Kew as that species. The present species is easily distinguished from *S. Carestianum* by the lack of probasidia and the irregularly bent basidia of the former. The species is characterized by the thin papery, small patches of growth, by the clay color of the newly formed top surface and the drab color of the old fruiting area. The whitish, thin subiculum and the absence of probasidia and the structure of the basidia are also characteristic.

Specimens examined:

South Africa: Natal, Pieter Maritzburg, on *Prunus domestica*, J. H. Spencer, coll., June 21, 1923, type in Herb. Division of Plant Industry, Pretoria, no. 17276, and U. N. C. Herb. Also near Hanover, on *Prunus persica*, Mr. Dale, coll., Sept. 9, 1928, Pretoria Herb., no. 21011, also in U. N. C. Herb.

Septobasidium Reinkingii Pat. (Unpublished, but so labelled in Pat. Herbarium.)

Plate 58, figs. 1–3, and pl. 96, figs. 1–4

Resupinate, covering extensive areas on the stems and leaves of *Citrus* and other plants, the larger specimens up to 15 cm. long and completely girdling the smaller limbs. Color of hymenial surface pale drab gray or sometimes almost white to light cinnamon drab, fading to a dark brown; in old specimens the surface may be mottled with whitish areas over the brown due to renewed growth; the context near Natal brown. Surface of hymenium smooth, papery, finely powdery when fresh, becoming shiny when old; with maturity the surface becomes broken by a few conspicuous cracks. Margin determinate, though the subiculum can only be seen with a lens. The marginal region is frequently dotted with minute pillars and in some specimens where the hymenium is not well developed shows a spongy appearance. In section about 0.8–2 mm. thick. Usually composed of a subiculum, a region of pillars, and the top layer over which the hymenium is formed. Frequently however there may be several stories by the formation of the subiculum, then pillars, and then another thin layer over the tops of the pillars and then pillars again and so on for three, four, or even five stories. Subiculum characteristically thin even in the older parts of growth, composed of brownish threads about 4.2μ thick and hyaline threads about 2.1μ thick, the hyaline threads arising from the brownish ones. The brownish hyphae extend mostly in the same direction and are usually separated by several microns from each other and the hyaline threads form a connecting network between the brown threads. From the subiculum arise the pillars, which may or may not be very distinct, i.e. the hyphae may arise in a fairly compact column or may be so loose as not to form pillars; pillars 30–150 x 130–200μ, hyphae of pillars 2.4–3.8μ thick. The pillars may branch out above to form a fertile top layer or a thin sterile horizontal layer may be formed, after which the pillars renew growth, growing up through this sterile layer and thus in the same way several (three have been seen) sterile horizontal layers may be formed before the hymenial layer is formed. Fertile top layer 200–300μ thick, hyphae rather loosely entangled, branched; hymenium 60–85μ thick, composed of densely packed and entangled hyphae, pale brownish or hyaline, wavy, sometimes circinate, branched, about 2–2.5μ thick. Basidia arising directly from the hyphae without a probasidial cell and forming a rather compact region on the outer side of the hymenium, irregularly curved, 5.4–6.7 x 35–50μ, 4-celled, sterigmata long, 2.5 x 10.5 or much longer. Spores bent-elliptic, 4–4.2 x 13.4–16μ.

Associated with scale insects some of which are parasitized by irregularly coiled haustoria while many are free from parasitism.

This species is easily confused with *S. bogoriense* Pat. The color, habit of growth, and microscopic characters are much the same. However, *S. Reinkingii* has a lighter colored and more papery hymenium and the pillars are more distinct than in *S. bogoriense.* Also *S. bogoriense* does not form several stories as does the present species. The surest means of differentiating the two is in the lack of a probasidial cell in *S. Reinkingii.* The present species is very close to *S. bogoriense* and is most likely a hybrid, with *S. bogoriense* as one of the parents.

Specimens examined:

Philippine Islands: Los Banos, college farm, on *Citrus* leaves and stems, Nov. 13, 1919, O. A. Reinking, coll., no. 6608; also no. 8848 from same location, Feb. 16, 1920, in Pat Herb., type; also another collection in Pat. Herb. as *S. bogoriense* marked "typique," on stems, from same location, Jan. 15, 1912, M. B. Riemonds, coll.; another collection, same location, Reinking, no. 8831, Feb. 16, 1920, on *Citrus,* determined by Patouillard as *S. Reinkingii* but is entirely distinct and perhaps a new species. It is, however, sterile. On *Citrus maxima,* same loc., August 28, 1933, Alvaro, coll., in U. N. C. Herb., 10429.

Septobasidium Westoni n. sp.

Plate 59, figs. 1–3, and pl. 82, figs. 3, 4

Resupinate, forming small, individual or confluent patches which vary in width from 1 to 5 mm. or more. Sometimes the patches may grow together forming a smooth surface as much as 3 cm. wide; wood brown in the older regions to pale vinaceous buff in the younger parts; surface smooth; margin determinate, smooth or fimbriate. In section 350–700μ thick, usually about 500μ thick, composed of three regions: (1) the subiculum which is very thin and which extends over the leaf surface; (2) the middle region composed of pillars from which branch out loosely entangled masses of hyphae so that the spaces between the pillars, empty at first become filled with hyphae, pillars about 20–40μ thick at base, about 320–420μ tall, threads of pillars brownish, 4.2–5.4μ thick, septate, without clamp connections, threads between pillars much branched, lighter colored than thread of pillars, about 3.8–4μ thick; (3) hymenial region of variable thickness usually about 35–45μ thick, considerably thicker in old regions; probasidia and paraphyses not formed, basidia one or one and a half times coiled or cork screw-shaped or irregularly twisted, 4–5 x 30μ, thrice septate; sterigmata not seen; spores bent-elliptic, 3.8 x 15.5μ (only 3 seen). Elliptic, rough-walled conidia formed in abundance over the lower surface.

Associated with *Diaspis* sp. some of which are parasitized by irregular coils of hyphae while others are free from fungal infection.

This fungus may be distinguished by its peculiar habit of growing in small roundish patches; by the fact that the spaces in between the pillars become more or less stuffed with hyphae; by the absence of paraphyses and probasidial cells and by the small closely coiled basidium. In habit of growth it is somewhat like *S. minutulum* Sydow which grows on the surface of leaves, forming minute roundish patches. *Septobasidium minutulum* Sydow has a probasidial cell whereas the present species has not.

Specimen examined:

Panama Canal Zone: Barro Colorado Island, on living leaves of pineapple (*Ananas magdalenae*), Nov. 28, 1928, Wm. H. Weston, coll., type in Weston Herbarium and U. N. C. Herb. 9960.

Septobasidium granulosum Sydow. Botan. Jahrb. **54:** 253. 1916. Ill. Petch. Trans. Brit. Myc. Soc. **12:** pl. 18, fig. 3. 1927. Excellent section view.

Plate 59, figs. 7–10, and pl. 96, fig. 7

Resupinate, extending out on the sides to form lateral wings which are minutely fimbriate on the margins. Hymenial layer grayish purple, densely granulose; subhymenial region deep chocolate brown; alveolate in the wing region. In section up to 1.5 mm. thick in the thickest parts, the wings thinner, about 0.5–0.7 mm. thick. Not divided into distinct top, bottom, and middle regions as in most species of the genus but composed of the context and hymenial regions. Context made up of deep brown, considerably branched and entangled hyphae, 2.1–4μ thick, most about 3.8μ thick, which are fairly compactly arranged in the older parts where the fungus is attached to the bark, but which are loosely packed and are more or less arranged in irregular anastomosing fascicles in the winged portion, thus giving the latter an alveolate appearance. Hymenium arising directly from the context, at first in scattered patches, which increase in size until finally many small oval or conical mounds are formed, the basidia covering the mounds. Hymenium at first not continuous, covering only the mounds but later more or less continuous as the mounds increase in size, becoming contiguous. Hymenium composed of densely packed, much branched (sometimes dichotomously), and coiled threads, which in the basal part of the hymenium are concolorous with the threads of the context; threads in the outer region hyaline, 1–3μ thick. Basidia arising directly from the threads without the intervention of probasidia, formed over the outer surface and not sunken in the tissue, curved, 3-septate, 5.8–7.8 x 40–50μ; sterigmata about 8μ long; spores 4.5–5 x 18.9–25μ, elliptic, sharply bent at the distal end.

Associated with scale insects, some of which are parasitized by irregularly coiled hyphae.

This species is very close to *S. alatum* Lloyd and *S. aligerum* Petch, but can readily be distinguished by the granular surface of the hymenial region, the peculiar shelving habit, and by the absence of pillars. For further comparison, see under those species.

Specimens examined:

New Guinea: On living branches of a tree, C. Ledermann, coll., Sepik Expedition, no. 12983.

Septobasidium rhabarbarinum (Mont.) Bres. Ann. Mycol. **14**: 240. 1916.

Daedalea rhabarbarina Mont. Ann. Sci. Nat. ser. 2, **13**: 205. 1840. Also in Sylloge Generum Specierumque Cryptogamarum. Paris, 1856. This is merely a reprint of the original description.

Corticium reticulatum B. & Br. Jour. Linn. Soc. Bot. **14**: 69. 1873.

Septobasidium reticulatum (B. & Br.) Pat. Herbarium name for above.

Corticium rhabarbarinum B. & Br. Jour. Linn. Soc. Bot. **14**: 69. 1873. *Pro parte.*

Hymenochaete frustulosa B. &. C. Jour. Linn. Soc. Bot. **10**: 334. 1868.

Septobasidium frustulosum (B. &. C.) Pat. Bull. Soc. Myc. Fr. **10**: 79. Pl. 3, fig. 4. 1894.

Peniophora citrina Henn. Hedwigia **36**: 192. 1897.

Septobasidium (*Hymenochaete*) *radiosum* (P. Henn.) Lloyd. Myc. Notes **5**: 722. 1917.

Septobasidium sulphurellum Sydow. Ann. Mycol. **17**: 33. 1919.

Plates 5, 7, 14, 58, figs. 11, 12, and pl. 96, figs. 5, 6.

Resupinate, forming extensive patches (15–20 square cm.) on the trunk and limbs of deciduous trees; olive ochre to old gold; surface very variable, in young specimens the marginal region may extend for some distance in the form of flattened, sinuous, rhizomorphs 0.2–0.8 mm. wide, having the gross structural appearance of a leafy liverwort; in other specimens the marginal region may show the rhizomorphs and numerous little elevated tufts of hyphae; in mature specimens the surface is cut up by numerous, narrow (0.1–0.3 mm. wide), sinuous, anastomosing cracks; rarely the surface may be nearly smooth or dotted with numerous pin holes formed by the incomplete development of the top layer. In section 0.6–1 mm. thick, composed of three layers: (1) a lower, compact, thin layer of entangled, matted threads which run more or less parallel to the substratum (this layer is often rather indistinct or entirely absent) about 35–50μ thick; (2) the middle layer 300–550μ thick, of loosely packed, much entangled and branched threads about 2μ thick; broken by numerous connecting channels; in old specimens the middle region is usually more or less compact; and (3) the hymenial region which varies from a few microns thick in the young parts up to 300–370μ thick in the older regions; hymenium stratose by the formation of densely packed layers of basidia; basidia also formed in between the strata, i.e. scattered throughout the hymenial region; hymenial region composed of densely packed, circinate hyphae which are many times dichotomously branched, threads 0.8–2μ thick, the ends of the branches curved backwards and often spirally coiled; probasidia usually lacking, the young basidia pyriform, about 6 x 15μ, borne laterally on the hyphae, developing into a curved 4-celled basidium, 6.3–7.1 x 50–68μ; probasidial cell present in some collections and remaining as a distinct empty cell after the basidium has been formed; sterigmata 3 x 25–34μ. Spores 4.2–5.4 x 21–25.5μ, elliptic and pointed at both ends or pointed at only the mucro end. There are also found in the hymenial region numerous two or three times septate, elongated bodies which may be terminal or intercalary.

Associated with scale insects (*Aspidiotus fabernii*) in the Jamaican material) some of which are parasitized by haustoria composed of spindled-shaped segments connected by fine threads; other insects are free from parasitism.

This species is easily distinguished by the yellowish color (olive-ochre to old gold), the sinuose margin, the usual absence of supporting pillars, the anastomosing cracks in the surface, the dichotomously branched and coiled threads in the hymenium, and the coiled basidia. Another very striking characteristic is that when some of the context is teased apart in a weak (5%) solution of KOH a deep yellow or orange color oozes into the fluid which changes in a few minutes to a pale or greenish yellow.

The nomenclature of this species is considerably confused. Unfortunately I have not been able to study the original type material of *Septobasidium* (*Daedalea*) *rhabarbarinum* (Mont.) Bres. The type specimen was sent to me by Dr. Heim in Paris but was lost in the Christmas mail. However Patouillard (1897), Bresadola (1916), and Petch (unpublished notes, very generously sent me by Mr. Petch) have compared the types of *Daedalea rhabarbarina* Mont. and *Hymenochaete frustulosa* B. &. C. and find them identical. Bresadola (1916) also recognized the identity of *Peniophora citrina* P. Henn. with the present fungus. Upon reading the original descriptions of *Corticium rhabarbarinum* B. & Br. and *C. reticulatum* B. &. Br. (Jour. of Linn. Soc. Bot. **14**) it appeared probable that these two species were incorrectly described as *Corticium* and might belong in the genus *Septobasidium*. Through the kindness of Dr. Hill, Director of Kew Gardens, and Miss Wakefield, I was permitted to study these two collections and found that both were identical with *S. rhabarbarinum* (Mont.) Bres. Miss Wakefield (1915) had already noted that *C. reticulatum* and part of *C. rhabarbarinum* were identical but did not note their identity with *Septobasidium*. Patouillard (Bull. Soc. Myc. Fr. **24**: 2. 1908) recognized that *Corticium reticulatum* B. & Br. was a *Septobasidium* but did not indicate that it belonged here. *Septobasidium sulphurellum* Sydow also belongs here, agreeing perfectly with the American material.

Specimens examined:

Cuba: Wright, coll., comm. by Curtis, as *Hymenochaete frustulosa*, nos. 225 and 244 in Kew Herb. and Curtis Herb.

Brazil: St. Catharina, on branches of a tree, leg., E. Ule, no. 741 in Herb. Bot. Gard. Berlin, also a fragment in U. N. C. Herb. (as *Peniophora citrina* P. Henn., type).

Ecuador: In Patouillard Herb. under *S. frustulosum* (B. & C.) Pat., de Lagerheim, coll., also in Herb. Bot. Gard. Berlin.

Guadeloupe: On living *Eugenia montana*, 1904, in Patouillard Herb. as *S. frustulosum* (B. & C.) Pat., no. 1458, also in N. Y. B. Gard. Herb. and Herb. Bot. Gard. Berlin.

Jamaica, B. W. I.: Two collections on living deciduous wood, J. N. Couch, coll., July 1926, U. N. C. Herb. Described (1929) as *S. frustulosum* (B. & C.) Pat.

Panama: C. W. Dodge, coll., July 10, 1925, in U. N. C. Herb. and Farlow Herbarium; also A. F. Skutch, coll., May 26, 1929, U. N. C. Herb.

Nicaragua: Castillo Viejo, Feb.–March 1893, C. L. Smith in Cent. Amer. Fungi 100, as *Thelephora retiformis*, transferred by Burt (1916) to *Septobasidium* as *S. frustulosum* (B. & C.) Pat.

Guatemala: A. F. Skutch, coll., U. N. C. Herb.

Mexico: Sanborn, Oaxaca, C. R. Orcutt, 3334, in Mo. Bot. Gard. Herb. as *S. frustulosum*.

Grenada: Grand Etang, R. Thaxter, coll., in Burt Herb. and Farlow Herb. as *S. frustulosum*.

Venezuela: Fendler, coll., 279, in Farlow Herb. and Mo. Bot. Gard. Herb.

Africa: Liberia, D. H. Linder, coll., September 1926, in Arnold Arboretum Herbarium, no. 680, and U. N. C. Herb.

Ceylon: Thwaites, no. 657, *pro parte*, July 1868, type material of *Corticium rhabarbarinum* B. & Br. Part of this collection was transferred by Miss Wakefield (1915) to *Stereum rhabarbarinum* (B. & Br.) Wakefield. Thwaites, no. 656, July 1868, type material of *Corticium reticulatum* (B. & Br.).

Java: Tjibodas, on living tree limbs, 1500 M., July 1898, E. Nyman, coll., in Pat. Herb. as *S. frustulosum* (B. & C.) Pat. variety *crassim*, also in Berlin Bot. Gard. Museum; Tjibodas, January 1899, leg., Massart, in Pat. Herb. as *S. frustulosum* (B. & C.) Pat. Form with pillars and with considerably coarser threads than is typical. This is the collection cited by Boedijn and Steinmann (1931).

Philippine Islands: Province Laguna, Luzon, M. Ramos, coll., Feb. 1913, in Bureau of Science Herbarium, no. 20616, N. Y. Bot. Gard. Herb., and U. N. C. Herb. Basidia up to 9.6μ thick.

Pelew (Palau) Islands: Babelthuap, on branches of trees, March 9, 1914, C. Ledermann, coll., no. 14546, in Sydow Herb., Herb. Bot. Gard. Berlin, and U. N. C. Herb.

New Guinea: Kaiser Wilhelmsland, on living tree trunk, Lauterbach, no. 1135d. Described by P. Hennings as *Hymenochaete*? *radiosa*. First recognized as *Septobasidium* by Bresadola who wrote on the label, "Ist ein *Septobasidium*, aber noch unreif." However, I have studied the same original material and find abundant probasidia, basidia, and a few spores, all of which are typical. This material differs from the American material in having probasidia; a smooth surface; distinct, though few and small, marginal pillars; and in having a less compact context. In gross characters this collection resembles the above mentioned material from the Philippine Islands. Original specimen in Herb. Berlin Bot. Gard.

Septobasidium subolivaceum Sydow. Philippine Jour. Sci. **9**: 158. 1914.

Plate 59, figs. 4–6, and pl. 95, fig. 1

Resupinate, forming closely encrusted, rather extensive, confluent patches 15 cm. or more long on the stems of living *Schizostachyum*. Near tawny olive. Surface smooth except for numerous, conspicuous, low flattened mounds about 2–4 mm. wide. The mounds are separated from the main body of the context by a narrow crack and usually have a dark center. Margin irregular in outline, determinate. In section 170–500μ thick, except through the mounds which are up to 1.5 mm. thick; composed except in the mound region of a very dense layer of entangled, much septate, branched, dark brownish threads, threads 3.2–4.2μ thick; towards the upper surface the threads become paler in color and are intermingled with numerous, pale yellowish or nearly hyaline threads which are sparingly septate, sometimes dichotomously branched and with recurved ends. Basidia are intermingled with the nearly hyaline threads. The structure of the context in the mound region is somewhat different. Under each mound there is one or several parasitized insects and occasionally a non-parasitized one. Each insect is covered by a dense layer of dark-colored hyphae which give rise to a distinct layer of pale yellowish, less compactly arranged, sometimes dichotomously branched, curved and sparingly septate hyphae (these hyphae are like the nearly hyaline ones mentioned above). This layer is about 100–170μ thick; the hyphae 2–2.3μ thick. The basidia arise from the pale hyphae, are 4-celled and frequently helically curved, 4.6–5 x 32–36μ; sterigmata long, spores not seen.

Associated with scale insects, some of which are parasitized by haustoria consisting of spindle-shaped segments connected by fine threads.

This is a very distinct species and is easily recognized by the tawny olive color, the thin compact context with the elevated mounds, each with a dark center; the curved (frequently making 1.5 turns) basidia; the spindle-shaped haustoria; and the habitat on *Schizostachyum*.

Specimens examined:

Philippine Islands: Taytay, Palawan, on living *Schizostachyum*, E. D. Merrill, coll., type in Herb. of Bureau of Science Philippine Islands, Sydow Herb., N. Y. Bot. Gard. Herb. and U. N. C. Herb.

Septobasidium Petchii n. sp.

Plate 63, figs. 12–15, and pl. 95, figs. 2–4

Resupinate, hypophyllous, forming thin extensive patches of growth sometimes covering nearly the entire under surface. Probably pure white when fresh, becoming dirty buff in the herbarium. Margin determinate but rather indistinct. The surface presents a very characteristic appearance, being closely dotted with large and small elevations. The small bumps are much the more numerous, averaging a millimeter or less apart, about 0.3–1.0 mm. wide; the larger bumps are 1–5 mm. wide and 1–1.5 mm. tall, rounded or slightly irregular in outline, and with a somewhat flattened surface. The small mounds are composed of hyphae which are cruciformly branched with one or several cross bars or the branches may be arranged like the rays of a star. When mounted in water these hyphae break up into T-shaped or cross-shaped or L-shaped or other variously shaped segments which float about in the water. The hyphae of this region as well as the hymenial region frequently bear long sterigmata-like processes on the ends of which there usually occurs a small globule of hyaline material. (The same type of structure is found in *S. apiculatum.*) The large mounds are composed toward the center of dark brownish, mostly straight and rather sparingly branched hyphae, 3–4.2μ thick; and toward the outside of hyaline, considerably branched hyphae, the branches arising mostly at right angles and frequently cruciform; hyaline hyphae 2–3μ thick. The fertile part of the fungus is formed on the very thin region in between the mounds, the hymenial surface apparently never extending over the mounds. The hymenial surface is finely granulose under a lens (10 x or more). In section the fertile layer is 50–90μ thick and is composed of a few, thick-walled, horizontal hyphae which extend over the leaf and in the central part is composed of densely entangled hyphae the branches of which arise at right angles or cruciformly. These hyphae when mounted also separate into small segments as described above. The hyphae of the middle region are 2–3.8μ thick and give rise on the outer surface to the basidia. Basidia closely coiled, formed directly from the hyphae without a probasidial cell, about 3.8–4.2 x 17–23μ, 4-celled; each cell giving rise to a sterigma which is about 10μ long. Spores elliptic, slightly bent, 3–4 x 10.5–11.7μ.

Associated with scale insects. The healthy non-parasitized insects are beneath the large mounds, usually one or two insects beneath each mound. Several of the healthy females were found giving birth to young. The parasitized insects are beneath some of the smaller mounds. It is well to note that relatively few of the small mounds cover insects. The typical parasitized insects were very difficult to find and it was only after an examination of a number of mounds that a few insects were seen with haustoria. These were of the spindle-shaped type connected by fine threads.

This species possesses very distinctive characteristics which readily separate it from all known forms. The habitat on leaves, the whitish color, the numerous small white mounds which give the surface a finely spotted appearance, the large, conspicuous, rounded mounds composed of sterile hyphae and the very thin, finely granulose hymenial region in between the mounds are distinctive gross features which characterize the present fungus. Another characteristic in which this fungus appears to be unique is the remarkable habit the hyphae possess of breaking up into straight, or L-shaped, or T-shaped, etc., segments when mounted in water. The small, closely coiled basidia and the small spores are also distinctive.

Specimens examined:

Ceylon: Nuwara Eliya, common on *Lasianthus* sp., September 11, 1927, T. Petch, coll., type in Petch Herbarium and U. N. C. Herbarium; Hakgala, on *Murraya exotica,* April 1917, T. Petch, coll. Also a young form or variety from the same locality, July 9, 1923, on *Murraya exotica.* The surface of the fungus is nearly smooth and brownish in color. Basidia as in type. Hyphae frequently curved and breaking up much less readily even when crushed than in the type.

Septobasidium tenue Couch. Jour. Elisha Mitchell Sci. Soc. **51:** 58. 1935.

Plate 37 and pl. 77, figs. 1–4

Resupinate, very closely adherent to the bark, forming irregular isolated or confluent patches which seldom exceed six to eight cm. in length. Color quite variable, mouse gray or drab or pale gray or with a buff or brownish tint. Surface usually inherently smooth, except for minute depressions and except for very minute cracks in the older regions and irregularities in the bark. Here and there on the surface are numerous, dark, elevated areas beneath which are parasitized and non-parasitized scale insects; in old specimens these elevated dots may not be evident. Margin irregular in outline; determinate. In section about 100–225μ thick, usually about 125μ thick, in very young regions not more than 25μ thick; compact, often stratose (no. 9183) in old regions, a new hymenial stratum forming each year (no. 8456 not stratose). Context composed of much entangled, branched, septate, mostly hyaline threads, without clamp connections; threads heavily encrusted with crystalline material; tips of threads at surface often bent like a shepherd's crook; 3.2–4.2μ thick; probasidia absent unless the sometimes thick-walled resting cells of the basidia may be considered as such. Basidia 6.3–7.5 x 29–37μ, up to 50μ long when nearly straight, but nearly always coiled and twisted, 4-celled, each cell usually giving rise to a sterigma on the end of which is formed a spore. Spores (of no. 8456) bent-elliptic, dividing into 4 or 8 cells, 2.5–4.2 x 13–22μ, most 3.2 x 16.8μ.

Associated with *Aspidiotus juglans-regiae* Comst. on bay and *Aspidiotus* sp. on oak, some of the insects being parasitized by haustoria which are in the form of miniature strings of sausages.

Very puzzling color variations occur in this species, sometimes even in the same collection. Number 8456 from South Carolina was growing on oak. In this lot of material of which there are over thirty patches of growth there is one patch nearly pure white which upon examination was found to be *S. apiculatum.* This latter species is also exceedingly thin and differs macroscopically from the species under consideration only in color and in the structure of the insect houses. The patches of growth in this collection vary in color from a drab gray to very light gray. In the lighter patches a microscopic examination shows the presence of thin-walled hyaline hyphae characteristic of *S. apiculatum.* Some of the color variation in this collection may possibly be due to the mixed growth of the mycelium of these two species.

This species may usually be recognized by the drab color, the small circular, dark elevations, and the coiled basidia. In no. 8456 there were large numbers of living algal cells embedded in the hymenial layer of the fungus.

Specimens examined:

South Carolina: Near Charleston, on *Quercus,* March 19, 1930, J. N. Couch, coll., 8456. Type.

Florida: Kissimmee, on *Quercus nigra,* March 13, 1932, J. N. Couch, coll., 9243, 9246; also near Gainesville, on *Quercus nigra,* Feb. 22, 1935, G. F. Weber and E. West, colls., Univ. Fla. Herb., 9965.

Mississippi: Near Pascagoula, on *Magnolia virginiana,* Jan. 1, 1932, J. N. Couch, coll., 9183, 9185, 9187, 9188, 9214.

Septobasidium rugulosum Couch. Jour. Elisha Mitchell Sci. Soc. **51:** 59. 1935.

Plates 36, 68, figs. 1–4, and pl. 77, figs. 5, 6

Resupinate, closely adherent to the bark, forming small (2–3 x 4–5 cm.) irregular, isolated or confluent patches. Color tawny olive to sepia. Surface minutely retiform, often with dark, circular dots beneath each of which is, as a rule, a parasitized or healthy scale insect, surface rarely nearly smooth. Margin at times abruptly determinate, more often minutely fibrillose. In section 60–250μ thick, compact, composed of a basal region of thick-walled, brownish, septate threads, which are deeply constricted at the septa, threads 4–10.5μ thick. These threads when crushed separate into short segments. Upper region composed of the same type of threads except smaller; tips of threads at surface often bent like a shepherd's crook, surface threads covered with crystalline material. Probasidia absent. Basidia 4–5μ wide, about 35–42μ in length (allowing for the curvature); twisted or coiled, 4-celled, each cell giving rise to a sterigma and a spore; sterigmata often very long; basidia usually difficult to see in section. Spores 3.9–4.6 x 12.6–18.9μ, most 4 x 14μ, rarely 5 x 21μ, bent-elliptic, becoming once or thrice septate.

Associated with *Chionaspis gleditsiae*(?), some of which are parasitized by haustoria which are in the form of strings of sausages.

This species occurs in the same region with *S. tenue* and sometimes on the same tree and is very closely related to that form. Generally the two species may be separated by color, *S. tenue* being drab (Ridgway) while *S. rugulosum* is a rich brown (tawny olive to sepia of Ridgway). The surface of *S. rugulosum* is generally retiform while that of *S. tenue* is inherently smooth. The surface of *S. tenue* generally shows numerous, dark, elevated mounds beneath which are scale insects, whereas in *S. rugulosum* the surface of the fungus immediately over the insects is not elevated. The surface of the latter, however, is dotted with the minute dark specks which form beneath the young scale insects. In microscopic section the two species are quite similar. Both have coiled basidia but those of *S. tenue* are shorter and thicker than are those in *S. rugulosum.* The spores are quite similar in shape but differ in size, those of *S. rugulosum* being slightly thicker and shorter. The threads in the context of the two plants are quite dissimilar. Those of *S. rugulosum* are composed of short segments, the threads showing rather deep constrictions at the joints. When a mass of these threads is crushed, the threads break into short joints. The threads are also generally coiled and twisted. In *S. tenue* the threads are composed of long, more or less straight segments, are not constricted at the joints, and do not separate into short segments when crushed.

Specimens examined:

South Carolina: Jacksonboro, on *Liquidambar styraciflua,* March 20, 1930, J. N. Couch, coll., 8472, type; near Myrtle Beach, on *Liquidambar styraciflua,* July 4, 1928, J. N. Couch, coll., 8314.

Florida: Jacksonville road near Ga. line, on *Liquidambar styraciflua* associated with *Chionaspis gleditsiae,* March 11, 1932, 9229; near Kissimmee, on *Magnolia virginiana,* March 13, 1932, J. N. Couch, coll., 9251; also 9244.

Louisiana: Near Baton Rouge, on *Quercus* sp., Dec. 31, 1931, J. N. Couch, coll., 9154.

Septobasidium arboreum (Couch) Boed. and Stein. Bull. Jard. Bot. Buitenzorg III, **11**: 168. 1931.

Helicobasidium arboreum Couch. Jour. Elisha Mitchell Sci. Soc. **44**: 257. 1929.

Plates 5, 12, 13, 14, 91

Resupinate, forming rather thin, cracked areas on the bark of living trees; largest areas up to about 6–8 sq. cm. in extent. Infertile areas blackish brown, myceloid. Hymenial surface light gray, forming a more or less cracked and granulated surface over the much darker-colored, subhymenial hyphae. In section composed of (1) the vegetative hyphae which penetrate throughout the cortex region developing particularly in the cells of the cork cambium and phloem, intracellular hyphae colorless; (2) the superficial vegetative hyphae, which are very dark-colored, septate without clamp connections, very loosely packed, 4–8.4μ thick close to the base, not arranged in pillars but arising individually and branching many times to form a dense cluster of hyaline, delicate, coiled and twisted hyphae. Hymenium composed of these clusters of hyphae among which are borne the spore-bearing bodies or basidia. Basidia not arising from an obviously preformed resting cell. Mature basidium one to one and one-half times coiled, and thrice septate, 4.8–7.4μ thick by 33–40μ long. From each of the four cells of the basidium a long sterigma grows to the hymenial surface and bears a long, narrow, elliptic spore, distinctly bent at the distal end and usually once septate but not rarely three times septate; spores 5–5.9 x 21–30μ, white in a spore print.

Collected only once and then on the bark of a living deciduous tree on the Blue Mountain Trail, Jamaica, June 1926. Associated with scale insects, many of which are parasitized by irregular, bent spindle-shaped segments of hyphae which are connected by fine threads. Many insects, however, are not parasitized.

This species can be recognized with the unaided eye by the gray color and peculiar granulated surface of the hymenium. In microscopic section the thick dark-colored hyphae arising individually and branching to form a dense cluster of hyaline, delicate, twisted hyphae and the coiled basidia are distinctive.

Specimens examined:

Jamaica: Blue Mountain Peak Trail, on bark of living deciduous tree, June 1926, J. N. Couch, coll., type in U. N. C. Herb.

Septobasidium pseudopedicellatum Burt. Ann. Mo. Bot. Gard. **3:** 327. 1916.

Plates 27, 35, 78, and 79

Resupinate, forming small to very extensive, conspicuous, foliose crusts on the bark and leaves of a great variety of trees and shrubs. Individual patches of growth up to 15 or 20 cm. in greatest dimension. Color very variable: light buff, or pale, smoke gray to cinnamon brown or chestnut brown. Hymenial surface smooth and usually shiny even under a lens, sometimes cracked and wrinkled in old specimens, in certain young specimens before the hymenium is formed the surface may be velvety or alveolate. Margin usually determinate, varying in appearance at different times of the year, subiculum of margin typically whitish, sometimes buff. Marginal region usually conspicuously marked by numerous upright pillars; however, in some conditions the top layer may extend out even with the subiculum obscuring the marginal pillars. Margin of upper layer sometimes fimbriate. In section 0.7–1.5 mm thick, composed of three distinct layers or regions: (1) the very thin whitish subiculum which extends over the bark, threads of subiculum 3.4–4μ thick, often covered with minute crystals; from the subiculum arise the pillars which compose the (2) middle region. Pillars simple or branched in the older parts, sometimes arising in concentric rows on the margin, erect, 0.5–1.2 mm. tall, usually about 1 mm. tall by about 40–60μ thick, composed of parallel, septate, mostly unbranched hyphae about 3.7–5.5μ thick; brownish with a tuft of whitish threads at the base, the white threads encrusted with minute crystals. The pillars branch out above to form (3) the top layer over which the hymenium is formed. Top layer stratose in old specimens by the formation of several hymenial layers; up to 300μ thick, composed of loosely packed, branched and entangled, brownish threads, 3.7–5.5μ thick; giving rise to the hymenium, which because of its densely packed and entangled, colorless paraphyses and probasidia is sharply differentiated from the underlying brownish threads, hymenium 35–50μ thick; paraphyses or sterile threads about 4μ thick at their origin and tapering down to about 2μ thick, wavy; probasidia usually subglobose or pyriform, sometimes globose, 11.5–13.8 x 16–22.2μ, giving rise to a straight, cylindrical basidium, 4.8–7 x 37–70μ, usually 6 x 60μ, thickest in the middle or toward the distal end, usually with a long stalk from which it frequently separates at maturity. Sterigmata 4–5μ long, lateral or sometimes terminal on the end cell. New probasidia may sometimes be formed within the old ones, such secondary probasidia giving rise to smaller basidia, which bear small spores. Spores white in print, bent-elliptic, 3.7–4.8 x 16–23μ (the few spores seen in the type material agree), rarely up to 5 x 30μ, becoming 4–8-celled.

Associated with a variety of scale insects on the leaves and bark of many kinds of trees and shrubs (*Fraxinus* spp., *Alnus* spp., *Pyrus* spp., *Quercus* many species, *Cornus*, *Hicoria*, *Carpinus*, *Liquidambar*, *Nyssa*,

Gleditsia, Acer, Crataegus, Betula, Magnolia, Ilex, Citrus, Taxodium, Staphylea trifolia, and other wild and cultivated genera). Some of the scale insects are parasitized by haustoria in the form of irregular coils; other insects are free from parasitism.

This species in its commonest and typical form may easily be identified by the buffy to cinnamon brown color of the top layer, and its smooth surface; the dark-colored erect pillars with the tuft of whitish threads at their bases, and by the thin, whitish subiculum. The pillars are always quite distinct in this species and it is unfortunate that it has the name *S. pseudopedicellatum*. The hymenial characters are also distinctive. The hyaline elements of the hymenium contrasting strikingly with the brownish underlying tissue, the subglobose or pyriform probasidia, the cylindrical, 4-celled basidium with its short sterigmata, and the wavy, tapering, hyaline threads help one to recognize this fungus.

This species may be distinguished from *S. Mariani* Bres., with which it frequently occurs, by the densely crowded, branched and entangled pillars and the minute, dark colored, marginal rhizomorphs of the latter. *Septobasidium Mariani* also frequently has an extra horizontal layer between the subiculum and hymenium. The reproductive structures of the two fungi are similar.

Widely distributed throughout the southeastern United States, reported as far west as Wisconsin and as far north as New Jersey. One of the commonest species of this region. Also occurring on *Citrus* in Brazil, perhaps imported into that country on diseased *Citrus* stock. Represented by sixty-seven collections in our herbarium.

Specimens examined:

North Carolina: Common in piedmont and coastal regions.

Virginia: Near Roanoke, on *Carpinus*, April 21, 1931, C. L. Shear, 9940; near Burkeville, July, 1928, on *Cornus, Liquidambar*, and *Hicoria*, J. N. Couch, coll.; Emporia, April 5, 1937, L. Barksdale, coll., U. N. C. Herb. 10412.

South Carolina: Common in piedmont and coastal regions. Near Myrtle Beach, on *Cornus amomum*, Aug. 28, 1931, J. N. Couch, coll., 9090; also on *Fraxinus*, 9092; near Charleston, on *Cornus florida*, March 19, 1929, J. N. Couch, coll., 8483; St. Helena's Island, on *Magnolia virginiana*, March 21, 1929, J. N. Couch, coll., 8462.

Florida: Common as far south as Sebring, J. N. Couch, coll.; Gainesville, on *Acer rubra*, Knight, coll., Herb. of the University of Florida, 9189; on *Fraxinus*, A. S. Rhoads and E. West, colls. Herb. of the

University of Florida, 9545; also two collections from Weber on *Fraxinus*, Herb. of the University of Florida, 9411 and 9413; on orange leaves, F. A. Wolf, coll.; Homestead, on *Citrus grandis*, Link, coll., Herb. of the University of Florida, 9187; also on *Citrus sinensis*, Herb. of the University of Fla., 9186.

Alabama: Nokomis, on *Liquidambar*, April 19, 1932, C. L. Shear, coll., in U. S. D. A. Herb. and U. N. C. Herb., 9937.

Louisiana: Opelousas, on *Taxodium*, April 16, 1932, P. R. Miller, coll., in U. S. D. A. Herb. and U. N. C. Herb., 9938, with *S. taxodii* and *S. Mariani*; also Honey Island, St. Tammany Parish, on *Liquidambar*, Dec. 31, 1931, D. P. Rogers, coll.; Covington, 1924, Arsene, coll., in Farlow Herb., 14891, and U. N. C. Herb.; Baton Rouge, three collections from C. A. Brown, on *Liquidambar*, no. 328, on *Ilex decidua*, no. 318, and on *Crataegus*, no. 331, in Brown Herb. and U. N. C. Herb.; near Baton Rouge, five collections, Dec. 31, 1932, by J. N. Couch, in U. N. C. Herb.

Tennessee: Gatlinburg, on *Carpinus caroliniana*, June 20, 1931, L. R. Hesler, coll.; another coll., same loc., July 19, 1937, both in Hesler Herb. and U. N. C. Herb.

Mississippi: Pascagoula, on *Magnolia virginiana*, Jan. 1, 1932, J. N. Couch, coll., in U. N. C. Herb. 9202; also on *Ilex americana*, in U. N. C. Herb., 9190.

Also reported by Burt from Wisconsin, Kentucky and Pennsylvania but I have not examined these collections.

Brazil: Estado de Minas Gerais, Viçosa. On *Citrus*, referred to *S. albidum* in Phytopathology **23**: 734–737, 1933. Albert S. Müller, coll.; Müller Herb., 44. Dr. Müller has very kindly supplied me with abundant material which is in fruit and agrees well with *S. pseudopedicellatum* Burt.

Campo Grande, Rio de Janeiro, abundant on stems, twigs and leaves of *Citrus*, June, 1934, Nearch Azevedo, coll., Dr. Silveira Grillio, comm. Referred to *S. albidum* (Estudo sobre *Septobasidium albidum* da Laranjeira. Agronomia **1**: 265–276, 2 pls. 1930).

Septobasidium Mariani Bres. Ann. Myc. **1**: 24. 1903. Diagnosis in Ann. Myc. **3**: 164. 1905.

Sense of Bresadola and Patouillard.

Plates 17, 26, and pl. 72, figs. 6–11

Perennial, resupinate, forming patches sometimes as much as 10 cm. long, often girdling the limbs. Composed of two distinct zones: (1) the older central part over which the hymenium has been formed and (2) the marginal region. Central region typically dark brown during the summer, becoming somewhat fleshy brown and later buffy brown in fall and retaining the lighter color until after the spore-bearing period in the spring. Surface of central region irregular and granulose in fall when forming, becoming smooth and shiny and finally becoming split by irregular cracks. Marginal region very distinctive and peculiar, deep brown in color (darker than central portion) except for the very narrow, outer, growing zone which is whitish; about 1 cm. wide in late summer, hardly over 2 mm. wide by late winter due to the formation of new hymenium; marginal region often higher than the central region, composed of a basal subiculum, over which extend numerous dark-colored rhizomorphs. From the upper sides of the rhizomorphs plates of horizontal threads grow outward. Where the plates come in contact with each other, they fuse, thus forming a layer over the subiculum. This condition is distinct at the outer growing margin. From the older portion of the marginal region a vast tangle of much branched and anastomosing pillars arise. The pillars branch out at the top giving rise to the hymenial surface. In section up to 2 mm. thick, stratose in older portions. In younger regions composed of (1) a basal subiculum which extends over the bark and over which numerous branched dark brown rhizomorphs extend, subiculum thin, about 25–30μ thick; (2) a region of very short stout pillars which arise from the subiculum and which with the rhizomorphs support (3) an incompletely formed layer parallel to the subiculum (the scale insects are located between the subiculum and the second layer), space between subiculum and second layer about 100μ high; the second layer and the rhizomorphs give rise to (4) the long, branched, slender, anastomosing pillars, often about 1 mm. tall, 15–60μ thick, threads of pillars 3.8–4μ thick, septate, without clamp connections, the pillars giving rise to (5) the top layer which bears the hymenium. Hymenium 50–90μ thick, composed of rather compact, branched threads, 2.2–3.5μ thick, from the sides of which the probasidia arise; probasidia pyriform, 6.7–8.4 x 10.9–18.9μ, giving rise to a long, straight, cylindrical, 4-celled basidium about 4.2–4.6 x 42–50μ; spores (no. 8400) bent-elliptic, white, 3.8–4.2 x 12.6–21μ.

Symbiotically associated with several species of scale insects some of which are parasitized by irregular, coiled hyphae, others of which are left unparasitized. In this species the threads which connect the coiled haustoria of the fungus within the insect with the fungal pad which envelops the posterior part of the parasitized insect pass through either the anus, the genital pore, or through the thin cuticle by the mouth.

The most remarkable characteristic of this plant is that it is composed of three distinct stories: i.e., there is the basal subiculum (the first story) covering the bark, next a region of short, thick pillars and rhizomorphs and then another layer (the second story) parallel with the subiculum, and arising from this layer is a region of long branched pillars on the tops of which is borne the top layer (the third story). This characteristic can easily be recognized with a hand lens in material collected from mid-summer until early winter and serves to distinguish this species from all but a few species of *Septobasidium.* Another characteristic by means of which this plant may be distinguished is the peculiar marginal condition of plants collected in the late summer and early fall. At this time the new hymenium for the following year has not begun to develop and the new zone of growth between the margin of the old hymenium and the outer margin is composed of a dense tangle of dark-colored pillars and rhizomorphs.

I have often found this species growing on the limbs along with typical *S. pseudopedicellatum* Burt and because of the similarity of color in some lots of material the two plants may easily be confused, as has been done by previous students. Typical *S. pseudopedicellatum* Burt is a rich, buffy brown color on top with much darker brown, usually straight, and unbranched pillars, and with a white subiculum. (For further distinctions between this species and *S. pseudopedicellatum* Burt, see under the latter species.)

This species like all the others which I have studied is distinctly perennial. We have collections from around Chapel Hill during every month in the year. Moreover I have kept several specimens on *Pyrus japonica* marked in the arboretum for several years. The fungus has a very different appearance in winter and spring from that in the summer and fall, so different, in fact, that one would almost certainly describe the forms collected during these two seasons as distinct species. It is only by keeping the same specimen under observation for several years that one may prove that such apparently distinct fungi are merely stages in the development of the same fungus.

In the Bourdot Herbarium there are two collections, no. 37453 and no. 37460, determined by Bresadola as *S. Mariani.* These are nearly identical with *S. Michelianum* (see that species for comparison).

Widely distributed and fairly common in some localities.

Specimens examined:

Italy: Near Rome, on *Pyrus communis,* 1902, Mariani, Patouillard Herb., Harvard Univ., and N. Y. Bot. Gard. Herb. Type or cotype.

Caucasus: Near Suchum, on *Pyrus communis,* July 1915, W. Siemaszko, 9954.

Brazil: Near Bono Principio, 1928, J. Rick, Farlow Herbarium, also U. N. C., 9358, sterile.

North Carolina: Chapel Hill, on *Pyrus malus*, 4157, and on *Quercus nigra*, part of 3936 (large piece), four other collections on *Quercus*, 8182, 8378, 8405, 9708. On *Pyrus japonica*, eight collections; on *Cornus florida*, 8253, 8395, 9898; on *Cornus amomum*, 8276, 9113; on *Ilex decidua*, 8377; on *Liquidambar styraciflua*, 8400, 9332.

South Carolina: Society Hill, on *Pyrus angustifolia*, July 22, 1922, J. T. Rogers, U. S. D. A. Herbarium and U. N. C., 9955.

Tennessee: Sevierville, on *Acer* sp., L. R. Hesler, two collections (Univ. Tenn. Herb. 538 and 540), U. N. C., 9709 and 9710.

Louisiana: Opelousas, on *Crataegus*, April 16, 1932, C. L. Shear, U. S. D. A. Herb. and U. N. C., 9707; on *Taxodium distichum*, April 16, 1932, P. R. Miller, U. S. D. A. and U. N. C. Herb.; near Baton Rouge, on *Crataegus*, Dec. 31, 1931, J. N. Couch, 9165 and 9173.

Arkansas: Near Little Rock, on *Quercus phellos*, April 12, 1933, Dow V. Baxter, 9376.

Septobasidium Mariani var. **japonicum** n. var.

Plate 110, fig. 7

The correct position of this fungus will have to remain doubtful until the probasidial and basidial stages can be found.

In color and superficial appearance without a lens the fungus exactly resembles the late fall collections of *S. Mariani*. The outer edge of the margin is whitish with a finely cottony appearance for a region about 1 mm. wide. Just behind this zone arise irregular plates of hyphae which spread out horizontally to form an incomplete horizontal layer parallel with the subiculum. Some of these plates turn upwards so that this zone has an irregular honey-comb-like structure. In some specimens these plates of hyphae are not evident but pillars arise from the region just back of the margin. Pillars are also present among the plates. The pillars are very densely arranged and arise in tufts, branch and anastomose freely, 40–200μ x 1–3 mm. Hyphae of pillars 3–4μ thick, branching out above in brush fashion to form the top layer (100–160μ thick) the upper hyphae of which extend mostly in a vertical direction. Hyphae of top layer, hyaline or pale brownish, 3–4μ thick, considerably branched, the branches arising laterally or at times dichotomously or sometimes three branches may arise at one point. No probasidia, basidia, or spores seen. Subiculum 80–250μ thick, composed of a densely compacted layer of brownish hyphae.

Associated with scale insects some of which are parasitized by irregularly coiled haustoria, such insects being imbedded in the subiculum, while others are free from parasitism and are not encased by a thick subiculum tissue.

The present species agrees with *S. Mariani* in having three horizontal layers and also the hyphae of the two fungi are similar. This fungus has considerably thicker and taller pillars, a thicker subiculum and top layer than has *S. Mariani*. In *S. Mariani* the pillars are very distinct in the marginal region while in the present form very few pillars are found on the margin, these being replaced by irregular plates of hyphae. Pillars, however, are conspicuous in the older parts. In *S. Mariani* the margin shows distinct rhizomorphs while such structures are entirely lacking in the present form. Also in the present fungus the hyphae of the hymenium extend in a more or less vertical direction while in *S. Mariani* the hyphae in the hymenium are entangled and extend diagonally or horizontally.

Specimens examined:

Japan: Sendai, on *Styrax japonica*, Oct. 1, 1912, Prof. A. Yasuda, coll., in Lloyd Collection, no. 39231.

Septobasidium Cokeri Couch. Jour. Elisha Mitchell Sci. Soc. **51**: 40. 1935.

Plate 39, figs. 11, 12, and pl. 85, figs. 3–6

Resupinate, covering an area of several square centimeters (up to 15 cm. in diameter) on the bark of a variety of trees and shrubs. Typically pure white throughout when fresh and frequently retaining the pure whiteness in the herbarium, sometimes creamy on the surface and becoming brown upon aging, sometimes becoming blackish in the older parts. Margin determinate. Surface smooth and usually free from cracks except in old specimens. In section 0.6–1.3 mm. thick, composed of three distinct layers, frequently stratose: (1) the subiculum which extends over the bark 40–70μ thick, threads of subiculum 3–3.7μ thick; (2) the pillars, which arise from the subiculum, arising singly or sometimes in tufts, each pillar with a circlet of short threads at its base, these threads encrusted with crystals; 18–185 x 370–510μ, composed of much septate threads sometimes constricted at the septa and frequently thick-walled. 4–5.5μ thick. The pillars often arise in concentric rows on the margin, each row being about 0.25 mm. apart. In the older parts this concentric arrangement of the pillars is not evident. Threads of pillars branching above to form (3) the top layer over which the hymenium is formed, 70–148μ thick; threads 4–5.9μ thick, frequently thick-walled, with a very small lumen and sometimes constricted at septa. Hymenium 25–50μ thick, composed of probasidia and hyaline paraphyses. Probasidia globose, subglobose, or more usually pyriform, 10 x 11–16μ, sometimes subtended by one or more large cells, probasidial wall often considerably thickened, up to about 2μ thick. Threads in hymenium 2–3μ thick, recurved but not wavy and of more or less even diameter throughout. Basidia long-cylindrical or sometimes curved,

4.4–6.2 x 40–55μ, 4-celled and with a long stalk which remains attached to the empty probasidial cell; sterigmata lateral, small, about 5μ long. Spores 3–3.7 x 14.8–22.4μ, bent-elliptic.

Associated with scale insects some of which are parasitized by irregular coils, while others are unharmed by the fungus. Most commonly found on *Quercus* spp., also on several other trees and shrubs. Distributed from New Jersey southward to Florida. Frequently found with typical *S. pseudopedicellatum*, *S. Mariani*, *S. Burtii*, *S. Curtisii*, and *S. apiculatum*.

This species has been included in *S. pseudopedicellatum* by Burt (1916) and Coker (1920) but may be distinguished as a rule by its almost snow white or creamy color throughout. This whiteness is very distinctive and by marking certain specimens and keeping them under observation through the seasons it has been found that the whiteness is not a mere seasonal condition but persists throughout the year. The very distinct unbranched pillars, which in the marginal region arise frequently in concentric rows, the pillars in each row often being so close to each other as to form a solid phalanx, are also characteristic. Even in older specimens the older surfaces of which are buffy or brownish in color the pillars will always be whitish or light buff and never dark brown as in *S. pseudopedicellatum* Burt. Another distinguishing character of this fungus is in the hyphae. The subhymenial hyphae are frequently thick-walled, constricted at the septa and with small lumina. Such hyphae resemble those of *S. apiculatum* with which this form sometimes occurs. The basidia and spores are also smaller than those in *S. pseudopedicellatum*.

Specimens examined:

New Jersey: Newfield, on living saplings of *Quercus tinctoria*, Jan. 1900, J. B. Ellis Herb., also in Farlow Herbarium. Burt included this under *S. pseudopedicellatum* but wrote on label: "This does not have exactly the structure of *S. pseudopedicellatum*. Perhaps it is weathering that has caused change."

Virginia: Mecklenburg Co., on *Cornus florida* with *S. filiforme*, July 13, 1927, J. N. Couch, coll., 8179.

North Carolina: Chapel Hill, on *Quercus rubra*, 8439, 8440, 8441, 9108, 9326, 9385, 10004, type; on *Quercus velutina*, 10000; on *Quercus palustris*, 8303A, 8563; on *Quercus phellos*, 8391, 9224; on *Quercus montana*, 9102; on *Cornus florida*, 8394, fertile; also on *Cornus amomum*; on *Fagus grandifolia*, 8187; on *Acer negundo*, 9072, 9109; Cary, on *Quercus phellos*, John R. Raper, coll., 9430; Pasquotank Co., on *Hicoria aquatica*, April 4, 1933, Willard Hewitt, coll., fertile, 9367; near Magnolia, on *Quercus*, Jan. 1931, A. C. Mathews,coll., 9966. Beautiful specimens.

Florida: Near McClenny, on *Quercus* sp., Jan. 18, 1933, W. L. Kersey, coll., Univ. Fla. Herb., 8923, also U. N. C. Herb., 9344.

Septobasidium obscurum Boed. and Stein. Bull. Jard. Bot. Buitenzorg III, **11**: 212, figs. 30, 31. 1931.

Plate 57, figs. 13, 14, and pl. 101, fig. 7

Resupinate, forming rather small, inconspicuous orbicular or irregular patches 2–6 mm. wide, context deep rich brown, hymenial surface paler, smooth, usually glabrous. Margin determinate. In section up to 1 mm. thick, composed usually of three more or less distinct regions: (1) the very thin subiculum 5–30μ thick; the subiculum does not extend between the patches of growth but with a binocular dissecting microscope (20 x or above) one can distinguish fine rhizomorphs and branched threads extending between the patches; hyphae of subiculum 3–5.2μ thick, with rather thick walls and a rich brown color, sparingly septate with septations rather inconspicuous; branched, the branches usually arising at right angles; from the subiculum arises (2) the middle region composed of scattered pillars; pillars 300–450μ tall by about 10–90μ thick, composed of rather loosely arranged hyphae which are similar to those of the subiculum; the hyphae of the pillars spread out above to give rise to (3) the top layer which is composed of very loosely arranged hyphae. Top layer 150–450μ thick, hyphae similar to those of pillars. Over the top layer is formed the very thin hymenial crust about 40–60μ thick; hymenial layer composed of very conspicuous, spherical, hyaline probasidia, 16–21μ thick, which remain as an empty cell after the formation of the basidia; and thin, hyaline, branched at right angles, long, straight or slightly wavy threads, 1.5–3μ thick. Basidia long-cylindrical, 8.2–9 x 67–75μ, four-celled, sterigmata 10–15μ long. Spores 5.8–6.3 x 30–32μ, elliptic, recurved at one end. In some specimens the pillars are not formed, the hymenium being formed directly over the lower layer, in fact in some specimens pillars may be formed in part of fungus but not in other parts.

Associated with scale insects, some of which are parasitized by haustoria in the form of very irregularly coiled hyphae, the connection between the haustoria and the external fungus being established through the vaginal pore; other insects are free from parasitism.

This species may be recognized by its color and occurrence in small patches, by the very loosely packed hyphae of context, the thin subiculum, thin hymenium with the large probasidia, basidia, and recurved spores.

Specimens examined:
Java: Priangan, Tjibodas, on bark of living (?) tree, November 1929, Boedijn, coll., no. 69, type.

Septobasidium triviale Boed. and Stein. Bull. Jard. Bot. Buitenzorg III, **11:** 210, figs. 28, 29. 1931.

Plate 57, figs. 8, 9, and pl. 99, fig. 6

Resupinate, forming small (1–6 mm. in diam.) orbicular or irregular patches on the lower surface of leaves; sordid wood brown or buffy brown, margin sharply determinate. In section 550–700μ thick, composed of three distinct regions: (1) the very thin (5–20μ thick) subiculum, composed of brownish, branched hyphae 2.5–3.5μ thick; (2) the middle region of pillars which are about 10–50μ thick and 270–450μ tall, hyphae of pillars 3.8–4.2μ thick, compactly or rather loosely arranged; the pillars branch out above to form (3) the top layer 84–210μ thick, usually about 125μ thick, composed of very loosely arranged, considerably branched hyphae, the branches arising at right angles; hyphae throughout top layer and pillars with rather conspicuous but incomplete septations; on the upper surface of the top layer is formed the hymenium, which is about 30–42μ thick and composed of rather fine (1–2μ thick) densely entangled, pale brown to hyaline hyphae which mostly extend in a horizontal direction. Probasidia globose or pyriform, 10.9–15μ thick, germinating to form a cylindrical, four-celled basidium 5.8 x 67μ. Spores not seen. Haustoria in the form of irregular coils. Connection through vaginal pore.

This species appears to be close to *S. pseudopedicellatum* Burt but can easily be distinguished from that species by gross characters. The habit of growth in small patches, the sordid brownish color, the inconspicuous pillars, the loosely packed hyphae in the subhymenial region, the very thin hymenium and the delicate, long, hyaline threads in the hymenium are characteristic.

Boedijn and Steinmann did not find the basidia but by tearing apart a number of hymenial fragments I was able to find several basidia, some of which showed short sterigmata.

Specimens examined:
Java: Res. Batavia, Depok, Van der Brink, coll., no. 5907, on leaves of *Litsea*, type in Herbarium Botanic Garden Buitenzorg and U. N. C. Herbarium.

Septobasidium pachydermum n. sp.

Plate 44, fig. 16, and pl. 102

Resupinate, covering large areas on the stems of bamboo, the specimen examined 8 x 45 cm., the surface mostly smooth and velvety but in places with bumps and depressions, near snuff brown. The margin determinate, somewhat strigose. In section 3–5 mm. thick, sometimes stratose by the formation of several hymenial layers. Composed usually of a blackish, basal subiculum which forms an exceedingly compact layer over the stem, 60–200μ thick, usually about 150μ thick; hyphae about 4–5μ thick. From the subiculum arises the context which in places consists of irregular, branched, blackish pillars while in other places the context consists of a tangle of threads. Usually the pillars and threads are irregularly intermingled to form a fairly compact region of entangled hyphae. Hyphae of context 3–8.4μ thick, most about 6.7μ thick, thick-walled, ascending, sparingly branched. Hymenial layer is formed over the surface of context, 60–100μ thick, composed of very compactly arranged, branched, entangled hyphae the hyaline ends of which are recurved like a shepherd's crook. Probasidia globose to subglobose, usually elongating upon germinating, frequently borne in clusters, 15–18μ thick, hyaline, remaining as an empty vesicle after the formation of the basidia. Basidia straight, cylindrical, 4-celled, 7.8–8.4 x 42–50μ, sterigmata about 3μ long. Spores bent-elliptic, 4.6–6.3 x 18–23μ, most about 5 x 20μ.

Associated with scale insects some of which are parasitized by irregularly coiled haustoria.

This species may be distinguished by its velvety snuff brown surface, very thick, rather compact structure, by the context which is composed of intermingled pillars and hyphae, by the thick hyphae of the context and the hyaline hyphae of the hymenium recurved like a shepherd's crook. Its occurrence on bamboo may also be distinctive.

Specimens examined:

Burma: Mawkmai, in teak and bamboo forest, elevation 3,000 ft., on a bamboo (*Dendrocalamus strictus*) about 3 feet above the ground, coll., W. A. Robertson, Jan. 1911, type in Kew Herb. and U. N. C. Herb.

Septobasidium hakgalanum Couch and Petch n. sp.

Plate 57, fig. 6, and pl. 97, figs. 6, 7

Resupinate, girdling a small limb for about 10 cm. Verona brown to bister in central part, margin vinaceous buff. Surface of top layer very irregular, due to its incomplete formation. In the marginal region consisting mostly of flattened areas with an ovoid or irregular outline, 1.5–2 x 2–2.5 mm. As the fungus grows the margins of these elevated surfaces anastomose with each other, the top surface, however, remaining broken into numerous sinuous patches. Margin determinate. On the subiculum in the marginal region there are "tent-like" structures each formed by two diagonally ascending plates of hyphae, the ends of the hyphae of each plate anastomosing above with those of the opposite plate to form the top of the tent. From the subiculum around the margin of each tent there grows up a number of pillars which extend up some distance above the top level of the tent; the hyphae of the pillars then spreading out and anastomosing to form the top surface and thus each tent is covered by its own covering and also by a separate part of the top layer. In section 700–1000μ thick. Composed of a subiculum, 160–220μ thick which is made up of compactly arranged, brownish hyphae, the lower ones (about 4–4.2μ thick) extending horizontally over the bark and giving rise to vertically extending, unbranched, hyaline, sometimes crystal-encrusted hyphae, about 3μ thick. From the subiculum arise the tent-like insect houses which persist even in the older parts of the fungus. Walls of insect houses exceedingly compact, usually 60–80μ thick but much thicker when the house contains a parasitized insect. Pillars 40–90 x 250–450μ, hyphae of pillars brownish, 4.2–5μ thick, sparingly septate. Top layer 60–160μ thick, composed of compactly arranged, much entangled brownish hyphae, 4.2–5μ thick. Hymenium not differentiated from underlying tissue. Probasidia formed near the top surface, globose or subglobose, 15–20μ thick, becoming empty in the formation of the basidia. Basidia long-cylindrical, 4-celled, the end cell pointed, 5.8–7.1 x 68–84μ, sterigmata 5–6μ long. Spores 4.2–5 x 25–27μ, hyaline, long-elliptic, slightly bent.

Associated with scale insects some of which are parasitized by irregularly coiled haustoria. Some of the insects are not parasitized and were in the act of giving birth to young when collected. A large number of insects are parasitized by the larval stage of one of the exoparasitic insects.

This species may easily be recognized by the very distinctive insect houses which are exceptionally large and symmetric and each of which is at first covered by a separate flat roof ovoid in outline.

Specimen examined:

Ceylon: Hakgala, April 1919, T. Petch, coll., in Petch Herbarium and U. N. C. Herbarium.

Septobasidium atratum Pat. Bull. Soc. Myc. Fr. **16:** 181. 1900.

Plate 57, figs. 1, 2, and pl. 101, fig. 6

Resupinate, fuscous or dark fuscous (Ridgway), lighter toward margin. Surface of top layer smooth and shiny with a fimbriated margin. Margin of subiculum determinate, the outer region being distinctly marked (under the lens) by rows of hyphae which grow up from the subiculum and are so arranged as to form minute cave- or tent-like structures. Just within this region are dense tufts of newly formed pillars. In section 1.0–1.6 mm. thick, composed of three distinct regions: the subiculum, the central region of upright pillars, and the top layer. Subiculum thin in the young parts, becoming 300μ thick in the older regions; composed of thick-walled hyphae which are 4–5μ thick. From the subiculum arise the pillars which are densely arranged, twisted, branched, 20–115μ thick; hyphae of pillars 4–5μ thick, septate and sparingly branched. Toward the top the hyphae of the pillars branch out horizontally to form the top layer, which is usually rather thin, 75–100μ thick. (In the material which I have examined there is a layer of amorphous, hyaline material about 20μ thick covering the top layer. This layer seems to be composed largely of spores, collapsed basidia, and hyphal ends.) Threads of top layer 4–5μ thick, extending in all directions but mostly obliquely, rather densely packed, branching sparingly and at right angles. Probasidia formed on the sides of the threads, sessile, usually near the hyphal ends; spherical, 11–13.4μ thick, giving rise to a long, narrow, straight, cylindrical, 4-celled basidium, 4.2–4.6 x 50–64μ; sterigmata about 4.5μ long. Spores long, bent-elliptic, 3.2–4 x 19.3–23μ.

Associated with scale insects (*Aspidiotus* sp.) some of which are parasitized by irregular coils while others are free from infection. There are two collections in the museums which are incorrectly labelled *S. atratum* Pat. One is in the Museum of the Botanical Garden in Berlin, on *Hevea brasiliensis*, Rio Jura, no. 2729; the other, which is apparently the same fungus, is in the Kew Herbarium on the same species of tree, Estado de Amazones, "Juruá Boni Fini," Nov. 1900. The latter species is being described as *S. heveae* n. sp. For a comparison see under *S. heveae*.

Septobasidium atratum Pat. is related to *S. Mariani* Bres., but may be easily distinguished by color, the latter being invariably brown.

Specimens examined:

West Indies, Guadeloupe: on living limbs of *Eugenia jambos*, R. P. Duss, coll., type in Pat. Herb. and also in Mus. Bot. Gard., Berlin.

Septobasidium suffultum (B. & Br.) Pat. Bull. Soc. Myc. Fr. **24:** 2. 1908. Also published as *Septobasidium* by Bresadola, Ann. Mycol. **14:** 241. 1916.

Thelephora suffulta B. & Br. Fungi of Ceylon. Jour. Linn. Soc. Bot. **14:** 63. 1873. Not *Corticium suffultum* B. & Br.

Plate 44, fig. 15, and pl. 101, figs. 1–5

Resupinate, covering rather extensive areas, one broken piece of type being 9 cm. long. Hymenial surface near cinnamon brown or darker, the marginal region of pillars a much darker brown, the subiculum near mummy brown. Top surface smooth to the unaided eye but with a lens numerous, minute (about 0.1 mm. wide), conical elevations are visible, these disappearing from the older regions as the top layer thickens. The surface in the older parts but where the top layer has not been formed is covered by an exceedingly dense tangle of pillars, the ends of which curve over. In the younger regions the pillars arise from mounds, leaving part of the subiculum free from pillars, but in the older parts the pillars appear to arise directly from the subiculum as well. Out towards the margin the surface is closely dotted with hollow mounds which are about 1.5 mm. wide and about 0.5 mm. high on the inside. These mounds begin their development as semicircular or horseshoe-shaped plates of hyphae which arise diagonally from the subiculum and growing inwards meet in a median line to form a roof leaving an open door towards the older part of the fungus. Surface of subiculum smooth except for mounds. Margin determinate. In section up to 1.5 mm. thick, composed in the older parts of three distinct regions: (1) the subiculum about 100μ thick, except beneath the mounds and here much thicker; threads of subiculum 3.8–5μ thick, so densely compacted and entangled as to appear pseudoparenchymatous in section, brownish in color; from the subiculum and particularly around the mounds arise (2) the region of densely arranged pillars, which are much entangled and anastomose freely with each other; pillars 30–85μ thick, sometimes as thin as 12μ; hyphae of pillars 3.8–4.2μ, up to 6.7μ near base of pillars; the pillars giving rise above to (3) the top layer which is 150–400μ thick, usually about 300μ thick and composed in the lower part of entangled, thick-walled hyphae which are up to 4.4–6.7μ thick and which extend mostly horizontally and from which arises the hymenial layer. This is composed of vertically-extending septate hyphae which are very closely arranged, 3.4–5.4μ thick; from these hyphae arise the probasidia which are usually spherical but may be pyriform, 12.6–13.8μ; apparently new probasidia may proliferate into the empty ones. These secondary probasidia are elongated and irregular in shape. Basidia straight, cylindrical, 4-celled, 6.5–7.1 x 42–48μ; easily separating from the probasidia (all basidia seen had broken away from probasidia), sterigmata not seen. Spores bent-elliptic or flattened on one side, 4–4.8 x 13–15μ.

Associated with scale insects, usually one or more beneath each mound in the older parts. Some insects are parasitized by irregular coils while others are free from fungal infection.

This species is distinguished by the deep brownish color, the mound-shaped insect houses which are very distinct in the marginal region, the exceedingly dense growth of pillars which grow up over the mounds, and the rather thin but very compact subiculum.

Petch has sent me a collection of material from Ceylon which he considers the young stage of the present species. Petch's material shows a very thin, brown, compact subiculum, the hyphae of which are 4–6.3μ thick. Spherical hyaline or deep brownish cells, 15.9–21.8μ thick, which are probably probasidia, are found abundantly in certain places on the subiculum and also on the pillars. These cells resemble the probasidia of *S. lepidosaphis* which are also borne on the subiculum. I am unable to find any probasidia on the subiculum of the type. Also in Petch's material the insect houses are scattered about irregularly over the subiculum and are of a different shape, being elongated instead of rounded and only half roofed over (Pl. 101, figs. 3–5). Also in Petch's fungus the top layer does not seem to be formed, the fungus consisting solely of subiculum and insect houses. Such variations, however, have been recorded for *S. Curtisii* and it would seem best until more fresh material can be studied in Ceylon to treat Petch's material as *S. suffultum.*

Specimens examined:

Ceylon: Southern part of Island, on bark July 1868, Thwaites, no. 669, type in Kew Herb.; also Sita Eliya, on *Eurya*, T. Petch, coll., February 1923, in Petch Herb. and U. N. C. Herb.

Septobasidium lepidosaphis Couch. Jour. Elisha Mitchell Sci. Soc. **51**: 35. 1935.

Plate 19 and pl. 74, figs. 13, 14

Resupinate, extending over the surface of the stem and leaves of *Citrus* sp., covering a relatively small area (6 sq. cm.) in my specimens, carob or Van Dyke brown (Ridgway). Surface covered with numerous pointed spines which may be 1.2 mm. high, spines usually anastomosing considerably, giving the surface a wefty or irregular honeycomb appearance. Margin usually determinate. In section composed of two or three regions: (1) the subiculum which is very thin, 10–20μ thick, threads of subiculum 3–6.3μ thick, septate without clamp-connections but with much anastomosing between threads; (2) from the subiculum arises a tangled weft of hyphae or hyphae which unite to form thin pillars which in turn unite to form larger upright spines, spines up to 75μ thick, usually thinner, hyphae of spines 4.2–5μ thick, septate without clamp connections, usually straight and rarely branching. (3) In some specimens a third region may extend over the tops of the spines, forming an irregular hymenial layer. Probasidia formed over the marginal region of the subiculum and over the top layer if a top layer is formed, and rarely on the spines, spherical 8.4–12.6μ thick, usually about 9.5μ thick, wall hyaline and rather thin (in my material). (Along with the probasidia are numerous enlarged bladder cells or perhaps aborted probasidial cells.) Probasidia germinating to form an elongated, straight cylindrical or sometimes curved, 4-celled basidium, 4.2–6.3 x 31.5–46μ, usually about 5.4 x 35μ; sterigmata small, about 4.2μ long; spores strongly bent-elliptic, 3.2–4.2 x 10.9–13.8μ.

Associated with the purple scale insect (*Lepidosaphes beckii*), some of which are parasitized by irregular coils while others are unharmed by the fungus. The fungus passes through the vaginal pore of the parasitized insects.

This species is very distinct from any other one so far described and may easily be recognized by the brownish color, dense tall spines, the sometimes coiled basidia, and most striking of all characters—the fact that the probasidia and basidia are formed on the flat thin margin as well as on the top layer when a top layer is present.

If infection of the young occurs here by the bud cells which come from the spores as in *S. Burtii*, the fact that the latter are borne on the margin would facilitate the infection of young migratory insects which happened to crawl out over the margin as the spores were being formed.

Specimens examined:

Florida: Okeechobee, on *Citrus grandis* with *Lepidosaphes beckii*, Jan. 20, 1933, L. S. Light, coll., in Univ. of Florida Herb., 8933, and U. N. C. Herb., 9345; near Venus, on *Citrus aurantifolia*, March 23, 1933, R. B. Linger, coll., in Univ. of Fla. Herb., 9053, and U. N. C. Herb., 9363.

Brazil: Viçosa, Estado de Minas Gerais, on *Citrus* sp., June 10, 1929, A. S. Müller, coll., type, in Müller Herb., 43 (see Phytopathology **23**: 734–737. 1933) and U. N. C. Herb., 9721; Lagoa Grande, on native tree, June 24, 1934, A. S. Müller, coll., in Müller Herb., 831, and U. N. C. Herb., 9961; also near Viçosa, on *Vernonia crotonoides*, July 10, 1934, A. S. Müller, 832, and U. N. C., 9962; also on *Mangifera indica*, A. S. Müller, 833, and U. N. C., 9963. Müller's collections also deposited in Cornell Univ. Herb.

Septobasidium castaneum Burt. Ann. Mo. Bot. Gard. **3**: 330. 1916.

Plate 25, figs. 1–3, and pl. 71, figs. 1–5

Resupinate, growing in large patches and often covering an area equal to several square feet; not rarely extending in effused patches from the base of a tree nearly to the top. Chocolate brown when dry, becoming nearly black when wet. Surface typically smooth and shiny, sometimes with cracks or pin holes. Margin distinctly fimbriate, mostly concolorous with the surface except in rapidly growing specimens and then nearly white in the growing region, determinate; typically with stout or slender pillars from near the bases of which arise horizontal hyphae which extend out and anastomose at the tips, forming a second layer parallel to the subiculum; from the tops of the pillars threads extend out horizontally forming the top layer; in the older parts this middle horizontal layer is usually quite indistinct. Thus there are three horizontal layers as in *S. Mariani*. Margin of top layer at times extending out over the margin of bottom layers; margin sometimes with an irregular honeycomb appearance. In section 1–1.5 mm. thick, often stratose by the formation of several fruiting layers in old specimens; composed of three regions: (1) the subiculum made up of a compact layer of threads which grow over the surface of the bark, 30–50μ thick; (2) the middle region which at first is composed of upright inconspicuous pillars or tufts of hyphae not grouped into pillars; pillars becoming indistinct as the fungus develops, the spaces between them being nearly filled by the growth of hyphae until in the older parts of specimens the middle layer is more or less continuous with the subiculum and top layer; (3) the hymenial layer which is about 30–40μ thick and is composed of probasidial cells, basidia, and simple hyphae; probasidia usually subglobose, quite irregular in shape, the large functional cell frequently subtended by a somewhat smaller swollen sterile cell 10–14.7μ, most about 12μ thick, thick-walled, sprouting under proper conditions to form a cylindrical basidium; basidia 5–6.7 x 37–50μ, becoming thrice septate, i.e. divided into four cells, each cell bearing a spore on short sterigma. Spores 3.8–4 x 11–14.2μ, bent-elliptic, usually not becoming septate.

This species shows considerable variation in color shade even on the same host, but can be recognized and distinguished from the related species by the following combination of characters: deep brownish chocolate color, inconspicuous pillars, peculiar margin, and comparatively small spores which do not become septate. For a comparison with *S. Leprieurii* (Mont.) Pat., to which this species is most closely related, see under *S. Leprieurii.*

This is one of the commoner species of *Septobasidium,* being by far the commonest species in the eastern part of South Carolina, so far as my observations go. By far the commonest host is the willow oak (*Quercus phellos*). The worst infected trees are the young ones with trunks from one to five inches in diameter. The trunks of such trees are often nearly girdled with the fungus from near the base almost to the top. Such trees are very unhealthy, with numerous dead branches.

Specimens examined:

Alabama: Near Montgomery, Dr. R. P. Burke, Mo. Bot. Gard. Herb., 20421 (type) and 20693, part of type in Farlow Herb.

North Carolina: Near Beaufort, very abundant on *Quercus rubra, Cornus florida, Ilex opaca;* specimens on *Ilex* associated with *Aspidiotus forbesi* Johns, J. N. Couch, coll., Aug. 3, 1929, several fertile specimens; near Wilmington, on *Quercus virginiana,* Dec. 28, 1921, Couch and Grant, colls., 5932, sterile; near Magnolia on *Quercus* sp., A. C. Mathews, coll., 9969.

South Carolina: Near Cheraw, Pee Dee River swamp, abundant on *Quercus phellos*, June 29, 1928, A. B. and J. N. Couch, colls., 8318; near Georgetown, on *Hicoria*, 8329; also on *Quercus phellos*, *Ilex opaca*, 8332 and 8330, *Magnolia virginiana*, 8331; on *Pyrus communis* in orchard, half dozen trees noted all infected, 8336; on *Nyssa sylvatica*, 8328; near Clemson College, on *Azalea* sp., Feb. 19, 1932, Mr. Lutken, coll., sent by G. M. Armstrong, 9225. Mr. Lutken reports that the fungus was abundant, killing entire plants.

Georgia: About 19 miles east of Valdosta, on *Quercus*, March 19, 1932, J. N. C., 9299.

Florida: Homestead, on *Albizzia* sp., with *Aspidiotus herculeanus*, March 9, 1933, spores forming, Univ. Fla. Herb., 9024; also common on *Quercus* sp., *Liquidambar*, *Ilex*, and other species as far south as Sebring, J. N. Couch, coll. Gainesville, on *Quercus nigra*, Feb. 22, 1935, G. F. Weber and E. West, colls., Univ. Fla. Herb., 9966.

Mississippi: Near Pascagoula, on *Magnolia virginiana*, 9201, mixed with *S. Leprieurii* (Mtg.) Pat., not typical; on *Quercus*, 9208; on *Fagus*, 9182, Jan. 1, 1932, J. N. Couch, coll.

Louisiana: Baton Rouge, on *Liquidambar*, Jan. 8, 1932, C. A. Brown, no. 327, not typical, apparently a cross between *S. castaneum* and *S. Mariani*; near Madisonville, on *Quercus nigra*, C. A. Brown, no. 320, typical; near New Orleans, several collections on *Cornus* and *Quercus*, J. N. Couch, coll., Dec. 31, 1931.

Septobasidium heveae n. sp.

Plate 57, fig. 10

Resupinate, forming an extensive growth over the bark of *Hevea*; blackish brown or grayish black or black in places. Surface irregular with cracks, holes, and bumps. Texture of top layer rather brittle. Margin with conspicuous, tough rhizomorphs. In section about 0.5–2 mm. thick, without a distinct subiculum but with numerous large and small rhizomorphs which extend over and between the fragments of bark. Rhizomorphs much branched, the largest ones up to 1 mm. thick, black or blackish, extending longitudinally and giving rise to blackish pillars which arise vertically or diagonally about 1 mm., forming a much entangled region, finally branching out above to form the top layer. Hyphae of rhizomorphs straight, very compactly arranged, deep brownish, about 3.7–5.4μ thick. Top layer 100–600μ thick, distinctly stratified in places, composed of densely entangled, brownish, branched, curved hyphae which usually have hyaline tips. Probasidia borne near top surface, ovoid, 9.4–12.6 x 12.6–16.8μ; sprouting and becoming empty to form the cylindrical, 4-celled basidia, which are 6–6.3 x 29–38μ. Sterigmata about 4.2–4.8μ long. Spores bent-elliptic, 3.6–4.2 x 12.6–13.4μ.

Associated with scale insects. Haustoria not seen. This species was put under the collections of *S. atratum* by Patouillard in his herbarium. Both are very dark-colored and might be confused. The present species is a much coarser plant than *S. atratum,* having large and conspicuous rhizomorphs beneath the top layers, while such structures are lacking in *S. atratum.* In the latter species a peculiar, pen-like insect house is formed in the marginal region; such structures are not formed in *S. heveae.* The two also differ in hymenial characters. In *S. atratum* the hymenium is thin and the sterile hyphae in it are mostly straight or slightly curved, and the basidia are long and narrow. In *S. heveae* the hymenium is thick, the sterile hyphae are curved, and the basidia are short and thick.

This species may be recognized by its blackish color, the large coarse rhizomorphs, the coiled threads in the hymenium, and the short thick basidia.

Specimens examined:

Brazil: Estado de Amazones, on *Hevea brasiliensis* Müll., Nov. 1900, in Kew Herb.; also Rio Jura, on *Hevea brasiliensis,* no. 2729, in Mus. Bot. Gard. Berlin, type.

Septobasidium alni Torrend. Broteria, Ser. Bot. **11:** 84. 1913.

Plate 21 and pl. 70, figs. 1–5

Resupinate, growing in medium sized patches, 2–10 x 3–20 cm., often girdling the limbs. Surface inherently smooth in mature specimens, becoming wrinkled and cracked when old; color, when fresh but dry, cinnamon to Prout's brown or Saccardo's umber; much darker, about burnt umber, when wet. Margin of growing region irregular, sometimes byssoid, whitish, buff colored or sometimes yellowish. In section up to 400μ thick, usually composed of three regions: (1) the subiculum or lower layer which extends over the bark, very compact, about 30–65μ thick, composed of thin-walled, pale brown to nearly hyaline hyphae about 3μ thick; (2) the middle region composed of a very few, short, upright pillars between which are very wide spaces; pillars 40–100μ tall, often up to 175μ thick, sometimes much thicker and not rarely as thin as 20μ, composed of brownish hyphae about 4.2μ thick, pillars branching out above and forming (3) the more or less papery top layer on the outer surface of which is formed the hymenium, top layer 200–250μ thick, composed of fairly densely packed, slightly branched and entangled hyphae which grow at a slight angle to the surface, hyphae deep brown in color except in the hymenial region, rather thick-walled,

3.5–4.2μ thick. Threads throughout with few septa and no clamp connections. Hymenial region very thin (in my specimens), composed of thin-walled, slightly branched, nearly hyaline threads, probasidial cells, and basidia. Probasidial cells globose to subglobose, sessile, 9–12μ thick before germinating, nearly hyaline and quite inconspicuous in the hymenium, probasidium sprouting into a long cylindrical basidium into which the probasidium empties its entire contents, empty probasidial cell elongated ovoid or ellipsoid, 8.8–10 x 18–24μ, thick-walled (1.5–2μ), new probasidia often formed in the old empty probasidial cells, elongated. Basidium 5–5.4 x 50–60μ, becoming thrice septate, and thus four-celled, each cell giving rise to a sterigma about 8μ long. Spores (of no. 8325) 3.4–4.2 x 15.9–21μ, smooth, bent-elliptic, becoming thrice septate, at times seven times septate soon after shedding.

Associated with scale insects some of which are parasitized by irregular coils while others are not parasitized.

This species shows remarkable similarities to *S. Cavarae* Bres. The American material has been compared with type material lent to me by Sydow (Torrend, no. 68) and the two lots of material agree in most respects. The type material does not show any basidia or spores, but does show vast numbers of empty probasidial cells and a considerable number of probasidia growing up into the older, empty probasidial cells. I have found more or less the same condition in my material collected during late summer, fall, or winter. I am convinced that Torrend mistook these probasidia for the basidia, since the measurements which he gives for the basidia agree with the size of the probasidia found so abundantly in his material. He does not mention the probasidia, but concerning the basidia he says, "at first globose, 10–12μ in diameter, at length ellipsoid, 3-septate, 20–25 x 7–9μ; spores not seen."

This species can be recognized by the thin fructification, by the more or less papery top layer, by the few, inconspicuous, short, scattered pillars, which, however, cannot be seen well unless the top layer is removed and turned over, in which case the pillars adhere to the top layer; and by the buff colored margin, contrasting strikingly with the brown older region (this color is conspicuous only at the end of the growing season—late summer and fall). One remarkable peculiarity of this species is the fact that new probasidia are often formed within the old empty probasidial cells. Such probasidia are present in large numbers in regions of old growth in the late summer and fall and if a specimen bearing such probasidia is kept in a damp chamber, the latter will germinate, forming basidia. Another peculiarity, in some specimens, is the presence of numerous clusters of green algal cells imbedded here and there in more or less regular order just below the upper surface of the fungus. Delicate fungal threads often surround some of the algal cells and seem at places to penetrate into the cells.

In typical specimens of this species the pillars arise in a row from the subiculum, at first about one millimeter behind the growing margin. These grow upward and quickly branch out at the top, developing first into an umbrella-like structure with a very short, thick stalk and an exceedingly wide, flat covering. After the subiculum has grown out a third or half a centimeter from the first row of pillars, another row arises and develops in the same way. The tops of the pillars grow in a horizontal direction until their margins come together and fuse. In the hybrid described below growth may stop before the margins all come together, thus leaving irregular spaces several millimeters wide between the margins. This condition gives the fungus a peculiar flaky appearance as though the top layer were peeling off. In my collections the flaky appearance is commoner in the sweet gum specimens, though it is also found in the willow oak, hickory, and American holly specimens. This flaky appearance has not been noticed on specimens occurring on pecan. In some specimens pillars may be entirely lacking in certain regions.

Fairly common in eastern North Carolina, South Carolina, Georgia, and probably in states farther south.

Specimens examined:

Portugal: Type, on *Alnus*, Torrend, coll., 373, in N. Y. Bot. Gard. Herb. and 68 in Sydow Herbarium, now in Upsala.

North Carolina: Near Beaufort, on *Liquidambar*, Aug. 4, 1929, J. N. Couch, coll., 8415, common.

South Carolina: St. Helena's Island, J. N. Couch, coll., on *Quercus nigra*, with *Aspidiotus osborni*, March 21, 1930, U. N. C., 8485; on *Quercus phellos*, with *A. osborni*, U. N. C., 8481; on *Magnolia virginiana*, with *Aspidiotus diffinis*, U. N. C., 8482; Springwood, on *Q. phellos*, with *Chrysomphalus obscurus* Comst., July 2, 1928, J. N. and A. B. Couch, colls., U. N. C., 8323; on *Liquidambar*, with *Aspidiotus* (*Cryptophyllaspis*) *liquidambaris* Kot., U. N. C., 8325; on *Ilex opaca*, with *Aspidiotus forbesi*, U. N. C., 8327; on *Hicoria*, with *Aspidiotus ancylus* Putn., 8324; near Clemson College, on *Hicoria pecan*, with *Aspidiotus ancylus* Putn., Feb. 20, 1930, Geo. M. Armstrong, coll., U. N. C., 8437, also Nov. 24, 1930, U. N. C., 8556.

Georgia: Barnesville, on *Hicoria pecan*, common, Oct. 14, 1930, B. B. Higgins, U. N. C., 8498, fertile.

Venezuela: Carabobo, Valencia, 430 meters, on *Citrus sinensis*, Chardon and Stelling, 806, August 6, 1932, also in U. N. C. Herb. and L. O. Overholts Herb.

Septobasidium alni var. **brasiliense** n. var.

Plate 44, fig. 7, and pl. 114, fig. 1

Resupinate, covering extensive areas on the bark, the largest patches up to 10 cm. in diameter. Texture tough, not easily torn with needle. Wood brown, except in the older parts where the color is a much darker brown, surface glabrous, almost shiny, with here and there a conical spine projecting up about a millimeter above the surface. These spines are the ends of rhizomorphs and give the surface a characteristic feel. They are never very abundant and are entirely absent from some pieces. Margin determinate, usually with a finely toothed appearance, with a narrow whitish or buff zone or concolorous with the surface. Margin of subiculum either even with margin of top layer or projecting up to 1.5 mm. beyond, very compact, usually with a row of indistinct hyphal tufts. In section about 200–600μ thick. Composed of subiculum, pillars and top layer in the younger parts, irregularly stratose in the older parts. The subiculum is exceedingly thick, compact, brownish to nearly black, sometimes with conspicuous, inherent rhizomorphs. Pillars much as in type, becoming inconspicuous in the older parts. Top layer, hymenium, basidia and spores as in typical material.

Associated with scale insects some of which are parasitized by irregularly coiled haustoria while large numbers of all ages are free from parasitism.

Specimens examined:

Brazil: Sao Leopoldo, Rick, coll., Dec. 16, 1937, on undetermined host, U. N. C. no. 9903; on *Melia* sp., Aug. 6, 1937, U. N. C. no. 10419; on undetermined host, Jan. 6, 1936, Rick, no. 7, U. N. C. no. 10083.

Septobasidium alni var. **squamosum** Couch. Jour. Elisha Mitchell Sci. Soc. **51:** 35. 1935. Apparently a hybrid between *S. alni* and *S. castaneum*.

Plate 25, fig. 4, and pl. 70, figs. 6, 7

Resupinate, the patches of growth being intermediate in size between *S. alni* and *S. castaneum*. Color about Saccardo's umber, sometimes brown with a distinct grayish tinge (no. 8485). Margin determinate or indeterminate. Surface flaky, due to the incompletely formed top layer. The flakes, which form all the top layer present in this species, vary from about 1 up to several square mm., crowded and anastomosing or scattered with 1 or 2 mm. between each flake. In section the entire plant is exceedingly thin, the flakes papery thin. Pillars scarce, the flakes usually arising directly from the basal layer. Pillars when present short and thick. Probasidia formed over the surface of the flakes and also over the basal layer except beneath the flakes; spherical, 9–11.3μ thick; basidia 4–5 x 40–50μ, cylindrical, 4-celled (no. 8323). Spores as in *S. alni* except slightly smaller. Associated with scale insects, some of which are parasitized by irregular coils.

This hybrid form can easily be recognized by its brownish color, exceedingly thin structure, and the flaky nature of the upper layer. It occurs with *S. alni* and *S. castaneum* and is possibly a hybrid formed by the crossing of these two species. The forms belonging here were first segregated under *S. alni*.

Specimens examined:

South Carolina: Near Myrtle Beach, on *Liquidambar styraciflua*, common, August 29, 1931, 9087, 9088, 9089; St. Helena's Island, on *Liquidambar styraciflua*, March 21, 1930, J. N. Couch, coll., 8479; Springwood, on *Liquidambar styraciflua*, with *Aspidiotus* (*Cryptophyllaspis*) *liquidambaris* Kot., July 3, 1928, J. N. Couch, coll., 8325; also on *Hicoria* with *S. alni*, 8324.

Septobasidium grandispinosum n. sp.

Plate 98, figs. 6, 7

Resupinate, forming tough, irregular patches up to 4 cm. long on the twigs. Surface mottled with dark and light-colored areas, from nearly white through avellaneous and wood brown to snuff brown. The surface as a whole is characteristically roughened by a few, irregularly scattered, large, conical, usually sharp-pointed spines, which are about 1 mm. thick at base by 1–1.5 mm. tall. The spines may be entirely absent from some parts. Surface very compact and smooth where there are no spines. Margin determinate, fimbriate or scalloped. In section 1–1.5 mm. thick, sometimes stratose. Composed of a white basal region which does not form a continuous subiculum but extends here and there over the insects in the form of rhizomorphs or plates of hyphae. The parasitized insects are surrounded by compact, thick, pseudoparenchymatous tissue. From this basal zone arises the context, the lower part of which may sometimes be arranged in pillars. Hyphae of context hyaline to pale brownish, thick-walled, most 4–4.2μ thick, compact, much branched, entangled; giving rise above to a lighter colored zone of still more compactly arranged hyphae. Hyphae of this upper zone 2.8–3.5μ thick, much branched, densely interlaced and entangled. The probasidia are formed irregularly near the upper surface of this zone. Probasidia characteristically irregular in shape, mostly subglobose, 12.6–18μ thick. Basidia long-cylindrical, straight or slightly bent, 4-celled, frequently attached to the empty probasidial cell by a long stalk, 5.4–7.5 x 54–68μ. Each cell of the basidium gives rise to a short sterigma on which is borne a spore. Spores flattened or slightly concave on one side, convex on the other, thickest in the middle, 4.5–5.4 x 20–22.2μ.

Associated with scale insects which are exceedingly abundant in the material examined by me. Some of the insects are parasitized by irregularly coiled haustoria.

This species is easily distinguished by its irregular appearance, by the few but conspicuous spines on the surface, by the mottled brownish and whitish color, and by its compact texture.

Specimen examined:

South Africa: Transvaal, near Pretoria, on *Gymnosporia buxifolia*, coll., L. Kresfelder, March 1925, in Herb. Div. of Plant Industry, Pretoria, no. 20350, and U. N. C. Herb., type.

Septobasidium Carestianum Bres. Malpighia **11:** 254. 1897.

Plate 20 and pl. 74, figs. 1–12

This species has been collected in the western hemisphere only by Dr. John Dearness in Canada and by Padre J. Rick in Brazil.

Perennial, resupinate, covering an area of several (2–6) square centimeters on the bark of living shrubs and trees. Color quite variable: about cream color in young specimens, becoming buffy brown and sometimes cinnamon brown. Surface inherently smooth but often with minute pin holes or irregular fissures due to the incomplete formation of the top layer, the latter condition being particularly noticeable in young specimens. Margin determinate or often indeterminate and irregular in outline due to the formation of isolated patches of hymenium. In section 250–700μ thick, often stratose, the primary stratum usually composed of three indistinct regions (in old, well developed specimens the subiculum, middle region, and top layer may be continuous): (1) the subiculum, usually exceedingly thin, hardly exceeding 10–15μ thick in young specimens, in older specimens 30–60μ thick, very compact, hyphae of subiculum about 3–4μ thick, minutely rough; (2) the middle region which is composed of loosely packed, ascending, septate, pale or nearly hyaline hyphae about 3–4μ thick without clamp connections and arranged in short cone-shaped tufts or in thick short pillars, the hyphae passing into and forming (3) the top layer over which is formed the hymenium; top layer 110–210μ thick, composed of branched septate hyphae, hymenium sharply demarked from the top layer by a region of densely packed, entangled, darker colored threads at the base of the hymenium, usually 35–40μ thick but often up to 170μ thick in old specimens, hymenium at first composed of hyaline, loosely packed, slightly curved threads, 2–3μ thick, and irregularly arranged probasidial cells, in old specimens exceedingly compact, composed of densely packed threads, probasidia, and irregular branched cells which perhaps are probasidia which failed to germinate normally. Probasidia usually with a distinct stalk, often borne in clusters, usually pyriform, 6–8 x 11–14μ, germinating to form a club-shaped basidium which is thickest at the distal end. Basidia 4.2–5 x 33.6–47μ, 4-celled. Spores about 4 x 14.7μ, bent-elliptic (only a few seen).

Associated with *Chionaspis corni* Cooley, some of which are parasitized by irregularly coiled hyphae, while others are left to reproduce.

In material collected in the fall, the healthy scale insects on the bark as well as beneath the fungus are exceedingly abundant. Beneath practically all the scales of the healthy insects there were vast numbers of eggs which are most likely intended for over-wintering. It would be interesting to see if the eggs unprotected by the fungus covering overwinter as well as those beneath the fungus. This insect has a comparatively thin scale and it is possible that eggs protected only by the scale could not resist either prolonged desiccation or severe freezings.

Since the fungus does not penetrate at all into the bark of the dogwood, it appears to be entirely dependent upon the insects for its food supply, drawing this indirectly from the shrub through the parasitized insects.

This fungus (Dr. Dearness, coll., 3396) was examined by Burt (1916) and placed under *S. pseudopedicellatum* Burt (Mo. Bot. Gard. Herb., 43802). The two plants, however, are very distinct in a number of ways and are easily separated by a casual examination. In *S. pseudopedicellatum* Burt the subiculum is white throughout, the pillars are dark colored, long and straight, unbranched, narrow, and are extraordinarily distinct, as distinct as in any species of *Septobasidium.* These characters, together with the reproductive organs, will serve to distinguish *S. pseudopedicellatum* Burt from *S. Carestianum* Bres.

The present species is very close to *S. alni* Torrend. In color characters *S. Carestianum* and *S. alni* may be separated, but only after an examination of a large amount of material representing all stages of development. The color of *S. Carestianum* is usually a lighter shade of brown than is that of *S. alni,* often being cream colored. In *S. alni* the margin is usually determinate and sometimes (especially in specimens on pecan) the new growth region is a very distinct yellowish clay color as contrasted with the older dark brown region of growth, whereas *S. Carestianum* usually has a more or less indeterminate margin and shows no distinct color zones. Another important point of distinction is in the way in which the hymenia are formed in the two plants. In *S.*

alni the hymenium as a rule extends almost out to the margin of the subiculum and does not show gaps and irregularities as in *S. Carestianum* in which the hymenium is incompletely formed in a number of areas during the period of growth. Microscopically the two plants are different in the size of the hyphae of the pillars and top layer and the size and shape of the probasidia and the size of the basidia and spores.

This plant is also related to *S. filiforme* but may easily be separated from that plant by the coiled basidia in *S. filiforme* and the presence of the abundant delicate threads with brownish contents in the hymenial region in the latter plant. In *S. Carestianum* there may be found a few threads in the upper layer with dark contents. Such threads are much thicker than the threads found in the hymenium of *S. filiforme.*

This plant may be distinguished by the thin fructification, the peculiar marginal characters, and the pyriform and frequently very irregular probasidia.

Specimens examined:

Italy: Type (?), on branches of *Salix,* 1897, part in N. Y. Bot. Gard. Herb.

Location ?: On *Salix incana* Schrank with *Chionaspis salicis* (L.) Sign., Aug. 15, 1917, Otto Jaap, leg., part in U. S. D. A. Herb. (v. Höhnel, 820).

France: Epinay, on *Cornus sanguinea* with *Chionaspis salicis,* July, 1906, in Patouillard Herb., Harvard University, spores present, 3.8 x 13.8μ; *S. Carestianum* Bres. forma *Fraxini* Pat., on trunk of *Fraxinus excelsior,* October, 1909; shows more distinct pillars but otherwise agrees with type.

Canada: North of London, Ontario, on living limbs of *Cornus stolonifera,* collected by Dr. Dearness, no. 3396, Sept. 25, 1911, not in fertile condition, no. 9115; also July 10, 1929, no. 9116, fertile; August 23, 1929, no. 9117, no basidia found; November 11, 1929, no. 9118, not fertile; March 22, 1930, no. 9119, not fertile; Sept. 11, 1931, no. 9120, not fertile.

Brazil: Sao Leopoldo, Dec. 16, 1934, J. Rick, coll., in U. N. C. Herb., 9902, 9903.

Septobasidium Carestianum var. **natalense** n. var.

Plate 44, fig. 8, and pl. 114, fig. 4

Resupinate on the smaller twigs and leaves of *Citrus,* largest patches up to 6 cm. long. Texture tender and fragile. Color of younger parts avellaneous to wood brown or Sayal brown, oldest parts near cinnamon brown. Surface with a fine bloom under lens, wrinkled and irregular in the older parts. Margin usually determinate but sometimes indeterminate, usually dotted with minute whitish tufts of hyphae. In section 200-630μ thick, stratose in the older specimens by the formation of more than one hymenium. With a thin white subiculum which becomes thick and compact forming a stromatic tissue where there are insects. The pillars which are indistinct in the marginal region are nearly entirely replaced in the older parts of the context by loosely arranged, obliquely ascending branched hyphae which are pale brownish, 2.1–4.2μ thick, rarely up to 5.4μ thick. In the specimens examined there is a narrow horizontal streak, the old hymenial layer, 20–45μ thick, composed of deep brownish hyphae and brownish probasidia. In some sections two of these old hymenial layers were evident. The context above the first hymenium is composed of more compactly arranged hyphae than below. Hymenium 30–63μ thick, not composed of a compacted layer, with dark and light-colored hyphae in the surface. Probasidia spherical, 11.3–15.1μ thick, becoming brownish in old specimens. Basidia long-cylindrical, pointed at the distal end, 4-celled, 5-6.7 x 62–71μ, with short sterigmata. Spores bent-elliptic, 4.2–5 x 21–23.1μ.

Associated with scale insects some of which are parasitized by irregularly coiled haustoria.

The variety differs from typical *S. Carestianum* in having a less compactly arranged hymenial layer and in having spherical probasidia instead of pyriform ones. Also the probasidia in the present form are borne singly and not in clusters and the peculiar elongated bodies found in the hymenium of typical *S. Carestianum* are lacking here. The basidia and spores are similar to those of the type but much larger.

Specimens examined:

Africa: Natal, Pieter Maritzburg, on *Citrus* sp., coll. by J. van der Vywer, May 19, 1932, in Herb. Div. of Plant industry, Pretoria, South Africa, no. 26321, and U. N. C. Herb. 10425.

Septobasidium alveomarginatum n. sp.

Plate 44, figs. 13, 14, and pl. 98, figs. 1–5

Resupinate, covering several (3–6) sq. cm. on the bark or leaves of *Citrus* sp. Color of current season's growth pale ochraceous buff to vinaceous buff (Ridgway), the older regions a dirty dark brown, near bister. Surface smooth where the hymenium is well formed, areolate where it is poorly developed; the older parts broken by cracks and hence having a scaly appearance. Margin determinate, pure white with a shiny appearance, dotted here and there with minute short pillars; margin of the top layer irregular in outline. The lateral margins show a distinct and characteristic, thick spongy or areolate condition. In section 450–625μ thick, usually about 550μ thick; becoming stratose with the second season's growth by the formation of a new context and hymenial layer over the old growth; composed at first of three layers: (1) the subiculum which extends over the bark, 25–50μ thick, composed of whitish hyphae; (2) the pillars which are rather few and short, 25–60 x 60–80μ; (3) the top layer, which is formed over the pillars, 230–420μ thick and thus making up the greater part of the context, composed of very loosely woven, branched, septate hyphae, 3.8–4.4μ thick, without clamp connections, pale olive brown under microscope; the hyphae of the context may extend more or less vertically to the hymenium or may be arranged in irregularly anastomosing columns to form the spongy or areolate areas mentioned above. Hymenial layer extending over the upper part of the top layer and contrasting strikingly with the context in color, the hymenium being hyaline under the microscope. Hymenium 33–42μ thick, composed of densely packed threads which are hyaline toward their tips, 1.8–2.4μ thick, and which bend over and intertwine with each other giving the surface its smooth texture; and probasidia and basidia in various stages of development. Probasidia 11.5–16μ, most about 12.6μ thick, subglobose, sprouting and becoming entirely empty to form the basidia. Basidia straight, 4-celled, 5.8–6.4 x 42–54μ, sterigmata short. Spores not seen.

Associated with scale insects, some of which are parasitized by haustoria which are in the form of irregularly coiled hyphae. Other insects are not parasitized and are in all stages of development.

Upon superficial examination this species may easily be taken for *albidum*, as has been done by Patouillard and Rick. In the buffy color and the narrow, whitish, marginal region of the subiculum the two plants are similar. However, the true *S. albidum* has a deeper, richer buff color. The surface texture of the two plants also differs. In the present species the surface of the hymenium when well developed is inherently smooth and almost shiny, whereas in *S. albidum* the surface is minutely pulverulent. In microscopic characters the two plants are very distinct: in true *S. albidum* the hymenial region is composed of much septate, more or less parallel, upright hyphae which are comparatively thick, while in the present species the hyphae in the hymenium are very delicate and intertwined. The hymenial layer in the latter species is very thin (33–42μ thick) and papery as compared to true *S. albidum*. The basidia in the two species are quite distinct. Also the two fungi occur on different species of plants, so far as the collection records show.

The present species is very closely related to *S. pseudopedicellatum* Burt, the hymenial characters, including the probasidia and basidia, being quite similar. The two species may easily be distinguished by macroscopic characters. The present species is much lighter in color, thinner in section, and has much less conspicuous pillars than has true *S. pseudopedicellatum.* Also the striking, honeycomb appearance of the lateral margins of the present species would easily serve to differentiate it from *S. pseudopedicellatum* Burt. Further evidence that this plant is not merely the South American form is that true *S. pseudopedicellatum* does occur in Brazil. This fungus is also close to *S. Carestianum* Bres. but can be distinguished by the macroscopic characters. In *S. Carestianum* the context is thinner and never shows the areolate or spongy condition frequently found in the present species, and *S. Carestianum* appears to be a smaller and more delicate fungus.

Specimens examined:

Brazil: Sao Leopoldo, on *Citrus*, Oct. 13, 1932, J. Rick, coll., type in U. N. C. Herb. 9364; also as *S. albidum* Pat. in Pat. Herb., on leaves and branches of *Citrus*, Rick, coll., in Farlow Herb. and U. N. C. Herb. 10424; also Parecy Novo, on leaves and stem of *Citrus*, Rick, coll., in Farlow Herb. as *S. albidum*; also two other collections from Rick, in U. N. C. Herb., nos. 10419a and 10422.

Septobasidium Galzini Bourdot. Matér. Fl. Myc. de France, Ass. fr. Av. des Sc. p. 575, 1921.

Plate 44, fig. 9, and pl. 100, figs. 3–5

Resupinate, covering extensive areas, frequently girdling the branches and extending for 20 cm. or more. Adhering very closely to the bark and sometimes appearing as a mere discoloration on the bark. Varying in color from wood brown in the lighter parts to bister or Vandyke brown. Surface smooth in places but mostly irregular due to the presence of numerous irregular small mounds or tufts or plates of hyphae. Mounds 1–3 mm. wide and up to 1 mm. thick, frequently with a finely spongy appearance under a lens. Under these mounds there are usually one or more scale insects. Margin determinate, but very thin and nearly concolorous with the bark and hence very indistinct. The marginal region is sparsely marked by minute horizontal plates of hyphae supported by one or more stubby pillars. In most specimens the plates anastomose along their margins, thus forming the top layer; in other specimens the top layer is formed directly over the bark without any distinct subiculum or pillars. In section either composed of only one layer, i.e. without subiculum or pillars, or composed of three distinct layers: (1) A very thin subiculum, (2) a few, short, stubby pillars, and (3) a top layer composed of one to several horizontal layers of hyphae. About 100μ thick when composed of only one layer, up to 650μ thick when composed of subiculum, pillars and top layer. Subiculum 20–65μ thick, composed of brownish hyphae 3–4μ thick; pillars

60–160 x 60–80μ, hyphae of pillars 4–4.2μ thick. The pillars branch out above to form the top layer which may consist of 1–3 horizontal strata of hyphae, 80–350μ thick; hyphae of top layer fairly compactly arranged, entangled, 4–5.4μ thick, deep brown in the lower part of the top layer becoming paler and densely packed and entangled toward the surface. Without a sharply delimited hymenial zone. Hyphae of the hymenial region are densely entangled and extend mostly in a horizontal direction. Probasidia spherical, 10–12.6μ thick, giving rise to a cylindrical, 4-celled basidium, 4–5 x 44–50μ; sterigmata short, about 3μ long, and inconspicuous. Apparently new probasidia may be formed within the old ones. Spores bent-elliptic, 3.2–4.2 x 12.4–15.4μ.

Associated with scale insects some of which are parasitized by haustoria in the form of irregular coils while others are free from parasitism.

This species appears to be closest to *S. alni* var. *squamosum*, agreeing with that species in most characters. The present species differs, however, in being darker in color and in having the surface roughened by numerous irregular mounds. Another striking characteristic of the present species is that the mounds show a finely alveolate or spongy structure.

The fungus also shows a relationship to *S. castaneum* Burt as noted by Bourdot and Galzin but *S. Galzini* is much thinner in structure than *S. castaneum.*

Specimens examined:

France: l'Aveyron, on *Calluna vulgaris* and on *Erica cinerea* when the latter is in contact with the *Calluna*, Galzin, coll., June 9, 1913, and May 12, 1914. In Bourdot's Herb. nos. 20212 and 20221, also in U. N. C. Herb.

Septobasidium ugandae n. sp.

Plate 57, figs. 11, 12, and pl. 99, fig. 1

Resupinate, forming small circular firm but easily separable patches up to 3 cm. in diameter, the lighter areas near cinnamon brown, the darker patches near fuscous or clove brown, surface smooth or finely pulverulent in some specimens, distinctly bumpy in others, and in the older specimens the central parts may be broken by minute cracks. Margin abruptly determinate, irregular in outline and thus minutely fimbriate, the white subiculum extending beyond the darker top layer about 0.5 mm. In section 200–450μ thick, composed of three distinct regions. In some specimens (the cinnamon brown ones) composed of a subiculum and short stubby pillars which give rise to the top layer over the surface of which is formed the hymenium, in other specimens (the clove brown ones) the context arises from exceedingly thick tufts of hyphae or directly from the subiculum. In the latter the context is cut up by anastomosing chambers and tunnels. Subiculum 21–84μ thick, composed of compactly arranged hyaline to pale brownish threads, which are septate and 3–3.8μ thick. Pillars 60–168 x 40μ, composed of compactly arranged, upright, pale brownish to hyaline hyphae, 3.2–4μ thick. The pillars branch out above to form the top layer, 96–140μ thick, considerably thicker in the specimens without distinct pillars, being up to 240μ thick. Hyphae of top layer 3–4μ thick, compactly arranged in the lower part of top layer, loosely arranged, in the middle region, ascending to form the hymenial region, the hyphae of which are indistinguishable from those of the context except they are more compactly arranged and darker colored. Hymenial region 40–50μ thick, hyphae compactly arranged, entangled, about 3.2–4μ thick, recurved at the tips. Probasidia globose to ovoid, when globose 12.6–14μ thick, when ovoid 10.5–13.4 x 13.8–28.5μ; germinating and becoming empty to form a four-celled, cylindrical basidium, the distal cell of which is pointed at the end, 7.4–8.4 x 44–54μ; sterigmata short; spores slightly bent-elliptic, 4.2–5.4 x 18–25.2μ.

Associated with scale insects some of which are parasitized by regular coils as in *S. Burtii*.

This species resembles *S. Burtii* in microscopic details and is undoubtedly closely related to that species. The two are readily distinguished by gross characters. The present species being much smaller and having as a rule a nearly smooth surface. On the basis of color characters the two are sharply distinct.

The present species looks somewhat like *S. Patouillardii* and *S. indigophorum* but may easily be distinguished.

Characterized by the thin structure, the small size, the cinnamon brown to clove brown color, the smooth or slightly bumpy or minutely cracked surface, the irregular probasidia, the pointed basidia, and the regularly coiled haustoria.

Specimens examined:

Central Africa: Uganda, coll., T. D. Haitland, on live *Loranthus* sp., 1915, type in Kew Herbarium and U. N. C. Herb.

Septobasidium leprosum Couch. Jour. Elisha Mitchell Sci. Soc. **51:** 42. 1935.

Plate 29, figs. 4, 5, and pl. 80, fig. 11

Resupinate, forming rather small irregular patches on the bark of the younger limbs of living trees, usually not more than two or three centimeters in diameter, rarely as much as four or five, white. Surface irregular in young specimens, the marginal region dotted with numerous, small (about 1 mm. in diameter) apparently isolated patches of growth, marked in the older regions by numerous sinuous, anastomosing channels; as growth proceeds these channels are more or less closed so that numerous unconnected round or irregular pin holes are left in the surface; in the still older regions of growth the pin holes are closed. On *Ilex* and *Hicoria* the top layer never becomes a smooth continuous surface but remains in the form of small patches 1 or 2 mm. wide (hence the name *leprosum*). Margin indeterminate. In section exceedingly thin, 60–130μ thick, rather indistinctly differentiated into layers or regions except where there are chambers or tunnels. In such regions composed of a very thin subiculum about 12–20μ thick, from which arise a few thick (usually about 65μ), short (about 45μ tall) pillars which partly support the top layer, the roof of the chambers or tunnels; top layer 40–50μ thick. Threads septate, without clamp connections, hyaline; about 4.2μ thick in subiculum, with thickened walls, threads thinner in hymenial region, 2.1–2.5μ thick, with thin walls. Hymenium formed over the entire surface, composed of probasidia and basidia and simple, slender threads. Probasidia globose or slightly pyriform, 10–11 x 10–13.8μ, hyaline, with slightly thickened wall; formed during the early winter and germinating in the early spring to form a long, cylindrical, four-celled basidium. Probasidium becoming entirely empty during germination and persisting as an empty sac. Basidium 6–6.5 x 50–56μ, straight, sprouting a short sterigma from each of the four cells and normally bearing four spores. Spores white, smooth, bent-elliptic, 4.2–5 x 12.6–15μ.

Symbiotically associated with scale insects, some of which are parasitized by irregular, gnarled coils of hyphae and others of which are not parasitized.

This species can be easily recognized by its white color, very thin structure, by the irregular holes in the plant body, and by the numerous, small, apparently isolated, patches of growth around the margin.

This plant shows a remarkable adaptation to a symbiotic relation with scale insects. In the plant body there are numerous chambers connected by tunnels which open to the exterior through the irregular holes described above. The floor of these chambers and tunnels is formed by the thin subiculum and the roof is formed of a thin but compact mat of threads which bear the hymenium. The roof of each chamber is supported by one or two thick short pillars. This adaptation can be better seen if the development of the fungus is followed. If a young specimen or the margin of an old specimen is examined one sees numerous, small, apparently isolated, patches of growth. If these patches are examined under a dissecting microscope, they are seen to resemble the snow huts of Eskimos in shape. These little huts arise by upgrowths of hyphae from the subiculum. As growth proceeds the tips of the hyphae bend over toward each other, finally coming in contact and forming an arch. These little huts are about a millimeter apart. As development goes on, the roofs of the huts branch out and grow toward each other until the space between the huts is closed, except for numerous tunnels which connect the huts with each other, and connect the older huts with the newer ones, and thence with the exterior.

These huts are the homes of the scale insects. The houses containing parasitized insects finally become completely closed while the houses of the healthy insects remain open, communicating with the exterior through the tunnels. As the young hatch out they crawl out to the new houses and settle down there.

This species is commoner than my collection data indicate. It is so inconspicuous as to be easily passed over.

Specimens examined:

Virginia: Mecklenburg Co., on *Cornus florida* with *S. filiforme*, 9179.

North Carolina: Chapel Hill, on *Ilex decidua*, several collections from same tree, 9150, 9352, 9372, type; on *Cornus amomum*, Jan. 20, 1928, J. N. Couch, coll., 8302A; on *Fraxinus* sp., March 8, 1929, P. A. Rhodes, coll., 8387; on *Crataegus* sp., Dec. 20, 1931, J. N. Couch, coll., 9149.

South Carolina: Near Charleston, on *Cornus florida*, March 19, 1930, J. N. Couch, coll., 8463; Springwood, on *Hicoria*, July 2, 1928, J. N. Couch, coll., 8324a.

Louisiana: Near Baton Rouge, on *Ilex decidua*, Dec. 31, 1931, J. N. Couch, coll., 9171.

Septobasidium Curtisii (B. and D.) Boed. and Stein. Bull. Jard. Bot. Buitenzorg III, **11**: 181. 1931.

Glenospora Curtisii Berkeley and Desmazières. Journ. R. Hort. Soc. **4**: 243. 1849.

Plates 16, 68, figs. 5, 6, pl. 73, figs. 2–4, and pl. 86

Resupinate, thin, growing in effused patches which often cover an area equal to 30 or 40 square centimeters. Surface nearly smooth in some specimens but more often dotted with minute areolations, considerably cracked in old specimens (in some specimens, especially those on oak, the surface is roughened by upright pillars of hyphae); usually black with a purplish tint (about plumbeous black of Ridgway) but often fuscous. Margin variable, sometimes thick, terminating abruptly and then concolorous with the main surface, usually thin during the growing season and of a lighter color than the main surface, whitish with a purplish sheen toward the outer edge, and sometimes indeterminate. Marginal region dotted with numerous patches of elevated hyphae which sometimes are star-shaped, umbrella-shaped, tent-shaped or shaped like an Eskimo hut. In section about 200–400μ thick, sometimes compact throughout but more often composed of three more or less distinct regions: (1) the subiculum which grows horizontally over the bark, very thin, hyphae of subiculum 4–5μ thick in marginal region with much anastomosing between threads; (2) the middle region which arises from the subiculum sometimes as short thick pillars, the threads of which quickly branch out to form the top layer; (3) top layer 60–300μ thick, the thickness depending upon the age, threads of top layer and pillars about 4μ thick; in young specimens composed of only one hymenial region while in old specimens the top layer may be stratified by the successive formation of several hymenial layers. Hymenium composed of probasidia and basidia. Probasidia formed during the fall and early winter (young, small, thin-walled Oct. 10, mature Dec. 10), germinating into the basidia during the spring and early summer; mature probasidia 10.8–16.8μ thick (most about 12μ thick), spherical, wall about 1.5μ thick, often with numerous minute furrows on the inner side which give the wall, in a section view, the appearance of being pitted; basidia 6.3–7.6μ thick by 62–70μ long, counting the stalk, straight, thickest in the middle, often breaking off from the probasidium and its stalk; stalk 10–20μ long; basidium becoming divided by three transverse walls into four cells, each cell sprouting a short pyramidal sterigma, which is very peculiar in that, after the formation of the basidiospore, the sterigma remains full of protoplasm and then buds several very minute sporidia; basidiospores hyaline, usually bent-elliptic, becoming once or thrice septate, 3–4.2 x 13–21μ, up to 29.4μ long.

Associated with *Chionaspis sylvatica* Sanders, *C. gleditsiae* Sanders, *Chrysomphalus obscurus* Comst., and *Aspidiotus* sp. on a large variety of trees and shrubs, being perhaps the most widely distributed and commonest species in the United States, also reported from Java on tea by Boedijn and Steinmann.

The present species because of similar habitat and color may be confused with *S. Patouillardii* Burt. Burt's material of *S. Patouillardii* was collected on living branches of *Fraxinus*, *Liquidambar*, and *Nyssa*, which hosts I have found to be very common ones for the present species. Burt (1916) describes his plants as being "aniline-black at first, becoming fuscous in the herbarium," colors which agree with the present species. The two plants may be separated by gross and microscopic characters. *Septobasidium Patouillardii* always shows a distinct differentiation into three layers, the top layer being supported by pillars. In the present species the layered condition is often quite indistinct and the plant is never distinctly three-layered throughout. In probasidial and basidial characters the two plants are quite distinct. In *S. Patouillardii* the probasidia are small and elongated in shape; in the present species they are large, distinct, thick-walled, and globose. In the former the probasidial cell elongates to form a two-celled basidium, the probasidial cell therefore does not persist after the formation of the basidium; while in the present species the four-celled basidium sprouts from the probasidial cell, the latter persisting for a long time as an empty cell after the basidium has matured.

The fungus-insect combination causes considerable damage to certain trees and shrubs, particularly species of *Fraxinus* and *Nyssa*. The commonest type of injury shows itself as a cracking of the bark with a marked hypertrophy of the underlying tissue and sometimes the injury may show itself in the form of witches' brooms.

Specimens examined:

New Jersey: Newfield, on *Nyssa multiflora*, winter 1874, J. B. Ellis, coll., probasidia abundant, 10–14μ thick, many 12.6μ, no basidia. Associated with *Chionaspis* sp., many of which are not parasitized and are giving birth to young. Typical. In U. S. D. A. Herb., De Thümen, Mycotheca Universalis, 292.

Maryland: Hyattsville, on *Nyssa*, F. L. Scribner, coll. Typical.

Virginia: Clarendon, on living limbs of *Nyssa sylvatica*, April 16, 1923, J. R. Weir and W. W. Diehl, colls. Two other collections from Va.; also near Roanoke, April 21, 1931, C. L. Shear, coll.; all in U. S. D. A. Herb.

North Carolina: Exceedingly common from mountains to coast, represented by 60 collections, the only species common in mountains, J. N. Couch, coll; also collected by W. W. Diehl, on *Nyssa*, by F. L. Stevens on *Juglans*, and by P. O. Schallert, U. S. D. A. Herb.

South Carolina: Common, represented by 14 collections.

Georgia: Near Vaughan, on *Ilex decidua*, Nov. 3, 1933, B. B. Higgins, fertile, 9336; near Valdosta on *Nyssa*, 9300; near Darien, Ravenel, coll., 333, probasidia very abundant, 10.5–13μ thick, often slightly rough, no basidia seen, typical; also on *Carpinus*, U. S. D. A. Herb.; near Cairo, on *Pyrus communis*, Ogara and Worth, colls., Feb. 2, 1903, U. S. D. A. Herb.; near Fort Valley, on *Viburnum*, Jan. 19, 1922; also March 18, 1922, J. C. Dunegan, coll., in L. O. Overholts Herb., 8114, also in U. S. D. A. Herb.

Florida: Common as far south as Sebring. Near Gainesville, on *Fraxinus*, October 4, 1933, G. F. Weber, coll., Univ. Fla. Herb., 9412; several collections on *Nyssa*, J. N. Couch; also on deciduous tree, Calkins, coll., winter of 1886, U. S. D. A. Herb.

Mississippi: Pascagoula, very common on *Nyssa* and *Fraxinus*, J. N. Couch.

Alabama: Near Auburn, on *Nyssa*, F. S. Earle and C. F. Baker, colls., 2240, Jan. 16, 1897, typical; also on *Nyssa sylvatica*, Dec. 5, 1922, L. E. Miles, coll.; both in U. S. D. A. Herb.

Louisiana: On *Magnolia* sp., Langlois, coll., Nov. 4, 1888, many empty probasidia, no basidia; near Pointe a la Hache, on *Carpinus*, April 20, 1888, Langlois, coll., many probasidia, no basidia; on *Gleditsia triacanthos*, Jan. 5 and 6, 1886, Langlois, coll., material abundant, probasidia 10–13μ, some minutely spiny, no basidia seen (Langlois, 179); near Monroe, on *Nyssa sylvatica*, Jan. 30, 1928, E. L. Dennison, coll., all in U. S. D. A. Herb.; also near Opalousas, on *Carpinus* (?), April 16, 1932, C. L. Shear, coll., fertile, the best fruiting material I have yet seen of this species, probasidia 12–16.8μ thick, spores up to 30μ long, in U. S. D. A. Herb.; near Madisonville, on *Nyssa*, Nov. 6, 1932, C. A. Brown, 553, also in U. N. C. Herb., 9953; near New Orleans, very common on *Nyssa* and *Fraxinus*, Dec. 31, 1931, J. N. Couch, coll.

Arkansas: Near Little Rock, on *Gleditsia triacanthos*, 9374, and *Quercus phellos*, 9375, April 12, 1933, D. V. Baxter, coll. The collection on *Gleditsia* is not typical, having a much better developed top layer than usual.

Guatemala: On angiosperm tree, Dec. 28, 1906, W. A. Kellerman, coll., pillars very abundant, about one millimeter tall, hymenium forming a smooth surface over tops of pillars and hence not typical, probasidia very abundant, spherical, 8.4–12.6μ, most about 9μ thick, color and wall characters as in Chapel Hill and Ravenel plants, basidia not seen. Before positive determination can be made, basidia should be studied; in U. S. D. A. Herb.

Guadeloupe: Near San José, on *Erythrina*, May 1915, Condert, coll., in Patouillard Herb. Harvard Univ., with *S. Leprieurii* (Mont.) Pat. This material is much like the above collection from Guatemala.

Java: Marywattie Estate, on *Thea*, July 1930, Steinmann, coll., Herb. Hort. Bot. Bog., 11821, also in U. N. C. Herbarium, probasidia present, no basidia, many upright spines, no smooth hymenial surface.

Septobasidium ficicolum Pat. (Herbarium name)

Plate 44, figs. 5, 6

Resupinate, forming very thin, closely adherent incrustations on the bark of the smaller branches of *Ficus*. About clove brown (Ridgway) in the older parts, lighter toward the periphery. Margin determinate, whitish, inconspicuously dotted with tufts of hyphae which branch out above to form the top layer, margin of top layer with irregularly fimbriate margin. In section 250–370μ thick, composed of three layers: (1) the subiculum, which forms a very thin (12–30μ thick) closely adherent layer over the bark; (2) the middle region composed of stubby pillars or a continuous region of upright hyphae; the pillars, where distinct, 50–210μ thick by about 70–90μ tall; (3) the top layer 100–150μ thick. Hyphae of context 2.1–4μ thick, usually about 2.8μ, rather straight, only slightly branched, deeply brownish in color. Probasidia formed in the upper part of the top layer, 8.4–12.6μ thick, spherical. Basidia long-cylindrical, 4.6–5.2 x 52–58μ, thrice septate; sterigmata (only a few seen) about 4μ long. Probasidia remaining as an empty cell after the basidia have been formed. Spores elliptic, slightly bent at the distal end, 4–4.3 x 14.8–17μ.

Associated with scale insects some of which are parasitized by irregular coils.

This species may be distinguished by its thin structure, color, and short pillars. The marginal region is also distinctive.

Specimens examined:

Philippine Islands: Luzon, Mt. Maquiling, on *Ficus*, J. Lebunas, coll., Sept. 30, 1920, Herb. Bureau of Science no. 10243, type in Pat. Herb.

Septobasidium Burtii Lloyd. Myc. Notes **7**: 1286, fig. 2902, plate 296. 1924. *S. retiforme* of American authors but not of Patouillard.

Plates 1–4, 28, 69, 73, and text-figs. 1–60

Resupinate, forming rather tough and hard, more or less circular patches of growth, varying from 0.1–15 cm. in diameter, sometimes even larger, the patches sometimes anastomosing so that large limbs or trunks of small trees may be almost covered by the fungus. Color typically brown, the younger parts drab, the older darker brown, frequently mottled with lighter and darker spots. In some specimens the color is grayish. Margin characteristically determinate and whitish during the growing season. Surface typically formed of branched ridges which radiate in an irregular fashion from the center outwards. Between the ridges are valleys which are partially covered by thin flaps. The surface is also frequently marked by fairly distinct annual, concentric growth zones which are usually 2–3 mm. wide. The outer margins of the concentric growth rings are usually bordered by the above mentioned flaps. In section up to 2 mm. thick, composed of three regions: (1) the subiculum, forming a thin (35–60μ thick), closely adherent mat of densely packed threads on the surface of the bark and from which arise (2) the numerous anastomosing ridges composed of compacted more or less upright hyphae (about 4μ thick) which give rise to and support the top layer. Between the ridges are tunnels which radiate outward from the center; (3) top layer which varies from a fraction of a mm. up to 1.5 mm. thick. The lower surface of the top layer is black. The entire layer is composed of compacted hyphae which are of rather uniform diameter, 3–4.2μ, usually about 4μ thick, branched and with frequent septa, extending in a more or less vertical direction. Threads throughout brownish except in the regions of new growth where they are hyaline. Hymenium formed over the top layer, 40–65μ thick, composed during the fertile season of probasidia and basidia in various stages of development and recurved hyaline hyphae. Probasidia hyaline, subglobose, ovoid to pyriform, most frequently ovoid, 8–12 x 16–24μ, sprouting to form a cylindrical, thrice septate basidium. Basidia usually with a long stalk 8–15μ long, pointed at the distal end, thickest in the middle, frequently separating from the stalk when mature, 6–8 x 50–65μ. Each cell of the basidium gives rise to a short sterigma on the end of which is borne a spore.

Symbiotically associated with scale insects (*Aspidiotus osborni* and sometimes *Chrysomphalus obscurus*), some of which are parasitized by haustoria in the form of very regularly coiled hyphae, each coil being connected by a fine thread; the coiled hyphae eventually connecting with hyphae composed of spindle-shaped cells joined by fine threads. Some insects are not parasitized and are to be found usually in all stages of development.

This species is quite distinct from *S. retiforme* (Berk. and Curtis) Pat. (Bull. Soc. Myc. Fr. **16**: 55, 1900-no. 384 in Cuban Fungi represented by two collections, nos. 288 and 214, in the Kew Herbarium and Curtis Herbarium, one of which, no. 214, is not a *Septobasidium* but either a sterile lichen or blue green alga), as was first noted by Lloyd (1924). *Septobasidium Burtii* is usually predominantly drab with a dark brown center, while *S. retiforme* is cinnamon brown with a powdery gray sheen. The shape of growth also differs: in *S. Burtii* the growth is usually circular, in *S. retiforme* of irregular shape. The surface of *S. Burtii* is reticulately veined, the veins are rounded and adjacent veins are connected by a rather thin layer of hyphae and thus the top layer forms a continuous covering except for numerous holes which are partly covered by flaps; thus the subiculum in *S. Burtii* is not exposed except at the margin. In *S. retiforme* the surface is also reticulately veined but the surface of the veins is somewhat flattened without rounded edges, and the subiculum between the veins is exposed. In texture *S. Burtii* is hard and tough while *S. retiforme* is soft and flocculent. Unfortunately no one has seen mature basidia in *S. retiforme* but in some of the pieces of no. 288 the probasidia have elongated to form immature twisted basidia, some of which have become once septate. The probasidia and basidia of *S. retiforme* are quite distinct from those of *S. Burtii.* Furthermore, in *S. retiforme* the hyphae are loosely packed and frequently curved somewhat circinately and extend upwards at an angle of about 45°; in *S. Burtii* the hyphae are closely packed, and more or less straight and vertical. The hyphae of *S. Burtii* are slightly thicker than those of *S. retiforme.* In section the thickness of *S. Burtii* is characteristically variable due to the elevated ridges and depressions, being up to 2 mm. thick; in *S. retiforme* the thickness varies from 250–400μ. A very striking difference between the two fungi is in the haustoria. In a small fragment of *S. retiforme,* no. 288 from Patouillard's herbarium, one parasitized scale insect was found and upon soaking in KOH and then crushing, the haustoria were seen to be in the form of irregular coils as in *S. pseudopedicellatum,* thus proving beyond possibility of doubt that the fungus which has been called *S. retiforme* in the United States is quite distinct from the *S. retiforme* (B. & C.) Pat.

This species is easily recognized by its circular shape, tough, hard texture, drab color, and reticulately veined surface.

It causes marked damage to infected trees, badly infected limbs being killed after three or four years. Numerous spraying experiments have been carried out with this fungus over a period of several years but the only satisfactory way of eradicating the fungus and insects is to prune the infected limbs.

Coker (1920) first found and illustrated the germinating probasidia, basidia, and spores. Other studies showing the symbiotic relationship between the scale insects and fungus in this species have been published by the writer (1929, 1931).

This is one of the commoner species of *Septobasidium,* having been reported from the District of Columbia southward into Florida and westward as far as Texas. It is such a common and distinct species that I have taken only a few specimens on my collecting trips, hence my collection data give no conception of its abundance.

Specimens examined:

North Carolina: Near Chapel Hill, over forty collections during all seasons on *Quercus palustris, Q. phellos, Quercus* sp., *Liquidambar styraciflua, Acer saccharinum, Pyrus communis,* and other unidentified hosts; Smith Island, on *Quercus virginiana,* Dec. 28, 1921, J. N. Couch, coll., 5927.

South Carolina: Near Charleston on *Quercus nigra,* March 1930, J. N. Couch, coll.; near Springwood, on *Pyrus communis,* July 2, 1928, J. N. Couch, coll., 8737.

Georgia: Near Wrens, on *Quercus nigra*, March 11, 1932, J. N. Couch, coll., 9227.

Alabama: Near Auburn, on *Prunus*, 1897, Earle and Baker, colls. In New York Bot. Gard. Herb. A variety with nearly smooth surface.

Louisiana: Near St. Martinsville, 1890, Langlois, coll. (Flora Ludoviciana, no. 2233), in New York Bot. Gard. Herb. Apparently this is the collection seen by Patouillard and identified by him as *S. retiforme* (B. & C.) Pat. (Bull. Soc. Myc. de Fr. **16**: 55. 1900).

Brazil: Near Sao Paulo, on Meliaceae, Nov. 6, 1934, J. Rick, coll., 9894. Agrees with typical material except spores and basidia larger. Spores 4.2–5.4 x 21–29.4μ; basidia 8.4–9.2 x 50–63μ.

Septobasidium Burtii var. **acerinum** n. var.

Resupinate, forming small (up to 2 cm. wide) circular patches, near cinnamon brown to clove or blackish brown, surface nodulose, sometimes broken into sinuous segments by the incomplete formation of the top layer. Margin of subiculum determinate, margin of top layer distinctly fimbriate. In section up to 0.6 mm. thick, composed of three layers: (1) the whitish subiculum from which arise, (2) thick tufts of hyphae which branch out above to form, (3) the top layer. The top layer, as in *S. Burtii*, is composed of thick and thin areas but unlike *S. Burtii* there are no breaks in the thin areas. Hymenium, probasidia, basidia, spores, and haustoria as in *S. Burtii*.

This variety may readily be distinguished from the species by the entirely different gross structure of the two. The present variety is doubtless a hybrid formed by the crossing of *S Burtii* with some other species. Good evidence for this is that some segments of growth show a striking surface resemblance to *S. Burtii*.

Specimens examined:

North Carolina: Chapel Hill, on bark of living *Acer saccharinum*, Feb. 28, 1930, probasidia, basidia and spores abundant, no. 8438.

Brazil: Sao Leopoldo, Rick, coll., Nov. 6, 1937, on Meliaceae, no. 9894. Described as typical in my 1935 publication, but agrees well with the variety.

Septobasidium Mülleri n. sp.

Plate 99, fig. 5

Resupinate, forming circular patches 3–5 cm. in diameter, frequently girdling the small twigs. The mature part is clove brown or mummy brown to dusky drab with lighter brownish areas here and there and a whitish determinate margin, 3–5 mm. wide. Surface with low bumps and sinuous ridges and with a few very inconspicuous (visible only with a lens), half-moon-shaped openings to the interior. In the marginal region the surface of the top layer curves downwards fusing with the subiculum, leaving a few minute openings which connect with minute tunnels and chambers. The margin has a slightly pleated or fimbriate appearance. In section 400–600μ thick, composed of a subiculum and a context over which the hymenium is formed. Subiculum 35–50μ thick, composed of nearly hyaline to brownish, compactly arranged (pseudoparenchymatous in section), longitudinally extending hyphae from which arises the context. Context composed of rather compactly arranged, entangled, ascending, hyaline to brownish, branched hyphae, which are 3.4–3.8μ thick. Threads of context pass insensibly into the hymenium, about 20–50μ thick, composed of dark-colored, much entangled, compactly arranged hyphae and a few probasidia. Probasidia subglobose, usually irregular in shape, about 16 x 21μ; basidia cylindrical, 4-celled, 6.7 x 42μ (only one seen).

Associated with scale insects some of which are parasitized by haustoria in the form of regular coils as in *S. Burtii*.

Even in dried material it is possible to note here some interesting points in the fungus-insect relationship. The young insects settle down on the floor of the tunnels. As the insect grows the chamber in which it is imprisoned enlarges and an opening is formed (the half-moon-shaped openings) through the fungal tissue to the top surface. Such a course of development apparently occurs only if the insect is a fertilized, non-parasitized female. As the young hatch they may crawl out through the tunnels or may emerge to the top surface through the half-moon-shaped openings. Several young insects were seen in the tunnels leading to these openings and others were seen which had dried up on the surface of the fungus. The formation of the holes in the top surface is related entirely to the presence of a healthy female insect beneath that region of the fungus. Indeed no such hole is formed over the body of a parasitized insect. It is possible even in the dried material to find numerous stages in the development of these openings. The first external sign that a slit-opening is going to form in the fungus is the appearance on the surface of an elongated tuft of white hyphae. If a needle is inserted in this region, one finds that the fungal tissue gives readily and thus it is easy to dissect away the brownish tissue and to follow the white tissue downwards. Upon so doing one invariably finds that the whitish tissue ends around a non-parasitized, half grown, female insect. Apparently the presence of the insect causes the fungal tissue to grow in such a way that the split openings occur.

This species may be recognized by the clove brown color mottled with lighter areas, the wide whitish margin which slopes down and unites with the subiculum, leaving a few minute exits from the tunnels, by the surface which would be smooth but for the minute bumps and ridges. The minute, half-moon openings in the surface and the lack of pillars are also characteristic. This species is easily distinguished from *S. Burtii* on surface characters and also it is much thinner in section. Furthermore the probasidia in *S. Mülleri* are more irregular in shape than in *S. Burtii*.

This species is probably a hybrid formed from *S. Burtii* and another species. In fact on the surface of one of the pieces there is what appears to be a second hybrid descendant of *S. Burtii* and another parent. This is shown in the photograph (Plate 99, fig. 5).

Specimens examined:

Brazil: Viçosa, A. S. Müller, coll., no. 683, in Cornell Univ. Herb. and U. N. C. Herb.

Septobasidium Cavarae Bres. Ann. Mycol. **3**: 164. 1905.

Plate 53, fig. 15, and pl. 100, figs. 6, 7

Resupinate, forming orbiculate patches. Color about Mars brown (Ridgway), bordered by a conspicuous, whitish, pubescent margin about 2 mm. wide, which is abruptly determinate. In section 600–1,000μ thick, compact, except for a series of low, labyrinthiform tunnels between the subiculum and the context. Composed of a thin subiculum from which hyphae arise obliquely to form the context over which is formed the hymenial region. Hyphae of context septate, branched, entangled, brownish, 4–4.6μ thick. Hymenial region 40–75μ thick, concolorous with the context, distinguished from the latter only by the more closely packed and slightly thinner hyphae and the probasidia. The outer threads of the hymenium, however, are hyaline. Probasidia sometimes nearly spherical, 12.6μ thick, usually elongated and irregular in shape much as in the type of *S. alni*. Only a few pieces of basidia seen, about 5.4μ thick. On Bresadola's label there is a figure of a long-cylindrical, 4-celled basidium showing four sterigmata and three spores. Bresadola gives the basidia as 8 x 60–75μ. Several spores seen; bent-elliptic, about 4.2 x 16.8μ thick (Bresadola's figures, 5 x 27–30μ).

Associated with scale insects some of which are parasitized by irregularly coiled hyphae, while others are free from fungal infection.

The type is very close to *S. alni* Torrend originally described from Portugal but differs from that species in the absence of pillars, in being over twice as thick, in having a whitish cottony margin, and in having larger hyphae, probasidia, basidia and spores. The specimen from Tunis on *Pistacia* agrees with the type only in being a shade of brown and in having the same surface characters. The margin is distinctly different in the two fungi, being distinctly fibrillose in the collection from Tunis. In section about 400μ thick with a thin subiculum, scattered, short pillars, and a top layer 120–370μ thick. Hyphae 3.6–4.4μ thick. Hymenial region not distinct from the top layer. Probasidia subglobose, pyriform or elongated, characteristically irregular in shape, often in clusters, about 12.6μ when subspherical, about 12 x 21μ when elongated. Basidia abundant, separating easily from the probasidia, cylindrical, 4-celled, 5–6.7 x 54–60μ, most about 6.2 x 58μ. Spores 4.2–5 x 16.8–21.8μ, bent-elliptic. Associated with scale insects as in type. This plant is closer to *S. alni* than to the type of *S. Cavarae*, differing from the former only in marginal characters. It will be necessary to wait until more material can be studied before a decision can be made about whether we are dealing with one or two species.

Specimens examined:

Sardinia: On branch of *Pistacia lentiscus*, October, 1900, Cavara, coll., type in Bres. Herb., Pat. Herb., N. Y. Bot. Gar. Herb., and U. N. C. Herb.

Tunis: On *P. lentiscus*, May 2, 1901, in Bres. Herb., Pat. Herb., and U. N. C. Herb. Perhaps a distinct species close to *S. alni*.

Septobasidium Leprieurii (Mont.) Pat. Bull. Soc. Myc. Fr. **16**: 55. 1900.

Plate 25, figs. 5–7, and pl. 71, figs. 6, 7

Resupinate, covering extensive areas, up to 10 or 15 sq. cm., on the bark of living trees. Bister, Rood's brown, or clay color to tawny olive (Ridgway). Surface inherently smooth and velvety but frequently with pin holes and numerous sinuous or anastomosing cracks due to the incomplete formation of the top layer; in old specimens the surface frequently becomes broken by numerous fissures. Margin determinate, in some specimens the margins of both the top and bottom layers end abruptly at the same place as in the type material, without pillars, but with irregular small (about 1 mm. wide) plates of hyphae which arise from the subiculum and grow upward and backward to unite with the margin of the top layer. These plates are partly supported by tufts of hyphae. In section 500–800μ thick, frequently stratose due to the formation of more than one hymenial layer. Composed of three regions in the younger parts: (1) a subiculum from which arise (2) the short, scattered, tufts of hyphae which in turn give rise to and support (3) the top layer. In the older parts the tufts of hyphae are inconspicuous, due to the growth of hyphae which more or less fill the space between the tufts. The middle region is thus composed of a spongy context. Hyphae of context brownish, septate, about 4.2μ thick. Hymenium 33–44μ thick, composed of recurved threads, and probasidia and basidia in various stages of development. Probasidia globose to pyriform, sometimes subtended by an enlarged cell, 9–10μ thick, giving rise to a cylindrical, 4-celled basidium 4.2–4.6 x 28–35μ, probasidium remaining as an empty cell after the formation of basidia. Basidia usually falling away from probasidia, leaving a short stalk attached to the latter. Sterigmata arising laterally, one from each cell; spores small, bent-elliptic, 3.8–4 x 11–13μ, rarely becoming septate.

Associated with scale insects (*Aspidiotus diffinis*), some of which are parasitized by irregularly coiled hyphae.

Until studying the type of *S. Leprieurii* (Mont.) Pat., I have considered the present species as a hybrid form of *S. castaneum* Burt and have segregated it with the latter species. Unfortunately the type material of *S. Leprieurii* is sterile (I found one basidium, 4-celled, cylindrical, 5 x 33μ) but some of my collections (nos. 8317, 9254) agree almost perfectly with the type in color and vegetative structure. The type material is bister to Rood's brown with a smooth shiny surface. There are no pillars but only short tufts of hyphae toward the margin. These tufts can be seen only by turning over the top layer. The margins of both top and bottom layers end abruptly at the same place as in some of my specimens. In the older regions the context is composed of loosely packed hyphae and is stratose in places by the formation of more than one hymenial layer. Hyphae of type about 4.2μ thick. From the similarities in appearance and vegetative structure of the type of *S. Leprieurii* and the present species it is almost certain that they are the same but this can finally be decided only by the collection of fertile material of *S. Leprieurii* from the type locality.

The present species is very closely related to *S. castaneum* but may be distinguished from that species by the larger size, thicker structure, darker color, and larger probasidia, basidia, and spores of the latter. Typical *S. castaneum* is chocolate brown with a faint tint of purple, while *S. Leprieurii* is bister or Rood's brown, sometimes almost clay color. The marginal conditions of the two plants are also very distinct. In *S. castaneum* there are pillars on the margin and usually three horizontal layers are evident towards the margin, while in *S. Leprieurii* no pillars are evident and there are only two horizontal layers near the margin. Also in *S. castaneum* the marginal region is frequently distinctly fimbriated or sometimes fibrillose, while in *S. Leprieurii* the margin is smooth or indistinctly fimbriate, except with a lens. The two species occur in the same region but usually on different trees, *S. castaneum* most frequently being found on *Quercus*, while *S. Leprieurii* is usually on *Magnolia virginiana*. In one of my collections on *Magnolia virginiana*, however, both species occur, in some places actually growing intermingled.

Specimens examined:

Guiana: Perhaps type, in Patouillard Herbarium, Harvard Univ.

Guadeloupe: Near San José, on *Erythrina*, May 1915, Condert, coll. There are two species in this material. The black one is *S. Curtisii* (not typical) and the chocolate brown one is perhaps *S. Leprieurii* In color this resembles *S. castaneum* Burt but its very thin structure and lack of pillars suggest *S. Leprieurii*; another collection from the same locality on one of the Loranthaceae, May 1909, no. 151, is quite distinct from the type material of *S. Leprieurii*, in Pat. Herb. at Harvard Univ.

South Carolina: Near McBee, on *Magnolia virginiana*, June 30, 1928, J. N. Couch, coll., 8317, fertile; near Bethune, on *Magnolia virginiana*, March 20, 1932, J. N. Couch, coll., 9299, probasidia and basidia as in 8317, clay color.

Florida: Near Homestead, on *Albizzia* sp., with *Aspidiotus herculeanus*, March 9, 1933, O. D. Link, coll., Univ. of Fla. Herb., 9024, U. N. C., 9362; near Avon Park on *Magnolia virginiana*, March 13, 1932, J. N. Couch, coll., 9254; near Kissimmee, March 13, 1932, J. N. Couch, coll., 9238.

Mississippi: Pascagoula, very common on *Magnolia virginiana*, Jan. 1, 1932, J. N. Couch, coll., nos. 9184, 9188, 9201 (with *S. castaneum*), 9205, 9207.

Septobasidium lilacinum Burt. Ann. Mo. Bot. Gard. **3:** 343. 1916.

Plate 53, figs. 10–12, and pl. 97, fig. 4

Resupinate, about 6 cm. long by $1\frac{1}{2}$ cm. wide; about tilleul buff (pallid smoke-gray according to Burt) with a very faint lilac tint. Surface inherently smooth in places but here and there with numerous small bumps and ridges. In the older parts with crescent-shaped or irregular cracks which may be formed either by the incomplete growth of the upper part of the fungus or by the splitting of the tissue. Margin determinate, concolorous with the other surface areas of fungus. The ridges arise about two mm. back from the margin. In section 300–630μ thick, composed of context and a distinct hymenial layer; a distinct subiculum layer is not evident. The context is streaked with an irregularly horizontal layer of darker colored hyphae. Hyphae of context much entangled, ascending in diagonal direction, considerably branched, usually hyaline and characterized by the narrow diameter, about 2.1–2.8μ thick. Hymenium rather sharply distinct from the underlying hyphae; composed of densely arranged, circinate hyphae and probasidia and basidia in various stages of development. Hyphae of hymenium hyaline and of about the same diameter as those of subiculum or slightly thinner. Probasidia spherical, hyaline, rather thin-walled, 5.4–6.8μ thick. Probasidial cell persisting as an empty sometimes partially collapsed cell after the formation of basidia. Basidia 5.4–6.3 x 17.6–28μ, straight or more often slightly bent, distinctly larger at the distal end, usually 4-celled but sometimes 3-celled. Spores elliptic, slightly bent or flattened on one side, 3.8–4.2 x 15.9–16.8μ. Sterigmata not seen.

Associated with scale insects which are very abundant in the material studied by me. Some of the scale insects are parasitized by haustoria in the form of hyphal segments connected by fine threads (the strings of sausage type of haustoria), while others are free from parasitism. The hyphae which pass through the insect's derm making the connection between the external fungus and the internal haustoria are arranged in unique fashion. The penetrating hyphae after passing through the insect's skin branch out radially in several directions and the chitinous deposit around these hyphae give this region a star-like appearance. This species may be distinguished by the buff or grayish color with a faint lilac tint, by the compact context, the fine hyphae, the distinct hymenial layer with its circinate hyphae and very small probasidia and basidia.

In addition to the type material there are three other collections in the Missouri Botanical Garden Herbarium identified by Dr. Burt as *S. lilacinum* Burt. None of these, however, agree with the type. Two of these collections are from Bayamon, Porto Rico, on the trunks of grapefruit trees, collected by Dr. J. A. Stevenson (nos. 6742 and 6291). These two represent a new species, which I am naming *S. Stevensoni.* For a comparison with *S. lilacinum* see under *S. Stevensoni.* The other collection from Sabana Grande, San Domingo, on *Citrus sinensis*, collected by Dr. Stevenson is *S. griseum.*

Specimens examined:
Trinidad: Maraval Valley, Port of Spain, Thaxter, coll. Type.

Septobasidium nodulosum Pat. Bull. Soc. Myc. Fr. **31**: 73. 1915.

Plates 53, figs. 13, 14, and pl. 114, fig. 2

Resupinate, forming irregular patches 5–40 mm. across. The smaller patches convex but as the fungus develops the mounds become connected by fungal tissue so that the older patches consist of a smooth layer with here and there a mound which rises slightly above the surface. Snuff brown or lighter in the younger parts, about Hays brown or darker in the older parts. Margin of subiculum indeterminate, margin of top layer determinate. In section up to 2.5 mm. thick; composed of an exceedingly thin subiculum, about 20–30μ thick, from which the mounds arise, starting as inconspicuous flaky upgrowths. The mounds spread out horizontally, the edges of adjacent mounds anastomosing. Between the mounds and the subiculum numerous shallow anastomosing tunnels are left. Hyphae of context brownish, thick-walled, 4–6μ thick. Hymenial region (poorly developed in all specimens) composed of upright, rather straight hyphae with hyaline tips, probasidia (only a few seen) borne on the sides of upright hyphae, globose or subglobose, 13–17μ thick; basidia (only one seen), cylindrical, 4-celled, 7.1 x 67μ, the probasidial cell becoming empty during the formation of the basidium.

Associated with scale insects, some of which are parasitized by irregular coils while others are free from infection.

Characterized by the thick, firm structure, the nodulations, and perhaps by its habit of growing on bamboo. Also the upright hyphae in the hymenial layer are distinctive.

Specimens examined:

China: Tonkin, Cho Ganh, on bamboo with scale insects, Duport, 209, Feb. 1919, type in Pat. Herb., Harvard Univ.

Philippine Islands: Luzon, Mt. Maquiling, on bamboo, Abesamis, coll., Oct. 26, 1920, no. 10634, in Pat. Herb.; also same location on bamboo, O. A. Reinking, coll., in U. S. D. A. Herb., no. 10921.

Septobasidium pinicola Snell. Mycologia **14:** 55–60, pls. 11–13. 1922.

Plate 18, figs. 9–13, and pl. 75, figs. 12–14

Resupinate, forming small (2–3 cm. in diameter) roundish patches in the angles of the limbs on *Pinus strobus*. Army to Natal brown in the central part with a lighter buff colored margin. Surface smooth at first but becoming strigose toward the marginal region, usually remaining smooth in the center. The margin 1–2 mm. wide and more or less smooth. In section 400–1500 mm. thick, lacunar, spongy, not distinctly divided into regions. Hyphae 3–7.1μ thick, usually about 4.2μ thick, much septate, without clamps, sometimes slightly constricted at the joints, walls of hyphae slightly thinner than usual in the genus. Probasidia borne over entire plant body in my specimens, usually globose, 14–16.8μ thick, but sometimes subglobose, 13–16 x 15–20μ, hyaline, germinating and becoming entirely empty to form a cylindrical, 4-celled basidium 6–7.5 x 58–70μ, sterigmata about 8μ long. Spores 3–4.4 x 13–21μ, bent-elliptic.

Associated with an aphid relative (*Chermes* sp.), some of which are parasitized by irregular hyaline coils, each coil being shaped like the glomerules of a kidney, while other insects are not parasitized.

Easily distinguished by its usual habitat on *Pinus strobus*, its color, small size, and general appearance.

Described from New England and New York where it appears to be common (Snell). For collection data from this region see Snell, 1922. Also on *Pinus monticola* from Idaho, collected by Weir.

Specimens examined:

New Hampshire: North Conway, on *Pinus strobus*, W. H. Snell, coll., cotype in Mo. Bot. Gard. Herb.

Massachusetts: Shirley, on *Pinus strobus*, April 13, 1937, D. H. Linder, coll., in Farlow Herb. and U. N. C. Herb.

Pennsylvania: Near State College, on *Pinus strobus*, L. O. Overholts, coll., October 27, 1928, in U. N. C. Herb.

North Carolina: Highlands, on *Pinus strobus*, J. N. Couch, coll., Aug. 4, 1931, in U. N. C. Herb.

Septobasidium siparium (B. & C.) Bres. Ann. Mycol. **18**: 59. 1920.
Hymenochaete siparia B. & C. Jour. Linn. Soc. Bot. **10**: 334. 1870.

Plate 57, figs. 4, 5, and pl. 97, fig. 1

Resupinate, covering areas 2.5–5 cm. wide; about bister or warm sepia in color, except where the hymenium is formed and there with a grayish tinge; the surface marked with minute ridges in the older parts and with a distinct honeycomb appearance in the young regions. Margin determinate, fibrillose. In section 1.5–2 mm. thick, composed of three distinct regions. (1) The subiculum which is 50–125μ thick and composed of densely packed hyphae; hyphae of subiculum thick-walled, 4–8μ thick; from the subiculum arises (2) the middle region composed of a dense tangle of hyphae, which may be single or arranged in fascicles. The fascicles of hyphae branch and anastomose, thus forming a dense irregular network. The hyphae of this middle region are very variable in thickness, 3–10.6μ thick, but by far the greater number are about 8.4μ thick; thick-walled, deep brownish and frequently septate. Arising from this tangle of hyphae and covering it is (3) the top layer, 70–130μ thick, composed of hyphae which are densely and irregularly arranged, hyphae of top layer 3.2–4.2μ thick, branched, thin-walled, brownish or hyaline. The probasidia are not abundant and are scattered through the upper part of the top layer. Probasidia spherical to pyriform, 9–11.8μ thick, germinating and becoming empty to form a cylindrical or slightly curved basidium, which is four-celled, 4.6–5.4 x 35–42μ. Spores about 3.8 x 12.6μ, bent-elliptic, only a few seen (there are numerous bent-elliptic or crescent-shaped spores present which I do not believe belong to *Septobasidium*, at least they do not have the appearance of basidiospores.)

Associated with scale insects, some of which are parasitized by haustoria in the form of irregular coils while other insects are free from infection.

This fungus may easily be distinguished by the minute ridges and pin holes in the surface, by the fine irregular network appearance of the marginal region, by the unusually thick hyphae of the context. The small probasidia and basidia are also distinctive.

Specimen examined:

Cuba: Wright, coll., no. 220, type in Kew Herb. and Farlow Herb. Berkeley and Curtis state that this species was on dead wood which is undoubtedly a mistake since scale insects were present.

Septobasidium fumigatum Burt. Ann. Mo. Bot. Gard. **3:** 340. 1916.

Plate 23 and pl. 72, figs. 1–5

Resupinate, covering extensive areas on the trunks and limbs of trees, sometimes extending from the base nearly to the top of a tree; mummy brown or snuff brown, or sepia (on maple) to nearly black (on *Cornus*) where the hymenium has not been formed. Hymenium light to dark gray (pallid mouse gray or light mouse gray to mouse gray of Ridgway), contrasting strikingly with the brown where the hymenium has not formed. Hymenial surface smooth except for a few cracks, some times with numerous holes 0.3–0.5 mm. in diameter; surface without hymenium characteristically spongy or alveolate. Margin determinate or indeterminate, sometimes with fine rhizomorphs. In section 600–1400μ thick, usually about 1 mm. thick, composed of loosely interwoven, branched, septate, non-nodose hyphae, 3.8–4μ thick, which arise from the substratum to hymenial region, not differentiated into pillars. Hymenium 33–50μ thick, composed of once or twice coiled, thin-walled, hyaline threads about 3μ thick, and probasidia and basidia. Probasidia spherical to subspherical, with a hyaline wall when young, brownish and punctate when mature, 12–16μ in the greatest dimension, sprouting and emptying their entire contents to form club-shaped basidia. Basidia distinctly thicker at the distal end, four-celled when mature, 7.5–9.2 x 40–56μ, each cell giving rise to a sterigma (about 8–20μ long) on the end of which is borne a spore; spores bent-elliptic, 4.6–8 x 15–21μ, most 6.3 x 16μ (no. 8458).

Distinguished by the mouse or pale mouse gray color of the hymenium and the sepia, mummy, or snuff brown of the vegetative part, the lack of pillars, the spongy structure of vegetative part, the coiled threads in the hymenium, the probasidia with roughened walls, and basidia which are distinctly enlarged in the distal part. Related to *S. alveolatum* in vegetative structure but may be separated from that species by the smaller reproductive structures and the straight basidium in *S. fumigatum.*

Associated with *A. tenebricosus* and *Aspidiotus* sp., some of which are parasitized by spindle-shaped hyphae connected by fine threads These hyphae are of two sorts: the main threads composed of long cells. from which arise coiled or twisted hyphae composed of shorter cells. Vast numbers of the insects of both species are free from parasitism, carrying on their life histories year after year beneath the thick spongy mass of fungal tissue.

This is one of the commonest species from South Carolina southward into Florida, being most frequently found on *Acer rubrum.* In Highlands Hammock near Sebring, Fla., it was by far the commonest species, occurring with the scale insects on practically every red maple tree in the hammock. I found one tree of *Acer rubrum* with a trunk three feet in diameter, the main trunk of which was heavily infected. Frequently the stems of smaller trees may be almost completely clothed by the fungus for several feet. Such heavily infected trees were always unhealthy, showing large numbers of dead limbs in infected areas. Also occurs on mulberry, *Cornus,* and *Tilia.* Extending northward as far as South Carolina and westward as far as Louisiana.

I have studied type material (Mo. Bot. Garden Herb., 43822) collected on the trunk of living *Acer rubrum,* Santee River, near Gourdin, South Carolina, and find that my material agrees with the type. It should be pointed out, however, that in Burt's drawings of the type, figure 12, he fails to show the probasidial cell at the base of the basidium. Such drawings along with Burt's statement in the text convey the impression that the basidium is formed by the elongation of the probasidial cell and that the latter does not persist after the formation of the basidium. An examination of the type material shows that the probasidial cell becomes empty during the formation of the basidium, the contents of the former passing into the latter. The probasidium usually persists as an empty cell at the base of the basidium though sometimes it may collapse, but even then can be made out.

Specimens examined:

South Carolina: Type, Gourdin, on *Acer rubrum,* Nov. 4, 1914, C. J. Humphrey, coll., in Mo. Bot. Gard. Herb., 43822, Humphrey Herb., 2588, and Farlow Herb.; near Charleston, on *Acer rubrum,* abundant, March 19, 1930, J. N. Couch, coll., 8458, fertile; near Myrtle Beach, on *Acer floridanum*, August 29, 1931, J. N. Couch, coll., 9098; near Jacksonboro, on *Acer rubrum*, J. N. Couch, coll., 8476.

Georgia: Near Valdosta, on *Acer* sp?, March 17, 1932, J. N. Couch, 9297 and 9301.

Florida: Near Brooksville, on *Tilia* sp., Feb. 1922, J. A. Stevenson, 1641, in U. S. D. A. Herb., det. by E. A. Burt; also on *Acer*, J. A. Stevenson, 1673, in U. S. D. A. Herb., det. by J. R. Weir; near Sebring, Highlands Hammock, on *Acer rubrum*, March 14, 1932, J. N. Couch, coll., 9261, 9255, 9260, 9269, fertile; also on mulberry-like plant with white latex, 9275; also on *Cornus amomum* and *C. florida*, 9267 and 9267a.

Louisiana: East Baton Rouge parish, on *Acer rubrum*, March 19, 1932, C. A. Brown, 339, and U. N. C. Herb., 9964; near New Orleans on *Cornus amomum*, Dec. 31, 1931, J. N. Couch, coll., 9158, fertile.

Brazil: Campo Grande, on orange variety "Pera", Grillo, coll., Dec. 15, 1936. This specimen should perhaps be described as a variety. Color characters as in type. Surface slightly roughened by bumps and irregularities. Structure in section as in type but hyphae of context are more delicate and show a stronger tendency to be circinately branched. Probasidia 9.6–11.7μ, spherical, mostly hyaline and with smooth wall, sometimes brownish, thick-walled and with a minutely roughened surface. Basidia 6.7 x 40μ, otherwise typical, spores as in type but smaller, 5 x 10–14.2μ. Haustoria typical.

Septobasidium cervicolor n. sp.

Plate 50, figs. 14, 15, and pl. 97, fig. 2

Resupinate, extending for several centimeters over the bark of living trees. Wood brown to fawn color in some specimens to deep brownish drab to dusky drab in others (Ridgway). Margin determinate but irregular in outline. Surface smooth except for irregularities in the bark, felty. In section up to slightly over 1 mm. thick, usually 500–800μ thick; often stratose by the formation of several hymenial layers. Composed in section of loosely packed hyphae, the larger ones of which extend in a horizontal direction, hyphae much branched and curved; about olive brown under microscope except for a few which are reddish brown, 2.5–6.3μ, most 3.8–5μ thick, septate without clamp connections, and with rather thin walls. Scattered through the context and in the lower part of top layer are numerous bladder-like, spherical, nearly empty, thin-walled cells, 6–10μ thick. Hymenial region 80–150μ thick, composed of hyaline, or pale brownish, fairly closely packed, branched and recurved hyphae, 2.5–3μ thick, septate without clamp connections but with much anastomosing between threads. Probasidia spherical, 9.5–10.5μ thick, usually rather thin-walled but sometimes the wall may be considerably thickened and minutely rough, wall not rarely thickened on one side in old empty probasidia, usually hyaline but sometimes pale brownish, new probasidia sometimes formed in the old empty ones. Basidia cylindrical, 4.2–5.3 x 40–50μ, straight, usually with a short stalk, collapsing early, becoming 4-celled by three transverse walls. Sterigmata short. Spores 3.8–4 x 12.6–13.4μ, elliptic, flattened on one side, becoming once septate.

Associated with scale insects (*Aspidiotus* sp.) some of which are parasitized by spindle-shaped hyphae connected by fine threads, as in *S. fumigatum.* These hyphae are of two kinds: the main threads composed of long cells from which arise coiled or twisted hyphae composed of shorter, spindle-shaped segments connected by fine threads. Some of the insects were free from infection and were giving birth to young when the specimen was fresh.

Distinguished by the wood brown to dusky drab color, lack of pillars, by the curved threads in the hymenium, the peculiar haustoria, and the bladder cells in the context. Closely related to *S. fumigatum* Burt, but differing from that species in having a light fawn colored hymenium; also the probasidia, basidia, and spores are considerably smaller than in Burt's species. Some specimens of *S. cinereum* resemble the present species in color and general appearance. Such specimens of *S. cinereum* may be distinguished from the present species by the much thicker, brownish hyphae of the context, by the absence of globose bladder cells in the context, and by the thinner, more tufted, and more wavy threads of the hymenium. The haustoria within the parasitized insects are also strikingly different.

Specimens examined:

Brazil: Sao Leopoldo, Rick, 1929, type (Harvard, no. 17) in Farlow Herb., Harvard Univ., and Univ. of N. C. Herbarium no. 10383; also 10087 and 10382 from Rick (Rick's no. 9).

Septobasidium cinereum n. sp.

Plate 50, figs. 16–19, and pl. 99, figs. 3, 4

Resupinate, covering extensive areas 30–40 square centimeters or more on the bark and frequently girdling the smaller branches for a distance of 10–15 cm. or more. Very variable in color, the sterile areas, i.e. the subiculum, dark-colored, near fuscous, frequently with a greenish tint due to the presence of algal cells; the upper surface mouse gray to light gray to light mouse gray before the hymenium is formed, becoming light drab or drab to cinnamon drab or in some specimens dusky drab as the hymenium develops (gray with a faint violaceous tint when fresh according to Rick). Surface finely hypochnoid at first, becoming compact as the hymenium is formed, very irregular with bumps and cracks, the older parts cracked up into numerous irregular areas 1–5 sq. mm. in extent. In several collections the hymenial surface has the appearance of having been partly eaten away by insects. Margin indeterminate, growing intermingled with lichens and extending into the cracks of the bark. In section the fertile parts are 0.16–1 mm. thick, sometimes stratose by the formation of successive hymenial layers. Without a distinct subiculum. Context composed in the lower part of loosely arranged, obliquely ascending, mostly straight, thick-walled, brownish threads, which are frequently slightly constricted at the septa; threads 4–6.3μ thick, rarely up to 7.1μ thick. In about the median region of the context the brown threads give rise to thin-walled, much branched and entangled, hyaline or pale brownish threads, about 3μ thick. The brown threads become fewer and fewer towards the top surface though a few extend up to where the hymenium starts. Toward the upper surface the thin hyaline threads become compacted to form the hymenial region. Hymenium 40–170μ thick, threads much branched and entangled, 2.2–3μ thick, sometimes with recurved tips. Probasidia spherical, 12.6–14.7μ, hyaline, germinating and becoming empty to form the 4-celled, cylindrical basidia, which are 4.6–7.1 x 58–80μ, most 6.3 x 64μ. Spores elliptic, hyaline, flattened on one side, 5–8.4 x 12.6–17.4μ, most about 6 x 15.7μ.

Associated with scale insects (two species of *Aspidiotus* and one of *Chionaspis*) some of which are parasitized by haustoria which are of the type found in *S. Burtii* (i.e. spindle-shaped segments and perfect coils). The fungus penetrates the derm, the hyphae branching out from the point of penetration forming a star-like structure. Some insects are not parasitized. It is of interest to note that the mode of penetration and the haustoria are the same in both species of *Aspidiotus* and in *Chionaspis*.

This species according to Rick is very common on *Citrus* in Brazil, causing much damage to the trees. *Septobasidium cinereum* is quite variable in color and surface characters but in the greater number of specimens is characterized by the light drab to drab color, by the irregular, somewhat felty surface which in old specimens becomes cracked (sometimes the surface has a distinctly worm eaten appearance), by the irregular and indeterminate margin, by the lack of a distinct subiculum and pillars, by the thick brownish hyphae in the lower part contrasting with the hyaline, much branched and curved hyphae of the upper part. The haustoria and the star-shaped arrangement of the hyphae within the insects' bodies just where penetration occurs are also distinctive. Indeed after one has become adept at recognizing and dissecting out the parasitized insects the haustorial characters are the ones by which this species can be most easily and certainly identified.

This species is related to *S. fuscum*, *S. cervicolor*, and *S. fumigatum.* It may, however, be readily separated from those species.

Specimens examined:

Brazil: Sao Leopoldo, Rick, coll., ten collections on *Citrus*, in U. N. C. Herb. nos. 9408, 9409, 9886, 9896, 10040, 10088 (type), 10089, 10090, 10402. Rick reports much damage to *Citrus.* Also from Rick, same locality on unidentified tree, nos. 9356, 10092, 10404, 10405. No. 9356 has a hymenial surface composed of irregular anastomosing bumps but otherwise agrees with typical material except for slightly smaller probasidia, basidia, and spores in this form. Field studies might indicate that it should be separated out as a variety. Material of two of above collections (nos. 9356 and 9804) are also in Farlow Herbarium, Harvard University.

Septobasidium separans n. sp.

Plate 53, figs. 1–3, and pl. 100, figs. 1, 2

Resupinate over an area equal to 50 or more square centimeters on the smaller limbs. Young hymenium light brownish drab becoming a deep rich chocolate color, Mars or Van Dyke brown (Ridgway). Surface of young specimens floccose with patches of light-colored, crystal-encrusted hyphae; in mature specimens when the hymenium is well developed the surface may be smooth and nearly glabrous, but of a very delicate and fragile texture. In some specimens the surface may be exceedingly rough with ridges, rhizomorphs, and holes, not rarely with a roughly reticulate appearance. Sometimes overgrowing epiphytic mosses and liverworts. Margin usually determinate but irregular in outline. In section up to 2.5 mm. thick, usually much thinner, frequently stratose in old specimens by the formation of fresh layers of context and hymenium over the old region. Composed of the context and hymenial layer and without a distinct subiculum except in certain regions. Hyphae of the context of very loosely packed, branched, and upward ascending hyphae, threads 4–4.6μ thick, sometimes with swellings here and there, septate, the septum wall often being rather thick, anastomosing freely; golden brown under the microscope in the older parts, sometimes purplish brown. Pillars are entirely lacking but not infrequently several hyphae may be joined in a fascicle which extends horizontally through the context. Hymenial region very thin, usually not more than 20μ thick; when mature it separates from the context and may be peeled off in large thin flakes, composed of criss-cross hyaline and brownish hyphae, on which are borne the probasidia. Probasidia spherical or slightly subspherical, 10.5–13.8μ thick, thin-walled, or rather thick-walled, hyaline, sessile on the hyphae or with a very short stalk, germinating to form the basidia and remaining as an empty vesicle after the formation of the basidia. Basidia long, cylindrical, 5–6.3 x 46–71μ, usually 5.4 x 54μ, becoming divided into 4 cells by three transverse septa. Basidia usually falling from the probasidia and forming the spores while lying prostrate on the surface of the hymenium. Sterigmata short and delicate, 6–8μ long. Spores long, elliptic, slightly bent, 3.8–4.2 x 15–21μ, becoming thrice septate.

Associated with scale insects some of which are parasitized by irregularly coiled hyphae. Some of the insects in a collection from Rick (Nov. 9, 1931) contain two types of haustoria: haustoria in the shape of irregular coils and in addition sausage-shaped haustoria held together by fine threads. Evidently this collection is made up of two species of *Septobasidium* though outside of the parasitized insects the hyphae and fruiting bodies appear to belong to only one species.

This species judging from the number of collections received appears to be the commonest species around Sao Leopoldo, Brazil. It is a very distinct species and may be distinguished in the early stages by the light-colored flocculent hyphae which overgrow the bark, by the absence of a continuous subiculum and pillars, by the very thin, smooth, papery, chocolate colored hymenium, which may be renewed several times by the formation of a new hymenial layer over the old one. The fungus is frequently found overgrowing liverworts and mosses and in some specimens the surface may have an irregular honeycomb appearance. This latter condition is very exceptional, however.

Specimens examined:

Brazil: Sao Leopoldo, 23 collections, several on *Citrus aurantium*, Rick, coll., in U. N. C. Herb.; also in Farlow Herbarium, Harvard, nos. 8, 12, 15, 16, 18; also St. Maria, province Rio Grande do Sul, leg. R. Niederaner Zancho, in Sydow Herb.

Septobasidium Stevensoni n. sp.

Plate 50, figs. 12, 13, and pl. 97, fig. 5

Resupinate, covering extensive areas on the main trunk and limbs of *Citrus* trees, frequently overgrowing epiphytic mosses and liverworts. Margin apparently indeterminate. Surface of young areas cottony; but in the older parts where the hymenium is formed the surface is finely granulose and compact. Color quite variable, context and subiculum deep brownish with a purplish tint, the hymenial surface is dirty pinkish or ochraceous buff in one collection and buff to deep brownish drab in another collection. In section 300–600μ thick, sometimes up to 1000μ thick, composed of three distinct regions. First a thick region (about 300μ thick) of dark-colored, purplish brown, loosely arranged hyphae which extend over the bark more or less parallel to its surface, hyphae 4–5μ thick. From this layer of horizontal hyphae arises the second layer, composed of loosely arranged hyaline or slightly colored, about 4μ thick, branched hyphae which extend outwards more or less vertically from the basal layer, giving rise to the third and top layer, the hymenium. Hymenial region about 40–65μ thick, composed of hyaline upright or slightly curved, branched, thin-walled hyphae, 2.8–3.5μ thick, and probasidia and basidia in various stages of development. Probasidia spherical, 9.6–12.6μ thick, hyaline, persisting as an empty partially collapsed cell after the basidium has been formed. Basidia 4-celled, cylindrical, thickest at the distal end, 6.3–7.6 x 37–50μ, projecting for about half its length beyond the hymenial surface. Spores 4.2–5.4 x 13.4–16.8μ, bent-elliptic. Sterigmata about 7–10μ long.

Associated with scale insects some of which are parasitized by haustoria in the form of spindle-shaped hyphal segments connected by fine threads. This species has been identified by Burt as *S. lilacinum* Burt but is quite distinct from that species. The two differ strikingly in the context, the hyphae in *S. lilacinum* being about the thinnest of those of any species of *Septobasidium* (about 2–3μ thick) while in *S. Stevensoni* the hyphae of the context are 4–5μ thick. In the color and arrangement of the hyphae, the two fungi also differ. Furthermore, in the hymenium of *S. lilacinum* the hyphae are circinate while in *S. Stevensoni* they are upright or only slightly curved. Moreover the probasidia, basidia, and spores of *S. lilacinum* though of much the same shape as in the present species are much smaller. The haustoria are alike in both species.

Specimens examined:

Porto Rico: Bayamon, on trunks of grapefruit trees, Oct. 1, 1917, J. A. Stevenson, coll. Two or three collections in Mo. Bot. Gard. Herb. under *S. lilacinum* Burt. (One collection has the number 55075.)

Brazil: Sao Leopoldo, on *Citrus*, Rick, coll. Received Feb. 3, 1936, U. N. C. Herb. 10086.

Septobasidium spongium (B. & C.) Pat. Bull. Soc. Myc. Fr. **16:** 181. 1900.

Thelephora spongia B. & C. Jour. Linn. Soc. Bot. **10:** 330. 1868.

Plate 57, fig. 7

Resupinate, spreading for many inches, near hair brown, the hymenial surface being near cinnamon brown; the surface of the non-fertile part alveolate, while that of the hymenium is nearly smooth or pulverulent. Margin with numerous bristles, determinate. In section about 1 mm. thick when dry, 1.5–2 mm. when wet. Apparently without a distinct subiculum but with numerous longitudinally arranged masses of hyphae which give rise to hyphae which pass obliquely upwards, sometimes being arranged in fascicles, at other times arranged in plates forming a dense spongy structure with vacant spaces about 0.5 mm. wide. Hyphae of context 4.8–5.4μ thick, fuscous under microscope. Over the context the hymenial layer is formed, 75–150μ thick; composed of pale hyphae, 2.8–3.8μ thick, irregularly branched and curved and sometimes coiled. Probasidia spherical, hyaline, 9.2–9.6μ thick, becoming empty to form a straight or slightly bent, cylindrical, 4-celled basidium. Basidia 5.8–7.1 x 25–30μ. Spores (only 2 seen) bent-elliptic. There are a considerable number of bladder-shaped cells in the context, each cell with a small columella-like structure projecting into it.

Associated with scale insects.

This species is distinguished by the thick spongy texture, the cinnamon brown hymenium with the small probasidia and basidia.

The type material is so mixed with other growth as partially to obscure its structure and color. There is an olive greenish growth which, when present, gives the surface a crustose appearance. There are also a good many filaments of a blue green alga and filaments of *Trentopholia.*

Specimen examined:

Cuba: On bark, Wright, coll., type in Curtis Herb. and Kew Herb. Also fragment in Mo. Bot. Gard. Herb. and Pat. Herb.

Septobasidium sabal-minor Couch. Jour. Elisha Mitchell Sci. Soc. **51:** 19. 1935.

Plate 22 and pl. 82, fig. 2

Fungus body forming small, brownish, cottony, resupinate growths on the stems, leaves, and fruits of *Sabal minor.* The patches on the fruit are from 3 to 10 sq. mm. in area and always surround the pedicel scar. Margin determinate, surface spongy. In section up to about 1 mm. thick, not composed of distinct regions. Threads of context 3.6–4μ thick. Hymenium rather indistinctly set apart from the context, composed of sparingly branched hyphae, probasidia, and basidia. Probasidium spherical to pyriform, 8.4–10.5 x 10–12μ, frequently with an enlarged subtending cell, germinating and emptying its entire contents to form the basidium. Basidium cylindrical, divided by three transverse walls into 4 cells, 4–4.6 x 33–42μ. Sterigmata 4–5μ long. Spores bent-elliptic, no septate ones seen.

Associated with scale insects, *Aspidiotus* (?) *sabalis* Comst., some of which are parasitized by irregular coils. This species somewhat resembles *S. sabalis* in gross structure and resembles *S. alni* in the structure of the probasidia, basidia, and spores, but can easily be distinguished from both of these species.

Specimen examined:

Florida: 15 miles north of St. Augustine, on fruits of *Sabal minor*, Nov. 20, 1933, Miss M. L. Bomhard, coll., type in U. N. C. Herb., 9719, and in U. S. D. A. Herb.

Septobasidium acaciae Sawada. Nôji Shikenjô Tokubetsu Hôkoku Taiwan (Formosa), no. 2: 103–104, pls. 11, 12, Nov. 1911 (Japanese). Translation in Mycologia **10:** 88. 1918.

Plate 97, fig. 3

Resupinate, forming an exceedingly thin, adherent, and extensive covering over the bark. Snuff to Sayal brown or darker. Surface conspicuously dotted with numerous oblong or rounded, dark, somewhat elevated dots, 0.3–1 mm. wide, under each of which there is a scale insect. These dots vary in color from black to concolorous with the rest of the fungus. The surface of the dots has a spongy appearance while the rest of the surface is more or less smooth. Margin determinate or indeterminate, whitish. On the marginal region there are formed numerous oblong or irregular "fungal houses" in which some of the young insects settle down. In section 75–150μ thick, composed of one layer of fairly closely packed, entangled, pale brownish threads, about 4.2μ thick. Hymenial region paler, about 16–25μ thick, composed of compact hyaline hyphae; probasidia 11–16.8μ thick, globose or slightly subglobose, wall hyaline and up to 2μ thick. Since I was unable to find the basidia and spores in material sent me by Tanaka, I am quoting from Sawada (1911) translated by Tanaka (1918): "Basidia arising from protobasidia, easily detached, cylindric, subacute above and truncate below, straight or slightly curved, hyaline, 1–5-septate, 52–81 x 4–6μ; sterigmata from each cell of basidia, 4–12μ long; basidiospore hyaline, oblong to oblong-obovoid, curved, 18–22 x 3–6μ, germinating with short tubes carrying sporidia of about the same appearance as basidiospores measuring 11–15 x 3–5μ."

Associated with scale insects some of which are parasitized by haustoria in the shape of irregularly coiled hyphae while other insects beneath the fungus are free from parasitism and are in all stages of development.

Through the kindness of Dr. Tanaka I have been able to study some specimens of this fungus from the type locality collected by Sawada. It has been reported (Tanaka, see *Mycologia*, 1918) as sometimes killing *Acacia* trees.

Easily distinguished from other species by the exceedingly thin structure and the rather conspicuous dark dots. However, this species is very close to *S. atropunctum* from Jamaica. The present species is thinner in section, and has a lighter colored surface than *S. atropunctum.* Also in the present species the surface is usually glabrous while in *S. atropunctum* the surface is, as a rule, pulverulent. The dark dots are different in the two species, in *S. acaciae* the dark color appears to be due mostly to the black insects' scales showing through the thin covering of fungal threads, while in *S. atropunctum* the dark dots are nothing more than the blackened fungal tissue which forms a roof over the insects' houses. In the latter each of the insects' houses in the early stages is open on one side. In *S. acaciae* there is no such opening.

Specimens examined:

Formosa: Taihokuchô Shakukô, Sept. 26, 1910, and Oct. 7, 1910, coll. by K. Sawada and Y. Fiyikuro on *Acacia Richii, Glochidion obovatum, Citrus* sp., also on *Prunus persica, Prunus. salicina, Thea sinensis, Salix grandulosa* var. *Warburgii* and *Melia Azedarach.*

Septobasidium atropunctum Couch. Jour. Elisha Mitchell Sci. Soc. **44**: 251. 1929.

Plates 5, 8, 13, 90

Resupinate, forming a thin, closely adherent crust over the bark, usually girdling the limb and extending for a decimeter or more along its length. Surface rather smooth in the younger areas, becoming cracked in older regions; cinnamon brown with numerous black dots (the coverings for the scale insects) which are 0.3–1 mm. wide. Margin 2–3 mm. wide, byssoid, whitish. In section, 175–265μ thick, composed of three layers: (1) subiculum 50–65μ thick, composed of a dense layer of hyphae which lie next to the bark; (2) the middle region 150–200μ thick, composed of upright hyphae which are grouped in large loose clusters forming thick pillars and leaving small spaces between the pillars; (3) hymenial region 30–60μ thick, composed of branched sterile hyphae and probasidial cells, probasidial cells 11.8–14μ thick, wall considerably thickened, spherical or slightly subspherical, sprouting under favorable conditions to form the basidia, the former emptying its entire contents into the latter during the process. Basidia 4.6–5.4 x 58–70μ, long, narrow, cylindrical, becoming thrice septate, each of the four cells thus formed giving rise to a sterigma which bears a spore. Spores 3–4.2 x 16–22.5μ, bent-elliptic, usually becoming once or thrice septate, though if kept moist often becoming divided into eight cells.

Associated with scale insects, some of which are parasitized by irregular coils of hyphae while others are not parasitized.

This is an exceedingly interesting form since it shows an obvious and rather high degree of specialization on the part of the fungus in adapting itself to the needs of the scale insects. The fungus forms a little, flattened, oval-shaped house for the scale insects. The floor is considerably depressed so that the roof is even with the surface of the fungus. The roof is made of an exceedingly compact layer of hyphae and is nearly black in color. One end of the roof is free, thus leaving an entrance. This free margin is fringed with whitish hyphae so that the opening is quite conspicuous under a binocular microscope. If the roof is lifted, nearly every house is found to contain a scale insect. It is only in the marginal regions where the fungus is developing that the houses with "open doors" are found. In the older regions the growth of the fungus has pushed up around the roof and thus more or less closed the entrance. The majority of these older houses are empty, containing nothing but the large, old insects' shells. In a few of these old empty shells, the dried bodies of young were found which were absolutely untouched by the fungus. The new houses contain only small, young scale insects.

It is obvious, although these observations were made on material which I had collected two years previously and on which I was therefore unable to follow the behavior of the living insects, that the young hatch out in the older houses and crawl out of the ones that are not completely closed, migrate out to the marginal region, enter the little houses, and there develop their sucking organs which penetrate the tree for food. In return for this protection the fungus parasitizes many of the insects.

Specimens examined:

Jamaica: Cinchona, several collections on bark of living deciduous trees, July 1926, J. N. Couch, coll. Type in Herb. U. N. C.

Septobasidium fuscum Couch. Jour. Elisha Mitchell Sci. Soc. **51**: 15. 1935.

Plate 24 and pl. 81, figs. 11–13

Resupinate, very closely adherent to the bark, covering areas equal to several hundred square centimeters, clove brown, fuscous or Brussel's brown (Ridgway). Surface characteristically smooth except for irregularities in the bark, minute holes and elevated patches of coarse, loosely entangled hyphae. Beneath each of these patches is a scale insect. Margin usually indeterminate, composed of branched threads. In section 125–200μ thick, composed of three indistinct regions: (1) the subiculum, composed of a few branched threads about 4.2–5.8μ thick, rarely up to 6.3μ thick, which adhere closely to the bark; (2) upright hyphae arising from the subiculum and branching out to form (3) the top layer; top layer 60–85μ thick, composed of branched hyphae recurved at the tips, hyphae 2.2–3.5μ thick, probasidia and basidia in various stages of development. Probasidium spherical or slightly subspherical, 10–12.6μ thick, sprouting and emptying its entire contents to form a cylindrical 4-celled basidium which is 4.2–5.4 x 35–40μ; sterigmata 4–5μ long. Spores 4–5 x 6–13.4μ, bent-elliptic.

Symbiotically associated with *Aspidiotus diffinis* Newst., some of which are parasitized by very regular coils which are connected by extremely fine threads and spindle-shaped cells, as in *S. Burtii* Lloyd; other insects are free from parasitism and are in all stages of development.

This remarkable species has appeared only twice but because of its thin structure, it may be easily overlooked as a discoloration on the bark. It may be distinguished from other species by its thin structure, clove brown color, and by the minute elevated patches of hair-like hyphae which cover the scale insects.

Specimens examined:

Mississippi: Near Pascagoula, Jan. 1, 1932, with *Aspidiotus diffinis* Newst., J. N. Couch, coll., 9192. Type.

Louisiana: Near Baton Rouge, on *Quercus* (?), Dec. 31, 1931, J. N. Couch, coll., 9177. Differs from the type in being wood brown to Brussels brown (Ridgway) and in the absence of the patches of loosely woven threads which cover the insects.

Septobasidium schizostachyi n. sp.

Plate 46, fig. 15

Resupinate on the smaller stalks of bamboo, covering extensive areas up to 15 or 20 cm. long and frequently girdling the stems. Texture firm and hard but breaking easily into small fragments. The color is a sort of dark dirty brown, the darker areas near clove brown. The surface is smooth, not shiny in the younger parts but irregularly bumpy in the older parts, the bumps being 2–3 mm. wide; in the older parts the surface is broken by numerous minute (visible only with a lens) cracks which divide the surface into irregular areas less than a square mm. There are also a few large cracks. Margin indeterminate or vaguely determinate. In section up to 600μ thick, usually about 400μ thick when mature, exceedingly thin in the younger parts, sometimes stratose. Composed of three layers. (1) A very compact subiculum, 40–80μ thick in the mature parts. The threads in the lower part of the subiculum are thinner than in the context and are so compactly arranged as to appear pseudoparenchymatous in section. (2) Context without pillars, 60–440μ thick, composed of very compactly arranged, entangled, obliquely ascending, thick-walled, sparingly septate, dark reddish brown hyphae, 4.2–7μ thick, usually about 5.2μ thick. From the context arises (3) the hymenium, 60–210μ thick, composed of an almost solid layer of upright hyphae. The hymenial hyphae are of about the same thickness as those of context and concolorous except for hyaline tips, much branched, the branches arising at a very acute angle. Probasidia spherical, 12.6–16.8μ thick. No good basidia seen but several fragments found. One fragment showed 2 good cells between which was a collapsed cell and on the proximal end was part of a fourth cell, basidia apparently 4-celled, about 6.7 x 40–50μ. Spores plentiful, bent-elliptic, 3.8–5.4 x 19–22.2μ, most about 5 x 21μ.

Associated with scale insects some of which are parasitized by irregularly coiled haustoria.

This species was recognized by Dr. Ocfemia as probably distinct from *S. Bakeri*. It may be easily distinguished from that species by its much thinner structure in section, by the absence of pillars, and by the rough, nodulose surface.

Specimens examined:

Philippine Islands: Laguna, Agricultural College, December 1933, G. O. Ocfemia, coll., no. 50, type in U. N. C. Herb. no. 10409. In this same collection there are two other new species but the material is too scanty to describe adequately. One of these occurs on the young leaf petioles and is buff colored, the other is on the stems and is reddish brown in color.

Septobasidium taxodii Couch. Jour. Elisha Mitchell Sci. Soc. **51:** 48. 1935.

Plate 42, figs. 6–13

Fungus body exceedingly thin, forming resupinate crusts on the bark of young twigs of *Taxodium*. Dirty Mars brown in color. Margin indeterminate. Surface usually roughened by cracks in the bark, or sponge-like under a lens where the fungus overgrows insects. In section about 100–250μ thick, not divided into strata and without a distinct subiculum. Hyphae penetrating into and between the dead cells of the outer bark. External threads densely entangled, 2.1–4.2μ thick, brownish, septate, branched, frequently recurved in the distal part. Probasidia 10–12.6μ thick, spherical, giving rise to a coiled (often helically) 4-celled basidium, 5–6.3 x 33–40μ (the coil about 16μ across). Probasidium remaining as an empty cell after the basidium has been formed. Sterigmata 5–6μ long on the distal cells. Spores not certainly seen.

Associated with *Aspidiotus forbesi* Johnson on *Taxodium distichum*. Many of the insects are parasitized by irregular coils, the fungus passing through the insect's body wall at the vagina and anus. Occurring with *S. pseudopedicellatum* Burt and *S. Mariani* Bres.

Easily overlooked because of its thinness and inconspicuous brownish color. Its thin structure, brownish color, and coiled basidia distinguish it from most other species of *Septobasidium*.

Specimens examined:

Louisiana: Opelousas, on *Taxodium distichum*, April 16, 1932, P. R. Miller, coll., type in U. S. D. A. Herb. and U. N. C. Herb., 9938.

Septobasidium Michelianum (Cald.) Pat. Ess. Tax. p. 10, fig. 2. 1900.
Micheli. Nova Plantarum Genera. Table 90, fig. 2. 1729.

Plate 44, figs. 1–4, and pl. 100, figs. 8–11

Resupinate, forming circular or elongated crustose patches; frequently about sorghum brown (Ridgway), i.e. brown with a purplish tint, sometimes buffy brown or sometimes with a grayish sheen over the brown. In the older parts the surface is inherently smooth except for numerous cracks which divide the surface into irregular patches 1–5 mm. in width. Towards the margin the top surface is divided into sinuose ridges which are usually less than a millimeter wide, the ridges resulting from the incompletely formed top layer. Margin determinate, whitish, with a pale lavender tint. In section 300–500μ thick, usually about 400μ thick, composed of three rather indistinct regions. The very thin subiculum about 40–50μ thick, composed of compact, horizontal hyphae 3.8–4.2μ thick, which give rise to the thick columns or ridges, the hyphae of which branch out horizontally to form the top layer. The anastomosing tunnels between the columns and ridges are rather narrow, much narrower than the ridges, about 200–500μ wide. Top layer 200–325μ thick, composed of compact entangled hyphae which are indistinguishable from those of the ridges, hyphae considerably branched, the branches often arising at right angles, 3.8–4.4μ thick. No distinct hymenial layer, the probasidia arising near the upper surface of the top layer, probasidia globose or sometimes pyriform, 12–14.7μ thick, germinating and becoming empty to form a cylindrical basidium. Basidium 5–5.6 x 46–63μ, straight, 4-celled, falling away from the probasidium sometimes and leaving a short stalk, sterigmata 5.5–8μ long. Spores bent-elliptic, 3.9–4.3 x 18–22μ. A few hyaline, thin-walled threads extend horizontally over the surface, about 3.2μ thick.

Associated with scale insects some of which are parasitized by irregular coils, while others are free from parasitism.

This appears to be a very distinct and easily recognized species, characterized by the brownish color with a purplish tint, the cracked surface of older parts and the retiform surface of the marginal region. It has been confused with *S. Mariani* Bres. and *S. Cavarae* Bres. but may easily be distinguished.

In Bourdot's Herbarium under *S. Mariani* there are two collections (no. 37453 on *Quercus Ilex*, Lamalon les Bains, May 23, 1924, also no. 37460 on *Cneorum tricoccum*, same location) which show a striking resemblance to *S. Michelianum* in the cracked surface of the older parts, the reticulate margin, and the probasidia, and basidia. They differ from type material in color, typical *S. Michelianum* having a distinct purplish tint while the two collections from Bourdot are mostly buffy brown. Moreover, in the latter two collections the fungus frequently occurs in small flaky patches, a condition which I have not observed in *S. Michelianum*. Whether these two collections belong under the present species or should be treated as a new species can only be decided after a study of more material.

Patouillard divided the genus into two groups, *A. Typicae*, in which the probasidia elongated to form the basidia, and *B. Gausapia* Fr., in which the probasidia persisted as an empty cell. The present species was put in the first group by Patouillard. However, I have examined all the material of this species in the Patouillard Herbarium and find numerous empty probasidia and stages in the germination of the probasidia which clearly indicate that Patouillard was in error on this point. It is easy to explain Patouillard's mistake. In the first place it requires much searching over large amounts of material to find the empty probasidia and the early stages in basidial formation and secondly since the basidia when mature fall off and are therefore in the mature condition seldom or never connected with the empty probasidia it would be easy to conclude that the probasidium elongates to form the basidium.

Specimens examined:

Italy: On *Laurus nobilis*, Caldesi, no. 66, type, Dr. Marcucci, coll., 1866; also on *Laurus nobilis*, Blide, April 1907; also on *Olea*, Rome, Mariani, coll.

France: Nimes, Jardin de la Fontaine, May 1922, on *Laurus nobilis* (?), Patouillard, coll. (?); same loc., Feb. 1922, on olive, not typical, being much lighter colored. All the above in Patouillard Herbarium, Harvard Univ.

This species appears to be very variable, several forms having been recognized. These are listed as follows in Saccardo **21**: 446. 1912: *a. genuinum* (on *Laurus*); *b. Alni*, on *A. glutinosa* in Tunis; *c. Carestianum* (*S. Carestianum* Bres.), on *Fraxinus*, *Salix*, *Cornus*, and *Laurus* in Italy and Morocco; d. *Phillyreae* (in Tunis); e. *Cavarae* (*S. Cavarae* Bres.) on *Pistacia*, in Sardinia and Tunis; f. *Ceratoniae*.

From this it would seem that Saccardo considers *S. Carestianum* and *S. Cavarae* as forms of *S. Michelianum*. However, I have studied typical material of *S. Carestianum*, as well as several other collections and find that it is quite distinct even in gross structure from the present species. The two differ in color and in structure: *S. Michelianum* is brownish with a purplish tint, the surface is retiform toward the margin and cracked in the older parts and characteristically without pillars, while in *S. Carestianum* the surface is papery and never cracked and there are pillars throughout. I have also studied type material of *S. Cavarae*. That species differs from the present in marginal characters, in color, and in the structure of the surface. It appears sufficiently distinct to justify its retention as a distinct species. Before a final decision can be made a study of numerous collections of fresh material should be made.

Form d. *Phillyreae,* of which I have studied a specimen collected by Otto Jaap in Dalmatia, no. 697, is typical except for lighter color. (Collected March 28, 1914, in Kew Herb., Patouillard Herb., and U. S. D. A. Herb.)

Form f. *Ceratoniae* agrees in all respects with *S. Cavarae* Bres.

Form *b. Alni* on *Alnus glutinosa* I have not been able to study.

Septobasidium fragile n. sp.

Plate 53, figs. 7–9

Resupinate, covering areas up to 10 sq. cm. in extent, bister to mummy brown, minutely pulverulent to the eye, surface very finely alveolate with a lens, irregular due to irregularities in the bark, in the young condition consisting of numerous, small, thin, brownish patches, margin indeterminate. In section usually 40-150μ thick, in the older parts where the bark is irregular and broken up to 700μ thick, without a distinct subiculum but with threads which penetrate throughout the dead layers of bark and thick, dark-colored, scattered threads which extend over the surface of the bark. From these arise a few darkly colored, loosely arranged, mostly horizontally extending threads, 4.2–6.7μ thick. From the latter arises the top layer or hymenium 40–84μ thick, composed of diagonally ascending, hyaline, branched, entangled, thin-walled hyphae, the ends of which extend more or less upright; hyphae of hymenium collapsing readily. Probasidia spherical or slightly subspherical, 10–12.6μ thick, hyaline, basidia cylindrical, sometimes slightly bent, 5.2–6.3 x 40–50μ; probasidia partially collapsing after the basidia are formed. Spores not seen.

Associated with scale insects some of which are parasitized by haustoria of the sausage-shaped type.

When first collected this species was put under *S. jamaicaense*, but though it shows the same color as a washed out specimen of the latter it may readily be separated by its much thinner structure and lack of pillars.

It is characterized by its very thin structure, indefinite margin, finely alveolate surface, hyaline and quickly collapsing hymenial surface. It is closest to *S. fuscum* but differs in surface characters and the haustoria.

Specimens examined:

Jamaica, B. W. I.: Blue Mountains, Hardware Gap, on deciduous tree, July 1926, J. N. Couch, coll., type in U. N. C. Herb. no. 8517.

Septobasidium subcarbonaceum (B. & Br.) n. comb.

Thelephora subcarbonacea B. & Br. Jour. Linn. Soc. Bot. **14**: 64. 1873.

Plate 53, figs. 4–6

Resupinate, forming a thin, closely adherent growth over the bark. Fuscous to fuscous black in color with a flocculent grayish sheen. The surface is irregular due to irregularities in the bark and has the appearance under a lens of being covered by a cobwebby material or dusted with whitish powder. In section 300–450μ thick, not divided into distinct regions; composed of rather thick-walled, fuscous colored hyphae which extend in an irregularly horizontal direction, hyphae 3.3–5.4μ, most about 5μ thick, septate. In the basal region and particularly around the insects are numerous, globose, bladder cells, each with a small columella-like structure. Intermingled with the hyphae are numerous algal cells. The hyphae of context extend into the cracks and crevices of the bark. Hymenial region rather indistinct, composed of hyaline, thin-walled, branched, wavy hyphae which arise from the thicker fuscous threads; hyphae of hymenium 2–3μ thick. From these hyphae arise the probasidia which are spherical, 9–12.1μ thick. The probasidia germinate, emptying their entire contents to form a cylindrical basidium. Basidia cylindrical, thickest at the distal end, 6.3–7.9 x 30–38μ, divided by cross septations into 4 cells. Sterigmata about 5μ long. Apparently new probasidia may be formed by proliferation within the old membranes. Spores bent-elliptic, 4.2–12.6μ, only a few seen.

Associated with scale insects, but due to the scarcity of material I was unable to determine the type of haustoria.

Petch (1925) considers this immature *Septobasidium arachnoideum* (B. & Br.) Bres., and the two fungi are strikingly alike superficially, both being much the same in color and of the same loose habit of growth. The two, however, may be distinguished. In *S. arachnoideum* the growth is more compact, the hyphae more branched, twisted and entangled than in the present species. Furthermore, in *S. subcarbonaceum* the basidia are straight while in *S. arachnoideum* they are curved.

This species shows a slight resemblance to *S. fuscum* in color, in the thinness of structure and the close adherence to the bark but the two differ otherwise in gross characteristics. Also in the present species new probasidia are formed within the old ones and the basidia are shorter and thicker than in *S. fuscum.*

Specimen examined:
Ceylon: Peradeniya, Feb. 1868, Thwaites, no. 538, type in Kew Herb.

Septobasidium Boedijnii n. sp.

Septobasidium molliusculum Sydow (sense of Boedijn and Steinmann). Bull. Jard. Bot. Buitenzorg III, **11:** 201, figs. 22, 23. 1931.

Plate 56, figs. 1–3, and pl. 106, figs. 6, 7

Resupinate, often surrounding smaller branches and extending along the mid-vein on lower side of leaves; forming small irregular patches up to 17 mm. but usually about 2–5 mm. wide. Surface light buff or pale ochraceous buff. Margin of top layer determinate, of lower layer indeterminate. In section up to 1 mm. thick, composed of three distinct regions: subiculum, pillars, and top layer. Subiculum very thin and inconspicuous, 5–27μ thick, whitish in the younger parts, becoming pale brown upon maturity, in the dried condition almost concolorous with the leaf surface. From the subiculum there grow up crater-like insect traps. Pillars characteristically long, straight, and slender, 500–900μ tall by 10–40μ thick, most about 20μ thick, hyphae of pillars unbranched except at the top, 3–5μ thick, sparingly septate; branching and spreading apart above to form the top layer which is 70–170μ thick (up to 375μ, Boedijn and Steinmann). Top layer composed of loosely packed, pale brownish, branched hyphae 3.8–5μ thick. Hymenium about 30–50μ thick (in some specimens the hymenial region may become stratose, Boedijn and Steinmann), composed of densely and irregularly packed, mostly hyaline hyphae 2.2–4μ thick, probasidia, and basidia, some of the ends of the hyphae extending up vertically slightly beyond the probasidia. Probasidia arranged at rather uniform level, spherical or more often pyriform or irregular, 16–19 x 20–23μ, rarely up to 33μ long; germinating and becoming empty to form a four-celled, curved basidium, remaining as an empty cell after the formation of the basidium. Basidia 8.8–9.3 x 32–38μ, sterigmata and spores not seen. Spores (according to Boedijn and Steinmann) elliptic, straight or curved, 6–7.5 x 20–26μ. Haustoria of the irregular coiled type.

For a comparison with *S. molliusculum* Sydow see under that species. This species is characterized by the growth in small but conspicuous patches, the buff colored top layer supported by tall, thin, straight, dark-colored pillars, by the very thin subiculum and by the four-celled curved basidia. This species appears to be closest to *S. minutulum* Sydow, both of which occur on the lower surface of leaves in small patches. There is a slight difference in color and shape of patches, *S. minutulum* having a slightly vinaceous tint and more rounded patches; in *S. minutulum* the pillars are much shorter than in *S. Boedijnii* so that in the former the patches frequently appear nearly or entirely sessile; and in *S. minutulum* the probasidia and basidia are slightly smaller.

Specimens examined:

Java: Res. Priangan, Tjibodas, on the branches and leaves of *Fagraea obovata*, Boedijn, coll., October, 1929, type. Also a collection by Raciborski near Salak on *Caryota* sp., labelled *S. vulcanium*, in Raciborski Herb. Cracow, Poland, and U. N. C. Herb.

Ceylon: Nuwara Eliya, October 23, 1937, on *Lasianthus* sp., Petch, coll., in Petch Herb. and U. N. C. Herb.

Septobasidium stratiferum Boed. and Stein. Bull. Jard. Bot. Buitenzorg III, **11**: 199, fig. 2. 1931.

Plate 56, figs. 9–11, and pl. 107, fig. 8

Resupinate, forming rounded or irregular patches which vary from less than a cm. up to 10 cm. wide. Margin whitish, myceloid, the younger growth light pinkish cinnamon to cinnamon, the older parts snuff brown to Verona brown. Surface very irregular due to the incomplete formation of the top layer and to the persistence of top layers of past seasons. Margin of top layer determinate. In section composed of three main regions: (1) the subiculum, (2) the pillars and (3) the top layer which may be stratose. The subiculum, usually 200–375μ thick, rarely as thin as 40μ and up to 500μ thick, hyphae of subiculum brownish or hyaline, compact, 3–6.3μ thick. The surface of the subiculum is usually whitish, being covered with a flocculent layer of hyaline, thin-walled hyphae, the interior deep brown to blackish. Pillars rather scattered but visible with the unaided eye, curved or straight, when straight 850–1500μ tall by 25–100μ thick, when curved up to 2 or 3 mm. long, hyphae of pillars 4–6.3μ, straight, unbranched, single hyphae pale brownish, rather sparingly septate. The hyphae of the pillars branch out above to form the top layer which may be single or stratose due to the formation of more than one hymenium over the top layer. Top layer up to 1 mm. thick, composed when stratose of successive horizontal layers (as many as five in the material observed) of dense brownish tissue (old basidia, probasidia, and hyphae) and loosely arranged supporting hyphae. Hymenium about 75μ thick, composed of probasidia, basidia, and hyaline, entangled, branched hyphae which mostly extend vertically. Probasidia globose or pyriform, 18–21 x 21–27μ (Boedijn and Steinmann describe the probasidia as 35–46μ long but I am unable to find any so long), germinating to form a 4-celled, curved basidium; the empty probasidial cell persisting as a partially collapsed vesicle with the wall usually thicker on one side and slightly roughened by granules; basidia 11.6–13.6 x 40–65μ (11–13.5 x 72.5–81μ, according to Boedijn and Steinmann), sterigmata very long, 3–5 x 30–60μ. Spores 8.2–8.6 x 21–31μ, usually 8.4 x 25.2μ, once, thrice, or seven times septate, bent-elliptic, frequently rather sharply recurved at the distal end.

Associated with scale insects some of which are parasitized by irregularly coiled haustoria while others are free from parasitism.

This species is easily recognized by its occurrence in small, very irregular patches, the thick structure, the whitish subiculum, the stratose top layer, and the rather large probasidia, basidia, and spores.

Specimens examined:

Java: Tjisaroea, on stem of *Ficus lepicarpa*, Boedijn, coll., July 1930, type in Herb. Hort. Bot. Bog. and U. N. C. Herbarium.

Septobasidium minutulum Sydow. Ann. Mycol. **12**: 195. 1914.
Septobasidium phyllophilum Sydow. Ann. Myc. **12**: 545. 1914.

Plate 55, fig. 1, and pl. 105, figs. 4, 5

Resupinate, hypophyllous, forming circular, ovoid or irregular, elongated patches 3–30 mm. in extent, the patches frequently elongated and narrow, extending lengthwise on the leaf; ochraceous buff, avellaneous or wood brown to cinnamon brown; surface at first flocculent, becoming nearly smooth or finely granulose under a lens; margin sharply determinate, the top and bottom margins usually ending at the same place though not infrequently the lower layer, either with or without a few upright pillars, may extend beyond the top layer. In section 250–420μ thick, composed of three distinct regions: (1) a very thin, pale buff subiculum usually about 15–22μ thick, hyphae of subiculum hyaline or pale yellowish, usually thin-walled, about 3–3.8μ thick, hyphae sometimes covered with granules; from the subiculum arises (2) the region of pillars, which, though visible only under a lens, are fairly abundant, 10–16 to a square mm.; pillars 20–126 x 140–270μ, composed of hyphae which are mostly unbranched, except near the top, hyphae of pillars brownish, 4–8.4μ, most 5–6μ thick, with thickened walls, thus contrasting in thickness and color strikingly with the hyphae of the subiculum: the threads of the pillars branch above extending diagonally and horizontally to form (3) the top layer, which is 60–150μ thick. In the lower part of the top layer the threads are 4.2–5.6μ thick and extend more or less horizontally. These thick threads give rise to a dense tangle of smaller (3–4μ thick), much branched threads which extend upwards to form the hymenium; threads hyaline at tips. Probasidia formed near the surface of the top layer, 12.6–17μ thick, globose, subglobose or pyriform, germinating to form a 4-celled, curved basidium about U-shaped, 9.2–10.5 x 35–43μ, probasidial cell remaining as an empty vesicle after the basidia are formed. Spores not seen.

Associated with scale insects some of which are parasitized by haustoria in the shape of irregular coils.

I have made a careful study of type material of the present species and also *S. phyllophilum* Sydow and it seems that the only bases of separation are (1) the occurrence of *S. minutulum* on the leaves of *Thea* and of *S. phyllophilum* on the leaves of *Dracaena*; (2) the patches of growth formed by *S. phyllophilum* are also generally larger and darker colored, but the shapes of the patches, as well as the color, shade into one another. In all other characters, macroscopically and microscopically, the two agree perfectly and undoubtedly are the same species.

The present species may be recognized and distinguished from other hypophyllous species by its occurrence in small rounded or linear patches, by the very thin subiculum, composed of delicate threads; by the pillars composed of very thick, dark-colored, mostly unbranched hyphae; and by the usually spherical probasidia and the bent U-shaped basidia.

Septobasidium minutulum may be confused with *S. molliusculum* since both occur on leaves. *Septobasidium molliusculum* however occurs in patches up to 6 cm. wide, apparently never growing in numerous small patches as does the present species. Also in the former the marginal region is quite distinct, being dotted with numerous small pillars. The pillars in *S. molliusculum* are taller and slenderer (as are also the hyphae of which they are composed) than in the present species. The probasidia in *S. molliusculum* are 11–13μ thick, while in the present species they are 12.6–17μ thick.

Specimens examined:

Philippine Islands: Luzon, Laguna, Mt. Maquiling, on leaves on *Thea montana*, Jan. 31, 1914, C. F. Baker, coll., no. 2786, type in Sydow Herbarium and Bureau of Science Herb., Philippines; also Prov. of Bulacan, on living leaves of *Dracaena angustifolia*, September 1913, M. Ramos, coll., Bureau Science, no. 21831, type in Sydow

Herb., Kew Herb., U. S. D. A. Herb., and U. N. C. Herb. Described by Sydow as *S. phyllophilum.*

Septobasidium Merrillii Bres. Hedwigia **64**: 303. 1915.

Plate 55, figs. 8, 9, and pl. 107, fig. 10

Resupinate, apparently covering areas equal to several square centimeters in extent. Color cinnamon brown or Prout's brown. Surface smooth and shiny. Margin apparently determinate. In section up to 2 mm. thick, composed of four distinct regions. First, the basal region, made up of a subiculum which extends over the bark, 65–100μ thick, threads of subiculum very compact, usually thin-walled, dark reddish brown. Arising from the subiculum is a zone of lighter colored thin-walled, loosely packed hyphae. About 150μ above the subiculum is a second horizontal layer about 50μ thick, which covers the scale insects. This layer is composed of dark-colored, densely packed and entangled hyphae, 3.8–4.8μ thick, and is easily separated like a membrane from the underlying tissue. Second, the region of pillars. The pillars arise from the subiculum, passing upwards through the loosely packed hyphae and the membranous layer extending vertically or diagonally upwards for 500–1000μ. Pillars not abundant. Hyphae of pillars about 4.2μ thick, parallel. Pillars branching out above to form the third region, the spongy context 500–700μ thick, which is composed of loosely packed, pale brown, branched, sparingly septate hyphae, 4.2–5μ thick. The fourth region, the hymenium, is formed over the top of the context. Hymenium 70–80μ thick, sharply delimited from the underlying tissue, composed of upright threads and probasidia and basidia in various developmental stages. Probasidium clavate, about 8.5 x 21μ, with the long axis vertical in the hymenium, giving rise to an elongated, curved, 4-celled basidium, the probasidium remaining as an elongated empty cell. During the process of germination the probasidium may elongate to over twice its original length. Basidium 8.4–9.2 x 40–50μ (Bresadola's figures for the length are 60–70μ. Doubtless he included the basal cell which would make his figures correct, according to my measurements). Sterigmata of variable length depending upon the position of the cell from which it arises. Spores not seen but Bresadola says, "oblong, bent, hyaline, 5–7 x 27–35μ."

Associated with scale insects (*Chionaspis* sp.) some of which are parasitized by irregular coils while others are free from infection. Insects parasitized through vaginal pore.

This species is easily distinguished by the color and glabrous surface, the complicated structure in section, the spongy context and the very compact hymenium. The probasidia and basidia are also distinctive. It is easily distinguished from *S. accumbens* (B. & Br.) Bres. with which it has been confused (Bresadola).

Specimens examined:

Philippine Islands: Luzon, Benguet, on living *Strobilanthes cincinnalis*, E. D. Merrill, coll., no. 6668, May 1909, type in Herb. Bot. Garden Berlin.

Septobasidium leucostemum (B. & Br.) Pat. Bull. Soc. Myc. Fr. **36**: 175. 1920.

Corticium suffultum B. & Br. Fungi of Ceylon, No. 651.

Plate 55, figs. 5–7

Resupinate, covering extensive areas, pure white at first, pale buff throughout in herbarium material. Surface smooth except for cracks or holes and with a compact membranous appearance in old regions. Texture very friable, perhaps somewhat waxy in the fresh condition. Margin indeterminate. In section 400–600μ thick, composed of three distinct regions: (1) a very thin white subiculum; (2) a region of pillars which are 21–105 x 250–340μ, composed of upright parallel hyphae, pillars arising singly or sometimes in clusters about 2 or 3 to a sq. mm., threads of pillars hyaline, about 4.2μ thick with thick walls; (3) the top layer about 60–90μ thick, composed of a lower part of loosely packed hyaline hyphae and a very thin upper crust of amorphous material in which the probasidia and basidia are imbedded. (It is quite likely that this upper amorphous layer is absent in fresh material and that it is formed by the disintegration or possible gelatinization of the upper part of the hymenium.) Probasidia ovoid, 9.6–12 x 14.7–20μ, hyaline, thin-walled, germinating to form a curved, 4-celled basidium. Probasidial cell persisting or collapsing after the formation of basidia. Basidia curved, 4-celled, 6-8 x 42–52μ. Sterigmata not seen. Spores hyaline, bent-elliptic, 6.3 x 21.8μ.

In material from China, associated with *Parlatoria proteus* (?) Curtis, the few I could find being free from parasitism by the fungus.

This species is nearest *S. Cokeri* but may readily be separated from that species by the smaller size of *S. leucostemum* and its curved basidia.

Specimens examined:

Ceylon: Central Province, Feb. 1868, Thwaites, no. 551, type in Kew Herbarium; also in Pat. Herb. Harvard Univ., sterile.

China: Province of Kwang Si, on living branches of *Citrus*, parasitized by scale insects, June 7, 1919, O. A. Reinking, coll., Patouillard Herbarium, in fruit.

Septobasidium laxum Pat. Philippine Journal of Science **10**: 85. 1915.

Plate 56, figs. 4–6, and pl. 107, fig. 9

Resupinate, covering extensive areas on the bark; color of hymenial surface about bister (Ridgway), fuscous-purplish according to Patouillard. Surface of hymenium nodulose, the nodules about 0.3 mm. wide. Marginal region of subiculum determinate, 5–7 mm. wide, grayish white, or pale slate color, surface smooth, the outer region of margin minutely fibrillose under lens. The older region of margin is covered by pillars which are unusually long and slender, 20–80μ thick by 1–2 mm. tall. Sometimes the fungus may develop rather extensively before the top layer is formed, i.e. show nothing but subiculum and pillars. On the subiculum there are developed numerous cave-like structures formed by the upright growth of v- or u-shaped plates of hyphae which bend over and anastomose at their tips; these structures are about 5 mm. wide by about 6 mm. long and average about 50 to each square centimeter. Pillars may sometimes arise from the roof of these cave-like structures. In section 1–2 mm. thick, composed of three distinct regions: (1) the subiculum which extends over the bark 100–400μ thick, hyphae of subiculum 3.8–5.4μ thick, deep brownish, densely packed, mostly rather straight, and with few branches and septations; (2) the middle region composed of pillars and the cave- or tent-like structures; pillars arising singly from subiculum or in clusters, bending over and forming a loose tangle; hyphae of pillars about 4.2μ thick, deep brown in color, branching little and with very faint apparently incomplete septations; (3) top layer arising from the ends of pillars, hence pillars usually are parallel with top layer for some distance, top layer 20–150μ thick, composed of very loosely packed, much branched and entangled hyphae which are deep brown and 4–5μ thick. No distinct hymenium formed but probasidia scattered over surface of top layer. Probasidia spherical, borne laterally or on the ends of hyphae, with a short stalk, 15–20μ thick; germinating at the side or top to form a 4-celled, bent basidium 9–11 x 35–50μ. Sterigmata about 8.4 long, about 4μ thick. Spores not seen.

Associated with scale insects, which, however, in the material studied by me were very scarce, only a few young ones seen in the newly formed insect houses.

According to Patouillard this species is characterized by the ashy gray subiculum, the scattered, distant, fibrous pillars and by the exceedingly thin, dark reddish purple hymenium. Other characteristics which should be mentioned are the long, frequently curved pillars and the numerous, minute, insect houses, the latter being visible only under a lens. Microscopically the exceedingly thin top layer, the lack of a definite hymenium, the unusually large scattered probasidia are distinguishing characters.

Specimen examined:

Philippine Islands: near Mont Maquiling, Los Banos, on branches of *Astronia Cumingiana*, Nemesio Catalan, coll.

Septobasidium filiforme Couch. Jour. Elisha Mitchell Sci. Soc. **51:** 49. 1935.

Plate 32 and pl. 84, figs. 1–8

Perennial, resupinate, effused, occurring in patches of varying size up to 20 cm. long by 5 cm. wide. Color very variable, the young margin white and quite striking and conspicuous when fresh or damp; older regions pale buff in comparatively young specimens, becoming wood brown to deep buffy brown or army brown or Natal brown, sometimes with a dusky purple tint; color fading considerably upon drying, surface usually becoming cracked after maturity into irregular patches about 1 cm. wide; the cracked edges often curling back somewhat and exposing the whitish subiculum; surface, except for cracks, very smooth. Whitish marginal region about 1–8 mm. wide, composed of young, undeveloped hymenium from beneath which a narrow border of the subiculum grows. Margin of subiculum often speckled with minute white dots,—the young pillars. In section 350–500μ thick in young specimens, up to 800 to 1000μ thick in old specimens, usually composed of three layers; in old specimens composed of a number of layers due to the annual formation of new pillars and an hymenium directly on top of the hymenium of the previous year: (1) the subiculum, which runs over the surface of the bark, white, 20–50μ thick; threads of subiculum hyaline under microscope, thin-walled, 2–3.2μ thick, septate without clamp connections; (2) the region of more or less vertical hyphae which arise from the subiculum and may adhere together to form the supporting pillars for the hymenium or may pass into the hymenium without forming the pillars; threads of this middle region 3.2–4.4μ thick, branched when not forming pillars, but threads of pillars sparingly branched, thick-walled and rarely septate; pillars very variable in diameter, composed of compactly arranged threads for only about one-third the distance between their base and hymenium, distance between subiculum and top layer quite variable (150–400μ); (3) hymenium thin, 42–60μ, rarely up to 90μ thick, composed of probasidia and basidia in various stages of development and hyaline, slightly bent threads about 3μ thick and numerous conspicuous narrow threads, with deep brown contents, which usually break up into pieces of irregular length; probasidia pyriform, borne laterally, 7.7–11.8 x 14–23μ, sprouting a long narrow twisted and often coiled basidium 4–5 x 28–40μ. Basidium usually 4-celled, of which only the first three cells form spores; not rarely 3-celled or only 2-celled. Sterigmata very long, 20–30μ, partially or completely collapsing after the spores are formed. Spores hyaline, bent-elliptic, narrowed at the mucro end, 3.4–4 x 13–21μ.

Symbiotically associated with scale insects, some of which are parasitized by irregular coils, while others are left unharmed. The scale insects are parasitized through the anus and sometimes through both the anus and the genital opening. Through these openings a bundle of brown, thick-walled threads pass and within the insect these branch out into hyaline, thin-walled threads from which arise enormous clusters of the coiled haustorial threads.

Abundant on a large variety of trees and shrubs, but easily overlooked because of its comparatively small size. This species has been labelled in our earlier collections as *S. pseudopedicellatum* but can easily be separated from that form. Distinguished by the wide whitish margin contrasting with the much darker brownish color of the older regions; by its whitish subiculum, exposed in old specimens by the cracking of the upper layers. In microscopic structure it may be distinguished by the pyriform probasidial cells, by the peculiarly twisted or coiled basidia and *above all by the narrow threads with deep brown contents which break up into rods that adhere end to end.*

Specimens examined:

Virginia: Mecklenburg Co., on *Cornus florida* and *Liquidambar styraciflua*, July 13, 1927, J. N. Couch, coll., 9179, type, with *S. apiculatum* and *S. Mariani.*

North Carolina: Chapel Hill, on *Alnus rugosa*, Jan 24, 1920, J. N. Couch, coll., 4015, with *S. pseudopedicellatum*; on *Ligustrum sinense*, Jan. 9, 1920, J. N. Couch, coll., 3923; on *Cercis canadensis*, Oct. 22, 1927, J. N. Couch, coll., 8273, very common; on *Ceanothus americanus*, also on *Hicoria* sp., October 20, 1927, J. N. Couch and W. L. Hunt, colls., 8263; on *Pyrus japonica*, 8446, 9077; on *Crataegus Marshallii*, 8290; on *Syringa vulgaris*, 9106; on *Acer Negundo*, 9105; on *Carpinus caroliniana*, with *S. Mariani*, and *S. apiculatum*, 3924, also 8190, 8237, 8291; on *Cornus amomum*, 8388, 8558, 9104; on *Cornus florida*, 8272, 8349, 9107, 9432; near Raleigh, on *Cornus florida*, May 15, 1929, B. B. Fulton, coll., 8409.

Septobasidium sp. Hybrid form between typical *S. pseudopedicellatum* and *S. Schweinitzii*. Jour. Elisha Mitchell Sci. Soc. **51**: 51. 1935.

Plate 35 and pl. 81, figs. 5–10

I have found several cases of hybridization between *S. Schweinitzii* and *S. pseudopedicellatum.* In collection no. 8319 on ash from Lynch's River bottom, S. C., we found an American ash tree mottled with *Septobasidium* from the base to the ends of the branches. Typical *S. pseudopedicellatum*, *S. Schweinitzii*, *S. Patouillardii*, and *S. leprosum* were present, the first three being more abundant. The *S. pseudopedicellatum* and *S. Schweinitzii* were hybridizing. The typical *S. pseudo-*

pedicellatum had a buffy to dark brownish surface, a whitish subiculum and long, distinct, individual pillars on the margin and between the upper and lower layers. The *S. Schweinitzii* material was not exactly typical, being lighter colored and with the sterile upright threads in the hymenium more wavy than typical. The margin of the subiculum showed the typical tent-like structures rather than upright pillars. The *S. pseudopedicellatum* was about twice as thick as the *S. Schweinitzii* and hence even ignoring the color and marginal characters it was possible easily to distinguish the two species. In a number of places on the bark the two species were growing in contact along their contiguous margins but in such pieces one could easily tell where one species ended and the other began although the margins were apparently in intimate contact. In such specimens the vegetative and reproductive structures were apparently unaffected.

In a large number of cases the two species were growing together, forming a new type of growth which combined the characters of both. Such organisms were distinct enough from both parents in vegetative and reproductive characters to justify their being described as a new species. I am considering this, however, as a hybrid form between *S. pseudopedicellatum* and *S. Schweinitzii.*

The hybrid form varies from a few millimeters in diameter to several centimeters and is about as thick as typical *S. Schweinitzii.* Color a dirty washed out buff. Marginal subiculum pale purplish and with both pillars and upright hyphae, each group of which is arranged at first in a plate with the outline of an incomplete ellipse. The sides of the plate come together above forming a miniature tent, the open part of which is always directed toward the older part of the fungus, these tent-like structures being typical of *S. Schweinitzii.* The "tents" and pillars are not scattered irregularly but a row of "tents" alternates with rows of pillars Hymenium hyaline, in some specimens composed of much twisted upright threads and probasidia and basidia. In some specimens the hybrid hymenium may be formed over the old hymenium of *S. Schweinitzii,* the hybrid threads apparently coming through the old layer of *S. Schweinitzii.* Other specimens have been seen in which the hymenium may show a resemblance in vegetative characters to *S. pseudopedicellatum.* Still other types of variation in the vegetative structure of the hymenium occur in other specimens. Probasidia spherical or subspherical, 8.4–12.6μ, averaging about 10μ thick (12 measurements), collapsing after emptying. Basidium 4.2–5.8 x 32–41μ, coiled, 4-celled, sterigmata about 3μ thick by 12μ long, usually straight, but sometimes coiled or twisted. Spores not seen.

I am unable to say whether these hybrids are formed by the grafting of the vegetative parts or by some kind of fusion of reproductive cells (bud cells from the spores). From studies on *S. Burtii* it seems very probable that hybrids are formed as follows: A scale insect hatches out beneath *S. pseudopedicellatum* or *S. Schweintzii* and accidentally picks up the spores on its body and thus becomes infected. It now crawls away from the *S. pseudopedicellatum* out on the bark and finally settles down beneath a colony of *S. Schweinitzii*. The infecting cells of the fungus grow within the insect, as it develops, and eventually after the insect has molted the last time the fungal threads of the *S. pseudopedicellatum* grow out through the anal or vaginal pore and anastomose with the threads of *S. Schweinitzii*, and thus the *S. pseudopedicellatum* is grafted onto the *S. Schweinitzii*. The course of events may proceed in exactly the opposite direction, i.e. the insects may become infected with the spores of *S. Schweinitzii* and crawl away and settle down beneath *S. pseudopedicellatum*, etc.

Septobasidium ussanguense (P. Henn.) Lloyd. Myc. Notes **5:** 722. 1917.

Hymenochaete ussanguensis P. Henn. Engler's Bot. Jahrb. **30:** 255. 1902.

Plate 60, figs. 4, 5, and pl. 114, fig. 7

Resupinate, extending for several square centimeters over the bark, girdling the smaller branches, having a tough, elastic texture and very easily separated from the bark. Surface roughened by wrinkles and cracks and in places by minute bumps. Sayal brown to snuff brown or bister. Margin determinate, whitish when fresh; with rows of pillars arranged longitudinally, the pillars being flattened with their edges pointing backwards and forwards. In section 0.8–1.5 mm. thick, sometimes stratose by the formation of a new layer of pillars and a new top layer over the old growth. Usually composed of three horizontal layers and two regions of upright pillars. The basal layer or subiculum is 40–160μ thick, usually about 80μ thick, composed of very compactly arranged hyphae, which extend longitudinally, about 3–3.8μ thick. From the subiculum arise numerous pillars which in the marginal region are arranged in longitudinal rows but which by the formation of new pillars become irregularly arranged. Pillars about 40–170μ thick by about 1 mm. tall, characteristically flattened, considerably branched, hyphae 4.2 x 5.4μ thick, closely packed. About 80μ above the subiculum the pillars give rise to a second but incompletely formed horizontal layer, the hyphae of which are rather loosely arranged, somewhat flocculent, and 4–5.4μ thick. The pillars branch out above to form the top layer, which at first is very thin, about 40μ thick, but in the older parts may be up to 330μ thick, sometimes stratified by the formation of a new hymenium directly over the old one; composed of closely packed, thick-walled, pale brownish or nearly hyaline, entangled threads, 4–5μ thick, and numerous spherical, subspherical or ovoid probasidia, about 12–18μ thick, frequently borne in clusters on the same hypha and frequently flattened by mutual pressure. Germinating to form an elongated, curved basidium and apparently becoming empty during the process. No mature basidia or spores seen. Hennings gives the spores as subglobose, 6–8μ thick, which undoubtedly is a mistake.

Associated with scale insects (*Aspidiotus*? sp.) some of which are parasitized by haustoria of the irregularly coiled type, penetration being through the vaginal opening, while other insects are not parasitized and are found in neatly arranged fungal houses between the subiculum and second horizontal layer.

Recognized by the snuff brown to bister color, the somewhat elastic texture of the younger parts, the rather thick structure, the three horizontal layers, the laterally flattened pillars, borne in longitudinal rows in the marginal region, and the irregularly shaped, clustered probasidia. The thick-walled hyphae in the hymenium are also distinctive.

The elastic texture of this fungus is remarkable but I am somewhat doubtful if this condition is natural or is due to the treatment of the type specimen by some poisoning fluid which has changed the composition of the cell wall.

Specimen examined:

East Africa: Ussangu, Sept. 18, 1899, W. Goetze, coll., type in Berlin Bot. Gard. Museum, also in U. N. C. Herb.

Septobasidium tropicale Burt. Ann. Mo. Bot. Gard. **3:** 326, fig. **2.** 1916.

Plate 60, figs. 1–3, and pl. 106, fig. 3

Resupinate, forming a thick, conspicuous, circular or ovoid growth on the bark, the one piece examined by me 2 x 4 cm. in area; color about avellaneous; surface smooth but not shiny; crustose, the central region broken by irregular cracks which may be up to 1 mm. wide. Margin sharply determinate; the outer zone, about 2 mm. wide, composed of rhizomorphs; the inner zone, also about 2 mm. wide, composed of a spongy base from which arise numerous closely set pillars. In section 1–2 mm. thick, the marginal region about 1 mm. thick; composed of three regions but without a distinct subiculum. The basal region is composed mostly of spongy tissue though in places the hyphae may be compactly arranged into a layer 350μ thick; hyphae of basal region hyaline or pale brownish, about 3–4.2μ thick. From the basal region arise diagonal plates of hyphae which extend upwards and outwards toward the growing margin at an angle of about 45°. In cross section these plates appear as sections of 5–7 distinct horizontal layers. Each plate is 30–40μ thick and several hundred microns wide. In some parts of the fungus the plates are connected here and there and thus in section show a spongy appearance. This region is from 0.5–1 mm. thick. The plates break up above to form pillars which extend vertically. Pillars abundant 30–126 x 500–1000μ, hyphae of pillars brownish, 3-5.2μ thick, septate, with thick walls. The pillars branch out above to form the hymenial layer which is 126–168μ thick. In the lower part of the hymenium the hyphae are densely entangled, about 5.4μ thick, deep brownish, and thick-walled. From these entangled hyphae arises a densely packed region of pale brownish to hyaline, upright or slightly wavy, branched hyphae which are 3–3.8μ thick and give rise to the probasidia. Probasidia apparently not passing through a resting period; at first ovoid, elongating and enlarging to become irregularly club-shaped, 8.4–10.5 x 14–34μ, germinating and becoming empty to form a curved (like a shepherd's crook), 4-celled basidium. Basidia 8.4–9.6 x 37–50μ; sterigma about 4 x 12μ, spores not seen.

Associated with scale insects some of which are parasitized by irregularly coiled haustoria; others are free from parasitism and can be found in all stages of development.

This species resembles *S. Mariani* Bres. in surface color and the superficial appearance of the marginal region. In *S. Mariani* the margin shows a distinct subiculum and the partially formed, middle horizontal layer, whereas in the present species the margin does not show a distinct subiculum and the beginnings of a middle horizontal layer. The marginal region is considerably darker in *S. Mariani* than in *S. tropicale.* In *S. Mariani* Bres. the surface is glabrous while in *S. tropicale* the surface is very finely pulverulent under a lens. This difference in appearance is due to an even more striking difference in structure. In *S. Mariani* the layer is composed of closely interwoven hyphae which extend more or less in a horizontal direction while in the present species the top layer is composed of vertical hyphae. The probasidia and basidia are also different. This species also resembles *S. ussanguense* (P. Henn.) Lloyd in superficial appearance, but the two can be distinguished. In *S. tropicale* the surface is smooth with a slight pulverulence under a lens, while in *S. ussanguense* the surface is bumpy. Also the arrangement of the pillars is quite different in the two species.

The present species is characterized by the avellaneous color, the peculiar margin with a thick, finely spongy structure, the lack of a distinct subiculum, by the hyphae arising diagonally in plate-like structures breaking up into pillars above, and by the elongated probasidia, the curved basidia, and the upright threads in the hymenium.

Specimen examined:

Mexico: Locality not stated, on bark of living branches of *Quercus,* C. G. Pringle, coll., type in Farlow Herbarium and Mo. Bot. Gard. Herb. 44590.

Septobasidium cinchonae Raciborski. Bull. Intern. de L'Acad. des Sci. de Cracovie **3:** 346–394. 1909.

Plate 65, figs. 1–6, and pl. 114, fig. 5

Resupinate, covering small areas on the young branches of *Cinchona* species; whitish at first, becoming pale buff or cream colored in herbarium; surface in places smooth but usually very irregular with numerous minute rounded bumps or sinuous ridges with here and there in the older parts a larger bump 1–2 mm. wide. Margin determinate, dotted with a considerable number of short pillars, the tops of which spread out umbrella-like. In section 300–600μ thick, consisting of three distinct regions: (1) the white subiculum which extends over the bark, 30–40μ thick, and from which arise (2) the pillars which consist of loose bundles of upright, mostly unbranched hyphae, pillars 50–100 x 50–200μ, usually about 50μ tall, sometimes absent in the older regions of growth; the hyphae of pillars branch out above to form the context of (3) the top layer, which at first consists of thick and thin areas, the thicker areas being the bumps; context at first 50–100μ thick, becoming 250–300μ thick, consisting of loosely arranged, entangled hyphae; hyphae of pillars and context 4–5μ thick with narrow lumens, easily breaking into short segments when crushed under a cover glass. Hymenium arising in scattered areas over context, 65–130μ thick, consisting of upright, hyaline, septate, thin-walled, slightly twisted hyphae, 2–2.5μ thick, and probasidia and basidia in various stages of development; probasidia formed laterally, spherical, thin-walled, hyaline, 12.6–20μ thick (18–21μ thick, Raciborski); basidia curved making about one half a complete turn, 8 x 45μ (Raciborski), 4-celled; probasidia remaining as an empty cell after germination, usually with the wall thickened on one side; sterigmata short (Raciborski); spores 4–5 x 19μ (Racib.), slightly bent.

Associated with scale insects some of which are parasitized by irregularly coiled hyphae while others are free from infection. Reported by Raciborski (1909) as being very injurious to *Cinchona* sp.

This species has been confused with *S. lichenicolum* but the two are quite distinct, as an examination of type material of both and a careful reading of Raciborski's excellent description prove. For a comparison of the two see under *S. lichenicolum.*

In the nodulose surface of the younger parts of the top layer and in basidial characters this species is related to *S. Thwaitesii.* The two may, however, be readily distinguished. The present species is a much smaller form than *S. Thwaitesii* and is nearly pure white in color when fresh.

Specimens examined:

Java: Soekanegara, on *Cinchona* sp., Raciborski, coll., type in Raciborski Herb., von Höhnel Herb., and U. N. C. Herb.

Septobasidium clethrae n. sp.

Plate 61, figs. 1, 2, and pl. 106, fig. 1

Resupinate, covering extensive areas on the smaller limbs of *Clethra.* Very friable in texture. Hymenial surface usually occurring in irregular patches which vary in size from small specks up to 0.5–2 square cm. in area; avellaneous, finely granulose under a lens; margin of hymenium often fimbriate, sometimes upturned. Subiculum not broken into patches but more or less continuous, margin indeterminate or very irregular, deep brown in color. Scattered here and there over the subiculum are the very peculiar and characteristic insect houses. These are about 0.5–0.7 x 3–4 mm. and consist of a long covered part with a flattened roof in which lives the insect and a shorter part which is only partially covered, there being an oval hole in the roof. Insect house surrounded by a row of upright pillars or not infrequently covered over by a dense tangle of anastomosing pillars. In section through the hymenial regions 0.5–1.5 mm. thick, composed of three layers. (1) The subiculum which is usually quite thin, 20–60μ thick, composed of fairly compactly arranged threads which are about 4μ thick. From the subiculum arise (2) numerous short thin tufts of hyphae which quickly branch out and extend diagonally upward to form the loosely arranged hyphae of context, which is 300–600μ thick, hyphae 4.2–5.4μ thick, pale brownish, thick-walled, sometimes breaking into short segments. These hyphae branch above in a very irregular manner to form (3) the top or hymenial layer, 50–100μ thick and composed in the lower part of thick threads, 4.2–5μ thick, which branch extensively and extend irregularly upward, tapering to form thin threads, about 2–3μ thick, which are much entangled and densely compacted. Probasidia spherical, 11.7–15.1μ thick, germinating and becoming empty to form a curved or twisted, four-celled basidium, 8.2–9.6 x 38–45μ, sterigmata long; spores elliptic, bent at distal end, about 8 x 25μ.

Associated with scale insects some of which are parasitized by haustoria of the irregularly coiled type while others are free from parasitism.

Recognized by the avellaneous hymenium which occurs in irregular anastomosing patches, by the peculiar insect houses which are frequently covered over by a tangle of pillars and rhizomorphs and by the coiled basidia.

Specimens examined:

Philippine Islands: Davao province, Mindanao, Mt. Apo, on stem of *Clethra canescens*, June 10, 1924, Clemens, coll., no. 5313; type in Sydow Herb., also in U. N. C. Herb.

Septobasidium bogoriense Pat. In Otto Warburg, Monsunia **1:** 138. 1900.

Septobasidium mompa (Tanaka) Rac. Bull. Intern. del'Acad. des Sci. de Cracovie **3:** 355. 1909. Not *Helicobasidium mompa* Tanaka.

Plate 61, figs. 3–10, pl. 103, figs. 4–7, and pl. 104, figs. 10–12

Resupinate on the branches of living trees. Pale smoke gray, or pale drab-gray to avellaneous, usually with a pale lilac tint. Surface inherently smooth where the hymenium is formed. Margin determinate. In section 600–1000μ thick, usually composed of three layers. (1) A thin subiculum, 20–100μ thick, from which arises (2) a region of rather stubby or sometimes slender and tall pillars which may be closely set, i.e. about 100μ apart or may be 1 mm. apart, pillars 50–100μ thick, 100–500μ tall, hyphae of pillars loosely arranged. In the type the hyphae of this middle region are grouped in irregular anastomosing columns to form a fine network, the meshes of which are hardly visible without a lens. The pillars branch out to form the top layer which is very variable in thickness in the different collections (160–400μ thick in the type, 800–900μ thick in collections from Tonkin, 80–130μ thick in material from Boedijn and Steinmann). Hyphae of context rather loosely arranged, ascending, brownish, 3–4μ thick. Hymenial layer of variable thickness, usually 40–85μ thick, composed of densely packed, much entangled brownish hyphae, which give rise to hyaline, branched, irregularly twisted, sometimes circinate hyphae about 2.1μ thick. The hymenial layer is sharply distinguished from the context by the more delicate, entangled, and densely packed hyphae of the former layer. Probasidia spherical or slightly subspherical, 8.4–10μ thick, sometimes oval, 8.4 x 13μ, sprouting and becoming entirely empty to form an elongated twisted basidium, 5.3–6 x 25–35μ. Probasidial cell remaining as an empty cell after the formation of the basidia. Basidia 4-celled. Sterigmata and spores not seen in type but in a collection from Tonkin (M. Demange, no. 186) sterigmata short, spores bent-elliptic, 4 x 18μ.

Associated with scale insects some of which are parasitized by irregular coils, while others are not parasitized and are in the stage of giving birth to young.

In the Patouillard Herbarium there are nine collections labelled *S. bogoriense* Pat., none of which is designated as type. According to the original description the type was collected by E. Nyman on the bark of a living tree in the "Hortus Bogoriense," Java. One of these nine collections was collected by E. Nyman, "6/3/98 on bark in Hortus Bogoriense." I am considering this the type. On the label there are two drawings showing a spherical probasidium and another with a basidium.

This species appears to be the commonest in the Dutch East Indies and Southern China, having been reported on *Coffea*, *Cinchona*, *Thea*, *Broussonetia*, *Morus*, *Citrus*, *Manihot*, *Ficus elastica*, *Solanum quitoense*, *Erythrina*, *Crotalaria* sp., *Tephrosia candida*, *Leucaena glauca*, *Sesbania aegyptiaca*, *Lantana*, *Stachytarpheta mutabilis*, *Paritium*, *Calosanthes indica*, *Fraxinus*, *Marsdenia*, *Polyosma*, *Rosa*, *Wigandia Kunthii*, *Macaranga tanarius*, *Hibiscus rosa sinensis*, and *Mangifera*.

The species is very variable, the variations having to do chiefly with the structure of the context, a fact also noted by Boedijn and Steinmann (1930). In some collections there are distinct pillars, in others the pillars are stubby with a very thick top layer, and in others the pillars anastomose to form an entangled network. In spite of such variations the fungus can be recognized by the color of the surface, this usually being a pale smoke or drab gray with a faint purplish hue. The very compact hymenium with the hyaline, branched, recurved threads; the small, persistent, probasidial cell and the twisted, four-celled basidium are distinctive, as are also the narrow, slightly bent spores, if they have been correctly described. For a comparison of this species with *S. Reinkingii* Pat. see under the latter.

Several important papers have appeared on this species, chief of which is the one by Gaumann (1922).

Specimens examined:

Java: "Hortus Bogoriense," on bark of living tree, June 3, 1898, E. Nyman, coll., type, in Pat. Herb.; also a collection in U. S. D. A. Herb., R. D. Rands, Nov. 19, 1923; Tapos Estate, Dec. 1929, on *Morus*, A. Steinmann, coll., in Buitenzorg Herb. and U. N. C. Herb.; "Hortus Bogoriense," Raciborski, coll., in Raciborski's Herb., Cracov, Poland; also a collection by Raciborski from same location labelled *S. mompa* (Tanaka) Rac., in Herb. Cracov and U. N. C. Herb.

Ceylon: Peradeniya, October 1915, T. Petch, coll., in Petch Herb. and U. N. C. Herb.

China: Tonkin, La Pho, Jan. 1, 1908, M. Demange, no. 186; same location and date on *Aleurites cordata*, no. 162, fertile; Cho Ganh, Feb. 1929, M. Duport, coll.; also same location, Duport, no. 962;

same location on branches of "Murier," very common, M. Duport, 764, all coll. from China in Pat. Herb.

Japan: On *Broussonetia papyrifera*, sterile, in Pat. Herb.

Australia: North Queensland, on lemon, July 17, 1936, in Herb. Plant Pathology, Brisbane, no. 4673, also in U. N. C. Herb. With *S. aligerum* and *S. crustaceum*. Spores about 3.8 x 15.5μ, elliptic or slightly bent-elliptic. On the hymenial surface of this collection and frequently embedded in small pockets were clusters of green algal cells. The algal cells were so numerous in certain areas as to impart a faintly greenish hue to the fungal surface.

West Indies: St. Lucia, on *Citrus* (lime), F. W. South, coll., April 11, 1912, labelled *T. pedicellata* Schw.

Panama: La Campana, on *Citrus sinensis*, Oct. 6, 1937, James Zetek, coll., no. 3944, in U. S. D. A. Herb. and U. N. C. Herb.

Septobasidium pedicellatum (B. & C.) Pat. Jour. de Bot. **6:** 61, with figs. 1892. *Thelephora pedicellata* B. & C. of Wright's Cuban Exsiccati No. 798, but not *T. pedicellata* of Schweinitz.

Plate 60, figs. 6–8, and pl. 107, figs. 3, 4

Resupinate, forming small, inconspicuous patches, 2–3 mm. in diameter, on the bark of living bushes, intermingled with small liverworts and lichens. Hymenial surface avellaneous, finely pulverulent under a strong lens, margin of hymenium irregular. Where the hymenium is not developed, the fungus consists of entangled, branched and anastomosing pillars which arise from inconspicuous rhizomorphs. Pillars and rhizomorphs nearly black. Marginal region not seen. In section 600–1000μ thick, composed of three regions. (1) The basal part which consists mostly of much branched rhizomorphs from which arise, (2) the region of entangled pillars, 300–600μ thick; pillars arranged in a very complicated network, 12–110μ thick, most 40–60μ thick, hyphae of pillars deep brown, 4–4.2μ thick. The pillars mostly extend upwards branching out above to form (3) the top layer, which is 315–400μ thick and stratose in the type material due to the presence of two hymenial layers. In the lower and older hymenial layer there are numerous collapsed probasidia and basidial cells, and also numerous probasidia (?) with greatly thickened walls and yellowish contents, as well as yellowish basidia which failed to form spores. The active hymenial layer is 190–220μ thick and consists of densely packed and much entangled hyphae, which are pale yellowish or hyaline, 2.8–4.2μ thick, numerous probasidia which are scattered throughout the hymenial region but are much more numerous toward the outer zone, 16–25μ, spherical or subspherical, and curved basidia which arise from the probasidia, leaving the latter as an empty cell. Basidia 4-celled, 12.6–13.8 x 40–50μ. Mature sterigmata not seen. Spores, bent-elliptic, 5.3–7.5 x 25.2–31.5μ, frequently divided into 8 cells.

Associated with scale insects (*Aspidiotus* sp.) some of which are parasitized by irregular coils.

Several species have at times been confused with the present one: *S. Leprieurii*, *S. Mariani*, *S. pseudopedicellatum*, *S. Henningsii*, *S. bogoriense*, and several undescribed species. The species appears, however, to be known only from the type collections made by Wright in Cuba. It is a very distinct species, being characterized by its occurrence in small patches, by the dense tangle of blackish pillars (conspicuous only under a magnification of 20 or more) which stand out in striking contrast to the lighter colored pinkish brown (avellaneous) hymenial layer, by the exceptionally large probasidia, the thick, curved, four-celled basidia and the thin threads in the hymenium. In the type material the fungus occurred intermingled with lichens and liverworts, a fact which is most likely of no specific significance.

Specimens examined:

Cuba: Coll., C. Wright, on living shrubs, no. 798, type in Farlow Herbarium, Kew Herbarium, Herb. Bot. Gard. Ber. and perhaps other herbaria. There is another collection of type material but I have not examined it.

Septobasidium alveolatum Couch. Jour. Elisha Mitchell Sci. Soc. **44**: 253. 1929.

Plates 5, 9, 13, 14, 91

Resupinate, forming irregular patches on the bark of living trees, 1 to 3 or more cm. wide by a few to 10 or 15 cm. long, blackish with a little brown over the vegetative portion, the hymenium grayish. Surface of vegetative part areolate, spiny, presenting in places an irregular, somewhat honeycomb appearance. Hymenium formed over the areolated areas, at first in very small patches, then sometimes anastomosing to form a continuous hymenial surface for several square centimeters; hypochnoid at first, becoming compact and smoother with age. Margin very indistinct and irregular, apparent only with a lens. In section up to 3 mm. thick when fresh, usually 1–2 mm. thick; composed of three regions: (1) a very thin, inconspicuous subiculum which creeps over the surface of the bark, following especially the cracks and penetrating by means of them as far as the cork cambium; (2) the alveolate middle region composed of very loosely packed, slightly wavy hyphae, which may or may not be arranged in anastomosing, upright or semi-prostrate pillars (in some collections the middle region is entirely lacking); threads dark brown under microscope, septate without clamp connections,

4–5.4μ thick; (3) the hymenial region which is exceedingly variable in thickness, when well formed from 60–120μ thick, but in old specimens up to 250–350μ thick, composed of probasidia and basidia in various stages of development and numerous hyaline, branched, wavy threads, 3.5–4.3μ thick, from the sides of which the former arise. Probasidia spherical or subspherical, 14–17μ thick, with a slightly thickened wall which is as a rule thicker on one side than the other, developing into the basidia by sprouting and emptying the entire contents into the latter; basidia 10–12 x 50–58μ, long-cylindrical, usually slightly curved or twisted, rarely straight, becoming divided by transverse walls into four cells, each cell sprouting a sterigma which bears on its end a large spore. Spores white in print, elliptic to bent-elliptic, 7.4–11.7 x 18–29μ, becoming once septate a few hours after formation.

Associated with scale insects, some of which are parasitized by very irregular coils of hyphae and segmented hyphae, the segments being connected by fine hyphal threads like strings of sausages. Many insects, however, are not parasitized.

This species in appearance to the unaided eye is near *S. cirratum* Burt. The color in some specimens of the two plants is practically identical. A careful examination serves, however, to show unmistakable differences in the two plants: *Septobasidium cirratum* in most specimens shows a much more effused, continuous and compact growth; young specimens, but old enough to be fruiting, strikingly resemble a small foliose lichen with lighter colored and upturned margins; young areas spreading vegetatively over the bark do not show the honeycomb appearance so characteristic of this present form. A comparison of sections under the microscope shows even more distinct differences. The context of *S. cirratum* is very compact, that of *S. alveolatum* very loose. In the surface of *S. cirratum* are numerous coiled threads; these are absent in *S. alveolatum*. The basidium in *S. cirratum* is formed by an elongation of the probasidium forming part of it and is 2-celled, while in *S. alveolatum* the basidium is 4-celled and sprouts from the probasidium which becomes empty and does not form part of the spore-bearing body.

It is also closely related to *S. pedicellatum* which occurs in the West Indies (Cuba), both having hymenia in small patches and similar probasidia and basidia. The two may easily be distinguished. In *S. pedicellatum* the hymenial surface is pinkish buff, the pillars, though entangled, are fairly distinct, the threads of the pillars are straight, 4–4.2μ thick, the probasidia 16–25μ, the basidia 12.6–13.8 x 40–50μ, curved, forming about one-fourth of a complete turn; while in *S. alveolatum* the hymenial surface is grayish, the context, sometimes arranged in pillars, more frequently not, nearly always distinctly alveolate, threads of context wavy, 4–5.4μ thick, the probasidia 14–17μ thick, the basidia 10–12 x 50–58μ, frequently S-curved. The spores are also distinct, as are the threads in the hymenia.

Characterized by the nearly black, alveolate context, composed of wavy hyphae, the small patches of grayish hymenium, the large probasidia, S-shaped basidia, the thick threads in the hymenium, and the peculiar haustoria.

Specimens examined:

Jamaica: Portland Gap, Blue Mountain Peak, Mossman's Peak, and Cinchona, nine collections by J. N. Couch, June, July 1926, on *Psychotria* sp. and other deciduous trees, often mixed with mosses and liverworts. Type, 8547, U. N. C. Herb.

Septobasidium rubiginosum Pat. Ann. Jard. Bot. Buitenzorg, Supplement **1**: 118. 1897.

Plate 60, figs. 9–11, and pl. 104, figs. 7–9

Resupinate, extending for several centimeters over the bark. Purple brownish or dark Indian red (Ridgway). Surface with numerous cracks and holes due to the incompletely developed hymenium. Margin indeterminate, with numerous minute rhizomorphs (conspicuous under lens). In section up to slightly over 1 mm. thick in dried specimens. Specimens which have just reached maturity are composed of three layers. (1) The subiculum, which is very thin in the marginal region but is considerably thicker in the older parts, varying from 100 to 300μ; (2) the middle region of rhizomorphs and pillars, the latter arising from the subiculum and rhizomorphs, sometimes branched, 50–100μ thick, minutely puberulent. Hyphae of context branched and often coiled, 3.8–4μ thick. In old specimens the pillars disappear, the context becoming compact except for numerous anastomosing channels. (3) Hymenial region 150–230μ thick in well developed specimens, composed of narrow (1.8–3μ thick) hyphae which are much branched and coiled (much as in *S. rhabarbarinum*), probasidia which are subglobose and which give rise to the long, cylindrical, irregularly coiled, tri-septate basidium. Basidia 7–8 x 60–75μ. Probasidial cell is usually left as an empty cell after the basidium is formed. Sterigmata long. Spores long-elliptic, narrower at the mucro end, 5.2–6.3 x 20–26μ.

Associated with scale insects some of which are parasitized by sausage-shaped cells held together by very fine threads, the "haustoria" being like those in *S. rhabarbarinum* (B. & C.) Pat. Other insects, old and young, are free from parasitism.

The collection no. 1323 (Massart) in Patouillard Herbarium shows the pillars, while the collection by Nyman does not show any. Both specimens have the same purplish brown color tones, with slight differences due doubtless to age, and agree microscopically. The two collections are undoubtedly the same species.

There is a striking structural resemblance between the present species and *S. rhabarbarinum* (Mont.) Bres. However, the two may be distinguished by color and in the presence of pillars in *S. rubiginosum* and in the more regularly branched and coiled hyphae in *S. rhabarbarinum* (Mont.) Bres.

Specimens examined:

Java: Tjibodas, Massart, no. 1323, Jan. 1898, type, in Pat. Herb.; also collected by Nyman from same loc., June 24, 1898, in Pat. Herb., Berlin Herb., Buitenzorg Herb., and U. N. C. Herb.

Septobasidium irregulare n. sp.

Plate 55, figs. 2–4, and pl. 103, figs. 1–3

Resupinate, covering extensive areas, the larger patches up to 5 cm. in diameter. Color of the hymenial surface near benzo brown, the marginal region much darker colored. The surface is strikingly irregular. At the outer margin there are closely appressed, blackish rhizomorphs visible only with a lens. About 2 mm. back from the edge there are numerous small (0.5–1 mm. long by about 0.3 mm. wide by about 0.3 mm. high) hollow mounds. These mounds are very closely spaced, frequently with contiguous sides and are irregularly arranged though as a rule the open end of the mound faces towards the older part of the fungus. The doors of the mounds are fringed with hairs or bristles, and long spines, some of which are a mm. tall, arise from the tops and sides of the mounds. These mounds are the insect traps and in numbers of them young insects were found. As the fungus grows older, the spines increase in height and several may anastomose to form a single, long, conspicuous (under lens) spine (1–4 mm. tall) which may stand upright but more frequently bends over in a curved position. In this region the top layer begins to develop. This starts as a grayish brown growth which covers the upper parts of the spines and extends out horizontally to anastomose with other horizontal hyphae, eventually forming the irregular patches of the top layer. Many of the spines project up through the top surface forming blunt or acuminate projections, which are covered with hymenial tissue like the top surface. In section 400–900μ thick. Composed of a subiculum which in the marginal region is made up mostly of rhizomorphs but in the older parts is a more or less continuous sheet, threads of subiculum extending into cracks of bark; subiculum 25–126μ thick except near the parasitized insects where it is much thicker; hyphae dark brown, rarely nearly hyaline, 3–4.2μ

thick, much septate. From the subiculum arise the mound-like insect houses, the pillars, and in places loosely entangled hyphae. Walls of mound 20-35μ thick at first and composed of dark brown, very compactly arranged, much septate, 4–4.2μ thick hyphae, becoming much thickened internally if it contains a parasitized insect by the formation of a pseudoparenchymatous layer over the insect. Pillars abundant in some places, absent in others, arising singly or in tufts, usually most abundant near the mounds, most 25–80μ thick, vertical or extending diagonally through the context, hyphae 3.6–4.2μ thick, much septate, dark brownish. Hyphae of context much entangled and branched, frequently septate, 3.6–4.2μ thick. Top or hymenial layer 70–100μ thick, composed of much branched, rather compact, entangled, much septate, brownish hyphae with hyaline upright or curved tips. Intermingled with the threads are numerous spherical or club-shaped brownish bodies which are probably probasidia which failed to germinate. Probasidia hyaline to pale brownish, usually pyriform, 8–9.2 x 12–16.8μ, germinating and becoming empty to form a curved 4-celled basidium, about 5 x 34μ. Spores bent-elliptic, 4 x 16.8μ (only one seen).

Associated with scale insects some of which are parasitized by haustoria of the glomerulus type.

This species may easily be recognized with the unaided eye by the benzo brown color and by the bumpy surface, the bumps being very abundant and conspicuous in the marginal region, and somewhat less abundant where the hymenium is formed. The elongated hollow mounds (visible only with a lens) which occur over the entire surface of the subiculum are very distinctive.

While the hollow mounds are exceedingly interesting as insect houses, it is hard to see how the thin covering of the mound could afford much protection to the scale insects against the hymenopterous wasps. It is quite possible, however, that the bristles which grow up around the traps might make it very awkward for the wasp to puncture the fungal covering and deposit its egg in the body of the scale insect. The roofs when dry are hard and firm, a condition which might make penetration by the wasp's ovipositor difficult.

Specimens examined:

Africa: Uganda Protectorate, Mukono Hill, altitude 4300 feet, on unidentified tree, Nov. 1914, R. Dummer, coll., type in Kew Herb., U. N. C. Herb., and U. S. D. A. Herb.

Septobasidium citricolum Sawada. Descriptive Catalog of Formosan Fungi, Part IV. Rept. Dept. Agri. Research Inst. Formosa **61:** 177. 2 pl. 1933.

Plate 62, figs. 4, 5, and pl. 105, fig. 1

Resupinate, frequently extending for as much as 15 cm. encircling the young branches and extending out on the under surface of leaves, forming a thick cream colored or pinkish buff encrustation. Surface smooth except for a few cracks. Margin determinate, sometimes the threads of this region show a network or spongy arrangement. In section 0.5–1.2 mm. thick, composed of four distinct regions. (1) The subiculum which is white and very thin in the young growing region but which is 80–300μ thick in the older parts where the scale insects are abundant; threads of subiculum hyaline, 2.8–3.5μ thick, arranged so compactly as to appear pseudoparenchymatous in section; with abundant fatty or glycogen content. The subiculum gives rise to (2) scattered stubby pillars 65–150 x 84–126μ; hyphae of pillars brownish, 3.8–4.3μ thick. Frequently the brownish threads may ascend to form (3) the central, loosely arranged context directly from the subiculum without uniting into pillars. Context 250–590μ thick, threads of context 3.8–4.4μ thick, brownish, septate, considerably branched and entangled, ascending obliquely and giving rise to (4) the hymenial layer which is sharply set off from the context by the exceedingly compact structure of the former. Hymenium 100–390μ thick, stratose in old specimens, the old layers brownish while the active hymenium is colorless. Composed of densely entangled, much branched, narrow, twisted, hyaline threads which have a generally upward direction, 2–2.5μ thick; probasidia which are formed in a rather compact layer at the outer surface, spherical, subspherical, or pyriform, 13–14 x 16.5–23μ; probasidia germinating and becoming empty to form a curved basidium 8.2–9.7 x 50–65μ, 4-celled; sterigmata thick in the basal half and narrowing abruptly to a fine tube. Spores bent-elliptic, 4.8–6.3 x 17.6–25μ.

Associated with abundant scale insects some of which are parasitized by haustoria of the irregularly coiled type while other insects are free from parasitism.

This species may be distinguished by the cream to buff colored, smooth hymenial surface, by the very thick hyaline subiculum, the few short pillars, the loosely arranged hyphae of the context, and the comparatively very thick hymenium. Combined with the above characters the probasidia and curved basidia are also distinctive.

Specimens examined:

Formosa: Hayashida-mura, March 22, 1931, on *Citrus poonensis* Hart., K. Sawada, coll., type in Sawada Herb. and U. N. C. Herb.

Septobasidium sinense n. sp.

Plate 46, figs. 8–11, and pl. 104, figs. 1, 2

Resupinate, forming small patches (up to 3 cm. long) on the smaller twigs and branches of *Citrus* and other trees, pale buff to cinnamon, surface nodulose or nearly smooth. There are two sizes of nodules, those formed on the margin which are very minute, 0.1–0.3 mm. wide and flattened above with a dark brownish center and whitish margin, supported by short, thick tufts of hyphae and those which form the main context and top layer of the fungus and are usually 1–2 mm. wide. In some specimens the larger area of the top surface may be smooth. Margin determinate, the subiculum conspicuously and characteristically marked under a lens by the minute nodules. In section 600-800μ thick in the central mature area where it is composed of five distinct regions. (1) The subiculum 35–80μ thick, composed of very compactly arranged hyphae 3.8–4μ thick, hyaline to dark brownish; from the subiculum arise (2) hyphal tufts which are about 80μ tall and are composed of pale brownish or nearly hyaline hyphae 4–5.4μ thick; (3) the tufts spread out above to form a very thin compact deep brown to nearly black horizontal layer which in section is the most conspicuous and striking characteristic of the fungus, 20–30μ thick. The scale insects are between these two layers and when the parasitized insects are very abundant the region between these two layers may be more or less filled with closely packed hyphae. From the thin black layer arise (4) vertical, pale hyphae 4–5.4μ thick, which become entangled and rather compactly arranged above to form the main part of the context, 200–600μ thick. The hyphae of context break up into short fragments when crushed. Over the context is formed (5) the hymenial layer 65–100μ thick which consists of thin-walled, upright, hyaline, sinuous, branched and anastomosing hyphae, 2–3μ thick; probasidia arising laterally, at first hyaline and nearly spherical, becoming pyriform, 12.6–14.7 x 14.7–23μ, remaining as a partially collapsed cell with the wall thickened on one side after the formation of the basidium. Basidia slightly bent or curved (like a shepherd's crook), four-celled, 8–10.5 x 42–50μ, sterigmata long, spores not certainly seen. (In the material examined nearly all the probasidia and basidia had become deep brownish; a ew hyaline germinating basidia, however, were seen. The latter, doubtless, are normal.) Hyaline, minutely rough conidia borne in somewhat penicillioid chains are formed over parts of the subiculum and certain parts of the top surface.

Associated with scale insects, some of which are parasitized by haustoria of the irregularly coiled type. Others are free from parasitism.

This species is characterized by its small size, nodulose surface, the peculiar margin, and in section it may be recognized by the very thin blackish middle horizontal layer. It is very close to *S. cinchonae* Racib. in the nodulose surface and in hymenial characters. The two also agree in having the three horizontal layers. They differ in that the present species lacks pillars while in *S. cinchonae* they are quite distinct; also *S. cinchonae* is white to pale buff while the present species is much darker in color, and in the present species conidia are formed abundantly while in *S. cinchonae* they are formed but very rarely.

Specimens examined:

China: Tang nen, Kwang Si, Reinking, coll., no. 4986, on living twigs of *Citrus limonia*, May 17, 1919, in U. S. D. A. Herb. and Pat. Herb.; Tonkin, Cho Ganh, on one of the Aurantiaceae, Duport, coll., August 1918, in Pat. Herb.; same locality, on *Citrus*, 1922, Duport, no. 107, in Pat. Herb.; all three of the above were labelled *S. albidum*, "Conidifere" by Patouillard; also a collection from Canton, Jan. 1932, on *Citrus sinensis*, Steinmann, coll.

Septobasidium Thwaitesii (B. & Br.) Pat. Bull. Soc. Myc. Fr. **24**: 2. 1908.

Thelephora Thwaitesii B. & Br. Jour. Linn. Soc. Bot. **14**: 64. 1873.

Plate 46, figs. 1–4, and pl. 104, fig. 4

Resupinate, occurring at first in small rounded or sinuous patches, later covering extensive areas, often girdling the limbs and extending for a length of 15–20 cm. Surface densely warted in the younger parts where the hymenium has not formed; the warts are rounded or elongated and closely set, about 3 to a mm.; this warted layer is about chestnut brown in color or sometimes almost black. The top layer is usually broken up into sinuous, anastomosing areas 2–5 mm. wide between which there may be spaces 1–3 mm. wide. The hymenium appears at first in buff colored to cinnamon brown pulverulent patches but these may later anastomose to some extent. The top surface, however, remains considerably broken up. In the type material the surface is pulverulent and cinnamon brown in color. Margin determinate but inconspicuous, dotted with minute whitish pillars. In section 270–1500μ thick, composed of five distinct layers. (1) The subiculum 50–85μ thick, composed of compactly arranged hyaline or pale brownish hyphae about 4μ thick; from the subiculum arise (2) scattered, stubby pillars 63–130μ thick by about 130μ tall, hyphae of pillars about 5μ thick, much septate, pale yellowish or nearly hyaline; the pillars giving rise to (3) a layer which is apparently light-colored at first but which becomes chestnut brown or nearly black due to the deposition of pigment, 40–84μ thick; hyphae of this layer compact, about 5μ thick. This layer has a distinctly bumpy surface, a bump occurring over each pillar. The scale insects are found between the subiculum and this first horizontal layer. From this second horizontal layer arises (4) a thick region, 500–850μ thick, composed of loosely entangled, obliquely ascending, branched, pale brownish to nearly hyaline (under microscope) hyphae, 4–5μ thick, rather thick-walled; hyphae yellowish brown to the unaided eye. Over the surface of this spongy region is formed (5) a thin crust, which is composed of about one layer of closely arranged hyphae embedded in a brownish material. The surface of this layer is glabrous as is also the surface of the second horizontal layer. This spongy region with the thin crust is formed in patches over the surface of the second horizontal layer. The hyphae of the spongy region push up through this surface layer to form the hymenium. Hymenium 80–420μ, composed of more or less upright, branched, nearly hyaline hyphae, 3.5–4μ thick, between which there is much anastomosing. Probasidia at first nearly globose or ovoid, 16–19 x 20–21.8μ, elongating to become pyriform, the enlarged part of the pyriform body becoming the functional probasidium, the basal part remaining as a stalk; frequently the globose probasidia may germinate without the formation of the pyriform body. Basidia curved like a shepherd's crook, 12–13.5 x 37–46μ, four-celled; hyaline at first but becoming brownish; sterigmata not seen. Spores bent-elliptic, 7–8 x 25–27μ.

Associated with scale insects, some of which were parasitized by irregularly coiled haustoria. Other adults were not parasitized and were giving birth to young.

This species is characterized by the minutely bumpy surface, by the flaky structure of the top surface, by the occurrence of several horizontal strata placed one over the other like thin, irregular sheets of paper, by the very unusual sectional structure and finally by the large probasidia, the basal part of which is frequently cut off as a stalk, and by the large curved basidia. The bumpy surface and mottled brown and lighter colored areas remind one of *S. cinchonae* Rac. but the two are easily distinguished.

Specimens examined:

Ceylon: Thwaites, no. 325, type in Kew Herb.; also three collections by Petch: Hakgala, no. 6178, April 1919; Hakgala, on *Pavzolzia* sp., Jan. 1914; Nuwara Eliya, on *Piper* sp., July 1927.

Septobasidium fusco-violaceum Bres. Ann. Mycol. **1:** 112. 1903.

Plate 45, figs. 4, 5, and pl. 107, fig. 11

Resupinate, forming effused patches on living bark. Cinereous or brownish drab in the younger parts, purplish brown (Van Dyke brown, Ridgway) in the older regions. Surface tuberculate or reticulate due to the incomplete formation of the top layer. Subiculum grayish with a striking, finely honeycomb appearance under a high powered lens (20x). The pillars which spread out above to give the tuberculate appearance are few and purplish brown in color. Marginal region indeterminate (?). In section about 500μ thick; usually composed of three regions: (1) the subiculum, (2) a middle region of inconspicuous, thick columns and (3) the top layer which arises from the tops of the pillars but does not form a continuous horizontal layer. Sometimes the middle layer may be lacking, the hymenial layer being formed directly over the subiculum; hyphae of context about 3–4.3μ thick. Hymenium 60–80μ thick, composed of densely packed, upright sinuous threads, which are 2–3μ thick. Probasidia spherical, 12.6μ thick (only a few seen); basidia 4-celled, one to 1.3 times coiled, 5.8–6.3 x 32–38μ. Spores not seen.

Associated with scale insects some of which are parasitized by irregularly coiled haustoria while many young and adults are free from parasitism.

In superficial structure this species somewhat resembles *S. retiforme* and *S. Langloisii* but can easily be distinguished from those two species by color and by microscopic details.

Specimens examined:

Poland: Type, on branches of *Salix cinerea.* In Bresadola Herbarium and Patouillard Herbarium.

Septobasidium cupressi n. sp.

Plate 62, figs. 12–14, and pl. 105, fig. 3

Resupinate, forming a thin whitish or slightly buff colored flocculent growth on the leaves of *Cupressus.* Very small, covering areas only a few square millimeters in extent. Surface minutely fluffy or cottony. Margin irregular, determinate. In section 200–350μ thick, composed of three regions: (1) a thin (20–40μ thick), white subiculum which extends over and between the scale-like leaves and gives rise here and there to (2) a few stubby inconspicuous pillars or tufts of hyphae, which branch out above to form (3) the top layer of very irregular thickness, frequently about 200μ thick. In the older parts the fungus may consist of only one layer, due to the filling up of the space between the pillars by hyphae. Hyphae of subiculum, pillars, and top layer hyaline throughout, 3–4.2μ thick, with swollen places here and there in the hyphae, septate but sparingly branched, in places with rather thickened walls. Toward the top of the upper layer the hyphae branch considerably, the branches extending obliquely upwards to form the loose hymenial region; hyphae thin-walled, about 3–3.5μ thick. Probasidia globose to pyriform, hyaline, 7.5–10 x 13–15μ, germinating to form a slightly curved basidium, persisting as a semi-collapsed empty vesicle after the formation of the basidium. Basidia 5–6 x 28–33μ, apparently 2-celled. Spores bent-elliptic, only a few seen.

Associated with scale insects (*Aspidiotus* sp.) some of which are parasitized by irregular coils.

This species is easily distinguished by being pure white throughout, by its very small size, loose flocculent texture, and by its occurrence on *Cupressus* sp.

Specimens examined:

Argentina: Buenos Aires, July 1936, on *Cupressus* sp., sent by Grodsinsky, type in Dept. of Agriculture Herbarium, Buenos Aires, and U. N. C. Herb. no. 10406.

Septobasidium piperis P. Henn. (Not published but so named on herbarium label.)

Plate 45, figs. 6, 7, and pl. 106, fig. 5

Resupinate, occurring on the smaller twigs at the nodes and also on the leaves in very small inconspicuous patches about a cm. in diameter; surface buff colored, becoming dark brown in old material; inherently smooth but with a few cracks and wrinkles and holes where the top surface is incompletely formed; margin determinate, indistinctly fimbriate. In section 197–300μ thick, composed of three distinct regions. (1) The very thin membrane-like subiculum, 12–42μ thick; threads of subiculum 3–4μ thick, brownish, except for hyaline, bent, crystal-encrusted threads which extend upwards for a short distance from the lower brown threads. The upright hyaline threads are usually so abundant as to give the subiculum a whitish appearance. From the subiculum arise (2) a few pillars 20–95 x 90–115μ; threads of pillars brownish, 3.8–4.2μ thick. The pillars branch out above, the hyphae extending horizontally to form (3) the top layer, 60–120μ, usually about 84μ thick, composed of a compact lower part of brownish threads which are 3.8–4.2μ thick and the hymenial zone which is nearly hyaline, about 25–36μ thick. Threads in hymenium 3–3.8μ thick, branched, the ends extending upright through the hymenium, encrusted with minute crystals. Probasidia 9.2–9.8μ thick, spherical or slightly irregular, giving rise to an irregularly curved or twisted basidium, remaining as an empty vesicle after the basidium has been formed. Basidia 5.6–6.4 x 42–54μ, 4-celled. Sterigmata and spores not seen.

Associated with scale insects some of which are parasitized by haustoria in the form of irregular coils while others are free from parasitism.

This is a very characteristic species, being distinguished by its very small size, buff color, thinness in section, the minute but distinct pillars, the hyaline, crystal-encrusted, curved threads arising from the subiculum, and by the small probasidia and twisted basidia. The fungus shows a superficial resemblance to *S. albidum,* but can easily be distinguished by size and by hymenial and basidial characters.

Specimen examined:

Location uncertain, perhaps Java: On *Piper nigrum,* August 23, 1901, Zimmermann, coll., type in Bot. Museum Berlin.

Septobasidium cremeum Couch. Jour. Elisha Mitchell Sci. Soc. **51:** 46. 1935.

Plate 31 and pl. 80, figs. 1–5

Resupinate, closely adherent to the bark, forming areas as much as 10 x 10 cm. Younger areas light buff, older regions cinnamon brown. Surface at first smooth, becoming cracked into small irregular areas as the plant ages. Margin very irregular, usually indeterminate. In section up to 225μ thick, usually about 210μ thick in well developed specimens, composed of three layers except in the marginal region: (1) subiculum usually very thin, up to 33μ thick, composed of hyaline, septate, non-nodose, hyphae; (2) middle layer composed of very short pillars; (3) top layer 60–130μ thick. Hymenium about 35–45μ thick, hyaline in color and hence sharply distinct from the lower part of the top layer which is brownish, threads of hymenium 2.2–3.5μ thick, colorless, densely packed, branched, coiled, and entangled. Threads of context and pillars 3.8–4.2μ thick, septate without clamp-connections, brownish under microscope. On the margin the hymenium may sometimes be developed not only on the top layer but also over the subiculum. Probasidia subglobose, about 8–9μ thick, the thread which bears the probasidium may be attached to it at the side or even top (i.e. the end toward the surface), instead of at the bottom; basidia usually coiled, 4.2–5 x 35–40μ, divided into four cells at maturity, each of which may give rise to a very long sterigma and bear a spore. Spores bent-elliptic, 3.8–5 x 11.5–16.8μ, not becoming septate on the slide.

Symbiotically associated with *Aspidiotus* (*Cryptophyllaspis*) *liquidambaris* Kot., some of which are parasitized by irregularly coiled hyphae, while others are not parasitized. The coiled hyphae, or haustoria, of the fungus connect with the main body of the fungus by means of hyphae which pass through the insect's derm by the mouth. These hyphae do not appear to interfere with the sucking operations of the insect, for the insect, though it may become heavily parasitized, continues to live and suck the plant juice. The insects which are not parasitized reproduce and carry on the life of the insects.

This species may be easily overlooked because of its inconspicuous color and thinnesss. It is related to *S. alni* and some of my specimens strikingly resemble that plant in color, in marginal characters, and in the pillars. The present species generally lacks the rich deep brown color of *S. alni*, being usually much lighter in color. It may also be distinguished from *S. alni* by the coiled hyphae in the hymenium and the coiled basidia.

It is possible that the present species is a hybrid between *S. alni* and *S. tenue*.

Specimens examined:

Florida: Near Jacksonville, on *Liquidambar styraciflua*, with nearly black variety of *S. Burtii, S. sinuosum, S. castaneum, S. alni*, and *S. tenue*, March 11, 1932, J. N. Couch, coll., type in U. N. C. Herb., 9232; La Cross, on *Ilex glabra* with *Aspidiotus juglans-regiae*, May 1, 1933, Geo. F. Weber, coll., in Univ. Fla. Herb., 9188, and U. N. C. Herb., 9395.

North Carolina: Near Chapel Hill, New Hope Creek, on *Ilex decidua*, Nov. 10, 1935, J. N. Couch, coll., in U. N. C. Herb., 10069.

Septobasidium humile Raciborski. Bull. Intern. Acad. Sci. de Cracovie **3**: 363. 1909.

Plate 62, figs. 1–3, and pl. 104, fig. 3

Resupinate, hypophyllous, forming small (usually less than 1 cm. in area), thin, irregular patches of growth. Drab colored, sometimes with a faintly purplish tint in the older places. Surface smooth where the hymenium is well developed, except for a few minute holes, finely pulverulent under a lens; in the younger parts the surface is very irregular, being dotted with innumerable umbrella-like tufts of hyphae, each of which is rounded when looked at from above, with an asteroid appearance, 200–400μ wide. As development proceeds the margins of the umbrellas fuse, forming at first a very irregular surface; margins of adjacent umbrellas remaining distinct even after the top surface is formed; margin of subiculum indeterminate, margin of top layer determinate but very irregular, usually fimbriate. In section 135–200μ thick (very rarely small mounds may be formed almost 1 mm. high) composed of three regions: (1) a very thin (5–10μ thick), waxy, whitish subiculum, composed of one layer of loosely arranged hyphae imbedded in a continuous layer of hyaline substance, hyphae of subiculum hyaline or pale brownish under microscope, 3.6–4.2μ thick; from the subiculum arise (2) very short small pillars, about 50–100μ thick by 40–90μ tall, composed of hyphae 4.2–6.3μ thick, brownish, thick-walled, branched, septate; the pillars branch out above to form the umbrella-like structures which anastomose to form (3) the top layer, 60–126μ thick, composed of a lower part of brownish hyphae which extend horizontally, branch mostly at right angles and taper toward the tips, hyphae about 4.2–5μ thick, giving rise to short, upright, branched, twisted, hyaline, loosely arranged hyphae about 3μ thick, which with the probasidia and basidia form the hymenium. Probasidia 13.5–15.4μ, spherical, germinating and becoming empty to form a twisted, 4-celled basidium. Sterigmata about 3 x 10μ. Spores elliptic, slightly bent, 5 x 18–20μ (Raciborski's measurements). Probasidia, basidia and spores are all hyaline at first but after maturity becoming brownish in herbarium.

Associated with scale insects some of which are parasitized by haustoria in the form of irregular coils while others are free from parasitism.

This species may easily be distinguished from other leaf-inhabiting forms by the drab color, the exceedingly thin, white membranous subiculum, the very small, short, knot-like pillars, the hyphae of which branch out above to form umbrella-like structures. The speckled appearance of immature specimens under a lens due to the numerous scattered umbrella-like structures and the extreme thinness of the structure even when mature are also distinctive.

Specimens examined:

Java: Near the base of Mt. Pangerango on the western side above Tjitwruk, on underside of leaves of *Tetranthera* sp., Raciborski, coll., type in Raciborski Herb., Cracov, Poland, von Höhnel Herb., Harvard Univ., Berlin Museum, and U. N. C. Herb.

Septobasidium Lagerheimii n. sp.

Plate 45, figs. 11, 12

Resupinate, forming small (3–30 mm. long), inconspicuous, irregular patches, the older, lighter parts where the hymenium is forming avellaneous to wood brown, the younger, darker regions about Verona brown. Surface at first roughened by numerous small tufts of hyphae. These tufts may remain single, the hyphae extending upward and branching out horizontally to form the top layer or they may branch and anastomose to form an irregular network, the top layer being developed from the ends of the branches. Top layer distinctly granulose under a lens. When mature the surface is broken by numerous iregular anastomosing cracks. Margin indistinctly determinate, whitish or pale buff; sometimes with distinct (under a lens) short pillars but more frequently the margin of the top layer extends downwards, apparently anastomosing with the subiculum. Here and there, however, are inconspicuous narrow openings where the top margin does not come in contact with the subiculum. In section 250–650μ thick, composed in places of three layers; in other places the context is not distinctly divided into layers. When three-layered, the subiculum is 40–100μ thick and composed of loosely packed, pale brownish or hyaline hyphae, 3–4μ thick; the middle region 120–450μ thick, composed of loosely arranged, branched and entangled hyphae which extend diagonally upwards or sometimes the hyphae of the middle region are arranged in short thick pillars, hyphae hyaline or brownish 3–4.6μ thick; top layer 40–125μ thick, not sharply differentiated from the underlying tissue, in places stratose by the formation of two regions of probasidia and basidia, the old probasidia and basidia becoming yellowish; composed of pale brown to hyaline hyphae, which extend perpendicularly but are somewhat wavy, septate, about 3–3.8μ thick, and with numerous spherical, thin-walled bodies, 3.8–10μ thick, arising laterally from the hyphae. These spherical bodies suggest young probasidia but it is unlikely that they are such structures since they also occur in lesser numbers on the older hyphae of the context. Probasidia 9.6–12.6μ, spherical, germinating to form a coiled, 4-celled basidium, 5.2–5.8 x 33.6–42μ, probasidial cell remaining as an empty vesicle after the basidia are formed. Sterigmata long. Spores not certainly seen.

Associated with scale insects some of which are parasitized by haustoria of the glomerulus type.

This species was identified by Patouillard as *S. pedicellatum* (B. & C.) Pat. but the two can easily be distinguished. In *S. pedicellatum* (B. & C.) Pat. the pillars are much more conspicuous than in the present species and though both occur in small patches, in *S. pedicellatum* the top layer is elevated on pillars while in the present species the pillars are so poorly developed that the fungus appears to be crustose. Moreover, in *S. pedicellatum* the probasidia and basidia are much larger than in the present species. The occurrence in small thin patches, the avellaneous cracked surface of the hymenium, the inconspicuous, scattered pillars which may be single, or in dark brown, anastomosing groups (or sometimes distinct pillars may be absent), the small probasidia and basidia, and the small globose cells borne on the sides of the hyphae particularly in the hymenial region are striking characteristics of the present species.

Specimens examined:

Ecuador: Near Quito, Jan. 31, 1892, de Lagerheim, coll., two collections in Patouillard Herb., both of which look alike.

Septobasidium macadamiæ n. sp.

Plate 45, figs. 1–3, and pl. 106, fig. 2

Resupinate, extending 2–5 sq. cm. over the smaller branches and frequently overgrowing the lower and upper surfaces of leaves. Color pale pinkish buff to avellaneous or wood brown. The older parts are frequently lighter in color than the younger patches, due to the fact that the threads of the hymenium are hyaline. Surface smooth in the younger regions of growth, becoming cracked by numerous, sinuous, anastomosing fissures, which start as minute irregular slits and open through the top layer until they may be 300μ wide and 2–3 mm. long. Margin determinate, the margin of upper layer somewhat fimbriate; the subiculum with a few inconspicuous pillars. In section 230–420μ thick, composed of three distinct regions: (1) the subiculum which is usually very thin; threads of subiculum about 3.8μ thick; from the subiculum arise (2) a few stubby pillars, 40–60μ tall, the hyphae of which branch out above to form (3) a horizontal layer. The hyphae of the pillars and the horizontal layer are 3–4.2μ thick, usually about 3.8μ thick, and are considerably entangled. The horizontal layer is sometimes indistinctly stratose. From the horizontal layer arises the hymenial layer the threads of which arise more or less at right angles and thus extend vertically. Hymenium 50–150μ thick, hyphae branched, hyaline or pale brownish, 2.5–3.5μ thick, septate and usually wavy toward the tips. Probasidium about spherical at first, 11–12μ thick, about 10.5 x 18μ when the basidia sprout, remaining as an empty cell after the formation of the basidium. Basidia 6.7–8 x 27–37μ, curved like a shepherd's crook, 4-celled. Sterigmata about 6–8μ long. Spores bent-elliptic, 4.2–5.4 x 15.9–22.2μ.

Associated with scale insects (*Aspidiotus* sp.) some of which are parasitized by irregularly coiled hyphae.

This species may be recognized by the pale pinkish buff to wood brown color, the cracked surface microscopically resembling baked clay (the surface is cracked only in the older parts), and the thin structure. The lighter colored, cracked areas resemble *S. albidum* Pat., while the darker colored, smooth region somewhat resembles *S. alni* Torrend. It may seem rather odd that the older region should be lighter in color than the younger but this is due to the fact that in the older parts the hyaline threads which make up the greater part of the hymenium are better developed. The hymenial characters are also distinctive.

This species might be confused with *S. Lagerheimii* but the two may be readily separated. *Septobasidium Lagerheimii* occurs in smaller patches and the pillars are not umbrella-like as in *S. macadamiæ*. Also in the former the pillars are not present throughout while in the latter they are. In *S. Lagerheimii* the cracks are left by the incomplete formation of the top layer; in *S. macadamiæ* they are formed by a splitting of the top surface.

Specimens examined:

Australia: Queensland, on smaller branches and leaves of *Macadamia ternifolia*, October 19, 1933, type in Herb. Plant Path. Brisbane, no. 2943, and U. N. C. Herb.

Septobasidium mexicanum Sydow. Ann. Mycol. **18:** 154. 1920.

Plate 44, figs. 10–12, and pl. 106, fig. 4

Resupinate, closely adherent, 1–2 x 1–3 cm. (Sydow's measurements), about Sayal brown in the marginal region, becoming grayish brown in the older parts, surface inherently smooth. Margin of top layer determinate, irregular in outline, in places fimbriate, margin of subiculum whitish and vague but determinate. In section 250–400μ thick, composed of three distinct layers: (1) the subiculum, pale brownish or whitish, 40–100μ thick, composed of closely packed hyphae 3.4–

4.4μ thick; (2) a region of very scattered short pillars, 60–100μ tall by 40–225μ thick, usually much thicker than tall, hyphae of pillars vertical, unbranched except at the top where they branch out to form the top layer; (3) the top layer 40–210μ thick, composed of considerably entangled, much branched hyphae which extend in a more or less horizontal direction, hyphae 3.8–4.4μ thick. In the older parts the top layer becomes so filled with numerous irregular crystals as to obscure the hyphae. No distinct compact hymenial layer is formed, but the hymenial region consists of probasidia, basidia, and hyphae which extend out beyond the probasidia and then curve back and extend horizontally along the surface. Probasidia borne in tufts, ovoid or subglobose, 10–12.5 x 14–17μ, germinating and becoming empty to form a four-celled, bent basidium, 6.2–7.7 x 35–40μ; the basidium may be bent like a shepherd's crook or may be nearly S-shaped. No mature sterigmata or spores seen.

No scale insects were seen by me but I studied only a small piece from the margin and a piece from the center. This species is undoubtedly associated with scale insects.

This fungus is very close to *S. alni* Torrend, differing from that species only in having shorter, thicker, and curved basidia. The occurrence on *Cupressus* may also be of specific value, though *S. alni* is known to have a wide range of hosts and its occurrence on *Cupressus* would not be surprising.

Specimens examined:

Mexico: On living branches of *Cupressus* sp., Reiche, no. 46, 1918, type in Sydow Herbarium and Herb. Bot. Gard. Berlin.

Septobasidium proliferum Boed. and Stein. Bull. Jard. Bot. Buitenzorg III, **11**: 179, figs. 5 and 6. 1931.

Plate 49, figs. 11, 12, and pl. 104, fig. 6

Resupinate, frequently surrounding smaller limbs for a distance of 20 cm. Not differentiated into layers but with a subiculum and numerous small tufts of pillars which are 1–3 mm. wide, and 0.5–1.2 mm. tall, nearly black when moist, brownish gray when dry except for the tufts of pillars which are nearly black at all times. Surface in between the tufts of pillars quite variable: smooth or irregular and alveolate. The pillars are very striking and characteristic. Sometimes they may arise in a circle but more frequently in an irregular tuft. Each pillar is usually composed, at the base, of several smaller pillars which anastomose and pass upwards to a sharp point thus forming an elongated cone-like structure. In section 35–200μ thick except where the pillars are and then up to 1.2 mm. thick counting the height of the pillars. Hyphae of context and pillars brownish, 2.5–5μ thick, septate. Hymenial region not sharply distinct from underlying tissue, composed of curved, branched, pale brownish or hyaline hyphae, 3–3.5μ thick, and probasidia and basidia. Probasidia spherical, sessile or with a short stalk, 12–17μ thick; basidium curved, 5–6 (rarely up to 8μ thick) x 26–42μ, 4-celled, sterigmata conical, about 3.5μ thick at base and 8–10μ long. Probasidia renewed by internal proliferation, the process being repeated several times until the probasidium has an exceedingly thick wall and a very small lumen. Spores bent in a half circle, pointed at both ends, the distal end being prolonged into a thin process.

Associated with scale insects some of which are parasitized by irregular coils.

This species shows a superficial resemblance to *S. Curtisii.* The two may be easily separated, however. In the present species the upright spines are more abundant than in *S. Curtisii* and are arranged in groups, each group as a rule covering an insect house. The probasidia, the curved basidia, and the strongly curved spores are also distinctive.

Specimens examined:

Java: Tjibodas, on stem of a small tree, April 1930, Boedijn, coll., type in Herb. Bot. Gard. Buitenzorg and U. N. C. Herb.

Septobasidium neglectum Boed. and Stein. Bull. Jard. Bot. Buitenzorg. III, **11:** 183, fig. 9. 1931.

Plate 49, figs. 9, 10, and pl. 104, fig. 5

Resupinate, extending for 10 or more centimeters along the smaller branches, often surrounding them, frequently forming inconspicuous soot-like patches, spongy under the lens, blackish brown in color. Surface exceedingly irregular, nodulose, spiny, alveolate or rarely nearly smooth, apparently most frequently alveolate. Very variable in thickness, up to 1 mm. thick in section, composed of a basal layer about 10–100μ thick, from which arise irregularly arranged entangled hyphae, to form the very irregular surface. Frequently the hyphae unite into strands, the strands often anastomosing to form an irregular net; hyphae of subiculum and context 4.6–6.4μ, deep brown, forming in some places a hymenial layer, which is about 40–90μ thick. In other places the probasidia are scattered here and there. The hymenium when well developed composed of upright, much septate, mostly hyaline hyphae and probasidia and basidia in various stages of development. The probasidia arise in unique fashion, laterally along a hypha, the younger probasidia being near the tip of the hypha; almost invariably just beneath each probasidium there arises a hyphal branch; probasidia ovoid or elongated, hyaline or very pale brownish, apparently developing immediately (in the type material) without a rest period into the basidia; when mature 9.6–11.7 x 14–24μ. Basidia 7–8 x 33.6–50μ, curved, frequently only slightly thinner than the probasidia, three- or four-celled. The probasidia elongate considerably as the basidia are formed. Most of the basidia showed only two sterigmata, a few were seen with three and some with four sterigmata, 8.4–12.6μ long by about 4.2μ thick at the base.

Associated with scale insects some of which are parasitized by haustoria in the shape of irregular coils. Other insects are free from parasitism.

This species is easily recognized by the brownish, black, sooty appearance, by the loose, irregular form of the fructification which lacks distinct pillars and only exceptionally has a distinct hymenial layer, and by the curved basidia with a narrow, elongated probasidial cell. The probasidium in the present species seems to be intermediate between those species in which this structure elongates and becomes transformed into a basidium and those in which the probasidium becomes empty, persisting as an empty cell.

Specimens examined:

Java: Tjibodas, on branches, April, 1930, F. W. Went, coll., type in Herb. Hortus Bot. Bog. and U. N. C. Herb.

Septobasidium gossypinum Pat. Bull. Soc. Myc. Fr. **41:** 341, fig. 2. 1925.

Plate 62, fig. 11, and pl. 107, figs. 5–7

Resupinate; about Prout's brown (Ridgway); surface finely granular under a lens. Marginal region distinctly and regularly areolate. In section 500–1000μ thick, not divided into layers; composed of loosely packed, much entangled, and much branched hyphae, the branches mostly arising at right angles. Hyphae rarely septate, light brown, 2.2–6.2μ thick, most about 4.2μ thick. No distinct compact hymenial layer formed. Fertile hyphae usually dichotomously branched, much septate, the joints of the hyphae only about 2–3 times as long as their diameter. Probasidia arising laterally or as a dichotomous branch, several arising on the same hypha at different levels, ovoid, 9.6–10.5 x 12–15μ, persisting as an empty cell after the formation of the basidia. Basidia 5–5.5 x 26–35μ, 4-celled, the distal three cells coiled. Sterigmata and spores not seen.

Associated with scale insects some of which are parasitized by irregularly coiled haustoria.

This species is close to *S. pannosum* (Fries) Bres. The two may easily be separated. In *S. gossypinum* the alveolations are much more regular than in *S. pannosum* and the basidia in the former are closely coiled and make a complete turn while in the latter the basidia make only a half turn.

The flocculent nature of the context, the lack of a distinct hymenium, the origin of the probasidia in series one above the other on the same hypha, and the closely coiled basidium are distinctive characters. The original collection (which appears to be the only one) shows a peculiarly regular areolated condition in the marginal region. The color of the hyphae in this region differs slightly from that of the fertile zone.

Specimen examined:

China: Cho Ganh, Tonkin, on branches of liana, November 1923, M. Petelot, coll., no. 715, type in Pat. Herb.

Septobasidium pannosum (E. Fries) Bres. Ann. Mycol. **14**: 242. 1916.
Corticium pannosum Fr. Nov. Symb. Myc. p. 114. 1851.

Plate 62, figs. 6–10, and pl. 107, figs. 1, 2

Resupinate, forming rather extensive patches up to 5 x 9 cm., perhaps larger. Cinereous over a brownish context, near Verona brown or cinnamon drab. Surface of the older parts broken by numerous sinuous cracks; surface of the marginal region alveolate or spongy. Margin sharply determinate. In section 1.5–2.3 mm. thick, composed of three regions. (1) The basal subiculum extending over the bark and in between the flaky parts; rather thin in the younger parts but up to 325μ thick in the older regions. (2) From the subiculum arise loose fascicles or plates of hyphae which unite to form a dense alveolate structure which is usually a mm. or slightly more in thickness; hyphae of context 3.8–5.4μ thick, most about 4.6μ thick, branched, septate, the cross wall at the septum usually being clearly doubled. (3) Over the surface of the context the top layer is formed, about 125–290μ thick, sharply distinct from the underlying tissue, composed of compactly arranged hyphae which are much branched, sinuous, considerably septate, and 3–4μ thick; between the hyphae there is much anastomosing. Probasidia arising from the ends or sides of hyphae, frequently showing a cymose arrangement, ovoid, 6.3–8.4 x 11.7–12.6μ, germinating and becoming empty to form a four-celled basidium which is curved like a shepherd's crook, 4.2–4.4 x 30–34μ. Sterigmata 2.2 x 10.5μ; spores (in Guadeloupe coll.) 4–4.4 x 14–16.8μ, bent-elliptic.

Associated with scale insects some of which are parasitized by haustoria in the form of irregular coils.

This is a very distinct species and easily distinguished by its large size, grayish brown color, cracked surface, with a spongy margin, and the spongy context. Microscopically the very compact hymenium with the small, cymosely borne probasidia and the small curved basidia are distinctive. It is close to *S. gossypinum*; for a comparison see under that species.

Specimens examined:

Mexico: On fallen limbs, Liebman, coll., type in E. Fries Herb. Bot. Museum, Upsala.

Guadeloupe: April 15, 1903, no. 772, in Patouillard Herb., Harvard Univ., as *S. spongia* (B. & C.) Pat. Basidia slightly larger than in type, 5–5.8 x 40–45μ.

Septobasidium Sydowii Couch. Jour. Elisha Mitchell Sci. Soc. **51**: 47. 1935.

Plate 30, pl. 67, fig. 3, and pl. 73, figs. 6–8

Resupinate, forming circular patches up to 5 cm. in width. Chestnut brown or darker in the older regions, light grayish brown in the younger or fruiting regions. Surface smooth near the margin over a zone about 6–8 mm. wide, the rest of the surface conspicuously cut up by many deep fissures which are up to 1 mm. wide. Margin sharply determinate, with a minutely scalloped appearance caused by minute holes which lead back into short tunnels in the tissue of the fungus. In section about 1 mm. thick, very compact and hard, up to about 1.4 mm. thick, differentiated into context and hymenial regions. Context faintly stratose, composed of densely packed, entangled, rather sparingly branched threads which have mostly a vertical direction. Threads deep brown under microscope, thick-walled, rarely septate, without clamp-connections, 3.6–5μ, mostly 4.2μ thick. Hymenium 90–115μ thick, stratose, light-colored and thus distinct in color from the context, threads vertical, thin-walled, collapsing easily, about 2.8–3.5μ thick; probasidia subglobose or ovoid, 9–10.5 x 13–15μ, germinating to form a once coiled basidium, collapsing soon after basidia are formed. Basidia about 6–7 x 30–40μ, 4-celled, sterigmata about 6–10μ long. Spores about 3.8–4.3 x 15.5–20μ, bent-elliptic, sometimes somewhat S-shaped, becoming thrice septate.

Symbiotically associated with scale insects, some of which are parasitized by irregularly coiled haustoria while others are free from fungal infection.

The specimens from Texas are light grayish brown and the fissures run out to the margin. The margin also differs from that in the Philippine material in showing short pillars or tufts of threads. The surfaces of the two plants are cracked in the same way and the microscopic characters of hymenium, probasidia, basidia, and spores agree.

This species may readily be distinguished from all other species of *Septobasidium* which I have studied by the peculiar cracked surface.

The fungus has the appearance suggesting a miniature bit of baked, finely cracked clay.

The fungus-insect relationship is obviously a very interesting one. Numerous healthy (i.e. uninfected by the fungus) adult insects were found partly buried in the cracks, the posterior part of the ventral and dorsal scales projecting up through the fissures, the scales reminding one of the bivalved shell of an oyster. These adults contained many young and, judging by the widely opened condition of the genital pore, were in the act of giving birth to young. In the tunnels near the margin many dried up, but non-parasitized young were found and farther back in the tunnels there were half grown insects. A considerable number of insects were infected, and such insects were always completely covered by the fungus.

This fungus was first recognized as a new species by Dr. Sydow and turned over to me for study.

Specimens examined:

Philippine Islands: Province of Sorsogon, on *Pterocarpus*, Elmer, coll., 14820. Type in U. N. C. Herb. and Sydow Herb.

Texas: Near Austin, on *Ulmus crassifolia*, Sept. 7, 1934, J. J. Taubenhaus, coll., U. S. D. A. Herb. and U. N. C. Herb., 9868.

Septobasidium murinum (B. & Br.) Pat. (So labelled in Pat. Herb. but unpublished)

Corticium murinum B. &. Br. Jour. Linn. Soc. Bot. **14:** 70. 1873.

Plate 61, figs. 11–12

The type material of this species is so scanty that it is impossible to describe accurately its gross appearance. The piece I have studied covers about a square centimeter, is resupinate but shows no marginal condition. Berkeley and Broome report the fungus as being arachnoid at first with a very indistinct whitish margin and extending for several inches over the bark. As a result of the growth of insects and fungus the bark is in a rather unhealthy condition and plates of hyphae can be traced down into the dead regions of the bark into the regions where the scale insects are located. Surface smooth, grayish with a lavender tint over the compact brown context. Texture very brittle. In section up to 1 mm. thick, usually 400–600μ thick, composed of a lower region of hyphae, which are sometimes united into small rhizomorphs or plates, extending down between the dead pieces of bark; hyphae about 3.8–5.8μ thick, sometimes crystal-encrusted; from these lower hyphae arises the middle region, 275–300μ thick, composed of much entangled, much branched, and fairly compacted hyphae, which are 3.2–4μ thick; from the middle region arises the hymenial layer 125–

170μ thick. The tissue of the hymenium is densely compacted, the threads are mostly collapsed and hence are difficult to make out. They are much branched, curved, extending generally in a vertical direction, hyaline, about 3–3.8μ thick, the hyaline tips of the threads extending out beyond the probasidia and basidia. Probasidia and basidia scattered throughout the hymenial region, the younger ones being toward the upper side, the older ones being buried in the tissue and usually yellowish in color. Probasidia very abundant, globose or ovoid, 14.7–16.8μ thick or 13–15 x 15–21μ, germinating to form an elongated, 4-celled, slightly twisted (open S-shaped) basidium, probasidial cell persisting as an empty vesicle after the formation of the basidium; basidia 8–10.5 x 36–51μ. Spores bent-elliptic, only a few seen.

Associated with scale insects some of which are parasitized by irregularly coiled haustoria.

Characterized by the grayish color, compact brittle texture, lack of pillars, by the compact hymenial region, the abundant probasidia and the open S-curved basidia.

Specimens examined:

Ceylon: Peradeniya, Thwaites, no. 32, type, Dec. 1868, in Kew Herb., Pat. Herb., and in Herb. Peradeniya.

Septobasidium arachnoideum (B. & Br.) Bres. Ann. Mycol. **14:** 241. 1916.

Plate 61, figs. 13, 14

Resupinate, forming irregular patches of growth which cover several square centimeters. Color of the surface about hair brown, the context darker. Surface finely arachnoid in the marginal region, granulose in the older parts. Margin irregular, indeterminate or nearly so. In section 200–1000μ thick, composed of a context and a hymenial zone. Threads of context very loosely arranged, entangled, considerably branched, septate, 4–5.4μ thick, and about fuscous in color under the microscope. Toward the upper surface the threads of the context become thinner and hyaline, giving rise to branched threads, the ends of which are irregularly curved, and between which there is considerable anastomosing. These hyaline threads give rise to the probasidia, thus forming the hymenial zone. Threads of hymenium about 3μ thick; probasidia 11.2–13.8μ thick, spherical or slightly irregular, germinating and becoming empty to form an elongated, irregularly curved basidium. Basidia 7–7.5 x 40–50μ, usually four-celled but sometimes apparently more than four-celled. Sterigmata about 6μ long. Spores not seen.

Associated with scale insects some of which are parsitized by haustoria in the shape of spindle-shaped segments connected by fine threads, some of the chains of hyphae being curved; other insects are not parasitized. At the point where the haustorial hyphae pass through the insect's skin to connect with the external hyphae they are arranged in a star-like group. This species is characterized by the finely arachnoid surface near the margin, by the loosely arranged, thick, fuscous hyphae of the context, and by the irregularly curved basidia. For a comparison with *S. subcarbonaceum* (B. & Br.) see under that species.

Specimens examined:

Ceylon: Habgalla, Thwaites, coll., no. 539, type, Feb. 1868, in Kew Herb.

Septobasidium formosense n. sp.

Plate 49, figs. 1–3

Resupinate, growing in thin, indefinite patches on the upper and lower surface of leaves and the surface of the smaller branches of *Citrus*. Consisting of small (1–5 mm. wide) circular or irregular, brownish, elevated patches of mycelial growth and a pinkish buff subiculum which extends indefinitely over the leaf and twig surface. In section the subiculum is 10–25μ thick and consists of irregular, matted, hyaline to pale brownish hyphae 3–5μ thick. The probasidia and basidia are borne only on the subiculum, not occurring on the elevated patches. Probasidia hyaline, spherical to pyriform, 9.6–11.7μ thick, frequently with a distinct stalk. Basidia curved, frequently forming one complete regular coil, at other times irregularly curved and twisted, 4-celled, 5.4–7 x 33–42μ. During the formation of the basidium the probasidial cell becomes empty and the wall becomes much thickened on one side while it collapses on the other side. Spores not seen. From the subiculum there arise in groups small loose tufts of hyphae, the ends of which branch out and anastomose to form the top layer. Top layer 400–500μ thick, composed of loosely packed, much entangled, deep brownish hyphae 3–5.4μ thick, sparingly branched, the branches usually arising at right angles. No probasidia formed on elevated hyphal patches.

Associated with scale insects (*Parlatoria proteus pergandei*) some of which are parasitized by irregular coils of hyphae, the hyphae on the outside making connection with the internal haustoria by the vaginal pore and the mouth region. The hyphae penetrating in the mouth region do not follow the sucking tube. Both parasitized and healthy insects occur under the elevated tufts.

Occurring on both the top layer and the subiculum are some peculiar elongated brownish hyphal bodies, which seem to act as conidia and whose hyphae can be seen to connect with those of the present fungus. These bodies are 20–50 x 200–500μ and are composed of 6–12 or more rows of cylindrical or barrel-shaped cells, 8.4 x 6.3–12.6μ, each cell containing a large globule of fatty-looking material.

This species is readily distinguished from other species which occur on leaves. The thin, indeterminate, pinkish buff subiculum, the brownish, circular or irregular, elevated patches of sterile mycelium and the coiled basidia borne only on the subiculum are characteristic of the present species.

Specimens examined:

Japan: Formosa, Trail to Rengichi from Gojyo, on *Citrus* sp., March 13, 1928, R. Kent Beattie, coll., type in U. S. D. A. Herb. and U. N. C. Herb. 10360.

Septobasidium punctatum n. sp.

Plate 45, figs. 8–10, and pl. 114, fig. 6

Resupinate on living bark of *Citrus* species; pinkish buff or pale pinkish buff over the hymenial region in dried material, darker, slate colored where the hymenium has not formed; surface retiform in young specimens, becoming smooth in older specimens except for cracks and numerous dark-colored, slightly elevated dots (0.5–1 mm. wide) beneath which are scale insects. Margin indeterminate, sometimes with irregular rhizomorphs. In section 120–260μ thick, usually about 200μ thick, very compact throughout, composed of context and hymenial region. Context about 100–200μ thick, composed of entangled and very compact hyphae, which are 3.8–4.2μ thick, septate, distinct from hymenium by the brownish purple color of the threads of the former. Hymenium 40–60μ thick, very compact, composed of probasidia and basidia in various stages of development and of hyaline branched thin-walled hyphae; probasidia 7.1–10.5μ, most about 9.2μ thick, globose or subglobose, sprouting and emptying the entire contents to form an irregularly curved basidium. Basidium 4.2–7.1 x 25–29μ, 4-celled. Spores not seen.

Associated with scale insects (*Parlatoria proteus* ?), some of which are parasitized by hyphae composed of spindle-shaped segments held together by fine threads, while others are free from parasitism.

This species was determined by Patouillard as *S. albidum* Pat. but is easily distinguished from that species by the pale pinkish buff or pinkish buff color in the older regions and the numerous dark circular dots, by the irregular retiform structure of the fungus in the younger regions, by the thin compact structure in section, by the small probasidia and the irregular four-celled basidia. The spindle-shaped haustoria are also distinct.

Specimens examined:

Philippine Islands: Los Banos, Laguna, on living twigs of *Citrus* sp.,

October 1920, O. A. Reinking, coll., type in U. S. D. A. Herb. and U. N. C. Herb.

Septobasidium crustaceum n. sp.

Plate 49, figs. 7, 8

Resupinate, forming small, thin, closely adherent but readily separable patches on the smaller branches. Color avellaneous, or wood brown to snuff brown. Surface mostly smooth, but finely granulose under a lens, with a few bumps and cracks. Margin determinate. In some parts of the margin distinct (under binocular dissecting microscope) but very short and exceedingly minute pillars are present while on other parts the margin is flat and without pillars. In the older parts the pillars have entirely disappeared. In section about 50–100μ thick, not distinctly divided into regions but composed of brownish, entangled and compactly arranged threads which are about 3.6–4μ thick. No distinct hymenium is formed but the probasidia and basidia are scattered over the surface. Old and collapsed basidia are found on the older surface while active probasidia and basidia are found in abundance a few millimeters back from the margin. Probasidia spherical, 8.2–9.2μ thick, germinating to form a curved (frequently making a complete turn), four-celled basidium, the probasidial cell remaining as a partly collapsed structure; basidia 5.6–6.3 x 28–34μ. In a good many instances the distal end cell of the basidium is considerably larger than the others. Sterigmata 8–10μ long. Spores slightly bent-elliptic, about 4.2 x 14.7–21μ. The reproductive structures contain conspicuous shiny oil (?) globules.

Associated with scale insects (*Chionaspis* ? sp.) some of which are parasitized by irregularly coiled hyphae; others are not parasitized.

This species may be recognized by its color, and by the very thin, compact, brittle structure. The small spherical probasidia, and the curved, 4-celled basidia with the distal cell usually larger than the others are also characteristic.

Specimen examined:

Australia: North Queensland, on lemon with *S. bogoriense* and *S. aligerum*, July 17, 1936, type in Herb. Plant Pathology, Brisbane, and U. N. C. Herb.

Septobasidium Bagliettoanum (Fr.) Bres. Ann. Mycol. **3**: 164. 1905.
Hypochnus Bagliettoanus Fr. Hym. Eur. p. 705.

Plate 49, figs. 4–6, and pl. 105, fig. 2

Resupinate, extending in irregular, small (0.5–3 cm. in diameter) anastomosing patches over the old bark and often overgrowing mosses and liverworts with the habit of a *Coniophora* or resupinate *Stereum*. Color about Sayal brown (Ridgway) or slightly darker. Surface smooth. Margin determinate and turned up. In section 200–500μ thick, compact, not divided into layers but composed of the exposed resupinate fruiting surface and large rhizomorphs (0.5–1.0 mm. thick) which extend from the lower surface of the fruiting layer down between the crevices of the bark, branching out into hyphae which run throughout the dead (?) parts of the bark. Hyphae of context and rhizomorphs 2.8–3.8μ thick, septate, without clamps but between hyphae there is considerable anastomosing. The hyphae extend through the context in a diagonal direction, are considerably branched and fairly closely packed. Hymenial region rather thin, 40–50μ thick, composed of curved pale brown to hyaline hyphae about 2–3μ thick and probasidia and basidia. Probasidia subglobose to pyriform, 11–14 x 18–22μ; germinating to form a long 4-celled basidium which is usually bent like a walking stick, the handle being attached to the probasidium; basidia 6–6.8 x 50–63μ (the measurements of Bresadola and Bourdot are both too small); each cell of the basidium giving rise to a sterigma, 8–20μ, usually about 10μ long, on which is borne a spore. Spores 3.8–5 x 16.8–23μ, the distal one-third bent almost at right angles to the rest of spore.

This species may easily be distinguished by its peculiar habit of growth like a *Coniophora* or resupinate *Stereum*, and by the large rhizomorphs which extend through the dead bark. It is quite distinct from all species which have yet been described.

Through the kindness of M. l'Abbe Bourdot I have had an abundant lot of material for study. However, with only dried material available, I have been unable to be certain about the relationship with scale insects. A scale insect (*Chermes* sp. ?) is present deep in the crevices of the bark. They are rather scarce and all I have found are parasitized by the fungus. The fungus forms an entirely unique type of haustoria within the insects' bodies. Instead of small regular or irregular coils of hyphae, or sausage-shaped hyphal segments the haustoria are in the form of large balls of coiled hyphae (50–75μ thick). No connection between the fungus within the insect's body and external hyphae could be made out. These connections in most species even with fresh material are very difficult to find since they are so easily broken. It would appear from the extensive growth of the mycelium through the dead (?) parts of the bark that food may be obtained from this source.

The biology of this species should certainly be worked out by some one who can study fresh material.

Specimens examined:

France: Bouisson, on trunks of *Quercus* with mosses, Feb. 18, 1906, and June 1918, two collections from Herb. Bourdot, nos. 4064 and 24415, both in U. N. C. Herb.

Italy: Tuscany, on *Quercus* sp., May 11, 1908, determined by Bresadola. in U. S. D. A. Herb., U. N. C. Herb., and Pat. Herb.

Septobasidium fisso-lobatum Lloyd. Myc. Notes **6:** 888, Pl. 131. 1920.

Hymenochaete fisso-lobata Henn. Fungi Amaz. I. Hedwigia **43:** 172. 1904.

Plate 49, fig. 13

Resupinate on the living branches of lianas, with a resupinate basal part from which extend numerous flattened and often branched lobes, the lobed parts strongly suggesting a *Thelephora* or *Hymenochaete.* Lobes may be 5 cm. long by 5 cm. broad. Color of hymenium between cinnamon drab and drab (Ridgway), the opposite side blackish brown. Hymenial surface pruinose, the opposite surface with numerous inherent rhizomorphs and irregularities. In section (in the small fragments examined by me) 1–1.5 mm. thick, compact throughout. The basal (?) region is composed of densely packed threads which are not coiled and are sparsely branched, threads about 4μ thick; the main part of the context or middle region is composed of densely interwoven, coiled, sometimes dichotomously but usually irregularly branched threads. A thin section of the context under the low power of the microscope at first suggests a section of dead wood. Context faintly stratified with dark lines. Threads 2.2–3μ thick. Hymenial region not sharply delimited from the context. Probasidia borne laterally on the curved branches, spherical, thin-walled, 8–9.6μ thick, giving rise to a curved, 4-celled basidium, the curve being mostly in the distal half. Basidia projecting beyond the threads and shaped like the handle of a shepherd's staff, 4.8–5.4 x 33–42μ. Probasidia becoming empty during the formation of basidia, but usually collapsing soon thereafter. Spores bent-elliptic, 3.9–4.2 x 12.6–13.6μ.

The specimen which I studied did not have any of the stem with it and therefore I could not tell if scale insects were present or not.

This fungus is easily distinguished from any other species so far described. The structure of the context suggests a distant relationship to *S. rhabarbarinum.*

Specimens examined:

Brazil: Amazon region, distributed in Mycotheca Brasiliensis no. 41, in Pat. Herb.

Septobasidium Henningsii Pat. in Otto Warburg, Monsunia **1:** 138. 1900.

Septobasidium tjibodense Boed. and Stein. Bull. Jard. Bot. Buitenzorg III, **11:** 194, figs. 17, 18. 1931.

Plate 63, figs. 3–7, and pl. 109, fig. 8

Resupinate, forming rather extensive confluent patches which frequently surround the smaller limbs for a length of several (3–12) centimeters. The younger parts about snuff or cinnamon brown or tawny olive, the older parts where the hymenium is formed are considerably lighter, near cinnamon. The surface, where the hymenium has not formed, is finely spongy or alveolate, due to the exceedingly dense tangle of upright anastomosing pillars. These pillars are so densely packed that individual pillars are barely visible except with a lens. Hymenial surface may be formed in small saucer-shaped patches or may be flat, covering wider areas; it is smooth at first but becomes cracked by numerous fissures into many small irregular areas, which are frequently less than one sq. mm. in extent. These cracks are 0.1–0.8 mm. wide and usually deep enough to expose the pillars and sometimes the subiculum. Margin determinate, fibrillose. In section 1–2 mm. thick, composed of three layers: (1) the subiculum 30–130μ thick, whitish or pale brown, hyphae of subiculum septate, 2-5μ thick; (2) the middle region 300–1100μ thick, of dense, entangled, anastomosing pillars, this region frequently being lacunose, in certain parts of the fungus particularly the younger parts the pillars arising in a slanting direction (slanting towards the direction of growth), the hyphae at the tops of the pillars branching out to form (3) the hymenial layer, 60–400μ thick, simple or stratified by the formation of a new hymenial layer over the older one. The older parts turn brownish and in these one may easily recognize the old basidia; composed of upright, densely packed, septate, branched, hyaline or pale brownish hyphae, 2–4.2μ thick, and hence because of the structure and arrangement of these hyphae sharply distinct from the underlying tissue; the basidia arising laterally from these hyphae without the formation of a probasidial cell; mature basidium straight or more usually slightly curved, 7–9.8 x 36–54μ, 4-celled, each cell giving rise to a long (3–4 x 14–34μ) sterigma; sterigma sometimes almost as large as the basidium. Spores hyaline, bent-elliptic, quite variable in shape, some being considerably more bent than others, 4.5–6 x 20–28μ.

Associated with scale insects (*Mytilaspis* sp.) some of which are parasitized by haustoria in the form of irregular coils. Other insects are free from parasitism.

This species is characterized by the dense tangle of pillars found in the younger regions, by the peculiar hymenial layer which is divided by conspicuous fissures into numerous irregular areas. The fibrillose margin is also distinctive. Microscopically the compact, frequently stratified hymenium composed of upright hyphae and cylindrical basidia which arise without the intervention of a probasidial cell is distinctive.

Septobasidium tjibodense, recently described by Boedijn and Steinmann, agrees in practically all details with the present species. Type material of *S. Henningsii* Pat. was not available to these two authors, which fact undoubtedly explains why this was not recognized by them. Through the kindness of Dr. Ulbrich of Berlin I have been able to study the original type of *S. Henningsii* Pat. and Drs. Boedijn and Steinmann have very generously sent me parts of their Javan specimens.

This species shows a striking resemblance to *S. albidum* Pat. (type material from Ecuador). The two agree in color, structure of hyphae, and in hymenial characters. They differ, however, in the shape of the spores and in certain gross characters. *Septobasidium albidum* is a smaller, thinner fungus and lacks the conspicuous dense tangle of pillars characteristic of the present species. It is remarkable that Patouillard who described both of these species did not mention their striking similarity. The present species also shows a striking relation to *S. flavo-brunneum* B. & S. in hymenial and basidial characters but that species lacks the robust pillars characteristic of *S. Henningsii*.

Specimens examined:

Java: Gedeh, altitude about 2600 meters, on living tree limbs, E. Nyman, coll., July 17, 1898, type in Museum Bot. Garden, Berlin, also in Pat. Herb., Harvard, and U. N. C. Herb.; Res. Priangan, Goenoeng Gedeh, altitude about 2500 meters, on branches of *Albizzia montana*, Dr. van Leeuwen, coll., 13215, in Herb. Bot. Gard. Buitenzorg and U. N. C. Herb. This collection is the type of *S. tjibodense*.

Septobasidium albidum Pat. Bull. Soc. Myc. Fr. **9**: 136, Pl. 8, fig. 1. 1893.

Pl. 64, figs. 1–3, and pl. 109, figs. 1–4

Resupinate, forming circular or irregular patches 3–5 cm. wide; clay color to wood brown or light pinkish cinnamon to avellaneous; surface smooth or deeply cracked, usually pruinose; margin determinate, with numerous ridges or elevated, irregular patches of mycelium arising from the whitish subiculum. In section 270–370μ thick but in old specimens often up to 850μ thick, measuring the hyphal pads overgrowing the insects. Composed of three regions: (1) the subiculum (50–60μ thick) which extends over the bark and is made up of closely packed, septate, brownish hyphae 3.8–4.2μ thick; (2) a region of short pillars which vary immensely in thickness, 20–320μ thick by 60–80μ high; (3) top region over which the hymenium is borne, 100–350μ thick with hymenium; hymenial region quite variable in thickness, 35–140μ, composed of closely packed, upright or slightly wavy, branched, much septate, hyaline or pale brown hyphae, and basidia in various stages of development. Basidia arising laterally on the hyphae as small pyriform cells which enlarge and elongate directly to form the mature basidium without the intervention of a probasidial stage; straight or slightly curved, thickest in the middle or distal end, 4-celled, 9.2-11 x 29-50μ, most 10.5 x 37μ; the basal cell apparently sterile, each of the others giving rise to a long sterigma, 3-4 x 20-25μ, on the end of which is borne a spore. Spores 4.8-5 x 21-27μ, sharply recurved at the distal end, several times septate.

Associated with scale insects (*Aspidiotus rapax* Comst.) some of which are parasitized by irregular coils of hyphae while others are free from parasitism. In the parasitized insects the connection between the haustoria and the external fungus is made by dark-colored hyphae which surround the mouth with a circle of hyphae, penetrating the derm in this region. Some of the non-parasitized females were giving birth to young; one male in the second stage was found, and several young which had just started development, i.e. had just completed the formation of the thin pellicle over their bodies.

In the Patouillard Herbarium (Harvard University) and in several other herbaria which have some duplicates of Patouillard material there are several distinct species which have been labelled *S. albidum* by Patouillard. None of the material in the Patouillard Herbarium was designated "type" by him. However, in the original description of this species Patouillard (1893) cites two collections which are in his Herbarium, the one collected in Feb. 1892 by de Lagerheim, Cotocollao, Vallée de Chilo, Ecuador, the other collected in July 1892 from the same place. The latter collection is on *Prunus salicifolia.* I have studied both of these collections and though they are in different stages of development they appear to be the same species. The collection made in February 1892 has a very thick hymenium representing growth of several seasons, while the collection of July 1892 has a much thinner hymenium representing only a few seasons' growth. I am considering these two collections as typical material.

The fungus may be recognized by its clay color, thin structure, short and inconspicuous but persistent pillars. Microscopically the hymenium with the perpendicular hyphae, the large, curved basidia, and the spores recurved at the distal end serve to distinguish the species.

The only other collections labelled *S. albidum* which resemble the type are three from the Philippine Islands collected by Reinking, Feb. 16, 1920. It is well to bear in mind that these three collections were labelled *S. albidum* by Patouillard twenty-seven years after the type was described. They may be distinguished from the type on the basis of color, the absence of pillars in fully developed specimens of the

Philippine plants, and on the basidial and spore characters. These three are being described as *S. philippinense*.

The fungus described as *S. albidum* by Grillo (1930) and Müller (1933) is *S. pseudopedicellatum* Burt.

Specimens examined:

Ecuador: Cotocollao, Vallée de Chilo, Feb. 1892, cotype, Lagerheim, coll.; also on *Prunus salicifolia*, July 1892, type; also reported as occurring on *Piper Kunthii*, *Salvia tortuosa*, and *Melastoma* sp., same locality.

China: Kweichow, Liang feng-yah, no. 244, S. Y. Cheo, coll., August 5, 1931. Agrees with type except basidia longer and narrower. In Farlow Herbarium, Harvard University, and U. N. C. Herbarium.

Septobasidium philippinense n. sp.

Plate 64, figs. 6–12, pl. 67, fig. 6, and pl. 109, figs. 5, 6

Resupinate over areas equal to several square centimeters. Cinnamon to wood brown (Ridgway) in the older regions with a cream colored margin; surface smooth in the younger regions and usually smooth to the unaided eye in the older parts but filled with minute cracks giving the surface a faintly pruinose appearance; in some specimens the cracks are large and conspicuous being easily visible with the unaided eye; in the specimens on *Derris* and *Tarchonanthus* the posterior ends of the insects' scales project up through the cracks. Margin determinate, in some specimens particularly young ones the margin is dotted with circular patches of mycelium supported on short pillars; in old specimens the margin usually has a folded or pleated appearance. In section 0.4—0.7 mm. thick, compact; composed of (1) a subiculum, which in old specimens often merges insensibly with the context, (2) context, and (3) hymenial regions. Subiculum 30–50μ thick, composed of densely packed hyphae which extend over the surface of the bark; context 300–450μ thick, composed of entangled, moderately compact hyphae which extend in a more or less upright direction, hyphae of context and subiculum 3.8–4.4μ thick, septate; hymenium 60–110μ thick, sharply distinct from the context by the upright hyphae, which are much septate and about 4μ thick; basidia formed without the intervention of a resting probasidial cell stage. Basidia at first pyriform or subglobose, elongating to form a straight or bent, club-shaped structure, 9–11.7 x 33–48μ, which becomes transversely septate and thus 4-celled. Each cell germinates to form an elongated, club-shaped body, the end of which is pointed and on which in turn is borne a spore. Spores 4–6.7 x 21–27μ, bent-elliptic (8838).

Associated with a species of *Chionaspis* (when on *Citrus*) some of which are parasitized through the mouth region by irregularly coiled hyphae; others are not parasitized.

The species may be distinguished from *S. albidum* by the thicker, more compact structure in the mature condition, by the thicker and longer sterigmata, and the slightly bent, elliptic spores which are not recurved at the distal ends. In color and surface characters the two also differ.

The *S. albidum* group, to which this species belongs, is a quite difficult group and will require much more study in the field before one can be certain of the limits of variation of the species. It is entirely possible that under the present name I am including four distinct species: (1) the pinkish buff form on *Citrus*; (2) the much darker form on *Antidesma* with very conspicuous cracks; (3) the grayish buff form on *Derris*; and (4) the Verona brown to sepia or sometimes buff to gray form from South Africa.

Specimens examined:

Philippine Islands: Laguna, Los Banos, on *Citrus* hybrid in college orchard, Feb. 16, 1920, Otto A. Reinking, coll., determined by Patouillard as *S. albidum*, Reinking, no. 8917, type in Herb. U. S. D. A. and U. N. C. Herb.; also Reinking, 8838, collection data as above, on *Citrus*, with pinkish buff surface, cracks very minute, margin abruptly determinate; same location on *Derris* sp., December 1915, C. F. Baker, coll., no. 3920, labelled *S. bogoriense* in Patouillard Herb. (reported to cause large swellings on the stem), in Herb. Univ. of Philippines, and U. N. C. Herb. (specimens on *Derris* sp. are grayish in color with very conspicuous cracks); same location, on *Antidesma* sp., December 1933, Ocfemia, coll., no. 51, in Herb. Univ. of Philippines and U. N. C. Herb. (surface dark brown with conspicuous cracks in the older parts, lighter brown near the margin with only a few cracks, the top layer in the marginal region almost papery, spores 4–6 x 22–24μ, bent-elliptic, not recurved at the distal end); San Antonio, on living *Ficus* sp., A. Abesamia, coll., no. 9590, in Pat. Herb. as *S. ficicolum*.

South Africa: Cape Province, Trapps Valley, on *Tarchonanthus*? sp., P. J. Pienaar, coll., 1912, in Herb. Div. Plant Ind., Pretoria, S. A., and in U. N. C. Herb.; surface irregular by renewed growth over the old areas and by cracks.

Septobasidium flavo-brunneum Boed. and Stein. Arch. Theecult. **4:** 46, Pls. 23, 24. 1930 (In English). Also described in more detail in French in Bull. Jard. Bot. Buitenzorg III, **11:** 192–194, figs. 15, 16. 1931.

Plate 63, figs. 10, 11

Resupinate, covering an area of several square centimeters. In dried specimens the younger marginal zone (about 1 cm. wide) is light buff in color (pure white when fresh), the older fertile region being avellaneous to snuff brown. Marginal zone composed of hyphae so loosely entangled as to have an hypochnoid or flocculent appearance, the surface of the older parts is membranous and broken here and there by cracks and spaces left by the incomplete formation of the top surface. Margin of subiculum indeterminate, sometimes dotted by minute pillars. In section 270–750μ thick, composed of four regions: (1) the subiculum which varies from 40–85μ thick and is composed of compact hyphae; (2) the pillars which are quite short and usually scattered and inconspicuous, 25–110 x 100–210μ; (3) the pillars branching to form a layer of loosely entangled, nearly hyaline, ascending hyphae (3.2–5.6μ thick), this region of loosely packed hyphae about 400–500μ thick; (4) the hymenial layer which covers the loosely packed hyphae and stands out in sharp contrast to it. Hyphae of hymenium upright, densely packed, septate, 2–3μ thick; hymenium about 65μ thick; basidia arising directly from the hyphae in the basal region of the hymenium without a probasidial cell. Mature basidia thrice septate, fusiform, often slightly bent, 6.7–8.4 x 42–60μ. Spores not seen.

Associated with scale insects some of which are parasitized by irregularly coiled haustoria, while others are free from infection.

The structure of the hymenium and the basidia of the present species suggest a close relationship to *S. albidum* but the gross structure of the two fungi readily distinguishes them. This species is also close to *S. Henningsii* Pat. in hymenial characters but may readily be separated from that species by color and gross structure. Furthermore in the present species the subhymenial hyphae are more loosely packed, more hyaline, thicker, and with smaller lumina than in *S. Henningsii*. Also the arrangement of the pillars and context in the two is quite different.

Recognized by the wide (about 1 cm.) marginal zone of flocculent or spongy, whitish hyphae and the brownish hymenial zone, the latter frequently occurring in small irregular patches. The hymenial and basidial characters are also characteristic.

Specimens examined:

Java: Tjikadjang, on *Erythrina*, Feb. 1930, Prillwitz, coll., type; also Kedoe, Col de Kledoeng, on *Erythrina*, Aug. 1929, Steinmann, coll. Type in Herb. Bot. Gard. Buitenzorg, Java, and U. N. C. Herb.

Septobasidium lichenicolum (B. & Br.) Petch. Ann. Bot. **25**: 842. 1911.

Plate 65, figs. 7–14, and pl. 108, figs. 6–9

Resupinate, forming rather extensive, circular patches several centimeters wide. Surface Natal brown or Mars brown in the older parts, snuff brown to pale buff in the younger parts, mottled with lighter areas due to renewed growth in the older regions. Surface irregular with numerous mounds and ridges. These ridges are about 1 mm. or less wide and are more conspicuous towards the margin than in the older region. In the type material there are numerous rounded corky outgrowths on the bark about 1–2 mm. wide by about 1 mm. tall. These are covered by the fungus, a condition which increases its irregular appearance. In the older parts where the hymenium is well developed the surface is smooth except for the mounds and ridges but in the older parts where the hymenium is not developed the fungus has an alveolate appearance. Margin irregular, fimbriate but distinct. Subiculum in marginal region about snuff brown, with a few inconspicuous irregularly placed pillars, the tops of which spread out to form the top layer. In section 300–850μ thick, usually about 400μ thick, frequently stratose in the older parts, composed of three layers: (1) the subiculum 80–120μ thick from which arise (2) the inconspicuous, scattered tufts of hyphae which branch out above to form (3) the top layer 125–300μ thick, composed of densely packed entangled hyphae. Hyphae of subiculum, pillars, and top layer brownish, rather sparingly branched, septate, breaking up into short segments when mounted and slightly crushed, quite variable in thickness, 3.8–8μ thick, most about 5μ thick, rarely up to 9.6μ thick, frequently with narrow lumina. Hymenium composed of hyaline, upright, sparingly branched, sinuous hyphae and basidia in various stages of development, and thus because of color and the direction of hyphae sharply distinct from the underlying tissue; 40–130μ thick; hyphae in hymenium 2.2–3μ thick. These hyphae are quite variable even in different places of the same specimen, most frequently rather short, in other places longer and in parallel arrangement. Basidia at first subglobose, 12.6 x 16.8μ, elongating and becoming irregularly curved, 4-celled, 9–11.7 x 33–40μ. Sometimes part of the base of the basidium may remain as an empty cell. Sterigmata and spores not seen.

Associated with scale insects some of which are parasitized by irregularly coiled haustoria while others are free from infection.

The fungus grows so closely adherent to the substratum as to have the habit of a crustose lichen, which fact undoubtedly accounts for its name. The predominantly chocolate or cacao brownish color mottled with lighter buff areas, and the crinkled surface particularly in the younger parts are also distinguishing characteristics. The pillars of which only a few were seen are visible only with a high-powered binocular dissecting microscope. However, the specimen which I examined showed very little of the marginal region. Microscopically the following characteristics are distinctive: the hymenial region with upright parallel hyphae, the twisted basidia without a persisting probasidial cell, and the hyphae with narrow lumina which break up into short segments when crushed.

Boedijn and Steinmann, apparently without studying the type of *S. lichenicolum* (B. & Br.) Petch, have considered this species the same as *S. cinchonae* Rac. I have studied material of *S. cinchonae* Rac. sent to me from Raciborski's Herbarium by Dr. Szafer, and though the two fungi are very closely related they differ sufficiently to justify the retention of both specific names. The two differ strikingly in color: *S. lichenicolum* is brownish while *S. cinchonae* is white. *Septobasidium lichenicolum* is much coarser and larger than *S. cinchonae.* Both fungi have a rugose surface but the ridges of *S. lichenicolum* are about one mm. wide and easily visible without a lens while in *S. cinchonae* they are much smaller and less conspicuous. Moreover the ridges in the former are due to crinkles or folds in the surface while in the latter the ridges are frequently due to the incomplete formation of the top layer. In the type material of *S. lichenicolum* there are very few signs of pillars on the margin while in *S. cinchonae* there are numerous upright tufts of hyphae. The subhymenial hyphae are of about the same thickness in both species and in both the hyphae show the striking peculiarity of breaking up into short segments upon being teased apart and crushed. In both species the hymenia have the same structure, though in *S. cinchonae* there is a sharper difference between the hymenial layer and the underlying tissue. However the two differ microscopically in that *S. cinchonae* has a persisting probasidial cell, while in *S. lichenicolum* the probasidial cell elongates to form the basidium.

Septobasidium lichenicolum (B. & Br.) Petch, as understood by Boedijn and Steinmann, agrees with the type material in hymenial and basidial characters but differs strikingly in practically all other respects. In the material from Boedijn and Steinmann the surface of the fungus is nearly smooth, i.e. free from bumps and ridges and the marginal condition appears to be quite distinct from the margin in the type material. Moreover the subhymenial hyphae in the two are distinct: in the type material the hyphae vary from 3.8–8μ thick, averaging about 5μ thick, and have very narrow lumina, while in the material from Boedijn and Steinmann the hyphae vary from 3.6–4.2μ, averaging about 4μ thick, and have wide lumina. The hyphae of the type, when crushed, break up into short segments, whereas in the specimens from Java they do not when so treated. It is very likely that the material from Java should be described as a new species but this should be done only after large collections of fresh material have been studied and compared.

Petch (1911) has observed this species a number of times in Ceylon and remarks that it frequently extends for two or three feet over the bark of mango trees.

Specimens examined:

Ceylon: Peradeniya, on lichens and bark, Thwaites, coll., no. 1026, Dec. 1868, type in Kew Herb.; also on living tree, T. Petch, coll., Oct. 1915, part of 4781, in Petch Herb. and U. N. C. Herb. This specimen has much the appearance of the type but the bumps here are hollow and are not due to irregularities in the bark as in the type. Two other collections from Petch on mango, August 1917, may possibly belong here. One of these has a nearly smooth surface, the other has a surface broken into flaky areas due to the incompletely formed top layer, but neither shows the bumpy condition of the type.

Java: Res. Batavia, Pondok Gede, on *Cinchona*, Nov. 1929, Boedijn, coll., no. 11044, in Herb. Hort. Bot. Bog., and in U. N. C. Herb.; Kopeng (Helling Merbaboe), on *Morus* and *Cinchona*, Feb. 1930, Steinmann, coll., in Herb. Hort. Bot. Bog. and in U. N. C. Herb. Also reported from Java by Boedijn and Steinmann on *Crotalaria*, *Erythrina*, and *Thea*.

Septobasidium crinitum (Fr.) Pat. See note below.
Corticium crinitum Fr. Linnaea **5**: 530. 1830.

Plate 66, figs. 10, 11, and pl. 108, figs. 1, 2

Resupinate, often shelving on the lateral margins, covering extensive areas on the limbs of trees, often causing the formation of witches' brooms (Rick). Pale to a rich cream color, the subiculum usually white. Surface smooth, margin determinate. In section up to 0.5 cm. thick. Composed of three distinct layers: (1) the basal layer or subiculum which is rather thin; (2) the middle region of pillars which are very peculiar in that they arise in clusters (a scale insect usually beneath each cluster). Pillars 1.5–4 mm. tall, passing upward and often uniting (i.e. those in each cluster) to form a thick column which branches out to form and support the top layer; hyphae of pillars about 3.8–4.4μ thick, little branched; (3) top layer up to 1 mm. thick, usually 100–200μ thick, flexible like soft chamois skin. Hymenium formed over the surface of the top layer composed of upright hyphae 2–2.5μ thick which are recurved at the tips like a shepherd's crook or several times irregularly coiled in old specimens, and basidia in various stages of development. Basidia straight or slightly curved, 10.5–11.7 x 57–64μ, becoming four-celled by transverse septa, sterigmata 5 x 20–60μ. Spores strongly curved into a half circle, 6.3 x 32μ, only a few seen, usually 4-celled.

Associated with scale insects which, in young specimens, are usually found beneath the clusters of pillars. Some of the scale insects are parasitized by irregularly coiled haustoria while others are free from parasitism. Rick (1906b) states that this fungus is a true parasite on *Myrsine*; however in its association with scale insects it appears to differ in no way from the other species.

This remarkable plant may be recognized by the pale or rich creamy color, by the shelving habit of the lateral margin, by the pillars arising in clusters, and by the unusually long pillars. The delicate recurved hyphae in the hymenium and the basidia and spores are also distinctive.

Note: The combination *Septobasidium crinitum* seems to have been only a herbarium name used by Patouillard, but was so published by Rick, with Patouillard as authority, in Annales Mycologici **4**: 311, 1906, and also in Broteria **5**, in the same year.

Specimens examined:
Brazil: Sao Leopoldo, Rick, coll., on *Myrsine* sp., often causing witches' brooms, in Farlow Herb., Berlin Museum, and U. N. C. Herb.

Septobasidium pallidum n. sp.

Plate 64, figs. 4, 5

Resupinate, growing in small circular patches, the one patch seen 3 cm. wide. Tilleul buff with a slightly darker margin. Surface smooth, even under lens, except for irregularities in the bark. Margin sharply determinate, minutely fimbriate under a lens In section up to 500μ thick, composed of three distinct regions: the thin (25–40μ thick) whitish or pale buff subiculum from which arise (2) the middle region composed of stubby pillars, 40–80 x 84μ, the hyphae of which are brownish, 3–4.2μ thick, septate; the pillars branching out above to form (3) the top layer which is 300–400μ thick and composed of a lower brownish region of loosely entangled, obliquely ascending threads and the whitish hymenium which stands out in sharp contrast to the underlying brownish tissue; hyphae of context 3–4.2μ thick. Hymenium 40–60μ thick, composed of hyaline, thin-walled, much branched, irregularly twisted hyphae, about 3μ thick. Basidia arising laterally from these hyphae without the intervention of a probasidial cell, at first ovoid, when mature 7–8.4 x 29–34μ, 4-celled, quite variable in shape, thickest in the distal end, usually slightly bent, sometimes irregularly shaped, rarely almost straight, all four cells producing sterigmata and spores, the sterigmata quite variable in length, sometimes almost as long as the basidium Spores bent-elliptic, becoming 8-celled by transverse septa, 4–4.6 x 19–22μ.

Associated with scale insects (*Aspidiotus* sp.) some of which are parasitized by irregularly coiled haustoria. The haustoria are connected with the external fungus by hyphae that penetrate by the mouth as in *S. albidum.*

This species is closely related to *S. albidum* Pat., one of the most striking resemblances being in the way in which the fungus penetrates the insects' derm. The basidia are also similar though smaller in the present species than in *S. albidum.* The two differ in the following characters. In *S. albidum* the threads in the hymenium extend upright and are parallel while in the present species they are irregularly arranged. Furthermore, in the spores of *S. albidum* the distal end is sharply recurved while such is not the case in the present species. Also in macroscopic characters the two may be separated. In *S. albidum* the margin shows rhizomorphs and the surface is cracked while in the present species there are no rhizomorphs and the surface is smooth. The two also differ in color.

The present species may be recognized by the tilleul buff color, by the smooth surface, by the hyaline hymenium contrasting strikingly with the underlying brownish tissue, and by the whitish subiculum.

Specimens examined:

Brazil: Sao Lourenco, Minas Geraes, Dec. 1936, on *Pyrus communis,* sent by Nearch Azevedo, type in Herb. Jardim Botanico, Rio de Janeiro, and U. N. C. Herb. no. 10407.

Septobasidium septobasidioides (P. Henn.) Lloyd. Myc. Notes **5:** 722. 1917. (Merely listed as *Septobasidium* by Lloyd.)

Hymenochaete septobasidioides P. Henn. Hedwigia **43:** 172. 1904.

Septobasidium papyraceum Couch. Jour. Elisha Mitchell Sci. Soc. **44:** 249. 1929.

Plates 5, 7, 13, 14, 66, figs. 1–3, pl. 87 and pl. 108, figs. 4, 5

Subresupinate, with free shelving margins which extend out laterally about 5–12 mm. forming lobed wings on both sides of the branch. Margin of subiculum rather indistinct. Hymenial surface smooth, near tilleul buff. The subiculum, the top surface of the non-fertile part, and the sterile surface of the wings about bister or warm sepia, sulcate. In section the fungus consists of three main regions: (1) subiculum, (2) pillars, (3) top layer. The very thin subiculum extends over the bark. From this there arise here and there little pen-like structures (traps for the young scale insects) which are roughly circular or sometimes quite irregular in outline, and are 1–1.5 mm. wide. Each "pen" is composed of a roughly circular plate of hyphae which extend diagonally upward toward a central point. As the "pen" is formed a number of upright, unbranched, black pillars (10–25 x 350–450μ) grow up around it. The hyphae in the ends of the upright pillars now branch out and anastomose to form a horizontal covering over the "pen." As development proceeds pillars are formed over the entire subiculum, the pillars in turn giving rise to the horizontal covering so that this layer becomes continuous. From this layer arises a third horizontal layer, 200–400μ thick and easily separable from the underlying layer. Composed in section of two very distinct regions—the dark brown context and the light buff hymenial layer. Threads of the brownish context extend horizontally, are rather compacted, thick-walled, sparingly branched, 4–4.6μ thick. From the brownish context there arise at right angles the hyaline, thin-walled, branched, sinuous threads with recurved tips, which with the basidia form the hymenial layer. The hymenial layer develops from the central part outward toward the margin of the wings, i.e., the oldest and thickest hymenial region is near the center, while the thinnest and youngest is near the margin of the wings. Hymenium 40–168μ thick, hyphae about 2.5μ thick, basidia arising directly from the hyphae, young basidia ovoid or pyriform, elongating and enlarging to form a curved, sometimes an S-curved, 4-celled basidium, 8.4–10 x 40–55μ; sterigmata about 3μ thick at the base by 12–21μ long. Spores 5–6.7 x 21–26μ, elongated, usually recurved and with a papilla at the distal end.

Associated with scale insects some of which are parasitized by haustoria of the irregularly coiled type; others are not parasitized.

This remarkable species can easily be distinguished by the winged margin, by the light buff, smooth hymenium contrasting strikingly with the deep brownish subiculum, pillars, and context, and by the peculiar insect houses. The basidia and spores are also distinctive. The specimens from Jamaica differ from the type material in rarely showing a shelving margin (out of over fifty pieces which I still have only two show the shelving habit; the margin of the top layer is commonly uprolled, however, showing the shelving tendency), and in having a thinner (40–50μ thick), more papery hymenium. In the form from Jamaica several hymenial layers may be formed one over the other, which may be peeled off one after the other like thin sheets of paper. This latter condition appears to vary in the Jamaican material, sometimes a single, thicker, compact layer being formed as in the type. The spores in the Jamaican material are thicker and shorter but show the same peculiar papilla at the distal end as in the type. In the Jamaican material the subiculum is reddish brown, contrasting strikingly with the blackish pillars, while in the type both pillars and subiculum are very dark brown. In spite of these differences the two are strikingly alike.

Hennings mistook the basidia of this species for setae and described the species as *Hymenochaete septobasidioides*.

Specimens examined:

North West Brazil: Rio Jurua, on the branches of one of the Sapindaceae, Ule, coll., 1901, type in von Höhnel Herbarium, also in Farlow Herbarium (E. Ule, no. 2734, Herbarium Brasiliense).

Jamaica: Blue Mountains, Portland Gap, on *Psychotria* sp., July 10, 1926, type of *S. papyraceum*, no. 8500; also 8501 and 8505 (basidia 7.5–8.4 x 48–60μ, spores 7–9.2 x 21–25.2μ) on *Psychotria* sp.; 8504 on vine, basidia 8.4–9.6 x 37–50μ, spores 5–8.8 x 17.6–23.5μ; 8503, on *Vimea*; 8502, on ?; 8524 with *S. jamaicaense* on ?. All from Blue Mountains. In U. N. C. Herb., J. N. Couch, coll.

Septobasidium cirratum Burt. Ann. Mo. Bot. Gard. **3**: 334. 1916.

Plate 66, figs. 4–9, and pl. 108, fig. 3

Resupinate, covering an area on the bark several centimeters square; surface very irregular due to cracks in the hymenium, its irregular margin, and the fact that it is incompletely formed in places; the hymenial surface, however, inherently smooth to the unaided eye, finely pulverulent under lens, cracked in the older parts into irregular areas, light cinnamon drab, the context dark brownish (about bone brown) contrasting strikingly with the lighter colored hymenial surface. Margin not seen distinctly in the type material, but apparently the context is somewhat sinuose in that region. In section 1–1.5 mm. thick. Indistinctly divided into four regions: (1) the subiculum which extends over the bark, composed of branched, fairly compact and straight or sometimes circinately curved hyphae about 3μ thick, septate, slightly constricted at the septa; the subiculum passing imperceptibly into (2) a region where the hyphae unite into very loosely compacted, irregular and indistinct columns, which can be seen only in the younger parts, hyphae of columns about 3μ thick, branched, straight or curved; in the older parts of the fungus the region between the columns becoming loosely filled with hyphae; (3) context giving rise above to a horizontal layer, the hyphae of which are of the same structure and color as the underlying tissue (in places there may be several horizontal layers); on this top layer arising (4) the hymenium which is sharply distinct from the underlying region by the greater compactness of the hyphae, their lighter color, and thinner diameter. Hymenial layer readily separating from the underlying tissue. Hymenium 80–150μ thick, usually about 100μ thick, composed of densely packed, coiled hyphae which extend in an upright direction, hyphae about 1.5–2μ thick. Probasidia at first ovoid, elongating to form a straight or slightly curved, cylindrical basidium about 8–8.6 x 40–45μ, apparently 4-celled. Only the end cell and the one just below have been seen with sterigmata, the others doubtless with sterigmata when mature. Spores bent-elliptic, 4.2–5.8 x 16.8–21.8μ.

Associated with scale insects, some of which are parasitized by haustoria in the shape of irregular coils and also hyphae in the shape of sausage-shaped segments united by fine threads.

This species can be recognized by the light cinnamon drab hymenium, contrasting strongly with the darker bone brown context, the cracked surface, the almost entire absence of pillars, and by the delicate frequently dichotomously branched and circinate hyphae. The spherical cells with columella and thin protoplasmic contents are also distinctive. These are scattered throughout the context. The same type of cell occurs in *S. lanosum* Pat., a species which in no other way resembles the present one.

Specimens examined:

Cuba: Omaja, C. J. Humphrey, coll., on living hardwood tree, December, type in Mo. Bot. Gard. Herb. 15836 (Humphrey, no. 2773)

Septobasidium verrucosum n. sp.

Plate 63, figs. 8, 9, and pl. 109, fig. 7

Resupinate, covering areas several square centimeters in extent (2 x 4 cm.), forming a hard, rather inconspicuous, warty growth. The little elevations or warts are blackish brown arising from an inconspicuous brownish subiculum. The warts are closely arranged, separated only by narrow spaces, rounded on top with irregular sides, about 0.5–1.5 mm. wide by 0.5–1 mm. tall. Margin indeterminate. The subiculum is composed of pale brownish, irregular threads, which are considerably septate and branched. The warts are easily broken off from the subiculum and are very hard and compact. In section they show a remarkable structure, each wart having one or more hollowed out cavities each of which contains a scale insect. Each wart is composed of two distinct regions of tissue: a basal region of threads so closely packed as to appear pseudoparenchymatous in section and an upper region of extremely compactly arranged vertical hyphae. These two regions are connected by a narrow darker colored zone. The cavities containing the scale insects are in the lower region of tissue and are lined with a coat of tissue about 3 or 4 layers of threads thick. Lower region of tissue about 400–1000μ thick. Threads of lower part 4–6μ thick. The upper layer of tissue forms the hymenium and is 160–450μ thick, threads of upper layer 3–4.2μ thick, much septate and considerably branched. Young basidia pyriform or cylindrical, 5.8–6.7 x 20–25μ, elongating to form a 4-celled (?) basidium. Sterigmata not seen. Spores bent-elliptic, 4–4.2 x 16–16.8μ.

Associated with scale insects some of which are parasitized by haustoria in the shape of irregular coils; other insects are free from parasitism and are giving birth to young. The fungus insect relationship here is of exceptional interest. Each little wart or hill contains one or several parasitized insects, which undoubtedly were alive when the fungus was collected. In addition to the parasitized insects there were a few healthy insects giving birth to young beneath some of the warts. The healthy insects instead of being completely surrounded by the fungus tissue and thus imprisoned in it are in cavities which communicate with the exterior, the roof being formed by the fungus and the floor of the cavity by the bark. Such insects are well protected from their enemies, chief of whom are the hymenopterous wasps, as is indicated by the fact that in none of the several hundred warts examined was there the characteristic and unmistakable hole left by the exit of the adult wasp. The detailed relationship between fungus and insects, which could be worked out only with fresh material, would undoubtedly make a fascinating story.

The fungus is easily recognized by the numerous crowded blackish warts. The structure of the warts in section is also very distinctive. Its relationship to other species of *Septobasidium* must remain uncertain until the structure of the mature basidium can be determined.

Specimens examined:

California: Riverside, H. S. Fawcett, coll., 1915, with material of *S. canescens* Burt on *Quercus agrifolia*, part of material in Mo. Bot. Gard. Herb., Farlow Herb., and U. N. C. Herb.

Septobasidium accumbens (B. & Br.) Bres. (Apparently unpublished by Bresadola)

Thelephora accumbens B. & Br. Fungi of Ceylon, Jour. Linn. Soc. Bot. **14**: 65. 1873.

Plate 46, figs. 5, 6, and pl. 113, figs. 9, 10

Resupinate, or slightly shelving, strikingly resembling some of the Thelephoras in habit. Consisting at first of thick, pulvinate, irregular, cushion-like patches 1–3 cm. square, which later may become confluent; color about Prout's brown; surface irregular with large ridges, bumps, and depressions. Margin determinate, frequently with thick (up to 0.5 mm. thick) distinct rhizomorphs, which are sometimes brush-shaped at the end. In section up to 2 or 3 mm. thick, consisting of three more or less distinct regions: (1) a thick subiculum which overgrows the bark and insects, forming over the latter thick, tough-walled, hollow mounds (walls 0.5 mm. thick), from which arise (in young material from Tonkin) (2) the distinct pillars and rhizomorphs (in the type material only indistinct rhizomorphs are evident); pillars, when present, up to about 2 mm. high, frequently arising in tufts up to 1 mm. thick, pillars almost or entirely absent in old material; hyphae of subiculum and context very distinctive, rich brown in color, sparingly branched, considerably septate, the septa being exceedingly thin, hyphae 5–8.4μ thick, most about 6.3μ thick; from the pillars arising (3) the top layer. Hymenial layer not developed in any material examined, but just below the surface there are scattered numerous rounded clusters of conidia, these clusters frequently up to 50μ thick and made up of a compact, somewhat intestiniform mass of subglobose, frequently angular conidia; conidia 4–5μ thick.

Associated with scale insects some of which are parasitized by haustoria of the "glomerulus" type.

This species is easily distinguished from others by its shape, the cushion-like patches, the unusual thickness (up to 3 mm. thick) of the fungus, the thick-walled insect houses, the thick hyphae of the context and the conidia. It has been considered by Bresadola probably the same as *S. Merrillii* Bres. but the two are quite distinct.

Specimens examined:

Ceylon: Hautani, elevation 3000 ft., trunks of trees, G. H. K. Thwaites, Nov. 1854, type, labelled *Thelephora accumbens* B. & Br.; Habgalla, two collections, 544 and 546, Thwaites, Feb. 1868. On the label of no. 544 Bresadola has written "Ne pas = au type. Certainement un *Septobasidium*, probablement = *S. Merrillii* Bres."

Tonkin: Cho Ganh, M. Duport, coll., no. 144, October 1924, forming large thick patches. This form bearing conidia. See Patouillard (1925) for illustrations.

Septobasidium theae Boed. and Stein. Bull. Jard. Bot. Buitenzorg III, **11:** 177, fig. 4. 1931.

Plate 48, figs. 2–5, and pl. 114, fig. 8

Resupinate, forming a very extensive, mummy brown to nearly black, soot-like (except for color) growth over the main trunk and branches. Surface usually irregular and bumpy, sometimes nearly smooth, flocculent under a lens, often appearing as a loosely entangled mat of fine hairs, and very peculiar in that if rubbed, though gently, with a piece of filter paper or a white handkerchief the white surface is darkened by the vast number of conidia. Margin indeterminate. In section up to 2.2 mm. thick in dried condition, usually about 1–1.5 mm. thick in the mature parts. In the younger parts the context is not differentiated into regions but is flocculent throughout, in the older parts composed of two more or less distinct regions: (1) a basal region of very closely compacted threads which appear almost pseudoparenchymatous in section and (2) the upper flocculent region. This basal, compact, pseudoparenchymatous region may be as much as 2 mm. thick but usually is only a fraction of a mm. thick. It is deep brown or almost blackish in color. Hyphae of context loosely packed, 3–5.4μ thick, of even diameter throughout or inflated at irregular intervals; usually single but sometimes borne in diagonally ascending fascicles, considerably branched, the branches usually arising at right angles. Scattered throughout the context are numerous hyaline globose cells, 16–20μ thick, with rather thin protoplasmic contents and with the end of the hypha protruding into the cell like a small columella; borne singly or in groups; frequently two cells may be formed so close to each other that their walls flatten out on the sides in contact. These cells are easily separated from the parent hyphae, the cell always separating with a short stalk. The same type of cell occurs in several other species, e.g. *S. lanosum*, *S. cirratum*, and are not probasidia as is indicated by the fact that in the latter two species true probasidia occur in addition to the bladder type of cell. In addition to this bladder type of cell there is formed laterally or on the ends of the hyphae a cylindrical or subglobose cell with an elongated apiculus. This cell may be hyaline or concolorous with the hypha and is of the same diameter and 3 to several times as long as thick. The apiculus is about 1μ thick and about 4–10μ long and usually has a slightly enlarged end on which a small bubble of hyaline material is sometimes found. Hymenial layer not formed but in some regions there may be a fairly compacted top layer. Toward the upper surface of the fungus there are borne vast numbers of conidia. These are formed in chains and frequently have the appearance of a *Nostoc* filament; sometimes the conidia may be in clusters of branched chains; conidia black or very dark brown in mass; single conidia deep brown under microscope, cylindrical, barrel-shaped or rounded, 5–6μ thick; when mature the chains break up into single conidia or segments of varying length.

Associated with scale insects some of which are parasitized by haustoria in the form of spindle-shaped segments of hyphae connected by fine threads, these peculiar hyphae being coiled as in *S. fumigatum* Burt and *S. lanosum* Pat.

This species, as noted by Boedijn and Steinmann, is close to *S. lanosum* Pat., a conidial form described from Tonkin, China. The two may be readily separated. *Septobasidium lanosum* is blackish brown with a grayish sheen and has a determinate whitish margin. The surface is also smoother than in the present species. Microscopically the two are easily distinguished. In *S. lanosum* the conidia are hyaline and are borne in branched tufts. Also in *S. lanosum* the globose cells containing the columella-like structures are considerably smaller than in *S. theae*; in *S. lanosum* probasidia are formed while in the present species they appear to be lacking.

The species is readily recognized by its sooty (except that it is brownish) appearance, by the catenulate conidia, and by the globoid, bladder-like cells in the context. It has been found only on tea, on which it is reported to be very common, and this habitat may be distinctive.

Specimens examined:

Java: Tengalengan, on stems of *Thea*, August 1932, in Herb. Hort. Bot. Bog. and U. N. C. Herb.

Septobasidium pilosum Boed. and Stein. Arch. Theecult. **4**: 48. 1930.

Plate 18, figs. 1–8, and pl. 75, figs. 8, 9

This remarkable species described from Java on tea has recently been found in Florida on *Mangifera indica* and in Louisiana on *Magnolia virginiana*.

Resupinate, forming cushion-like patches of growth 1–10 mm. wide by 1.5 mm. high. Color deep brown. Surface minutely setose; margin determinate. In section not divided into layers, about 0.8–1.5 mm. thick. In the basal region rather few, thick hyphae extend over the surface of the bark, penetrating into the cracks; from these hyphae arise rather closely packed, entangled, upright and mostly straight, frequently septate, dark-colored, unbranched threads, 6–10μ, usually about 8.4μ thick. In the type these upright threads may be single or several may arise together, forming a small loose fascicle 10–50μ thick; threads of type 5–8μ thick, tips of threads hyaline. From the sides of the hyphae, particularly in their basal parts, just below the septum, are borne clusters of conidia, the individual cells of the cluster are hyaline at first but later become dark and thick-walled, 8–11.7μ thick. The individual cells of the cluster do not separate but the entire mass falls off. On these "multicellular conidia" or conidial bodies a distinct stalk cell is evident; conidial bodies irregular in shape but usually rounded, 25–65μ thick. Probasidia not formed in material from Florida but in the type material the probasidia are formed on the ends of short septate branches which apparently start off as conidia. Probasidia spherical, sometimes minutely rough, usually 16.8μ thick but varying from 12–17μ thick. Germination not seen.

Associated with scale insects, some of which are parasitized by hyphae the main branches of which are composed of spindle-shaped segments held together by fine threads. From these branches numerous coiled masses of spindle-shaped cells connected by fine threads arise. The sausage-shaped haustoria are connected with the external fungus through the derm pores as in *S. Burtii.*

The material from Florida differs from the type in having thicker hyphae and in the fact that the upright hyphae arise singly and not in fascicles as frequently happens in the type. Also the Florida material lacks the probasidia. Except for these slight differences the two fungi are strikingly alike. The basidia have not yet been found but the presence of the very highly specialized and distinctive haustoria typical of certain other species of *Septobasidium* is sufficient evidence that the two collections belong to that genus. The collection from Louisiana on *Magnolia virginiana* is producing probasidia, though I was unable to find any basidia. In color and other characters it resembles the type. The conidial form of *Septobasidium* studied by Petch (Trans. Brit. Myc. Soc. **11**: 62–65, 1926) on tea from Formosa also doubtless belongs here, as suggested by Boedijn and Steinmann (1930).

Specimens examined:

Java: Type, on tea, Steinmann, coll.; on *Artabotrys*, Dakkus, coll.

Formosa: On tea, Yaizw, Province Sugura, K. Hara, coll., studied by Petch.

Louisiana: Near Baton Rouge, on *Magnolia virginiana*, Dec., 1931, J. N. Couch, coll.

Florida: Delbray, on *Mangifera indica*, May 29, 1934, L. S. Light, Jr., coll.; also on *Mangifera indica*, near Oneco, Feb. 12, 1935, comm. by E. West. With *Leucaspis indica.*

Septobasidium conidiophorum n. sp.

Plate 48, figs. 6–8, and pl. 75, figs. 10, 11

Resupinate, extending for 15–20 cm. on the smaller limbs of *Citrus*, sometimes on the leaves, overgrowing solid patches of scale insects, fuscous to blackish brown, surface velvety or soot-like, margin determinate. In section about 400–700μ thick, composed of a very compact subiculum, 60–300μ thick, which overgrows the insects. Threads of subiculum mostly dark-colored but sometimes hyaline, very compactly arranged and intricately entangled, 3.5–5μ thick; in the lower part of subiculum there are formed, on the ends of certain short threads, a considerable number of small, 5.4–6.5μ thick, bladder-shaped, thin-walled, hyaline cells. From the subiculum arise two kinds of threads. One kind is dark brownish, straight or curved, septate, 3.8–5μ thick, sometimes slightly thickened and darker colored at the septa, extending irregularly upward and accumulating at the top surface to form an irregular mat. The other threads are hyaline, thin-walled, much curved and branched, with ends that are cut up into chains of short cylindrical or barrel-shaped cells, these cells being about as long as wide. From these chains of cells vast numbers of minute, oval, hyaline, bud cells are formed, about 1.5 x 4μ. Probasidia and basidia not seen.

Associated with scale insects (*Lepidosaphes beckii*) some of which are parasitized by haustoria in the form of regular coils as in *S. Burtii*. Others are free from infection. The fungus appears here to overgrow colonies of scale insects but just what the relationship is cannot be determined without further study of fresh material.

This species may be recognized by the dark brownish to blackish color with a moth-eaten, velvety appearance, by the two kinds of hyphae, the dark and light-colored, by the abundant minute conidia, by the lack of pillars, and by the regularly coiled haustoria.

Specimens examined:

Florida: Canal Point, on *Citrus* sp., Jan. 26, 1933, O. D. Link, coll., type in Herb. Agric. Exp. Stat., Univ. Fla., no. 8952, and U. N. C. Herb. 9347.

Septobasidium elatostemae n. sp.

Plate 113, fig. 7

Resupinate, hypophyllous, occurring in small (0.5–2 mm. wide), gregarious, rounded or irregular, cushion-like patches. Surface of larger patches smooth or finely granulose, dirty dark brown in color. Margin of subiculum indeterminate, of top layer determinate. In section up to about 1 mm. thick, composed of a subiculum which extends more or less indefinitely over the leaf surface and which gives rise here and there to loose, tangled mats of hyphae which extend upward clinging to the very long, dense, leaf hairs. Over these loose mats of hyphae the cushion-like patches are formed. Hyphae of subiculum pale brownish to hyaline, much branched and anastomosing freely, about 2.2–4μ thick; hyphae of context and top layer darker in color than subiculum, 3–4μ thick. Probasidia, basidia, and spores lacking. The top layer and subiculum are evenly dotted with dark-colored, three-celled, thick-walled, elliptic spores, about 8 x 24μ, which appear to arise on elongated, septate conidiophores, which in turn appear to be connected with the *Septobasidium* mycelium.

Associated with scale insects some of which are parasitized by haustoria of the irregularly coiled type.

Though this species is undoubtedly a new *Septobasidium* its position among the other species in the genus will have to remain doubtful until more and fertile material can be studied and the relationship of the conidia to the present fungus more accurately determined.

Specimens examined:

Philippine Islands: Mindanao, Mt. Apo, on leaves of *Elatostema*, June 18, 1924, Clemens, coll., no. 5622, type in Sydow Herb. and U. N. C. Herb.

Septobasidium lanosum Pat. Bull. Soc. Myc. Fr. **40**: 29, fig. 1. 1924.

Plate 47, figs. 3–8

Resupinate, covering several square centimeters. Blackish brown or dull slate color, with a determinate, narrow, grayish margin. Surface flocculent. In section 800–1100μ thick, not distinctly layered except for a narrow region of loosely arranged, horizontal hyphae close to the top; hyphae of context loosely packed, much branched and septate, frequently united into fascicles, deep olive brown to hyaline. In most parts of the context the brownish hyphae stand out in rather striking contrast to the hyaline ones. Brown hyphae thicker than hyaline ones, 4.2–5.8μ; hyaline hyphae 3.6–4μ thick. Probasidia spherical, 10–14μ, borne on hyaline recurved hyphae. Basidia not observed but a considerable number of bent-elliptic spores were seen which had the characteristic appearance of *Septobasidium* spores. Also borne toward the top layer but apparently originating from other hyphae than those which bear the probasidia are numerous tufts of conidiophores. These are branched, the branches being usually straight and upright, giving the conidiophore a brush shape. Conidia pinched off by a constriction in the hypha, about 3 x 6μ, hyaline, usually falling away in 2-celled units, sometimes in 3 or 4-celled units. Separate conidia were rather scarce in the material which I studied. In the basal region there are a considerable number of spherical cells about the same size or slightly smaller than the probasidia but with more nearly hyaline contents. There is also a peculiar type of cell which I have noticed most frequently on the conidial mycelium and on other threads which may or may not have been the conidial mycelium. These cells were on the ends of hyphae and were of equal diameter with the hypha or slightly swollen and were extended at the distal end into a long narrow tip. This same kind of cell has been seen before in *S. apiculatum*. Their function, if they possess any, is unknown.

Associated with scale insects some of which are parasitized by haustoria in the shape of short hyphal segments connected by fine threads, some of these hyphae being coiled as in *S. fumigatum* Burt.

This fungus is easily recognized by the dark brown color with a faint dust of gray, by the loose flocculent context devoid of a distinct top layer and by the conidia. One of its most remarkable characteristics is the apparent occurrence of two distinct types of mycelia, the conidia-bearing and the probasidia-bearing mycelia. It is close to *S. paulense* and *S. theae*. For comparisons see under those species.

Specimen examined:

China: Cho Ganh, Tonkin, on living branches of *Citrus decumans*, M. Duport, coll., February 19, 1922, type in Patouillard Herbarium, Harvard University.

Septobasidium paulense P. Henn. Hedwigia **43**: 197. 1904.

Plate 47, figs. 9–14, and pl. 113, fig. 8

Resupinate, covering rather extensive (?) areas on the limbs of trees. Buffy brown or drab. Surface smooth and velvety. Margin not seen. In section 600–1000μ thick, context composed of much entangled, much branched hyphae, which are not united into pillars or rhizomorphs; context differentiated into four rather indistinct zones. (1) The subiculum which extends over the bark and insects and from which arises (2) a zone of yellowish brown hyphae 250–400μ thick, hyphae 4–5μ thick, rarely up to 8μ thick, giving rise to numerous racemosely branched conidiophores, the primary branches frequently arising at almost right angles from the main hypha; conidia hyaline, cylindrical at first but becoming barrel-shaped, formed simply by the septation of the hyphal branches, with rather full protoplasmic content, 3.8–4 x 5–5.4μ. Very few conidia were seen which had been broken off from the conidiophores and, when separation does occur, the conidia break away in chains of 2 to 4 at a time. Scattered through this basal zone are numerous round bladder cells, with thin cytoplasmic contents, the end of the hypha projecting into the bladder like a columella, 8–12μ thick. Also in this region there are a considerable number of cells borne laterally or sometimes terminally, the ends of which are drawn out into a narrow tube the distal end of which is frequently slightly enlarged, 4.2–5 x 15–20μ, including the narrow part, which is less than a micron thick and 6–10μ long. These peculiar cells as well as the bladder cells are found on threads intermingled with the conidial hyphae and not infrequently the cells may be borne on branches of the conidiophore. From the zone of yellowish brown hyphae there arises (3) a zone of hyaline or subhyaline hyphae; this zone is about as thick as the darker zone. Hyphae 3–3.8μ thick, ascending diagonally to form (4) the hymenial layer which is 42–85μ thick, and composed of densely packed, much branched, entangled, hyaline threads, 2.2–3.5μ thick, the ends of whch are recurved like a shepherd's crook. From the sides of these recurved hyphae arise the probasidia, which are hyaline, spherical, 12.6–16μ thick, with rather thick, slightly roughened walls. Some of the basidia were seen in the early stage of germination but no mature basidia or spores were seen.

Associated with scale insects, some of which were parasitized by hyphae composed of spindle-shaped segments connected by fine threads, some of which are coiled. Other insects are not parasitized.

This species shows a striking resemblance to *S. lanosum* Pat. in the two kinds of hyphae, the conidial and the probasidial. Moreover the probasidia in the two differ only in that those of *S. paulense* are slightly larger and rough-walled and both are borne on recurved hyphae. The conidiophores are strikingly similar but in *S. paulense* the branches of the conidiophore arise more nearly at right angles to the main threads than in *S. lanosum.* The bladder cells and the cells with the spine tip are identical in both species. The two differ in gross appearance, in color, and in structure. In *S. paulense* the surface is smooth, in *S. lanosum* the surface is flocculent. Also in *S. lanosum* the conidia-bearing hyphae are deep olive brown and are frequently united into fascicles while in *S. paulense* the conidia-bearing hyphae are yellowish brown and are never (so far as I saw) united in fascicles.

This species is also related to *S. theae* Boedijn and Steinmann and *S. tuberculatum* B. & S. in having a conidial stage and in having the bladder-shaped cells and the cells with the pointed ends.

This species is very distinct from *S. velutinum* to which it was said by Hennings to be close.

Specimens examined:

Brazil: Sao Paulo, Alto da Servia, on limbs of trees, type in Herb. Bot. Gard. Berlin and a fragment in U. N. C. Herb.

Septobasidium molle n. sp.

Plate 48, fig. 9

Resupinate, covering areas up to ten or more sq. cm., forming a dry, loose, fluffy, irregular, russet-colored growth. Margin indeterminate. In section up to about 1 mm. thick in places, not stratified but composed throughout of the same type of loosely entangled, pale brownish to hyaline hyphae which penetrate deeply into the cracks of the dead parts of the bark; hyphae 4–6.3μ thick, branched, the branches arising usually at right angles, septate, the cross wall clearly double at septum. Old hyphae thick-walled and with narrow lumina, sometimes with hyaline, gelatinous (?) sheath. Giving rise here and there to conidia which occur in clusters, each conidium about 8μ thick. Probasidia also formed in groups but not arranged in a definite hymenium, with a one- or several-celled stalk, spherical, pale to dark brown in color, with a distinctly warted wall, 18–23.2μ thick. Basidia and spores not seen.

Associated with scale insects some of which are parasitized by haustoria of the spindle-shaped type, the spindles being connected by fine threads.

This species may easily be recognized by the soft, fluffy, russet-colored growth, which looks like very fine intertwined brownish hairs. The conidia, and the large, brownish, rough-walled probasidia are also characteristic.

Specimens examined:

Philippine Islands: Agricultural College, Laguna, on *Piper* sp., September 25, 1933, Domingo Altamirano, coll., type in Herb. Agric. College Laguna, no. 47, and U. N. C. Herb.

Septobasidium tomentosum n. sp.

Resupinate, covering the bark of small twigs for ten or more centimeters with a thin, rather inconspicuous, powdery, benzo brown growth. Surface under binocular dissecting microscope has a matted woolly appearance, also uneven and bumpy in places. Margin indeterminate. In section varying in thickness from a few threads up to 600μ thick, usually about 175–300μ thick, composed of two layers. A very thin, basal subiculum which does not form a continuous layer and may in places be so thin that the bark is visible; when well developed 40–100μ thick and composed of compactly arranged, dark brown hyphae. From the subiculum arises a loose tangle of branched hyphae which are 3–4.2μ thick, frequently septate, slightly constricted at the septa. Toward the upper surface the hyphae give rise to loosely arranged, branched, slightly twisted chains of hyaline, crystal-encrusted conidia, which are elliptic, cylindrical, or barrel-shaped, 4–5 x 6.3–12.6μ. Probasidia and basidia not observed.

Associated with scale insects, some of which are parasitized by irregularly coiled haustoria; others are not parasitized.

This species is somewhat like *S. lanosum* Pat. but differs from that form in color and in the structure of the conidiophores. The haustoria in the insects are also different. The color, the thin tomentose surface, the absence of probasidia and basidia, and the conidia are characters by which the fungus may be distinguished.

Specimens examined:

Australia: North Queensland, on *Citrus* sp. with *S. pilosum* B. & S., July 1936, no. 4516, type in Herb. Plant Pathology, Brisbane; also in U. N. C. Herb. 10319.

Septobasidium tigrinum Boed. and Stein. Arch. Theecult. **4:** 47, Pl. 25, 26. 1930. Also in Bull. Jard. Bot. Buitenzorg III, **11:** 189, figs. 12–14. 1931.

Plate 46, fig. 7

Resupinate, forming small (1–5 mm. wide), irregular, thin, tan, reddish brown or nearly blackish patches which may follow the branch for 30 cm. or more; patches frequently anastomosing. Margin determinate. In section 130–540μ thick, stratose. A very thin (12–40μ thick) basal layer extends over the bark, giving rise to a few, short, thick, very inconspicuous pillars, which are about 60μ tall and 60–250μ thick. These pillars give rise to a thin brownish, crust-like, compact layer of hyphae, the threads of which are 3–5.4μ, rarely up to 8.4μ thick, and extend mostly parallel with the surface. From this layer arise hyphae which give rise to another compact layer the hyphae of which are densely entangled and have a more vertical direction. Several horizontal layers may be formed. The top layer consists of pale brownish hyphae 3–4.2μ thick which extend mostly in a vertical fashion except for an upper zone of compact, entangled threads; this upper zone varies from about 125–400μ thick. Only a few probasidia (?) seen, spherical, about 16μ thick (10–12.5μ thick, according to Boedijn and Steinmann), with rather thick walls. Another collection from another locality in Java was bearing abundant conidia, which were formed either laterally or in chains, the laterally borne conidia 3.5–5μ thick, while the ones formed on the ends of the hyphae 3.5–5 x 5–9.5μ; single conidia nearly hyaline, yellowish in mass.

Associated with scale insects some of which are parasitized by irregular coils while others are free from infection.

This species is readily distinguished by its occurrence in small, irregular, single or anastomosing patches and by its stratose structure, the layer next to the subiculum being exceedingly thin and compact.

This species shows a striking resemblance to *S. Thwaitesii* in its sectional structure (see fig. 13, Boedijn and Steinmann 1931) and also both have top layers which occur in broken anastomosing patches. Having only a small sterile fragment of *S. tigrinum* for study, it is impossible at present to do more than call attention to the apparent close similarity between it and *S. Thwaitesii.*

Specimens examined:

Java: On living branches of *Thea,* July 1927, Steinmann, coll., in Herb. Bot. Bog. Hort. and U. N. C. Herb.

Septobasidium scopiforme Pat. Ann. Mycol. **5**: 364. 1907. Also Bull. Soc. Myc. Fr. **24**: 1–2, fig. 1. 1908.

Plate 113, fig. 6

Fungus body composed of a group or tuft of upright spines, each tuft consisting of many (perhaps 50) spines and having a diameter of 1–1.5 cm. Spines 1.3–2 cm. tall and 200–800μ thick, usually about 400μ thick. The spines usually arise by the anastomosis of several upright columns and not infrequently may branch once or several times. Each spine is usually thicker near the middle region and about 2 mm. from the tip, tapering abruptly to a fine needle-like point, brownish with a black tip. The tuft of spines arises from a very indistinct subiculum, which appears to consist merely of the bases of the spines. Beneath the tuft there was the shell of an insect (not a scale insect but perhaps related to *Phylloxera*) with which most likely the fungus is associated. Sometimes some of the upright spines may bend over touching the bark, in which case the hyphae from the tip region grow out spreading over the bark in the shape of a palm leaf fan. Hyphae of the spines olive brown under microscope, rather thick-walled, straight, and mostly unbranched in the central part, wavy and branched toward the outside, septate, 4–6.3μ thick. On the outer side of the spines and also over the fan-like structures the hymenial layer is formed. This layer consists of hyaline threads about 4μ thick and probasidia in various stages of development. Probasidia hyaline, spherical or pyriform, sessile or stalked, about 11–13.5μ thick. There were some structures which appeared to be straight, four-celled, empty basidia but were so collapsed that it was impossible to be certain. No spores seen.

The specimen studied by me had been broken off the wood and hence I was unable to determine if scale insects were present. There is no reason, however, to doubt that this species is associated with insects.

This species appears to be most closely related to *S. Clelandii* and also shows a slight resemblance to *S. pteruloides* but is easily distinguished from either of those species by its habitat on the stems of Bambusaceae and by the lack of a distinct subiculum. The spines in the present species are quite similar to those of *S. pteruloides* (Mont.) Pat. but in that species the subiculum is very distinct and characteristic. In *S. Clelandii* the subiculum consists of a distinct, thick, tough pad of tissue and the spines are shorter than in the present species and usually with blunt tips.

Specimens examined:

Brazil: Prov. Sao Paulo, Campinas, Noack, coll., no. 792, type in Pat. Herb., Harvard Univ.

Septobasidium pteruloides (Mont.) Pat. Bull. Soc. Myc. Fr. **41:** 337. 1925.

Hydnum (?) *pteruloides* Mont. in Bélanger: Voyage aux Indes Orientales. Champignons. p. 150–153, Pl. 14, fig. 3. 1834.

Lachnocladium rameale B. & Br. Jour. Linn. Soc. Bot. **14:** 67. 1873.

Septobasidium rameale (B. & Br.) Petch. Ann. Bot. **25:** 843. 1911.

Plate 113, figs. 1–5

Resupinate, having the appearance of a resupinate *Hydnum* with very long spines, extending for a length of 15 cm. or more over the smaller branches, usually completely surrounding the branch. "In color violet or purple black when fresh, with a narrow, white margin" (Petch), the dried type material close to Van Dyke brown. Composed of a thin basal layer, the subiculum. From the marginal region of the subiculum there arise the "insect houses" or traps. These "trap" initials show a rather regular arrangement, those of the same age being about equally distant back from the margin of the subiculum. Each trap is initiated as an incompletely oval plate of hyphae, the incomplete part of the oval always being toward the older part of the fungus. The hyphae forming the wall of each trap grow diagonally upward, eventually meeting in a longitudinal line to form a roof over the trap and thus the trap is closed in everywhere except at the end next to the older fungal growth. Many of these traps contain non-parasitized insects. Some of the hyphae extend upwards above the side walls of the traps to form numerous small bristles. As growth proceeds the small bristles which outline any one trap extend upwards diagonally, bending toward a central point and there anastomosing to form one or several large spines over each trap. Spines when mature 0.1–0.5 mm. thick by 6–14 mm. tall. Frequently the spines from adjacent traps anastomose, thus forming a dense tangle, resembling, under the binocular microscope, the roots of a mangrove swamp. The spines may be single, i.e. unbranched, or branched once or several times. The branching may occur near the middle but is more frequent near the tip. The spines are thickest near the middle and taper to a sharp point. In the young condition they are covered by a grayish sheen but in the more mature type material they are a deep brown. Hyphae of spines 4.2–6.7μ thick, straight or somewhat wavy, frequently septate, brownish. No probasidia, basidia, or spores of any kind seen.

Associated with scale insects (*Chionaspis* sp.) some of which are parasitized by haustoria in the form of irregularly coiled hyphae; other insects are not parasitized.

This species can be easily recognized by the long, branched, sharply pointed, dense spines. The marginal condition is also characteristic.

A species of fungus collected by Dr. Cleland in Australia has been confused with the present fungus by Patouillard and Lloyd. The two resemble each other in having the hydnoid processes but aside from this they differ in practically all other respects. In the first place *S. Clelandii* is not associated with scale insects as is *S. pteruloides* but is associated with a peculiar gall-forming insect (related to the Phylloxeras) which lives within the woody tissue. As a result of this association with the two different kinds of insects, the subicula of the two fungi are entirely different: in *S. Clelandii* the subiculum is a simple, thick, hard, flat pad of hyphal tissue while in *S. pteruloides* the subiculum in the marginal region shows the conspicuous insect traps specially adapted to catch the young scale insects. Also in *S. Clelandii* the spines are darker colored, being blackish brown, are usually shorter, and are rather rarely branched, never showing the peculiar much branched condition at the base as found in *S. pteruloides*. All the material of *S. pteruloides* which I have examined is sterile. The haustoria of the two fungi are different, those in *S. Clelandii* being of the glomerulus type while those of *S. pteruloides* are of the irregular coil type.

Specimens examined:

East Indies, Island of Madura: On the bark of trees in the mountains, January 1826, Belanger, coll. Type in Herb. Crypt. Mus. Paris.

Ceylon: Ambagamowa, Sept. 1862, Thwaites, no. 595, on living branches, also on the petioles of leaves, and sometimes on the leaves, as *Lachnocladium rameale* B. & Br. in Kew Herb.; also Nuwara Eliya, Nov. 1908, Petch, coll., in Patouillard Herb. and Petch Herb.

Septobasidium spiniferum Burt. Ann. Mo. Bot. Gard. **13:** 333. 1926.

Plate 111, fig. 3

Resupinate, covering areas 4 x 9 cm. on the smaller branches, blackish brown. The surface is veined, the veins extending longitudinally from the center outward toward the margin and usually arising obliquely from the surface at an angle of 5°–10°, to end in a sharp-pointed spine. Spines about 0.2–0.3 x 1–5 mm., apparently always pointed towards the margin. Except for the veins and a few minute holes the surface is inherently smooth. Margin determinate but obscured by the overlapping of the spines of this region. In section 200–1000μ thick, composed of three regions: (1) a thin and inconspicuous subiculum which forms an incomplete layer over the bark; over the subiculum extend (2) numerous rhizomorphs, the thicker ones being the cores of the veins, visible on the top surface and extending up above this layer to form the spines; from the sides of the rhizomorphs arise hyphae which form (3) the top layer which is 150–400μ thick. Between the top layer and the subiculum there are numerous anastomosing tunnels. Hyphae of subiculum, rhizomorphs, and top layer olive brown under microscope, considerably septate, 4–5μ thick. The hyphae of the rhizomorphs extending parallel, those of top layer densely entangled and branched. Probasidia (?) 10–16.8μ thick, usually globose but sometimes flattened, pale brownish in color, with minutely punctate walls. Basidia and spores not seen.

Associated with scale insects (*Aspidiotus* sp.) some of which are parasitized by haustoria in the form of irregular coils while others are free from parasitism.

This species is easily distinguished by the veined, blackish brown surface with the hydnoid teeth which extend outward almost parallel with the surface and by the minutely rough brownish probasidia. In the absence of basidia the best evidence that the species is a *Septobasidium* is the association with scale insects and the typical haustoria. Its relationship to other species must remain doubtful until the basidia are found. The colored, punctate probasidia suggest a possible relationship to *S. Curtisii*.

Specimen examined:

Chile: San Pedro, Concepcion, R. Thaxter, coll., dry scrub, Nov. 1905. Type in Farlow Herb. and Mo. Bot. Gard. Herb., 57895.

Septobasidium Bakeri Pat. Leaflets Philipp. Bot. **6**: 2239. 1914.

Plate 110, figs. 4–6

Resupinate, covering extensive areas up to several square decimeters on the living stems of *Schizostachyum* sp., very easily separable from the substratum when dry (perhaps also when wet). Bister in the younger regions, becoming clove brown to brownish black (Ridgway). Surface smooth in the younger parts, cracked in the older regions, the cracks being up to 2 mm. wide and a cm. or more apart. Margin determinate, the margin of the top layer fimbriate. The exposed region of the subiculum is covered, except for the outer two or three millimeters, with a dense growth of pillars between which there is much anastomosing. In some specimens in which the top layer has been formed out even with the subiculum the pillars are not evident on the margin. In section 1.5–4 mm. thick, in the younger parts composed of three distinct regions, subiculum, pillars, and top layer; in older parts the fungus though somewhat stratified is more or less compact throughout. (1) Subiculum 50–250μ, usually about 100μ thick, buff colored on the lower surface (i.e. when removed from substratum), the hyphae from the buff region being nearly hyaline under the microscope, main hyphae of subiculum brownish, about 4.2–5μ thick, septate without clamps. (2) The middle region 0.7–2.0 mm. thick, composed of thick columns or plates or pillars. The pillars never extend as individuals from bottom to top layers but unite with each other freely or the hyphae of a pillar may spread out in a fan shape and mingle with hyphae from other pillars, which have spread out in the same way. Hyphae of pillars about 4.2μ thick, characteristically much septate, the distance between septa being about 20μ, septa without clamps. (3) The hyphae of the pillars branch out to form the top layer which varies in thickness from 75–250μ. Hymenial layer 50–130μ thick, composed of brownish, entangled, densely packed hyphae and a layer of closely packed, ovoid, hyaline probasidia (?) 8.4–12.6 x 14–17μ. The subhymenial hyphae are considerably entangled and loosely packed, about 4.2μ thick. Basidia not seen, but there were a good many bent-elliptic spores about 3.8 x 13μ, which perhaps belong here.

Associated with scale insects, *Odanaspis schizostachyi* Ckll., some of which are parasitized by irregular coils while others are free from parasitism.

This is indeed one of the most remarkable of all species of *Septobasidium* which I have yet examined. The entire stroma is easily separable from the slick bamboo stem, exposing underneath and sticking to the bottom layer the scales of numerous scale insects. The stem beyond the fungus is free from scale insects.

Specimens examined:

Philippine Islands: Prov. Laguna, Los Banos, Mt. Maquiling, C. F. Baker, coll., no. 87, type in Pat. Herb., Bot. Herb. Berlin, N. Y. Bot. Gard. Herb., Univ. Philippines Herb., and U. N. C. Herb.; also C. F. Baker, coll., no. 73, Sept. 22, 1912, same loc., in Pat. Herb.; also same loc., Reinking, coll., no. 9975, Oct. 1920, on dead bamboo, fungus in bad shape; also a large collection sent by Ocfemia, from same loc., Dec. 1933, Francisco Gomez, coll., no. 49, in Herb. Univ. of Philippines and U. N. C. Herb. All on *Schizostachyum* sp.

Septobasidium myrsinae n. sp.

Plate 47, fig. 16, and pl. 111, fig. 4

Resupinate, covering extensive areas (some pieces up to 20 cm. long) on the limbs of trees; surface mostly smooth but with a few cracks and irregularities and minute holes or alveolations where the top surface is immature but with a smooth feel like kid, color whitish with splotches of buff or uniformly buff colored in some specimens. Margin sharply determinate, very peculiar and characteristic. The subiculum in the marginal region is exceedingly compact, thin (20–40μ), and closely appressed to the substratum. From the subiculum little tufts of hyphae (50–100μ tall) rise; the hyphae of each tuft branch out above to form a compacted knob of hyphae; from the lower edge of each knob a membrane of hyphae grows out horizontally, forming a wide-brimmed, hat-like structure, the membranes from adjacent knobs finally anastomosing to form a continuous layer of tissue. The knobs remain conspicuously (with lens) evident above the surface of this second layer. Subiculum at first whitish in marginal region, becoming pale brownish, knobs deep brownish. From this second layer arises a dense tangle of branched, anastomosing, dark brown pillars which form in section an irregularly honeycomb-like structure, pillars 16–80μ thick, this region being 1–2 mm. thick; hyphae of pillars compactly arranged, septate, 3.8–4μ thick, giving rise above to the thin, papery, compact top layer which is composed of closely interlaced, mostly horizontally extending, hyaline hyphae which vary from 3–4μ thick. Top layer 50–80μ thick. Probasidia, basidia, and spores not seen.

Associated with scale insects which are very abundant and in all stages of development. Some are parasitized by irregularly coiled haustoria, others are free from infection.

This species was first distributed by Rick as *S. albidum* (Rick, no. 157) but may readily be distinguished from that species. It seems to be closest to *S. Mariani* Bres., resembling that species in having three instead of two horizontal layers. It may easily be distinguished from *S. Mariani* by the marginal condition. Indeed the peculiar margin of the present species would distinguish it from any of the previously described species.

Specimens examined:

Brazil: Sao Leopoldo, distributed by Rick in Fungi Austro-Amer. no. 157, as *S. albidum* Pat. (see *S. Rickii* below), on living branches, 1906, in Museum Bot. Gard. Berlin and U. N. C. Herb.; also two other collections from Rick, 10321 on *Myrsine*, and 10427, both in U. N. C. Herb.

Septobasidium Peckii Couch. Jour. Elisha Mitchell Sci. Soc. **51**: 75. 1935.

Plate 85, figs. 1, 2

Resupinate, forming small inconspicuous patches of growth (3 x 6 cm. in the one specimen seen). Tawny to russet or hazel. Surface composed of usually flat, sinuose plates, i.e. the top surface is incompletely formed. Margin determinate but subiculum is indistinct, with a few very inconspicuous (visible only with lens) pillars. In section 500–1000μ thick, composed of a very thin subiculum which forms a continuous layer next the bark, from which arise very short pillars the hyphae of which branch out above to form the top layer which may be 500μ thick. Hyphae of top layer upright, frequently grouped into columns between which are small spaces, so that the top layer in section has a somewhat spongy appearance. Hymenium immature, composed of upright, closely compacted threads which give rise laterally to the probasidia, the latter being immature. Hyphae of subiculum hyaline, in younger parts thin-walled, about 3μ thick, hyphae of pillars, and subhymenial region pale brownish, 3.8–4μ thick, hyphae of hymenium 3.8–4μ thick.

Associated with scale insects (*Chionaspis* sp.), some of which are parasitized by irregular coils while others are free from parasitism.

The species is easily distinguished from all others by its striking tawny to russet color, surface appearance, and the peculiar hymenium with the upright threads. The fungus, collected in October, does not show mature fruiting structures but the structure of the immature hymenium and the old hymenium of the preceding spring suggests a relationship to *S. albidum*, which occurs in Ecuador and *S. flavo-brunneum*, which occurs in Java.

Specimens examined:

New York: Near Indian Lake, on living branches of *Alnus incana*, October, Chas. H. Peck. Type in Patouillard Herb., Harvard Univ., labelled *T. pedicellata* Schw.

Septobasidium Rickii n. sp.

Plate 48, figs. 10–12, and pl. 112, figs. 1, 2

Resupinate, growing in regular, circular patches up to 6 cm. wide. Color Sayal brown in the central part (hymenium) through clay color to pinkish buff in the marginal region. Hymenial surface smooth, almost glabrous in places, crustose; the surface where the hymenium has not been formed is nearly smooth to the unaided eye but is of a very fine woolly appearance under a lens. Margin abruptly determinate, the margin of the subiculum being white and about 1 mm. wide. In section up to 2.5 mm. thick, composed of five more or less distinct regions. (1) The white subiculum which extends over the bark, 30–85μ thick, hyphae very closely packed, in the older places appearing pseudoparenchymatous, 3.8–5μ thick, thin-walled. From the subiculum arise (2) a few stubby tufts of hyphae which branch out to form (3) the very thick layer of spongy tissue. In the older parts the tufts of hyphae and the tunnels between the tufts disappear, the spongy region arising directly from the subiculum. The spongy layer is 1–1.5 mm. thick, and is composed of loosely arranged, entangled, hyaline, thick-walled hyphae, 4.6–5.2μ thick. The hyphae anastomose rather freely and are glued together at numerous places by a brownish material. These hyphae are very brittle, breaking into short pieces, several of which are usually glued together by the brownish substance. The scale insects are embedded in this part of the fungus, the parasitized ones completely covered by the fungus, the healthy adults in cavities which open to the top surface of this layer. From this spongy region (0.5–1.5 cm. back from the margin) there arises (4) a region of pillars, which are about 4–8 to a sq. mm. Pillars usually unbranched, 40–160 x 1000μ, most about 80μ thick, hyphae compactly arranged, hyaline or pale brownish, thin-walled, about 4μ thick. The pillars branch out above to form (5) the top layer which is 160–250μ thick, friable when mature, and composed below of loosely entangled, hyaline to pale brownish hyphae, about 4–4.2μ thick, which extend diagonally upwards to form the hymenium. Hymenium 60–80μ thick, composed of closely packed threads the ends of which extend mostly vertically ending in one to three ovoid to subglobose cells. No structures could be found which could be certainly identified as probasidia. There were, however, numerous collapsed, brownish, somewhat bell-shaped, apparently gelatinized bodies, borne singly or frequently in moniliform series through which new hyphal growth had usually penetrated. Sometimes as many as five of these bodies on the same hypha in a row, the hypha penetrating through each one. No basidia or spores seen.

Associated with scale insects some of which are parasitized by haustoria of the glomerulus type; others are free from infection.

The unique structure of this species will readily distinguish it from all other species of *Septobasidium*. The very thick spongy layer from which the pillars arise and the thin top layer are characteristic. Indeed this is the only species known in which the context is below the pillars instead of above.

The fungus-insect relationship in the present species is immensely interesting and although the material is dried the main outline of the story can be pieced together. Back about 1 cm. from the margin where the pillars are beginning to form there are numerous lunulate openings formed by the slight separation of the dorsal and ventral scales of scale insects. These openings are so small that they can be seen only with a binocular microscope. If the two scales are pressed apart and this tunnel is followed downwards, it is seen to open into a large chamber which contains either the dried remains of a once healthy insect giving birth to young or the shell of a parasitized scale containing the body of a parasitic hymenopterous wasp. The young scale insects emerge from this chamber through the lunulate openings and crawl about over the surface of the spongy layer.

In the marginal region of the fungus between the subiculum and the thick layer of loosely woven hyphae there is a labyrinth of connecting tunnels. By dissecting away the covering of these tunnels a considerable number of insects are discovered. Some of the insects still possess their antennae and legs indicating that they had been crawling about when the fungus was collected. Others are further along in their development. In fact, from the marginal region backwards it is possible to find in these tunnels and chambers all stages from the crawling nymphs up to the mature adult female giving birth to young.

The curved, ascending tunnels through which the young insects emerge are so peculiar as to deserve special attention. They are formed by a great elongation upwards of the ventral and dorsal scales. In most scale insects which I have observed growing with or without a fungus, the insect or its ventral scale is flattened out against the bark unless the insects are so crowded that such a position is impossible. In the present species the insects settle down on the floor of the labyrinthiform tunnels but instead of keeping an open connection with the tunnels or instead of breaking through to the tunnels at maturity, the insects build the channels upward through the fungal tissue. In the material which I have studied it is possible to find numerous stages in the formation of the tunnels. In species of *Septobasidium* in which the fungal tissue is tough it would be very difficult, if not impossible, for the insects' scales to be built up through the fungal tissue. In the present species, however, it is very likely that the structure and texture of the fungus is the deciding factor in determining which direction this tunnel shall take. In the lower parts of the fungus the hyphae arising from the subiculum are closely arranged and extend vertically upwards for some distance, then branch repeatedly to form the loose spongy tissue. The hyphae of the loose spongy tissue are exceedingly friable. The scale is slowly built of waxy material excreted and deposited by the insect and when the scale encounters the tough vertical hyphae it turns upwards and the insect, finding the construction of the scale very easy going through the friable hyphae, continues to build the tunnel to the exterior in this direction.

A considerable number of scale insects were found parasitized by hymenopterous wasps. In order to escape from the shell of its former host as well as the fungal covering the parasitic wasp must chew its way out. Numerous minute holes (0.3–0.5 mm. wide) in the fungal tissue as well as several wasps were seen. It was noticed that the scale insects parasitized by the wasps were not buried so deeply in the fungal tissue as were those free from such parasitism.

The scale insects which become parasitized by the fungus are surrounded by a thick pad of hyphal tissue.

Specimens examined:

Brazil: Sao Leopoldo, Rick, coll., in Lloyd Collection no. 46400, also in U. N. C. Herb., type, distributed by Rick in Fungi Austro-

Americani, no. 157, as *S. albidum* Pat., in Berlin Museum and perhaps other herbaria. Rick's no. 157 included two species, neither of which was true *S. albidum*.

Septobasidium robustum Boed. and Stein. Bull. Jard. Bot. Buitenzorg III, **11:** 204, fig. 24. 1931.

Plate 111, fig. 6

Resupinate, covering comparatively small areas on the bark, up to 10 cm. long by about 2 cm. wide; frequently completely circling smaller branches; snuff brown or darker to almost brownish black in specimens over a year old; surface at first densely villose and alveolate, becoming nearly smooth or finely granulose with the development of the hymenial layer; surface glabrous in the older specimens; margin irregular in outline but determinate. In section up to 2.5 mm. thick, composed when mature of three regions: (1) the subiculum about 100μ thick except where it covers scale insects and then up to 480μ thick, hyphae of subiculum 2–4μ thick; from the subiculum arises (2) the middle layer which at first is composed of rather densely arranged, distinct pillars, which branch and anastomose eventually forming a dense network as (3) the top layer develops; the pillars meanwhile becoming inconspicuous, the structure in section thus has a lacunar appearance; over the upper surface of this network the hymenium is formed, and as the latter develops the spaces in the nets are closed. Hyphae of context brown, 3–4μ thick; hymenial region poorly developed (in the type material), about 40–150μ thick, composed of spherical or pyriform probasidia (very few seen) 12–15μ thick, and hyphae the ends of which are hyaline and about 2μ thick. Haustoria of the irregularly coiled type, connected with external fungus through vaginal opening.

Even though the reproductive structures are immature this species can be recognized by means of vegetative characters. The snuff brown hymenial region, the darker colored, network appearance of the less mature growth, and the lacunate structure of the context are all distinguishing characters. It shows a rather striking superficial resemblance to *S. Mariani* Bres. but in the latter the pillars are more distinct and there are three horizontal layers instead of only two; moreover, in *S. Mariani* Bres. the context does not show the lacunar structure as in *S. robustum.*

Specimens examined:

Java: Res. Priangan, Kamodjan near Garoet, on branches of *Litsea angulata*, April 1930, Dr. van Leeuwen, coll., no. 13263. Type.

Septobasidium Simmondsii n. sp.

Plate 111, fig. 5

Resupinate, forming small, thick, whitish or creamy patches on the smaller twigs, usually extending for two to three centimeters and as a rule not girdling the twigs. Surface smooth, like kid, dry, not shiny. Margin determinate. In section 1–1.6 mm. thick, composed in the younger parts of three layers: (1) a very thin, inconspicuous, whitish subiculum from which arise (2) a few scattered short and stout pillars and (3) the top layer. As the fungus matures the spaces between the pillars are mostly filled with hyphae, leaving, however, a few tunnels. Pillars 0.1–0.2 mm. tall, giving rise to the top layer which is up to 1.4 mm. thick and makes up by far the greater part of the fungal body. Hyphae of top layer loosely packed, hyaline, thick-walled, sparingly branched, the branches arising mostly at right angles, and infrequently septate, ascending diagonally but individual hyphae mostly straight; hyphae 3.6–4.2μ thick. At the upper surface the hyphae become densely compacted and entangled to form the hymenial region which is 65–120μ thick; hyphae of hymenium 3–3.8μ thick, hyaline. Probasidia spherical or sometimes pyriform, when spherical about 12μ thick, when pyriform 12–13.8 x 12–16.8μ, hyaline, rather thin-walled. Basidia and spores not seen, due to the fact that the probasidia had not germinated.

Associated with scale insects some of which are parasitized by haustoria in the form of irregular coils while others are free from infection.

This species is easily distinguished by its pure white to creamy color throughout, by the usual absence of pillars in the older parts, by the small but exceedingly thick, smooth patches.

Specimens examined:

Australia: Queensland, Yarraman, on *Milletia megasperma*, Jan. 9, 1934, type in Herb. Plant Pathology Brisbane, no. 3431, and U. N. C. Herb., 10357.

Septobasidium sublilacinum (Ellis and Ev.) Burt. Ann. Mo. Bot. Gard. **3**: 331. 1916.

Plate 112, fig. 5

Resupinate, forming small rounded or irregular patches (3–10 mm. in diameter in the material examined by me). Surface of top layer usually slate color or dark neutral gray, sometimes with a lighter margin or lighter throughout, subiculum usually with a brownish tint; smooth. Margin of subiculum usually indeterminate, margin of top layer usually determinate and smooth but sometimes fimbriate. In some specimens the top margin extends downward, merging with the subiculum. In section 200–600μ thick, usually about 375μ thick, composed of three distinct regions: (1) the subiculum which may extend over the bark for several mm. before the formation of pillars begins, 40–100μ thick, threads of subiculum compactly arranged 3.8–6.3μ thick, penetrating freely into the cortical layer of the bark; (2) the pillars which may arise singly or in a tufted group of four to eight, 160–400μ tall, usually about 250μ tall, by about 40–85μ thick, rarely single pillars about a mm. tall, threads of pillars 4–5μ thick, compactly arranged from base to top; (3) the top layer about 30–50μ thick, composed of exceedingly compactly arranged hyphae, 4–5.8μ thick, mostly dark-colored, and extending in a horizontal direction, top layer characteristically brittle. Probasidia, basidia, and spores lacking. There are numerous very fine (about 1μ thick) branched hyaline threads in the hymenium, the origin of which I could not determine.

Associated with scale insects (*Aspidiotus* sp.) some of which are parasitized by haustoria in the shape of irregular coils while others are free from infection.

This species though not in fruit, possesses several distinctive characteristics which enable one to readily separate it from all known species. The small patches with a slate colored top surface, the thin brittle top layer, which frequently anastomoses with the subiculum, the pillars which arise in tufts, and the dark brownish, indefinite, compacted subiculum are distinguishing characteristics.

Specimens examined:

Nicaragua: Castillo el Viejo, B. Shimik and C. L. Smith, 1893, on living branches, nos. 92 and 415 in Herb. State Univ. of Iowa, part of type also in N. Y. Bot. Gard. Herb. no. 108.

Septobasidium prunophilum n. sp.

Plate 48, figs. 13, 14

Resupinate, forming small circular patches or frequently completely covering the smaller branches of plum and cherry for a length of 10–15 cm. Surface smooth and velvety, rarely flaky; rich velvety brown in color (Natal brown or warm sepia to considerably darker). Margin lighter colored, determinate, the top and bottom margins ending at the same line; the margin may be even or somewhat fimbriate under a lens. In section 450–700μ thick, usually about 500μ thick, composed of four distinct regions. (1) The basal subiculum, about 40μ thick, of compact, mostly dark-colored hyphae, which are 3.6–4μ thick; from the subiculum arise (2) the rather scattered short and stubby pillars, 85–130μ thick by about 100μ tall, hyphae 3.8–5μ thick. In the older parts the spaces between the pillars become filled with hyphae so that the context appears to arise directly from the subiculum. The pillars give rise above to (3) the top layer which makes up the greater part of the context, 300–400μ thick, composed of rather closely packed hyphae, 3.8–5μ thick, sparingly branched, ascending diagonally or extending almost horizontally. Over these hyphae (4) the hymenial surface is formed, 40–80μ thick; composed of brownish, upright, mostly parallel, branched, straight or wavy, closely arranged hyphae and structures which appear to be the early stages of probasidial or basidial formation, hyphae of hymenium 3.6–4μ thick, septate. No basidia or spores seen. (Collected in fall of year and hence sterile.)

Associated with scale insects, *Aulacaspis pentagona* (Targ.), the "West Indian Peach Scale" on Japanese flowering cherry. Some insects are parasitized by irregularly coiled haustoria. The fungus enters the insect by growing up around the suctorial tube, forming a compact cylinder of hyphae which, following the suctorial tube, enter the retreat pouch for this tube and after passing backwards down this tube for some distance, the column of hyphae branches, the ultimate ramifications of which extend to most parts of the insect's body. Sixteen parasitized insects were examined and in each one the fungus penetrated the insect by following the seta. Surrounding the parasitized insect is a thick, compact pad composed of pseudoparenchymatous, storage hyphae, which, however, is not applied directly to the insect but passes over the dorsal and under the ventral scales. The cone-shaped beak through which the sucking tube emerges from the insect's body is surrounded by a funnel-shaped layer of hyphae. Also the hyphae which pass through the mouth by the seta enlarge immediately after entering the beak and thus the beak is anchored so firmly to the ventral fungal pad that when an insect is lifted up from the bark the beak is torn off and left adhering to the fungal pad. This species and *S. suffultum* (B. & Br.) Pat. (young material collected by Petch in Ceylon) are the only ones in which I have found the fungus entering the insect by following the seta.

A few non-parasitized insects were found beneath the fungus, each of which was giving birth to young.

This fungus may be distinguished by its rich brown color, by the lighter colored and finely fimbriate margin, by the very thin subiculum, the few stubby pillars, the thick top layer, and by the upright hyphae in the hymenium.

Specimens examined:

Japan: Shino Mura, Tamba Kyotofu, on Japanese flowering cherry (*Prunus*), no. M. 154, R. Kent Beattie, coll.; also on *Prunus*, no.

M. 151, both in U. S. D. A. Herb. and U. N. C. Herb.; on *Prunus*, found at Philadelphia, Pa., Feb. 16, 1938, by U. S. D. A. Plant Quarantine Inspectors in the salon of a Japanese ship, the *Prunus* branches being used as decorations. The inspectors turned the material over to Dr. J. A. Stevenson of the U. S. D. A., who reports the branches as very heavily infected. This material agrees with the specimens from Beattie except for a slight difference in the marginal condition, the present specimens having a lighter margin composed of yellowish somewhat fluffy hyphae. In the surface just below the hymenium is a layer of algal cells, the presence of which is not uncommon in certain species of *Septobasidium*. In this species and particularly this collection there is a peculiar type of hyphal anastomosis not observed in any other species. Two hyphae in contact may excrete a brownish substance which first glues the two together, after which a true anastomosis is formed.

Septobasidium indigophorum n. sp.

Plate 48, fig. 15

Resupinate, forming small (1–2 cm. wide) circular patches or patches which girdle the smaller limbs. Surface smooth, bluish black in color. Margin sharply determinate, minutely fimbriate. In section 330–550μ thick, composed of three regions. (1) A very thin subiculum, usually 16–20μ thick, sometimes up to 60μ thick, composed of a mixture of white and dark-colored hyphae about 3.8μ thick. From the subiculum arise (2) a few, short, stubby pillars, 90–150μ thick by about 75μ tall, threads 3.8–4μ thick, dark brownish in color. The threads of the pillars branch out to form (3) the top layer which is 190–365μ thick, composed of very dark brownish, loosely arranged, entangled threads 3.8–4μ thick, septate. No hymenial layer was formed in the material examined, doubtless due to the fact that the material was collected in the fall instead of the spring.

Associated with scale insects some of which are parasitized by irregularly coiled haustoria; others are free from infection. The fungus appears to enter the insects through the derm pores.

The present species resembles *S. Patouillardii* Burt but the two may readily be distinguished. This form appears to be distinguishable by the blackish color, the rather thin structure, the exceedingly thin subiculum and the comparatively thick context.

Specimens examined:

Japan: Shino Mura, Tamba Kyotofu, on *Prunus* (Japanese flowering cherry), Nov. 15, 1927, R. Kent Beattie, coll., with *S. prunophilum*, type in U. S. D. A. Herb. and U. N. C. Herb., 10358; also Senda, on *Morus alba*, Sept. 24, 1913, A. Yasuda, coll., no. 44, in N. Y. Bot. Gard. Herb. and U. N. C. Herb.

Septobasidium molliusculum H. and P. Sydow. Leaflets Philipp. Bot. **6**: 1920. 1913.

Plate 110, fig. 1

Resupinate on the lower surface of leaves, forming patches up to 6 cm. wide. Older region about snuff brown, margin about benzo brown (Ridgway). Margin indeterminate in places, speckled with fine dots, the young developing pillars. In section 450–550μ thick, composed of three layers: (1) the bottom layer or subiculum (10–60μ thick) which spreads over the leaf surface and from which the middle layer of pillars arises, hyphae of subiculum septate without clamp connections, much branched and irregular in diameter, 2–4μ thick; (2) pillars long and slender, 10–25 x 350–400μ, hyphae of pillars septate, without clamp connections, 4–4.4μ thick, branching out at the top to form the hymenial layer; (3) hymenial layer at first exceedingly thin, becoming 40–60μ thick in mature regions, main hyphae branched, septate without clamps, about 4μ thick. Numerous thin hyphae, about 2μ in diameter also present in the hymenium. Probasidia spherical, 11–13μ thick. Basidia and spores not seen.

Associated with scale insects some of which are parasitized by irregularly coiled haustoria while others are free from parasitism.

This species can readily be distinguished from other species growing on leaves by the fact that though extremely thin and delicate it has a two-storied structure, the top layer being supported by the abundant, delicate pillars. The speckled effect on the margin produced by the young pillars is also a distinguishing characteristic, though the dots may be overlooked unless the specimen is examined with a lens.

Boedijn and Steinmann (1931) have described a fungus from Java under the name of *S. molliusculum* Sydow and have kindly sent me material from their collection for study. Their fungus, however, is quite distinct from the type, large pieces of which I have been able to study through the kindness of Dr. Sydow. The two may easily be distinguished by their gross appearance. The fungus described by Boedijn and Steinmann as *S. molliusculum* from Java grows in small but conspicuous patches, the top layer being supported by conspicuous pillars which are frequently one millimeter tall, while in the type material the fungus is very inconspicuous and the pillars are only half so tall. The two also differ strikingly in color, the form from Java having a buff colored hymenium while in the type of *S. molliusculum* the hymenium is about snuff brown, the margin about benzo brown. The hymenium in the material from Boedijn and Steinmann is thicker and more compact than in the type and the probasidia are much larger, being 12–19 x 22–31μ, while in the type the probasidia are spherical and 11–13μ thick. The fungus from Boedijn and Steinmann is being described as a new species, *S. Boedijnii*.

Associated with this fungus and apparently belonging to it are numerous moniliform conidiophores. These arise from the basal layer apparently from the threads of the *Septobasidium*. The conidiophores are most abundant near the scale insects. The conidia are oval in outline with slightly roughened walls. The fact that they are most abundant near the scale insects would suggest that they might offer another source than the spores for the infection of the young.

The fungus lives symbiotically with scale insects. A large number of healthy, non-parasitized insects were dissected out from beneath the fungal covering. Some were giving birth to young while others were found in all stages of development. A large number of the insects, however, were parasitized by the fungus. Due to the fact that the insects were so thoroughly dried, it was extremely difficult to demonstrate the haustoria within the insects' body. They were made visible, however, by heating some parasitized insects for several minutes in a 7% solution of KOH and then allowing the material to soak in the same solution over night. The haustoria are in the form of much twisted intestine-like coils of hyphae.

Specimens examined:

Philippine Islands: Cabadbaran, Mt. Urdantea, Prov. of Agusan, Island of Mindanao, on old drooping leaves of *Litsea cassiaefolia*, August 1912, Elmer, coll., type, no. 13417 in Sydow Herb., Buitenzorg Herb., Herb. Berlin Bot. Gard., and U. N. C. Herb.

Septobasidium sp. hybrid between *S. castaneum* and *S. Patouillardii* (?). Jour. Elisha Mitchell Sci. Soc. **51:** 75. 1935.

Plate 48, fig. 1

I have six collections of a species from Louisiana and Alabama which I at first put under *S. Patouillardii* but a more detailed study of these forms has indicated their distinctness from that species.

Resupinate, forming thin, closely adherent, irregular, confluent, patches up to 10 cm. long. Bister (Ridgway) or slightly lighter. Surface smooth except for the wrinkles in the bark and distinct cracks in older parts. Margin abruptly determinate, the margins of the top and bottom layers usually ending at the same place, irregular in outline. In section 160–520μ thick, usually 210–315μ thick, composed of three regions: (1) the subiculum which is very thin, 20–70μ thick, usually about 40μ thick, composed of densely packed threads, 3–4.2μ thick, which extend in a more or less horizontal direction, subiculum concolorous with top layer or slightly lighter; (2) the middle region composed of scattered pillars or very thick columns of hyphae, 100–1000μ thick and only 50–80μ tall, threads of pillars 3.6–4.2μ thick, concolorous with threads of subiculum; (3) the top layer supported by the pillars or columns, 55–150μ, usually about 65μ thick, composed of closely packed and entangled hyphae the ends of which are not arranged vertically but are recurved, hyphae 3.8–4.2μ thick, concolorous with the pillars and subiculum. Probasidia and basidia not observed.

Associated with scale insects some of which are parasitized by irregular coils of hyphae.

This species is related to *S. castaneum* Burt, *S. Patouillardii* Burt, and *S. Leprieurii* (Mont.) Pat. From *S. castaneum* it differs in being much thinner, in having a simple margin, in having two instead of three horizontal layers, in having a middle region composed of distinct pillars or columns of hyphae and empty spaces instead of branched pillars and flocculent hyphae. The color of the two is also different, *S. castaneum* showing more of a purplish tint in the brown than the present species. From *S. Patouillardii* Burt it differs in color, being bister without a velvety sheen instead of blackish brown with a faint purplish tint and with a velvety sheen, in having very thick, stubby pillars instead of thin slender ones, in having a subiculum concolorous with the top layer instead of distinctly lighter and in having the ends of the threads in the top layer recurved instead of vertical. From *S. Leprieurii* it differs in having pillars throughout, in being much thinner, and having a tough, crusty texture when torn with a needle instead of a soft, flocculent one. The color of the two species is somewhat similar, though *S. Leprieurii* has a shiny instead of dull surface. *Septobasidium Leprieurii* is also somewhat lighter colored.

This species has so far been found only on *Liquidambar styraciflua*, and although the material is abundant it is all sterile.

Specimens examined:

Louisiana: Near Baton Rouge, Dec. 31, 1931, with *S. Mariani*, *S. pseudopedicellatum*, and *S. castaneum*, J. N. Couch, coll., 9181, 9193; also near Baton Rouge, Dec. 26, 1931, C. A. Brown, coll., 326, U. N. C., 9946; near Covington, Jan. 10, 1932, with *S. castaneum* Burt, C. A. Brown, coll., 321, U. N. C., 9949.

Alabama: Near Nokomis, April 19, 1932, C. L. Shear, coll., in U. S. D. A. Herb. and U. N. C., 9948.

Septobasidium capnodes (B. & Br.) Bres. Ann. Mycol. **14:** 241. 1916.

Resupinate, color near carob brown (considerably darker than bister), surface smooth and shiny except for a few pin holes, margin determinate. In section 168–300μ thick, not divided into layers or regions but of the same structure throughout. There is no distinct subiculum as in most other species but numerous fine hyaline hyphae penetrate the upper region of the cork. Hyphae of context 3–5μ thick, deep brown under microscope, thick-walled, much entangled and branched, the branches usually arising at right angles. No distinct hymenial layer is formed but the probasidia are borne on the upper surface of the context. Probasidia spherical, 12.6–14.8μ, wall 1.5–2μ thick, hyaline or pale brownish. Basidia and spores not seen.

Associated with scale insects some of which are parasitized by haustoria of the irregularly coiled type; others are not parasitized.

This fungus shows a rather striking superficial resemblance to *S. Leprieurii* (Mont.) Pat. Unfortunately, however, the small amount of material available and the absence of basidia and spores in the present species do not permit a satisfactory comparison. The present species differs from *S. Leprieurii* in being darker colored and much thinner in section.

Specimen examined:

Ceylon: Central Province, on sticks, Thwaites, coll., no. 547, type in Kew Herbarium.

Septobasidium scabiosum Couch and Petch n. sp.

Plate 111, figs. 1, 2

Resupinate, epiphyllous or hypophyllous, occurring in small (0.5–1 square mm. in area) scattered or closely arranged patches. When the patches are closely arranged, they show a roughly concentric grouping, usually anastomosing along their lateral margins and thus forming a circular patch of growth up to 6 cm. wide. Sayal brown to snuff brown or sepia. The subiculum is very inconspicuous with an indeterminate margin, composed of pale brownish to hyaline hyphae; hyphae 2.8–3.5μ thick, extending as a rule in radial direction but interlaced by lateral branches. From the subiculum arise small flakes of hyphae the ends of which grow back toward the older part of the fungus. The adjacent margins of the "flakes" may anastomose or as frequently happens the flakes may remain slightly separated from each other. Each flake forms a sort of "lean-to" or cave-like structure, the front edge being attached to the subiculum while the posterior edge is 0.5–1 mm. above the subiculum. Hyphae of flakes brownish, the main hyphae extending backwards from the point of basal attachment while numerous lateral pale brownish to hyaline branches arise at right angles forming an interlacing network. Main hyphae 6–8.4μ thick, the branches being considerably thinner. No hymenium, probasidia, or basidia seen.

Associated with scale insects some of which are parasitized by irregularly coiled haustoria.

Though not in fruit this is a very distinctive species and may be readily recognized by its occurrence on leaves in numerous little gregarious or widely separated patches. The somewhat concentric arrangement of the patches when they are closely gregarious is also distinctive.

Specimens examined:

Ceylon: Peradeniya, 1910, T. Petch, coll., on *Codiaeum variegatum*, no. 3168, type; also same loc., January 1912, no. 3353, in U. N. C. and Petch Herbaria.

Septobasidium abnorme (P. Henn.) v. Höhnel & Litsch. Beiträge zur Kenntnis der Corticieen (II. Mitteilung). Sitzungsberichten der kaiserl. Akademie der Wissenschaften in Wien. Mathem-naturw. Klasse. **116**: 740–741. 1907.

Corticium abnorme P. Henn. Hedwigia **43**: 186. 1904.

Plate 47, figs. 1, 2

Resupinate, covering several square centimeters on the surface of leaves, buff colored, surface smooth except for numerous circular mounds which are from 0.3–0.7 mm. broad and about 0.3 mm. high, margin determinate. In section 100–350μ thick, compact, composed of dichotomously branched and coiled hyphae, the method of branching and coiling resembling that of *S. rhabarbarinum*. Hyphae of context delicate, 1.5–2.5μ thick, hyaline; hyphae in the very thin and inconspicuous subiculum slightly thicker. No probasidia or basidia seen.

This species of *Septobasidium* is particularly interesting because it is the one under which scale insects were first discovered in 1907 by von Höhnel and Litschauer. It appears to have been collected only once and then in the sterile condition. It may be recognized by its color, the minute mounds, the dichotomously branched and coiled hyphae, which are unusually thin. In the structure of the vegetative hyphae the species is very like *S. rhabarbarinum*. The two, however, are quite distinct in gross structure and color.

Apparently a similar relationship between the fungus and scale insects exists here as has been described for *S. Burtii* Lloyd. Beneath some of the mounds are insects parasitized by haustoria which are in the form of short segments of hyphae held together by fine threads; other insects are free from parasitism. At the bases of a considerable number of the mounds there is a minute slit-like opening about large enough to permit the exit of a young crawling scale insect. Such openings are found at the margin of the mounds which cover the non-parasitized insects.

Specimens examined:

Brazil: Rio Jurua, on lower surface of living leaves of Myrtaceae, von Höhnel's Herbarium Brasiliense, no. 2939, Amazon Expedition, type in Museum Bot. Garden Berlin.

Septobasidium makilingianum Sydow. Ann. Mycol. **15:** 170. 1917.

Plate 47, fig. 15, and pl. 110, fig. 3

Resupinate, forming an exceedingly thin, almost powdery growth on the lower surface of leaves. Surface smooth except for the minute mounds of fungal tissue which cover the insects. Patches of growth of variable size, 1 mm. up to 4 sq. cms. in extent, irregular in outline and without a distinct margin; about mummy brown in color (Sydow says cinnamon brown) with darker regions here and there, particularly immediately over or close to the overgrown insects. In section 50–110μ thick, composed of much entangled, rather compactly arranged hyphae which break up into short curved and irregular segments when mounted and covered with a cover glass. The hyphae extend into and between the epidermal cells, destroying this region of the leaf. Hyphae in the lower part of fungus about 6.8μ thick, deep brown in color, becoming thinner as they ascend, much septate and branched, frequently circinately branched, the ends of the hyphae in the upper region sometimes hyaline and about 3μ thick. The hyphae covering the insects are 3–5μ thick, elongated and with occasional swollen places up to 7μ thick. Probasidia, basidia, and spores not seen.

Associated with scale insects some of which are parasitized by haustoria in the form of spindle-shaped segments held by fine threads (sausage-shaped type), while other insects are free from parasitism.

It is interesting that Sydow placed this species in *Septobasidium* solely on the basis of vegetative structure and association with scale insects. The haustoria, typical of *Septobasidium,* add further evidence that it is correctly placed.

Septobasidium makilingianum Sydow is related to *S. tenue* and *S.*

rugulosum in the general structure of the context and the structure of the haustoria. *Septobasidium makilingianum* may be distinguished from these two species by its thinner structure, different color, less compactly arranged hyphae and perhaps by its occurrence on leaves. It is quite distinct from *S. molliusculum* Sydow though the two have been confused by Patouillard (see C. F. Baker, no. 2389, on *Celtis luzonensis* in Herb. Univ. of Philippines and Patouillard Herb.) and Neuhoff (see notation on label of type of *S. makilingianum*). In *S. molliusculum* there are distinct though minute pillars while in *S. makilingianum* Sydow there are no pillars. There are also other striking differences.

Specimens examined:

Philippine Islands: Province Laguna, Mt. Makiling, near Los Baños, on living leaves of *Astronia* sp. associated with coccids, Feb. 1914, C. F. Baker, coll., type no. 2808 in Sydow Herb., U. S. D. A. Herb., and U. N. C. Herb.; same location, on leaves *Celtis luzonensis*, Dec. 26, 1913, C. F. Baker, coll., no. 2389 in Herb. Univ. of Philippines and Patouillard Herb., also in U. N. C. Herb. from Ocfemia. This latter collection differs from the type in color, being much lighter, and in having hyphae which are thinner and more noticeably recurved.

Septobasidium carbonaceum Pat. Bull. Soc. Myc. Fr. **36:** 175. 1920.

Plate 110, fig. 2

Resupinate, covering several square centimeters. Surface color hair brown (Ridgway), the context darker. Surface pulverulent, broken by a few irregular cracks and holes. Margin apparently indeterminate. In section 400–750μ thick, composed of (1) a very densely packed layer of dark colored threads which extend over the bark forming the subiculum; (2) the middle region, composed of ascending, much branched, fairly compact, entangled hyphae some of which are deep brown and 4.2–5.2μ thick, while others are hyaline, considerably collapsed and thinner; (3) the top layer (80–130μ thick), composed of very densely packed, much branched, entangled and curved hyphae. Probasidia, basidia, and spores not seen.

Associated with scale insects some of which were parasitized by coiled, sausage-shaped haustoria while other insects were not infected.

According to Dr. O. A. Reinking's label this plant was collected on dead branches of *Citrus* sp. and perhaps was not in a very healthy condition itself since it depended for sustenance upon the insects who in turn depend on the living tree. From Patouillard's brief description and mine both drawn from the same dead and perhaps abnormal material it may be difficult to identify this species again in the field.

The color, the compact structure, the thick dark-colored hyphae intermingled with hyaline hyphae, and the sausage-shaped haustoria in the insects are distinctive characteristics.

Specimen examined:

China: Kwangtung Province, Pumela Park, on *Citrus*, Reinking, coll., May 17, 1919, type in Pat. Herb.

Septobasidium perforatum n. sp.

Plate 112, fig. 3

Resupinate, closely adherent and of a firm, compact but friable texture, covering extensive areas on the stems of a woody vine. The younger parts wood brown, the older parts bister. Margin determinate, whitish. Surface of the younger parts very remarkable in that it is dotted with numerous circular, oval, crescent-shaped or irregular holes, the mouths of which are 0.1–0.3 mm. wide and so distinct as to have the appearance of minute pin pricks. The holes are usually about 2–4 to a square millimeter but may be as few as one or as many as eight. In places they are formed in rows parallel with the margin with slightly more than a millimeter of smooth fungal surface between each row. When in rows, the holes in the same row may be so close that they anastomose along their margins, forming an irregular crack. Over the greater part of the surface, however, the holes are irregularly scattered and not arranged in rows. In the older parts the fungus changes abruptly to a dark brown (bister) and the surface shows numerous irregular cracks, the cracks having been formed by the widening and anastomosing of the holes. The holes extend down to the subiculum, connecting with an anstomosing system of chambers and tunnels in which live the scale insects. In section 400–650μ thick, composed of three regions. (1) The subiculum 63–125μ thick, which however in most parts is not sharply differentiated from the context, subiculum about 40–125μ thick, composed of compactly arranged, densely entangled hyphae which extend more or less longitudinally over the bark, hyphae of subiculum hyaline to pale brownish, rather thin-walled, 3–4μ thick. From the subiculum arises (2) the context, 300–420μ thick, composed of compactly arranged, irregularly curved and twisted, much entangled and much branched, ascending hyphae, which are of rather uniform thickness, about 3μ thick, pale brownish to hyaline. In some places the context is stratified by the apparent formation of several hymenial layers, the upper limit of the old hymenium being indicated by a narrow zone of more compacted hyphae arising from a horizontal layer of brownish amorphous material. (3) Hymenium 30–50μ thick, sharply differentiated from the context by being composed entirely of hyaline hyphae, which are much entangled and 2.2–3μ thick. No probasidia, basidia, or spores seen.

Associated with scale insects, some of which are parasitized by irregularly coiled haustoria; others are free from parasitism. Haustoria connected with external fungus through vaginal opening.

This is one of the most distinct and easily recognized of all species of *Septobasidium.* The light brown, firm, closely adherent surface with the minute, regular holes and the dark brown older region with cracks are characteristic. Except for the holes and cracks the surface is remarkably smooth and even.

Specimens examined:
Brazil: Sao Leopoldo, Rick, coll., in U. N. C. Herb., no. 10395, type.

Septobasidium lacunosum n. sp.

Plate 112, fig. 4

Resupinate, covering small areas (2–3 sq. cm. in the specimens examined by me), deep brownish in the older parts, slate gray in the younger regions. Margin irregular but determinate. Surface very peculiar and distinctive in that, here and there, spaced at rather regular intervals are numerous circular or oblong depressions 0.3–0.6 mm. wide. Each depression is partially covered by a sunken flap of fungal tissue, a sort of trap door. It is these regularly spaced pits which give the fungal surface a reticulate appearance. In section 400–600μ thick, without pillars and without any structural peculiarities except the pits. Hyphae brownish to hyaline, 2.1–4.2μ thick, most 3.5–4μ thick, septate, rather sparingly branched, resembling the hyphae of *S. Burtii* Lloyd. Probasidia formed near the upper surface, ovoid or pyriform, about 10.5 x 18μ, basidia and spores not seen.

Associated with scale insects some of which are parasitized by haustoria of *S. Burtii* type, the fungus entering by setae as in the latter.

This unique species may be easily recognized by its small size, slate gray color, and above all by the reticulate surface, with the rounded pits each partially covered by a trap door.

Specimen examined:
Costa Rica: Villas del Rio Torres, Feb. 26, 1911, in Patouillard Herb., Harvard Univ. as *S. Burtii* Lloyd (*S. retiforme*, sense of most American authors) and in U. N. C. Herb.

Septobasidium tuberculatum Boed. and Stein. Arch. Theecult. **4:** 45, Pl. 20–22. 1930. Description in English and Dutch. Also in Bull. Jard. Bot. Buitenzorg III, **11:** 185, figs. 10, 11 and Pl. 16. 1931. (Figures same in both papers.)

Plate 46, figs. 12–14

Resupinate, extending for a length of 15 cm. on the smaller limbs; "ashy-grey with here and there a purplish brown hue" (Boedijn and Steinmann). Scattered about on the thallus are conspicuous rounded elevations, which are about 2 mm. high and 2.5–7 mm. wide. Not infrequently these mounds may fuse forming a small distinctly elevated area. Mounds concolorous with the main part of fungus but in old specimens becoming a dirty brown. The surface of the main part of fungus mostly shows a network of small cracks. In section 180–300μ thick, context more or less compact throughout, threads of internal part of context with an old gold color, giving to a 5% solution of KOH a reddish or orange color which disappears in two or three minutes, threads hyaline or pale brownish under microscope, straight or circinate, 3–4.4μ thick, frequently swollen at intervals. Scattered throughout the context are numerous spherical bladder cells with thin walls and thin protoplasmic content, borne singly on the stalk or sometimes in groups of several, and sometimes two-celled, 7–11μ thick, the hypha projecting into this cell in columella-like fashion. These structures can hardly be probasidia and their function is at present obscure. No distinct hymenial zone is formed but the hyphae toward the surface are thinner, about 2–3μ thick, and nearly hyaline. Probasidia, basidia, and spores not observed.

Associated with scale insects some of which are parasitized by rather irregular spindle-shaped segments held by fine threads. Other insects are free from parasitism and some were in the act of giving birth to young when the material was collected.

This species even though the probasidia, basidia, and spores are lacking possesses characteristics distinct enough to enable one to recognize it. The finely cracked surface, the ashy gray or brownish color, the compactness of the context, and above all the conspicuous mounds are distinctive. Microscopically the circinate hyphae mixed with the straight ones and the bladder cells are characteristic.

Specimens examined:

Java: Buitenzorg, Jardin de Culture, on *Thea*, Nov. 1926, Steinmann, coll., type in Herb. Bot. Gard. Bog. and U. N. C. Herb. Also reported by Boedijn and Steinmann from Sumatra on *Thea*.

NOTES ON INCOMPLETELY KNOWN SPECIES

I have been unable to secure any material of the following three species and am therefore including the authors' original descriptions.

Septobasidium foliicolum Torrend. Fungi Selecti Exsic. no. 127. Broteria, Bot. Ser. **11:** 169. 1913.

Effusum, resupinatum, tenue, primum ex brunneo-cinerascens, deinde brunneo-fuscum, vel cacainum, incrustans, margini tenui et colore dilutiori, interdum in lacinias radiatim superficiem substrati invadens; hymenium continuum, villosulum, ex hyphis ramosis, ochraceis, ad extremitatem hyalinis et multifidis; basidiis primum globulosis, continuis, nucleatis, 12–14 x 8–10μ.; basidiis evolutis et sporis non visis.

Ad folia viva *Lauri nobilis.* Funchal. Leg. J. Barreto.

Obs. Comme je l'ai fait remarquer dans la Contribution précédente (no. 161) c'est je crois la première fois qu'on trouve une espèce de ce genre sur les feuilles. Peut-être est-ce une adaptation foliicole de *Sept. michelianum* si commum en Portugal sur les branches et jeunes troncs de la même plante nourricière.

This species is previously mentioned and even named *foliicolum* in Broteria **10:** 35, 1912, but not described.

Septobasidium Cabralii Torrend. Broteria, Bot. Ser. **11:** 83. 1913.

Effusum, tenue, arcte adnatum, purpureo-cupreum (Klinck. C. C. 113) rimosum, ad marginem brunneo rubidulum, et pulchre fimbriatum ex hyphis ramosis, 4–5 μ. crassis, crebre septatis et dense in fibras conglutinatis; basidia primum subglobosa, hyalina, continua, 10–12 μ, dein piriformia, ovoideo-oblonga, tandem septata, variae longitudinis, ochracea; sporae hyalinae, proteiformes, plerumque curvulae, saepe sigmoideae, continuae, vel vix in media parte subconstrictae, (an signum septi futuri?) 8–16 x 5–8μ.

Hab. Ad truncum *Quercus Ilicis.* Serra de Arrabida. Aprili.

Leg. Franciscus Cabral, oculatissimus et egregius Collegii Campolitensis alumnus, cui grato animo dicavi.

Obs.—Cette espèce semble rare, puisque malgré toutes mes recherches et celles de plusieurs jeunes compagnons à l'oeil très observateur, nous n'avons pu trouver qu'un seul pied de *Quercus Ilex* attagué par *S. Cabralii.* Il se présente sous l'aspect d'une plaque soyeuse, rouge cuivre, à marge plus claire ou brun rougeâtre pâle. La couche de l'année précédente prend une teinte rouge bruni, presque noire (Klinch. C. C. 105).

De plus les hyphes de diverses couches hyménifères voisines se croisent à travers la surface raboteuse de l'écorce du substratum, et forment comme une série de cordons rougeâtres qui sillonnent le substratum. Ce caractère rapproche beaucoup notre espèce de *S. Mariani* (Bres.) Peut-être y aurait-il lieu de les identifier. Les caractères différents des basides et spores ont moins de valeur étant donné la variabilité si connue de ces caractères chez les *Septobasidium.*

Septobasidium lanatum Chow. Annales de Cryptogamie Exotique **5**: 68, pls. 1, 2. 1932.

Resupinatum, longe lateque effusum, tenue, lanatum, fuscopurpureum, subnigrum, leviter passim granulosum, margine indefinito, structura simplici; hyphis ramosis, septatis, 5–6μ crassis, brunneis, extremitatibus aliquoties pallidis vel subhyalinis; conidiis multis, 5–7μ diam., brunneis, moniliformibus vel raro glomerabilibus; probasidiis subglobosis, hyalinis, ad 20μ diam., extremis pediculis columellaribus; basidiis in probasidiis ex extremis columellis natis; sporis non visis.

Ce champignon a été récolté en Cochinchine, à Núi Chúa Chan (prov. de Bienhoa), par M. Poilane, en octobre 1931, sur une branche d'un théier planté en 1919 (Herb. Cryptog. Muséum de Paris, Poilane no. 19494). D'après les caractères que nous donnons ci-joint, il ne peut être identifié avec aucune espèce décrite. Nous le considérons donc comme une espèce nouvelle.

Le thalle est mince, largement étalé, à marge indéfinie, laineux, d'un brun pourpre noirâtre uniforme sur toute son étendue, et recouvert d'une poussière abondante de conidies. Il se développe tout autour du rameau de théier dont il provoque une légère hypertrophie.

De place en place se trouvent des groupes de granulations de consistance ferme et de taille atteignant parfois 1 mm. de diamètre. Des coupes pratiquées perpendiculairement à ces nodules montrent toujours la présence de coccides à leur intérieur.

La structure du thalle est très simple. Il ne présente pas plusieurs couches différenciées superposées. La trame est homogène, composée d'hyphes entrelacées lâchemont. Celles-ci sont ramifiées, cloisonnées, larges de 5 à 6μ, à paroi épaisse de 1 à 1, 3μ et de couleur brune au microscope; quelques-unes sont plus pâles ou presqu'incolores à leur extrémité.

Seuls les nodules ont une structure plus compacte. Les hyphes s'y montrent plus serrées, surtout autour des insectes. A la surface du thalle, les hyphes peuvent être simplement végétatives et porter soit des conidies, soit des probasides, soit à la fois les unes et les autres.

Les conidies se forment le plus souvent en chapelets, quelquefois en petits glomérules. Elles sont aussi de couleur brune au microscope, à paroi épaisse d'environ 1, 3μ, et mesurent de 5 à 7μ de diamètre.

Dans la partie superficielle du thalle, surtout à la surface des petit nodules, on peut trouver des sphères hyalines, plus ou moins longuement pédicellées, à paroi mince, mesurant jusqu'à 20μ de diamètre. L'extrémité du pédicelle entre dans la sphère à la maniere d'une columelle. Boedijn en a signalé de semblables chez le *Septobasidium tuberculatum* Boed. et Steinmann. La découverte de plusieurs stades de dévelopement de la baside dans ces sphères nous permet d'assimiler ces dernières aux probasides. La cellule initiale de la baside se forme par une cloison à l'extrèmité de la columelle. Les basides adultes et les basidiospores n'ont pas été observées.

NOTES ON EXCLUDED SPECIES

Septobasidium album Burt. Ann. Mo. Bot. Gard. **13:** 332. 1926.

Resupinate, forming small (2–6 square cm. in area) white, thickish patches of growth; the surface has somewhat the appearance of kid but is pulverulent under a lens; broken here and there by irregular cracks, margin sharply determinate. In section 600-700μ thick, whitish throughout, composed of two rather indistinctly differentiated regions, a lower and an upper, each of about the same thickness. The lower region is composed of diagonally ascending, entangled, compact hyphae which are about 4–5μ thick and are furnished with conspicuous clamp connections at the septa. In the lower part there are numerous thick, dark-colored hyphae scattered among the hyaline ones. The upper part of the fungus is composed of hyphae which are frequently coiled, are very compactly arranged and are 2.8–3μ thick. Hyphae rather sparingly septate and usually with clamps at each septum. At the upper surface of the fungus the ends of the hyphae are spirally coiled. The basidia arise from and are intermingled with these coiled hyphae. The basidia arise from the sides or ends of hyphal branches and usually a clamp connection can be made out at the base of each basidium. Probasidia lacking. Basidia characteristically twisted, 6–8 x 70–80μ, four-celled, the basal cell becoming empty before the septa are formed. Spores apparently are formed only on the three distal cells. Sterigmata 5–6μ long. Spores hyaline, ovoid, 7–9 x 10–15μ.

Occurring on dead bark, not associated with scale insects.

This species may easily be recognized by its white color; occurrence on dead bark without scale insects; by the compact context with the coiled hyphae near the surface; by the elongated, twisted, four-celled basidium the basal cell of which is sterile; by the ovoid spores, and by the clamp connections.

In spite of the cylindrical septate basidium it is very doubtful if this species belongs in the genus *Septobasidium*. The close association of that genus with scale insects, its invariable habitat on living plants, the absence of clamp connections and the narrow, bent-elliptic spores serve to separate it from the present species.

In the material which I have studied there are numerous dark-colored thick-walled hyphae scattered among the hyaline hyphae in the basal part of the subiculum. Burt seems to have overlooked these hyphae. They may, of course, belong to another fungus.

Specimen examined:

New Zealand: Queenstown, Otago. December 3, 1919. On fallen branches of *Nothofagus*, G. H. Cunningham, coll., no. 542. Type in Mo. Bot. Gard. Herb. 59315, also in U. S. D. A. Herb.

Septobasidium cinnabarinum Syd. Ann. Mycol. **12:** 195. 1914. A translation of Sydow's Latin description follows:

Living on branches, resupinate, flat or convex in center, irregular, forming patches 1–2 cm. long and ½–1 cm. broad, scarcely fimbriate on the margin, vividly cinnamon, hymenium continuous; context made up of yellowish or brownish yellow, sparingly septate branched hyphae, raised thence in agglutinated erect fascicles 200–400μ high and 3–4μ thick; basidia not seen.

Habitat on dead corticated branches, Mosumbue, province of Tabora, Dutch East Africa, 10/6/1913, leg. Braun (Bot. Institut Amani, no. 5644).

This plant has the apparent gross vegetative structure of a *Septobasidium*, there being a bottom and a top layer, the latter supported by pillars. In its habitat on dead corticated wood, and the absence of scale insects, and the lack of probasidia and basidia to confirm its taxonomic position, it seems best to consider the plant as a doubtful species of *Septobasidium* until further information can be secured.

The color of the fungus is very striking, the "scarlet" of Ridgway being nearest. The color is due to a crystalline material excreted by the hyphae. This crystalline material is very abundant, forming a broom-like deposit on the tops of the pillars and composing the greater part of the top layer. This material has a distinctly reddish color under the microscope, and can be seen on the hyphae in the form of reddish bands. The hyphae, however, except where the reddish bands occur, are hyaline.

Specimen examined: Mosumbue, Province of Tabora, Dutch East Africa, 10/6/1913, leg. Braun. Type in Sydow Herb., U. N. C. Herb., and in Bot. Inst. Amani, no. 5644.

Septobasidium polypodii Couch. Jour. Elisha Mitchell Sci. Soc. **44:** 255. 1929.

Plate 10

Resupinate, overgrowing and completely covering the under-surface of fertile fronds of *Polypodium* sp. and parasitizing the sori. Nearly white when fresh, becoming cream colored upon drying. In section 250–700μ thick, usually about 500μ thick at the thickest part, whitish throughout, composed of (1) a very thin lower layer of hyphae which extend over the lower surface of the fern leaf but do not penetrate into the epidermal tissue; and (2) a layer of compact, much branched, coiled and entangled hyphae arising from the lower layer, often stratose by the successive formation of two or three fruiting regions. No pillars present. Probasidial cells formed in the outer surface of the compact layer, oval, 12.5–14.8 x 16–23μ, wall slightly thickened, germinating into a 4-celled, coiled (about in the form of a semicircle) basidium, 8.2–12.6 x 50–70μ, which in turn bears four spores. Spores elliptic, 6.5–8.5 x 19–25μ, becoming once septate.

This species is being excluded from *Septobasidium* because of its growth on fern sporophylls and the absence of scale insects beneath the stroma. It should probably be placed in the Uredinales near *Goplana*.

When the material was collected it was passed over at the time without much study, and I am therefore unable to make a full and certain report as to the fungus-host relationship. A study of sections made of several infected sporophylls has shown no scale insects underneath the fungus. It starts development on the naked sori, covering the developing sporangia and their stalks with a mat of hyphae. In several instances hyphal penetration into the pro-sporangial cells could be made out. The fungus does not penetrate at all into the vegetative parts of the fern sporophyll. In spite of the obvious parasitism, a good many of the sporangia mature, forming apparently good spores.

Specimens examined:

Jamaica: Blue Mountains, on *Polypodium* sp., W. R. Maxon, coll., June 1926, type in U. N. C. Herb.

LITERATURE LIST

BALLY, W. Nieuwe wortelschimmels bij koffie. De Bergcultures **3**: 1669–1675, 6 figs. 1929.

BERKELEY, M. J., AND BROOME, C. E. Enumeration of the fungi of Ceylon. Part II. Jour. Linn. Soc. Bot. **14**: 64. 1875.

——— AND CURTIS, M. A. Fungi cubenses (Hymenomycetes). Jour. Linn. Soc. Bot. **10**: 280–341. 1869.

——— AND DESMAZIÈRES, J. B. H. J. On some molds referred by authors to Fumago, and on certain allied or analogous forms. Jour. Roy. Hort. Soc. **4**: 243–260. 1849.

BERNARD, CH. *Septobasidium bogoriense* op jonge kinaboomen. Cinchona **2**: 46–48, 2 pl. 1925.

——— De grijze dadapschimmel (*Septobasidium bogoriense*). De Thee **6**: 82–85, 1. pl. 1925. (Abst. in Rev. Appl. Myc. **5**: 253. 1926.)

BLANCHARD, E. E. Principales insectos y enfermedades que perjudican los cultivos citricos en la Repúblic Argentina. Min. Agric. Nac. (Buenos Aires) Secc. Prop. e Inform. Circ. **815**: 114 pp. 8 pl. (7 col.), 42 figs. 2 diag. 1930.

BOEDIJN, K. B. A new Septobasidium on tea. Comm. Gen. Exp. Stat. A. V. R. O. S., Gen. Ser. **26**: 6–10. 1926.

——— Beitrag zur Kenntnis der Pilzflora von Sumatra. Recueil des Travaux Bot. Néerland. **26**: 409–410. 1929.

——— AND STEINMANN, A. On the species of Helicobasidium and Septobasidium occurring on tea and other cultivated plants in the Dutch East Indies. Arch. voor de Theecult. **4**: 37–59, 29 pl. 1930.

——— Over de roetdauwschimmels van de Thee. Arch. voor de Theecult. Nederl.-Indië **5**: 25–57, 9 pl., 1 fig. 1931.

——— Les espèces des genres Helicobasidium et Septobasidium des Indes Néerlandaises. Bull. Jard. Bot. Buitenzorg III, **11**: 165–219, 5 pl. (2 col.), 31 figs. 1931.

BOURDOT, H., AND GALZIN, A. Hyménomycètes de France, pp. 6–9. Paris, 1927.

BRESADOLA, J. Fungi Polonici. Ann. Mycol. **1**: 112. 1903.

——— Hymenomycetes novi vel minus cogniti. Ann. Mycol. **3**: 164. 1905.

——— Basidiomycetes Philippinenses. Hedwigia **56**: 303. 1915.

——— Synonymia et adnotanda mycologica. Ann. Mycol. **14**: 232–242. 1916.

——— Selecta mycologica. Ann. Mycol. **18**: 59. 1920.

——— AND SACCARDO, P. A. Enumerazione dei funghi della Valsesia raccolti dal Ch. Ab. Antonio Carestia. Malpighia **11**: 241–325. 1897.

BRITON-JONES, H. R., AND BAKER, R. E. D. Notes on some other fungous diseases in Trinidad, 1933. Trop. Agriculture **11**, No. 3: 67–68, 2 pl. 1934.

——— Control of Grapefruit diseases by cultural methods in Trinidad. Trop. Agriculture, Trin., **12**, No. 5: 119–125, 4 pl. 1935.

BUCHNER, P. Tier und Pflanze in Intrazellularer Symbiose. Berlin, 1921. 2nd ed. 1930.

BURT, E. A. The Thelephoraceae of North America VII. Septobasidium. Ann. Mo. Bot. Gard. **3**: 319–343, 14 figs. 1916.

——— The Thelephoraceae of North America XV: Conclusion with supplement and general index. Ann. Mo. Bot. Gard. **13**: 173–354. 1926.

CHARDON, C. E., AND TORO, R. A. Mycological explorations of Venezuela. Monog. Univ. Puerto Rico, Ser. B., Phys. and Biol. Sci. **2**: 353 pp., 33 pl. (1 col.), 1 graph, 1 map. 1934.

CHOW, C. H. *Septobasidium lanatum*, nov. spec. Ann. Cryptog. Exot. **5**, No. 2: 68–69, 2 pl. (1 col.). 1932.

COKER, W. C. Notes on the lower Basidiomycetes of North Carolina. Jour. Elisha Mitchell Sci. Soc. **35**: 113–182, 38 pl. 1920.

COMSTOCK, JOHN HENRY. Reports on scale insects. Cornell Univ. Agric. Exp. Stat. Bull. No. 372. 1916.

COUCH, J. N. A monograph of Septobasidium. Part I: Jamaican species. Jour. Elisha Mitchell Sci. Soc. **44**: 242–260, 16 pl. 1929.

——— The biological relationship between *Septobasidium retiforme* (B. & C.) Pat. and *Aspidiotus osborni* New. and Ckll. Quart. Jour. Micr. Sci. **74**: 383–437, 5 pl., 59 figs. 1931.

——— The biological relationship between Septobasidium and scale insects. Proc. Fifth Intern. Bot. Congress, 1930, p. 369. Cambridge, 1931.

——— Basidia of *Septobasidium (Glenospora) Curtisii*. Jour. Elisha Mitchell Sci. Soc. **49**: 156–160. 1933.

——— Structure of Septobasidium in relation to the association with scale insects. Proc. Sixth Intern. Bot. Congress, 1935, **2**: 154–156. Leiden, 1935.

——— Septobasidium in the United States. Jour. Elisha Mitchell Sci. Soc. **51**: 1–77, pl. 1–44. 1935.

——— A new fungus intermediate between the rusts and Septobasidium. Mycologia **29**: 665–673, figs. 1–30. 1937.

GALLOWAY, B. T. A new pear disease. Jour. Mycol. **6**: 113–114. 1890.

GÄUMANN, E. Über das *Septobasidium bogoriense* Pat. Ann. Mycol. **20**: 160–173, 9 figs., 1 pl. 1922.

——— Enkele gegevens omtrent de grijze dadaschimmel. De Thee **3**: 48, 2 pl. 1922.

——— Vergleichende Morphologie der Pilze. Jena, 1926.

——— Über den Parasitismus des *Septobasidium bogoriense* Pat. Ann. Mycol. **25**: 175. 1927.

——— AND DODGE, C. W. Comparative Morphology of Fungi. New York, 1928.

GIDDINGS, N. J., AND WOOD, JESSIE I. Diseases of fruit and nut crops in the United States in 1924. Plant Disease Reporter Suppl. **39**: 1–105. 1925.

GRILLO, H. V. DA S. Estudo sobre o *Septobasidium albidum* da Laranjeira. Agronomia (Ann. Soc. Brasil. Agron.) **1**: 265–276. 1930.

HENNINGS, P. Beitrage zur Pilzflora Südamerikas II. Hedwigia **36**: 190–246, 1 pl. 1897.

HENNINGS, P. Fungi monsunenses. *In* Warburg, Monsunia **1**: 137–174. 1900.

——— Champignons observés sur divers caféiers par E. Laurent. *In* De Wilderman, Mission E. Laurent. Fasc. **3**: 316–318. 1906.

HÖHNEL, F., AND LITSCHAUER, V. Beiträge zur Kenntnis der Corticieen. (II Mitteil.) Sitzungsber. d. kaiserl. Akad. d. Wissensch. Wien. Math.-naturw. **116**: 739–852, 4 pl., 20 figs. 1907.

JAAP, OTTO. Beitrage zur Kenntnis der Pilze Dalmatiens. Ann. Mycol. **14**: 27. 1916.

JACKSON, H. S. The nuclear cycle in *Herpobasidium filicinum*, etc. Mycologia **27**: 553–572, 4 figs. 1935.

KILLERMANN, S. Auriculariaceae. *In* Engler and Prantl, Naturl. Pflanzenfam. 2nd Aufl., **6**: 105–108. 1928.

KÜHNER, R. Contribution à l'étude des Hyménomycètes et Spécialement des Agaricacés. Le Botaniste **17**: 17–23, figs. 2, 3. 1926.

LLOYD, C. G. The Genus Septobasidium. Myc. Notes **5**: 720–722, fig. 1077. 1917.

——— The Genus Septobasidium [*S. alatum*, *S. rameale*, *S. fisso-lobatum*]. Myc. Notes **6**: 887–888, figs. 1538–1540. 1919.

——— *Septobasidium pteruloides* and *Septobasidium pungens*. Myc. Notes **7**: 1195, figs. 2429, 2430. 1923.

——— *Septobasidium cladoderris*. Myc. Notes **7**: 1245, fig. 2658. 1924. (Not a *Septobasidium*.)

——— *Septobasidium retiforme*. Myc. Notes **7**: 1286, figs. 2901, 2902. 1924.

——— *Septobasidium protractum*. Myc. Notes **7**: 1336, figs. 3088, 3089. 1925.

MARCHIONATTO, J. B. Argentine Republic: fungous parasites of scale insects of fruit trees. Intern. Bull. Pl. Prot. **9**: 102–103. 1935.

MONTAGNE, C. Champignons. *In* Bélanger, C., Voyage aux Indes Orientale pendant les années 1825–1829. 1834.

MÜLLER, A. S. Observations and notes on Citrus diseases in Minas Geraes, Brazil. Phytopath. **23**: 734–737. 1933.

Mycological Notes. Ann. Rept. Seychelles Dept. Agric. for the year 1926, p. 24. 1927.

NEWSTEAD, ROBERT. Monograph of the Coccidae of the British Isles. Vol. 1, 1901, vol. 2, 1903. London.

ORTON, C. R., AND WOOD, JESSIE, I. Diseases of fruit and nut crops in the United States in 1923. Plant Disease Reporter Suppl. **33**: 1–147. 1924.

OVERHOLTS, L. O. Mycological Notes for 1921–22. Mycologia **16**: 233. 1924.

PATOUILLARD, N. Note sur un genre nouveau d'Hyménomycètes (*Helicobasidium*). Bull. Soc. Bot. Fr. **32**: 171–172. 1885.

——— Septobasidium, nouveau genre d'Hyménomycètes hétérobasidiés. Jour. Bot. **6**: 61–64, 2 figs. 1892.

——— Espèces critiques d'Hyménomycètes. Bull. Soc. Myc. Fr. **10**: 79. 1894.

——— Énumération des champignons récoltés à Java par M. Massart. Ann. Jard. Bot. Buitenzorg, Suppl. **1**: 118. 1897.

——— Description d'une nouvelle espèce d'Auriculariacés (*Septobasidium Langloisii*). Bull. Soc. Myc. Fr. **16**: 54–55. 1900.

——— Essai Taxonomique des Hyménomycètes. Lons-La-Saunier, France. 1900.

Patouillard, N. Septobasidium. *In* Hennings *in* Otto Warburg, Monsunia **1:** 137–138. 1900.

——— Champignons de la Gaudeloupe. II ser. Bull. Soc. Myc. Fr. **16:** 181. 1900.

——— Basidiomycètes nouveaux du Brésil recueillis par F. Noack. Ann. Mycol. **5:** 364–65. 1907.

——— Champignons nouveaux ou peu connus. Bull. Soc. Myc. Fr. **24:** 1–2, 1 fig. 1908.

——— Note sur un Septobasidium conidifère. Comp. rend. Acad. Sci., Paris, **156:** 1699. 1913.

——— Champignons des Philippines. I. Leaflets Phil. Bot. **6:** 2239–2256. 1914.

——— Quelques champignons du Tonkin. Bull. Soc. Myc. Fr. **31:** 73. 1915.

——— Champignons des Philippines communiqués par C. F. Baker. Philipp. Jour. Sci. C. Bot. **10:** 85. 1915.

——— Quelques champignons du Tonkin (Suite) (1). Bull. Soc. Myc. Fr. **36:** 174–177. 1920; **40:** 29–37, 3 figs. 1924; **41:** 337–342, 2 figs. 1925.

——— and Lagerheim, G. de. Champignons de l'equateur. Bull. Soc. Myc. Fr. **9:** 136–137. 1893.

Petch, T. Note on the biology of the genus Septobasidium. Ann. Bot. **25:** 843. 1911.

——— Fungi parasitic on scale insects. Trans. Brit. Myc. Soc. **7:** 18–40. 1921.

——— Diseases of the tea bush. 220 pp., 69 figs. 3 pl. Macmillan. London, 1923.

——— Entomogenous fungi and their use in controlling insect pests. Bull. Dept. Agric. Ceylon, Peradeniya, **71:** 1–40, 2 pl. 1925.

——— Notes on Ceylon Thelephoraceae, etc. Ann. Roy. Bot. Gard. Peradeniya **9:** 259–298. 1925.

——— Studies in entomogenous fungi IX. Aegerita. Trans. Brit. Myc. Soc. **11:** 62–65, 1 fig. 1926; XII. Ibid. **12:** 47. 1927; XIII. Ibid. **12:** 105–113. 1927.

——— *Septobasidium rameale.* Trans. Brit. Myc. Soc. **12:** 276–282, 2 pl. 1927.

Prillwitz, P. M. Vergelijkende beschouwingen over de grijze dadapschimmel en hare bestryding. De Thee **7:** 91–94, 1 pl. 1926.

Raciborski, M. Parasitische und epiphytische Pilze Javas. Bull. Intern. Acad. Sci. Cracovie **3:** 346–394. 1909.

Rick, J. (a) Fungi Austro-Americani. Ann. Mycol. **4:** 311. 1906.
(b) Pilze aus Rio Grande do Sul. Broteria **5:** 5–6. 1906.

Ritzema, Bos J. *Septobasidium bogoriense* Pat. op jonge Kinaboomen. Tijdschr. over Plantziekten **32:** 303–304. 1926.

Saccardo, P. A. Septobasidium. Sylloge Fungorum **11:** 118. 1895; **14:** 215. 1899; **16:** 184–185. 1902.

——— Notae mycologicae. Ser. III. Ann. Mycol. **1:** 24. 1903.

——— Septobasidium. Sylloge Fungorum **17:** 203–204. 1905; **21:** 445–447. 1912; **23:** 559–567. 1925.

Sawada, K. Descriptive Catalogue of Formosan Fungi. Spec. Bull. Agric. Exp. Stat. Formosa **19:** Part 1. 1919.

SAWADA, K. Descriptive Catalogue of Formosan Fungi. Part VI. Rept. Dept. Agric. Res. Inst. Formosa **61**: 117 pp., 2 pl. 1933.

SIEMASZKO, W. Badania mykologiczne w górach Kaukazu. Discip. Biol. Arch. Soc. Sci. Varsaviensis **1**, Fasc. 14: 27. 1923.

SNELL, W. H. A new Septobasidium on *Pinus strobus*. Mycologia **14**: 55–60, 3 pl. 1922.

STEINMANN, A. Voorloopige mededeeling omtrent het optreden van *Septobasidium bogoriense* en *rubiginosum* op Thee. De Thee **7**: 52–53, 1 pl. 1926.

——— Over het optreden van de grijze Dadapschimmel en andere op Thee voorkomende Septobasidium—soorten. De Thee **7**: 156–158. 1926.

——— Eenige beschouwingen omtrent voorkomen en bestrijding van de op kina optredende Septobasidium-soorten. Cinchona **3**: 123–128, 2 figs. 1926.

SUBBA RAO, M. K. Report of the mycologist. Adm. Rep. Tea Sci. Dep. Unit. Plant. Ass. S. India, 1935–36, pp. 46–54. 1936.

SYDOW, H., AND P. Beschreibungen neuer südafrikanischer Pilze. Ann. Mycol. **10**: 33. 1912.

——— Notes and descriptions of Philippine fungi. Leaflets Phil. Bot. **6**: 1920. 1913.

——— Fungi from northern Palawan. Philippine Jour. Sci. C. Bot. **9**: 158–159. 1914.

——— Novae fungorum species XII. Ann. Mycol. **12**: 195. 1914.

——— Diagnosen neuer philippinischer Pilze. Ann. Mycol. **12**: 545. 1914.

——— Fungi papuani. Botanische Jahrb. **54**: 246–261. 1916.

——— Beitrag zur Kenntnis der Pilzflora der Philippinen-Inseln. Ann. Mycol. **15**: 170. 1917.

——— Mykologische Mitteilungen. Ann. Mycol. **17**: 33. 1919.

——— Novae fungorum species. XVI. Ann. Mycol. **18**: 154. 1920.

SYDOW, H. Bemerkungen zu einer Anzahl in letzter Zeit als neu beschriebener Pilze der Philippinen-Inseln. Ann. Mycol. **20**: 66. 1922.

TANAKA, N. A new species of hymenomycetous fungus injurious to the mulberry tree. Jour. Coll. Sci. Imp. Univ. Tokyo **4**: 191–204, 4 pl. 1891.

——— New Japanese fungi; notes and translations VI. Mycologia **11**: 80–86. 1919.

TORREND, C. Les basidiomycètes des environs de Lisbonne. Broteria, ser. bot. **11**: 73–98. 1913.

——— Troisième contribution pour l'étude des champignons de l'ile de madère. Broteria, ser. bot. **11**: 165–181. 1903.

TUNSTALL, A. C. Some observations on stem and root diseases. Quart. Jour. Indian Tea Assoc. **3**: 86–91. 1923.

ULTER, A. J. Verslag over de werkzaamheden van het Proefstation in het jaar 1928. Meded. Proefstat. Malang **69**: 64 pp. 1929.

VAN DER GOOT, P. Ziekten en plagen der cultuurgewassen in Nederlandsch.-Indië in 1935. Meded. Inst. Pl. Ziekt., Batavia **87**: 106 pp. 1936.

WILSON, G. W. Notes on three limb diseases of apple. Thelephorose. N. C. Agric. Exp. Stat. Rept. **25**: 53–55. 1912.

PLATES

Plate 5

All figures on this plate are semi-diagrammatic. All figures ×33, except Fig. 1.

Fig. 1.—Section through very thick region of *S. jamaicaense* Burt, where four hymenial regions had been formed. × 6.

Fig. 2.—*S. arboreum,* showing the tree-like clusters of hyphae.

Fig. 3.—*S. atropunctum,* showing the thin hymenium and the short pillars and threads penetrating into the crevices of the bark.

Fig. 4.—*S. areolatum,* showing holes (areolations) in context covered here by the hymenium. Note the scale insect with its sucking apparatus penetrating the wood.

Fig. 5.—*S. septobasidioides,* showing thin erect pillars and part of a scale insect to right.

Fig. 6.—*S. rhabarbarinum,* showing exceedingly thick columns of hyphae, and hymenium with dichotomously branched, coiled hyphae.

Fig. 7.—*S. purpureum,* showing context without pillars, etc. Note scale insect to right with sucking apparatus extending down into the wood.

Fig. 8.—*S. alveolatum,* showing an alveolus covered by the hymenium. Note empty scale at bottom of fungus.

PLATE 5

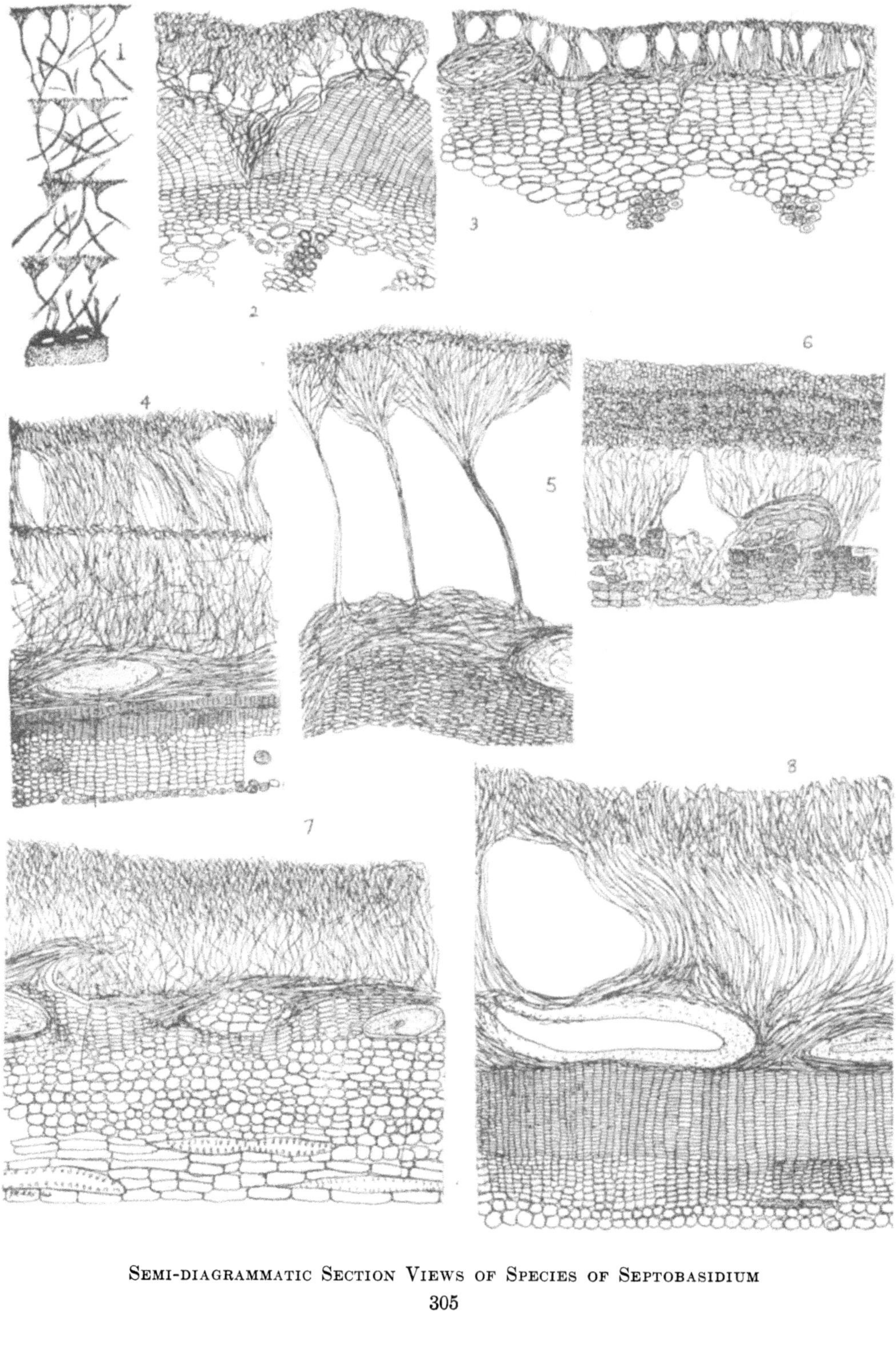

SEMI-DIAGRAMMATIC SECTION VIEWS OF SPECIES OF SEPTOBASIDIUM

PLATE 6

S. jamaicaense

Figs. 1–4.—× 873. Basidia in various stages of development. Fig. 4 shows an empty basidium with two long sterigmata. Fig. 5. Spores with sterigmata and one sporidium, abnormal. × 720.

S. areolatum

Figs. 6–13.—Development of basidium from the stage when it becomes septate to the empty basidium.

Figs. 14, 15.—Spores, fig. 14 an abnormal one. All figs. × 873.

PLATE 6

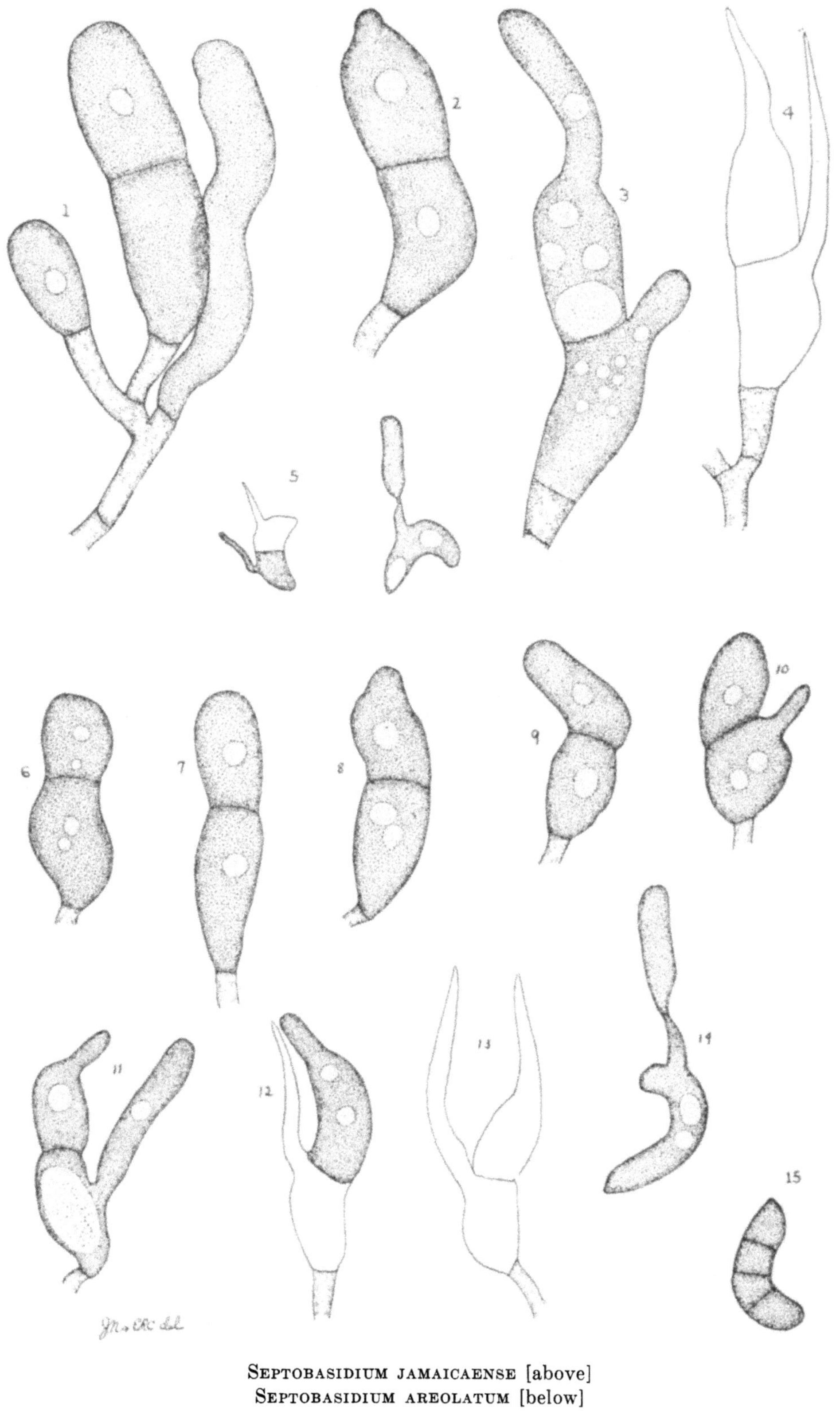

Septobasidium jamaicaense [above]
Septobasidium areolatum [below]

Plate 7

S. septobasidioides

Figs. 1–14.—Development of basidium from the probasidium up to the formation of sterigmata. Fig. 14. A basidium which has fallen from the parent hypha. All × 847.

S. rhabarbarinum

Fig. 15.—Basidium with four sterigmata. × 847.

Fig. 16.—Part of the hymenium showing dichotomously branched hyphae, probasidia, and coiled basidia, one with a spore. Note also the septate bodies, perhaps gemmae. × 847.

PLATE 7

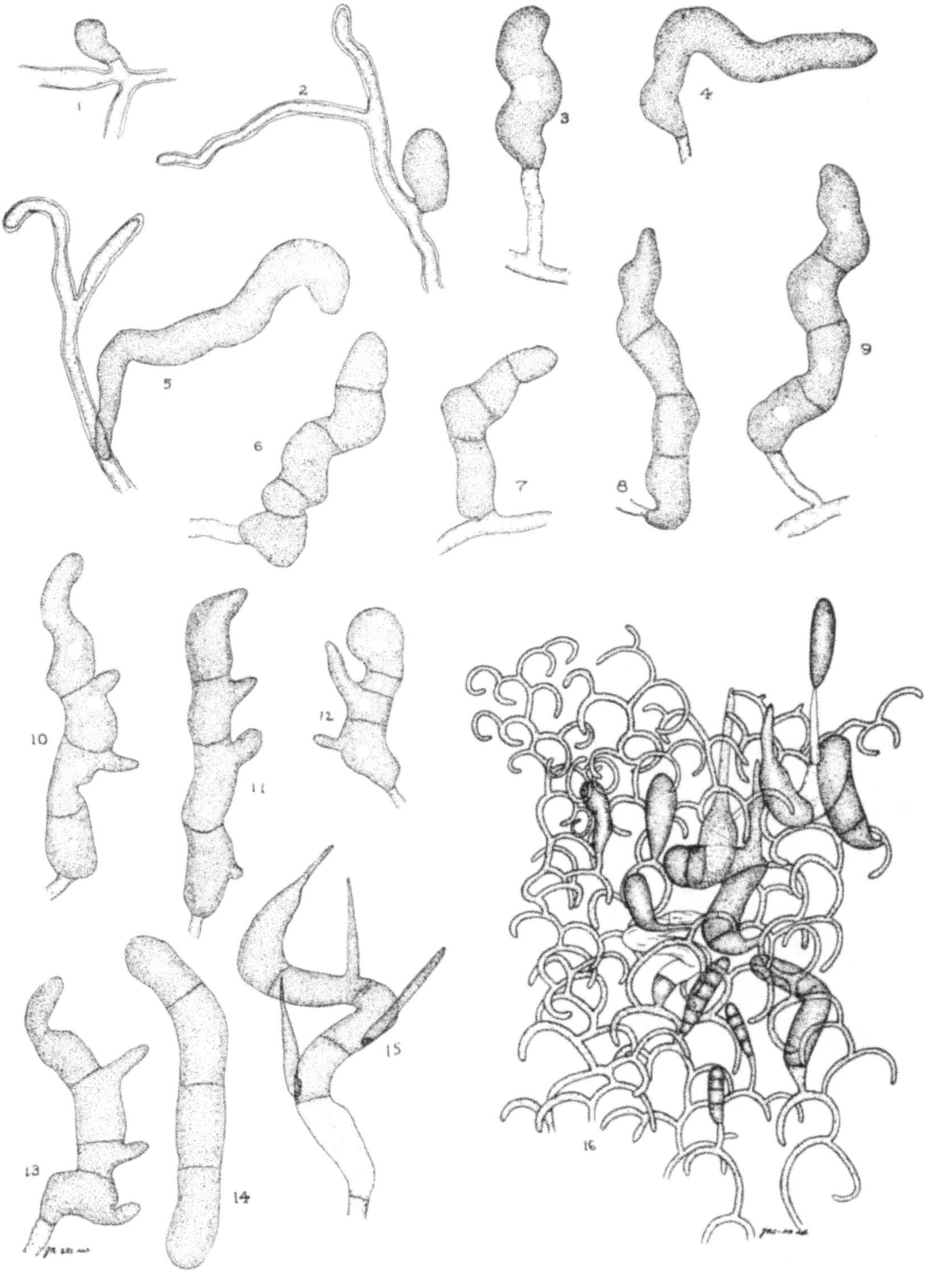

SEPTOBASIDIUM SEPTOBASIDIOIDES [Figs. 1–14]
SEPTOBASIDIUM RHABARBARINUM [Figs. 15, 16]

PLATE 8

S. atropunctum

Probasidia and basidia in all stages of development; pillars and subiculum. × 558.

PLATE 8

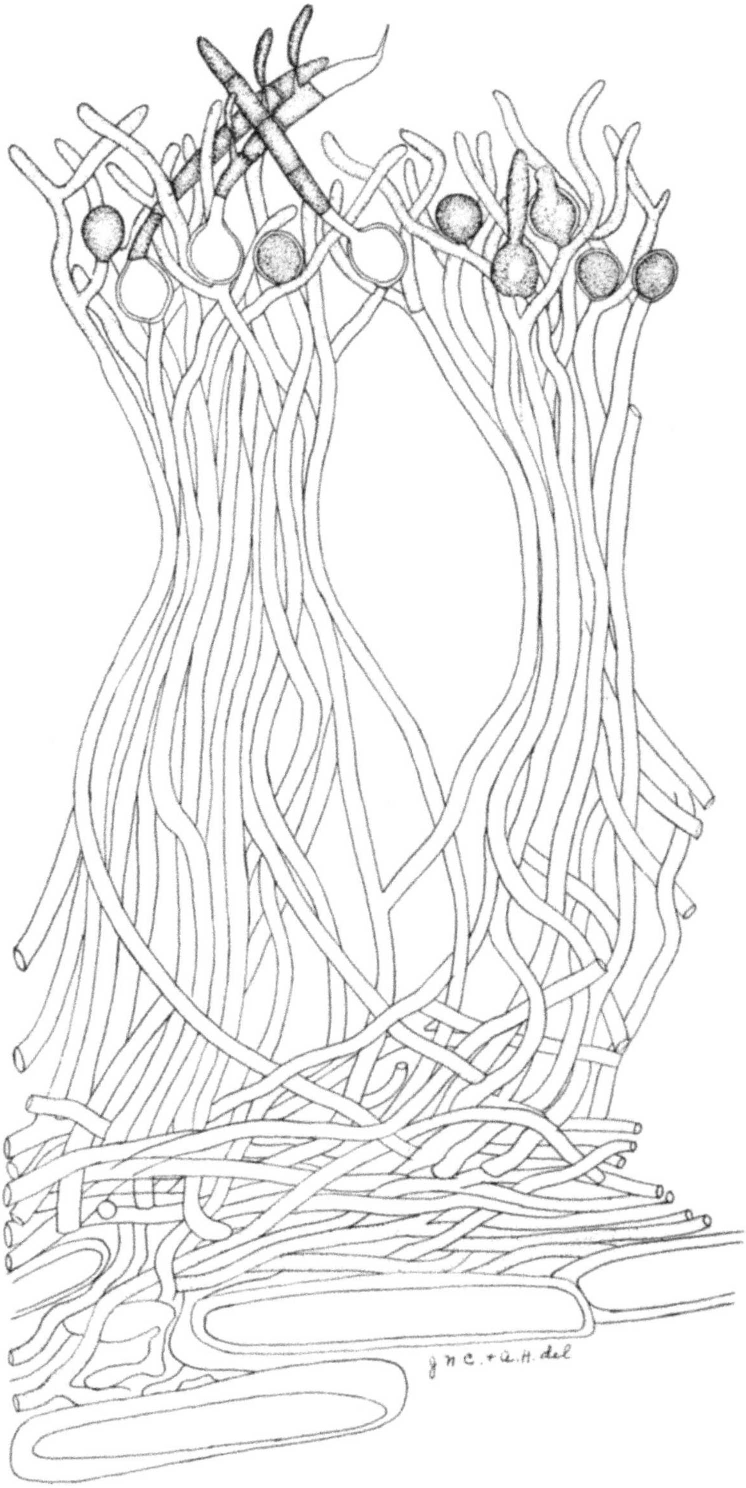

SEPTOBASIDIUM ATROPUNCTUM

PLATE 9

S. alveolatum

Development of probasidium into the basidium and the formation of spores. All × 873.

PLATE 9

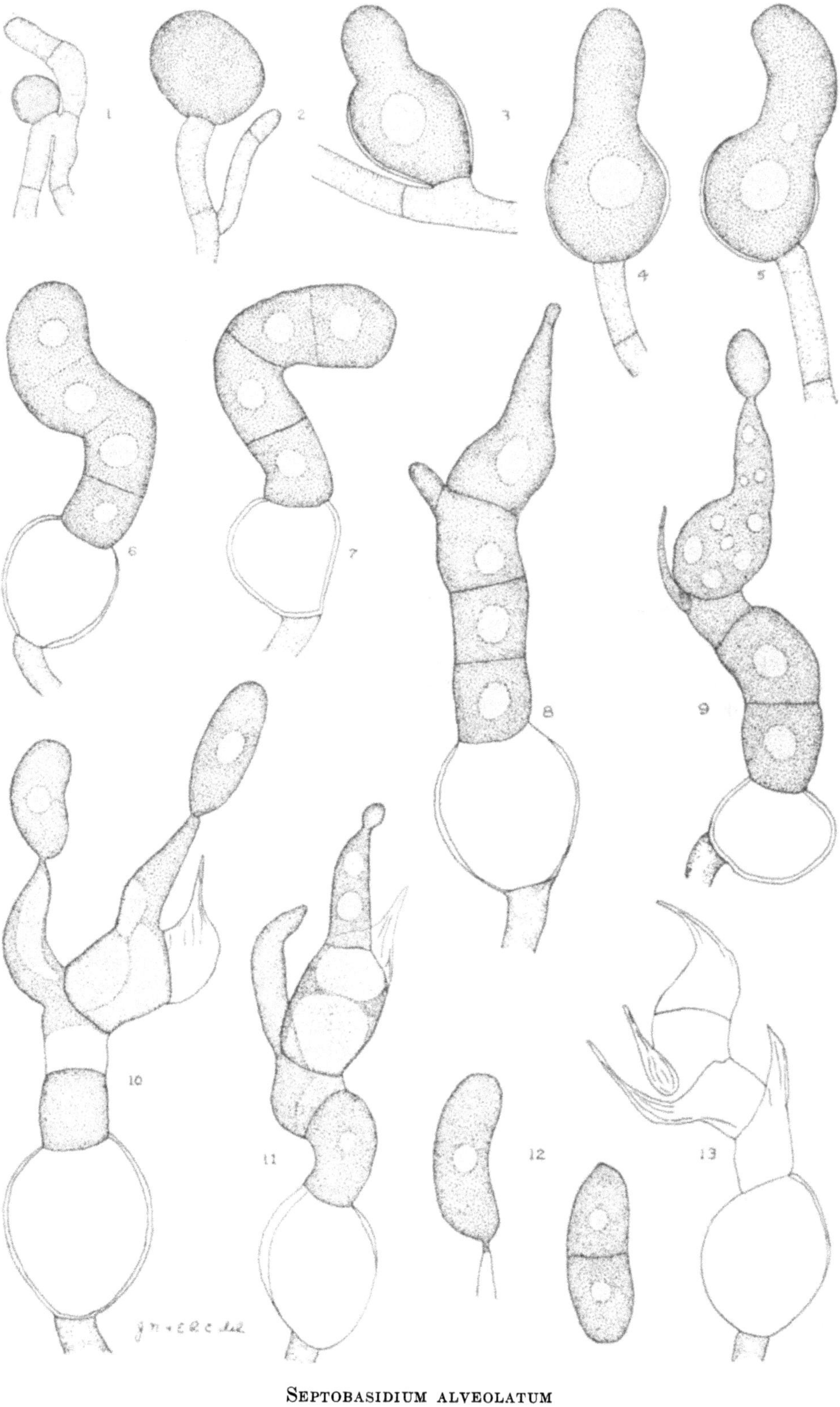

SEPTOBASIDIUM ALVEOLATUM

PLATE 10

S. (?) *polypodii*

Section of infected sporophyll, below, showing how the fungus grows in and around sporangia. × 37. Above, section of hymenium showing probasidia and basidia. × 1012.

PLATE 10

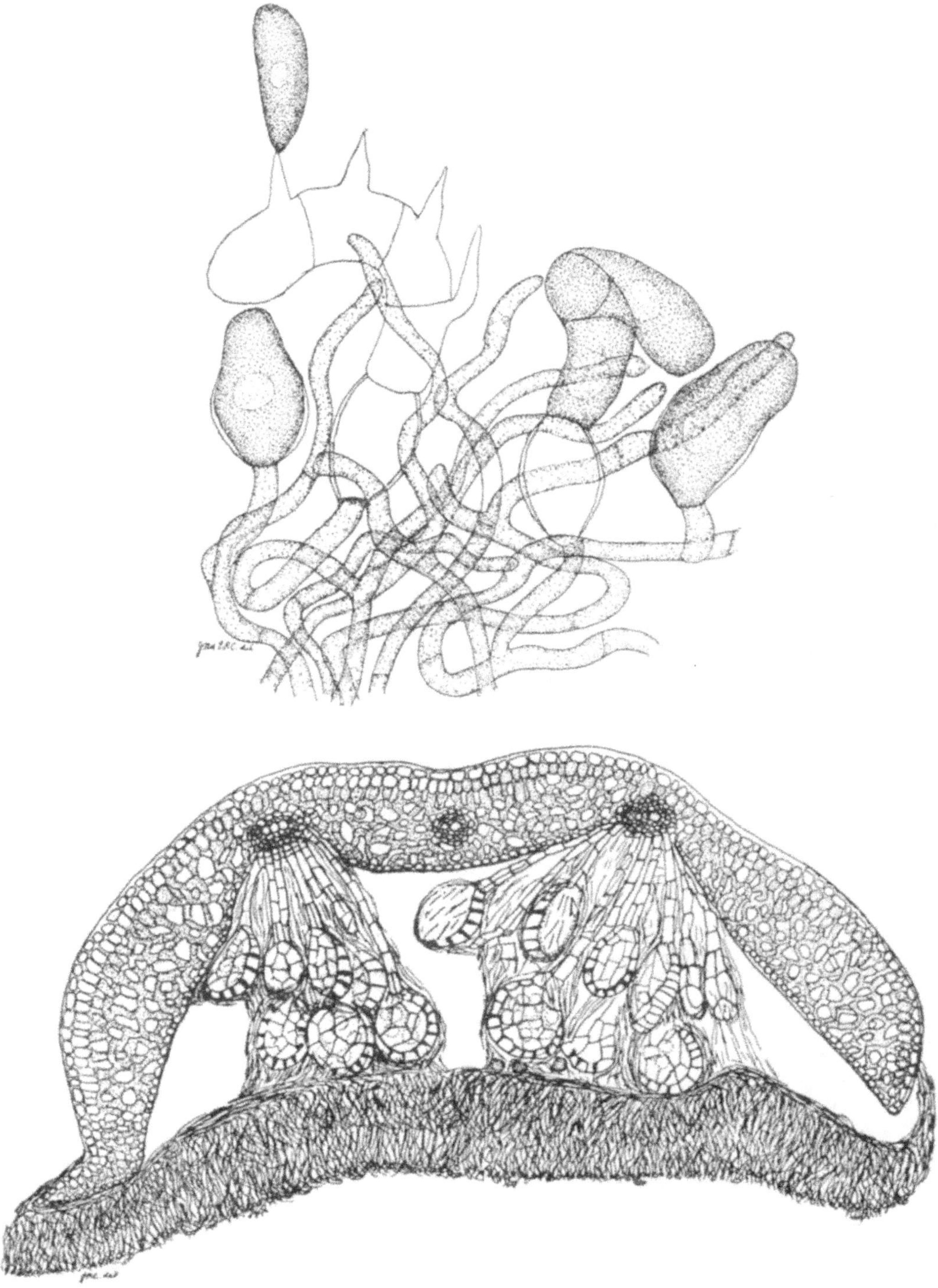

SEPTOBASIDIUM (?) POLYPODII

PLATE 11

S. purpureum

Figs. 1–11.—Development of probasidium into basidium which bears a single spore. × 1016.

Figs. 12–14.—Showing how spores if kept moist may divide into eight cells. × 1280.

PLATE 11

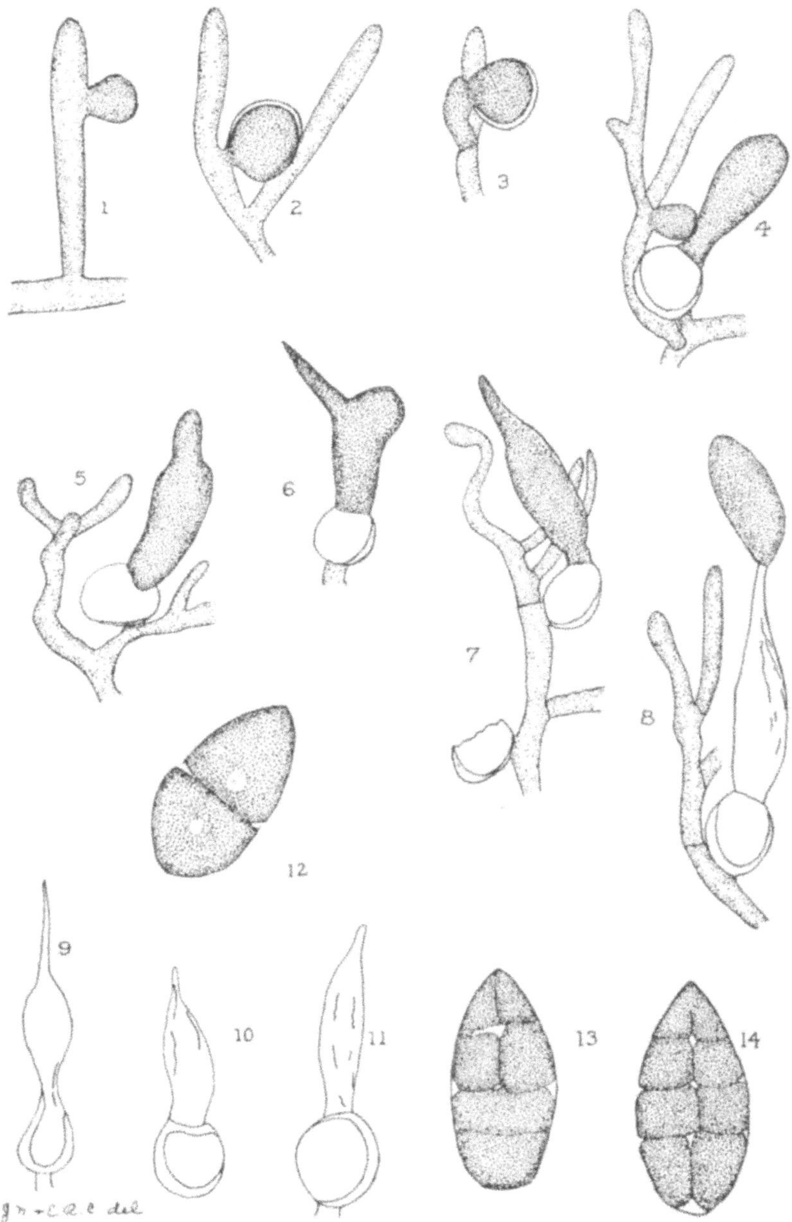

SEPTOBASIDIUM PURPUREUM

PLATE 12

S. arboreum

Fig. 1.—Basidium and coiled threads borne on very thick hypha.
Figs. 2-6.—Basidia, figs. 4 and 5 showing sterigmata.
Fig. 7.—Spore sprouting two sterigmata, which is unusual. All × 1310.

PLATE 12

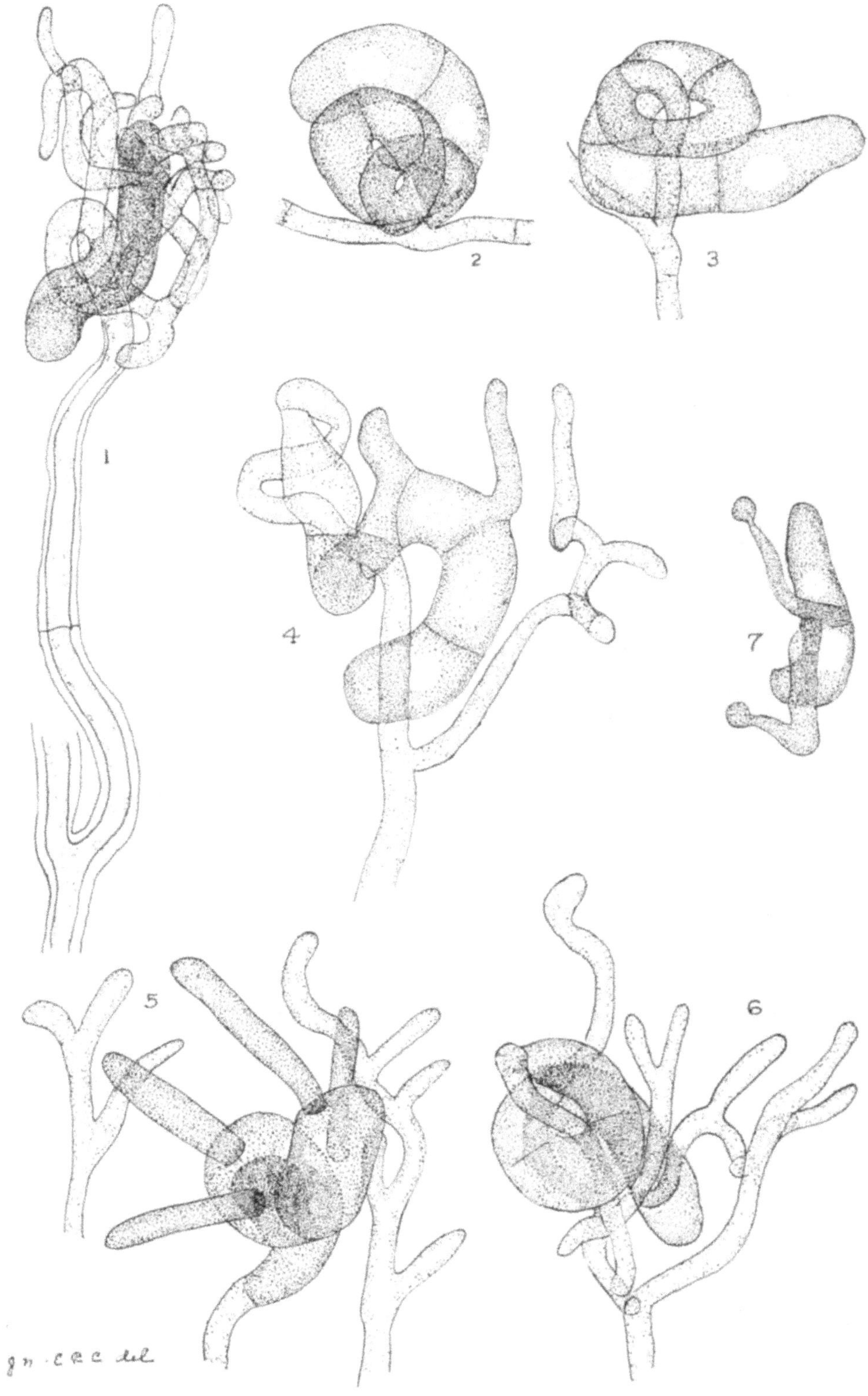

SEPTOBASIDIUM ARBOREUM

PLATE 13

Spores

Fig. 1.—*S. atropunctum* × 578; Fig. 2.—*S. arboreum*; Fig. 3.—*S. alveolatum*; Fig. 4.—*S. jamaicaense*; Fig. 5.—*S. septobasidioides*; Fig. 6.—*S. arcolatum*. Figs. 2–6, × 982.

PLATE 13

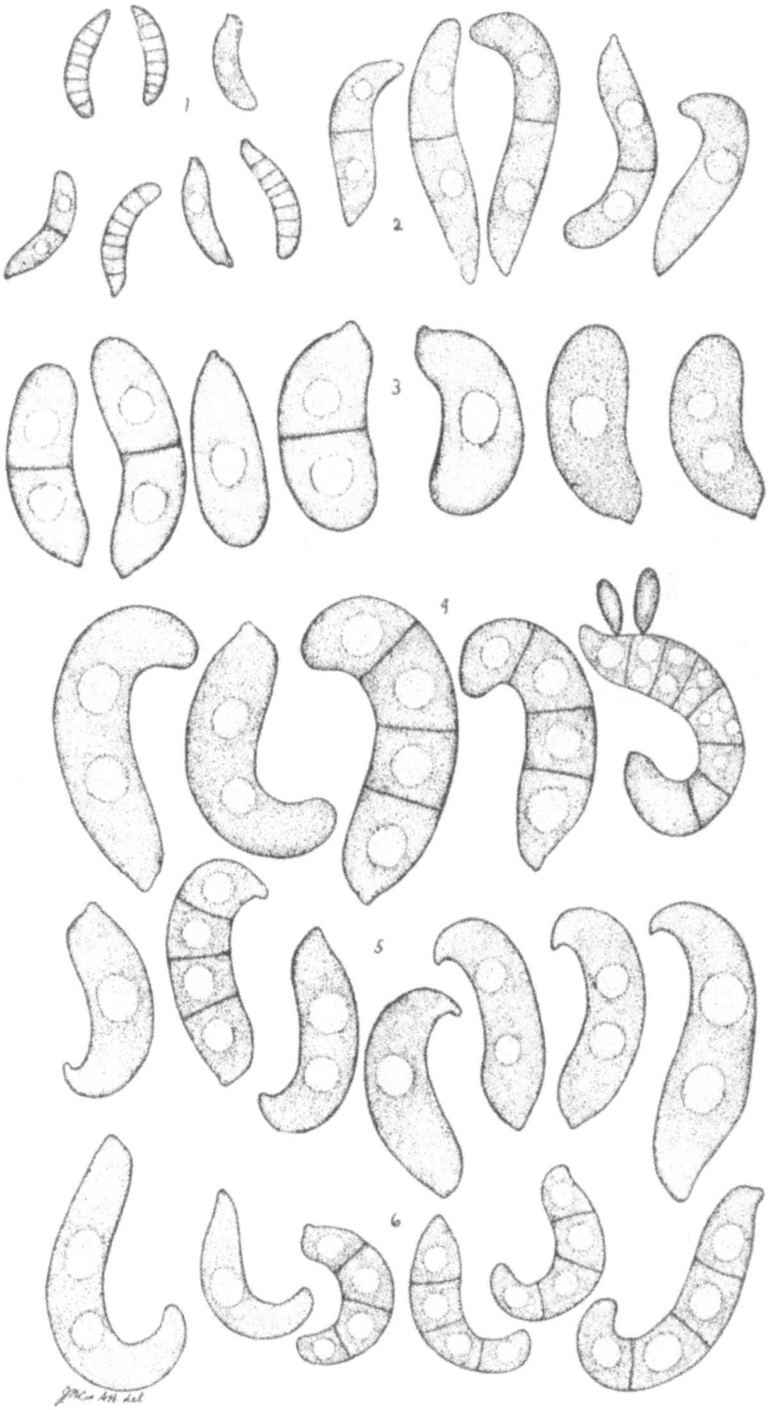

SPORES OF VARIOUS SPECIES OF SEPTOBASIDIUM

PLATE 14

Several types of haustoria from scale insects

Fig. 1.—Haustoria from within body of a parasitized scale insect beneath *S. purpureum.* Haustoria consisting of spindle-shaped segments of hyphae connected by fine threads, like strings of sausages.

Fig. 2.—Haustoria from within body of a parasitized scale insect beneath *S. alveolatum,* consisting of sausage-shaped hyphae and irregular coils of hyphae.

Fig. 3.—Haustoria from within body of a parasitized scale insect beneath *S. rhabarbarinum.* Note cross wall at the smaller end of the segments.

Fig. 4.—Haustoria from within body of a parasitized scale insect from beneath *S. arboreum.* Note how the segments are entangled and how almost every segment is bent.

Fig. 5.—Haustoria from within body of a parasitized scale insect from beneath *S. septobasidioides.* Haustoria consist of more or less regularly coiled hyphae.

Fig. 6.—Haustoria from within body of a parasitized scale insect from beneath *S. jamaicaense.* Haustoria consist of irregular coils of hyphae. All figures × 952.

PLATE 14

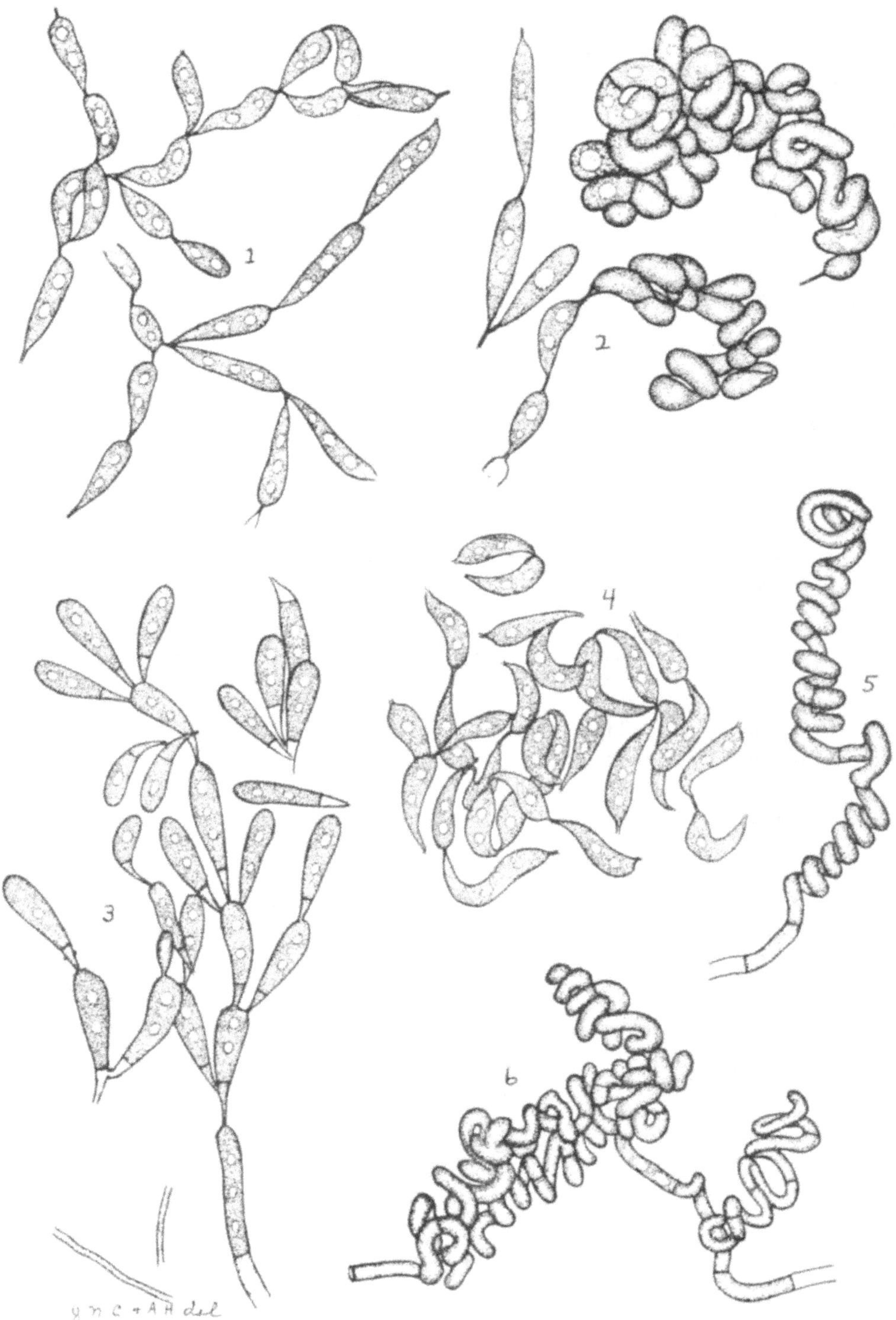

Several Types of Haustoria from Bodies of Scale Insects

Plate 15

S. purpureum

Semi-diagrammatic view of section of *S. purpureum* showing a parasitized scale insect over which the fungus has developed a dense pad of hyphae and the host plant into which the insect's suctorial apparatus penetrates. *P.B.*, probasidia; *B*, basidia; *B. S.*, basidiospores; *O.B.*, old collapsed basidia; *H.P.*, hyphal pad; *D.S.*, insect's dorsal scale; *V.S.*, ventral scale; *I.*, insect; *P.A.*, insect's pumping organ; *S.*, insect's suctorial "tube" represented by the two middle solid lines; *H.S.*, sheath of cellulose material secreted around the suctorial apparatus; *M.R.*, medullary ray cells; *N.*, nuclei of host plant; *H.*, haustoria of the fungus inside the body of living insect. × 135.

PLATE 15

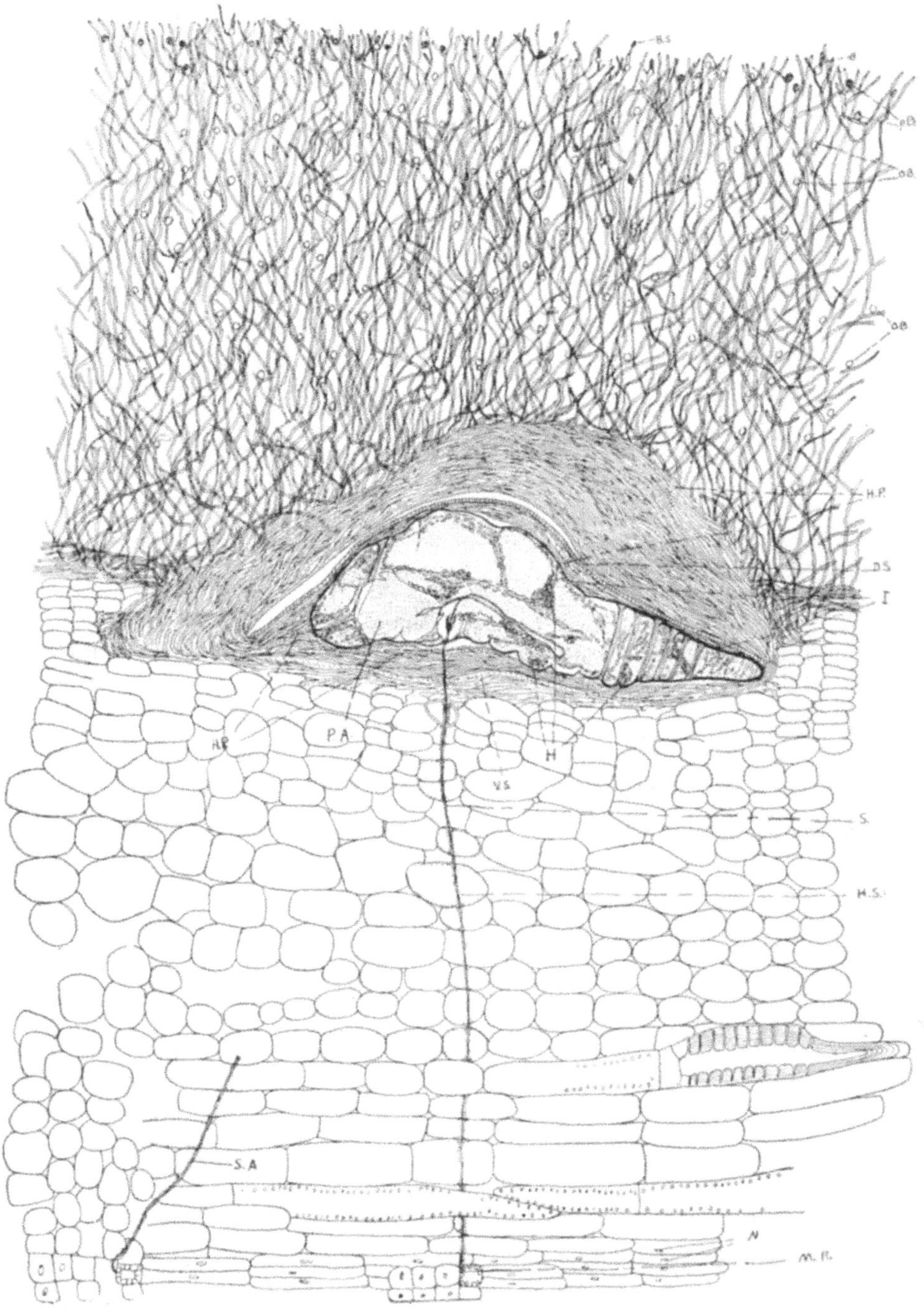

SECTION OF SEPTOBASIDIUM PURPUREUM AND TYPICAL PARASITIZED SCALE INSECT AND PART OF THE BARK OF HOST PLANT, SHOWING SUCTORIAL APPARATUS PENETRATING THROUGH BARK TO MEDULLARY RAY CELLS

PLATE 16

S. Curtisii

Fig. 1.—Section through fungus showing pillars, hymenium with probasidia, basidia, spores, and conidia. × 234.

Fig. 2.—An upright spine on which are several probasidia and basidia. × 100.

Figs. 3, 4.—Probasidia, the latter showing the shiny globules mistaken for nuclei by Berkeley. (Ellis, no. 292). × 835.

Fig. 5.—Empty probasidial cells showing stalk left after basidium has fallen off. × 835.

Figs. 6–10.—Basidia in various stages. Figs. 6 and 7, young basidia. Fig. 8 shows minute bud cells on sterigma after spores have been formed and fallen off. × 835.

Fig. 11.—Spores. × 835.

PLATE 16

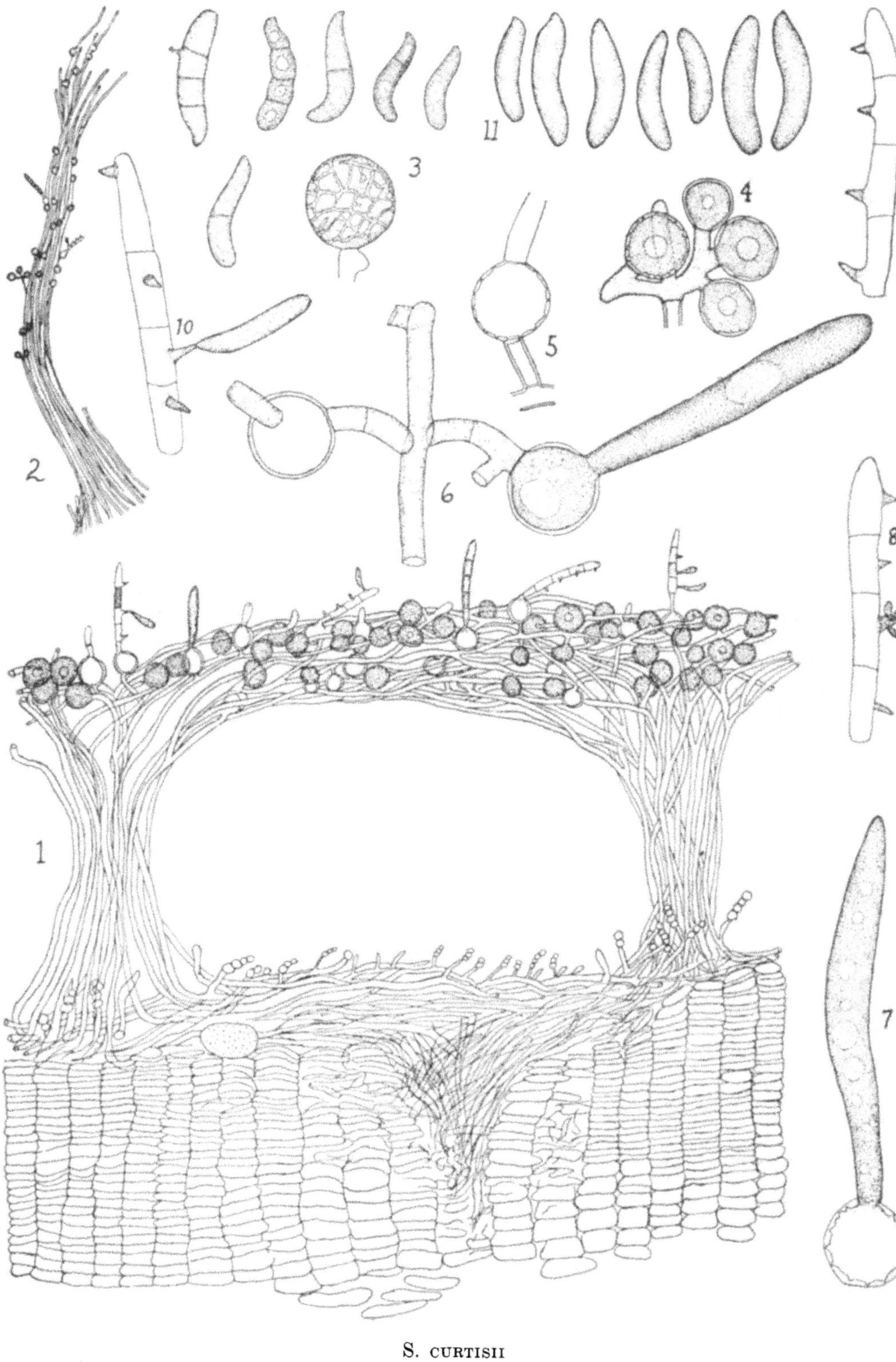

S. CURTISII

PLATE 17

S. Mariani

Section and surface view of fungus at time when young are being born. Note three horizontal layers: (1) subiculum (*sub*), (2) middle layer (*m.l.*), and (3) top layer (*t.l.*) which may sometimes be stratified. Scale insects inhabit the region between middle layer and subiculum, if insects are parasitized, or between middle layer and bark, if insects are healthy. One healthy adult female (*s.c.b.*) is giving birth to young, several of which can be seen crawling about over the fungus (*n*). Some have settled down on the subiculum (*n.r.*) Rhizomorphs (*r*) are distinct, also pillars (*p*). Several insects are parasitized by fungus (*s.c.p.*). × 20.

PLATE 17

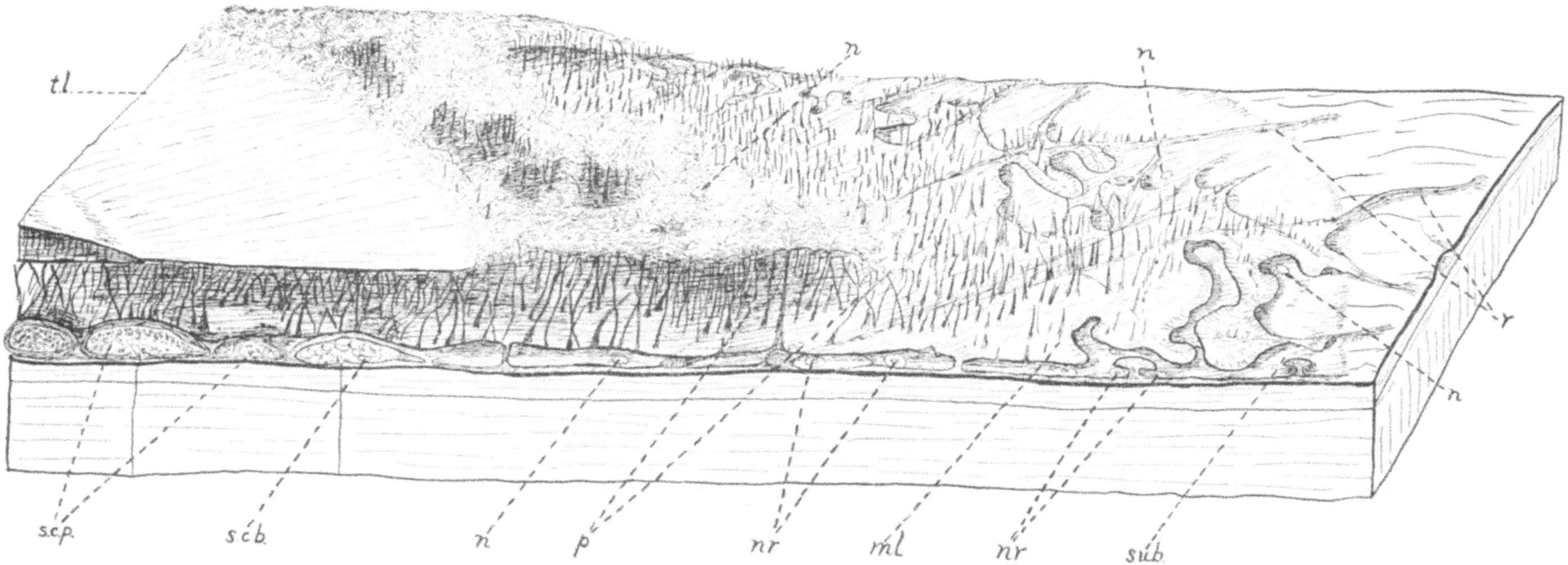

PLATE 18

S. pilosum

Fig. 1.—Diagram sketch through a tuft of growth, showing upright hyphae bearing conidia, and beneath the fungus in the center a parasitized scale insect. Note the coiled threads connecting insect with hyphal pad. × 18.

Fig. 2.—Several upright hyphae bearing conidia. × 110.

Figs. 3 and 4.—Conidia. × 525.

Fig. 5.—Thick-walled hyphae found in basal region. × 525.

Fig. 6.—Haustoria from parasitized insect. × 525.

Fig. 7.—Probasidia of type. × 525.

Fig. 8.—Showing connection between haustoria and external thread. × 525.

S. pinicola

Fig. 9.—Sketch of section of *S. pinicola*, showing very irregular surface, parasitized and healthy insects. × 18.

Figs. 10 and 11.—Basidia and spores. × 525 and × 580.

Figs. 12 and 13.—Haustoria of "glomerulus" type. × 580.

S. Hesleri

Fig. 14.—Sketch of margin showing a few scattered pillars arising from subiculum, several "insect houses" from one of which roof has been lifted, exposing insect. × 18.

Fig. 15.—Hymenial section showing dichotomously branched paraphyses, coiled basidia, and spore. × 580.

Fig. 16.—Spores, one of which has become septate, each cell forming a sterigma. × 580.

Fig. 17.—Crystal-encrusted, upright threads from subiculum. × 335.

PLATE 18

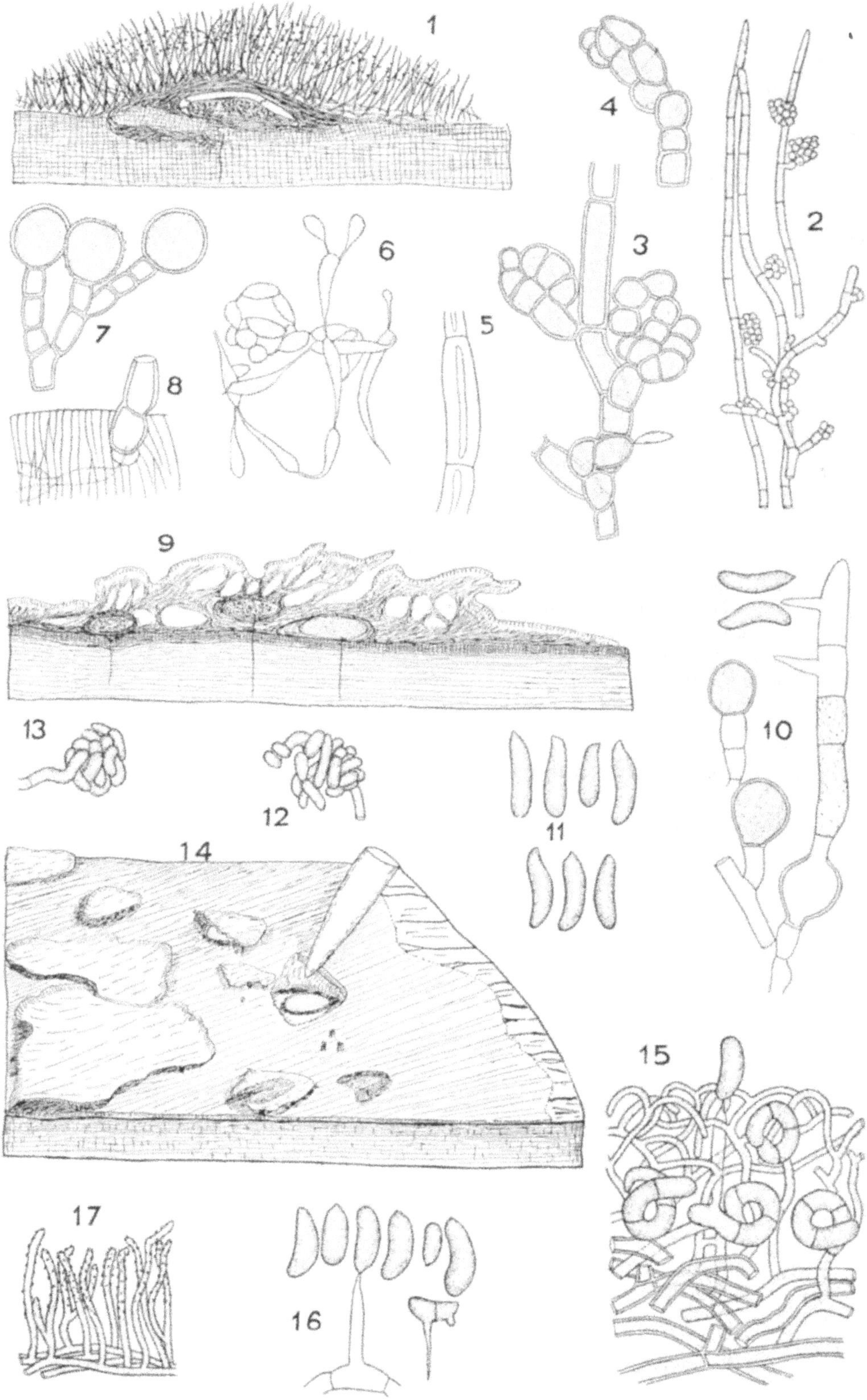

PLATE 19

S. lepidosaphis

Fig. 1.—Sketch of marginal region, showing subiculum which in this species bears probasidia, basidia, and spores. Note anastomosing and branching pillars, lop layer, and part of a scale insect. × 23.

Fig. 2.—Spores. × 833.

Fig. 3.—Large cells filled with protoplasm in subiculum. × 833.

Fig. 4.—Surface view of subiculum showing hyphae, probasidia, basidia and spores on epidermis of leaf. × 833.

Fig. 5.—Irregularly coiled haustoria. × 833.

Fig. 6.—Adult healthy scale insect, *Lepidosaphes beckii*. × 100.

PLATE 19

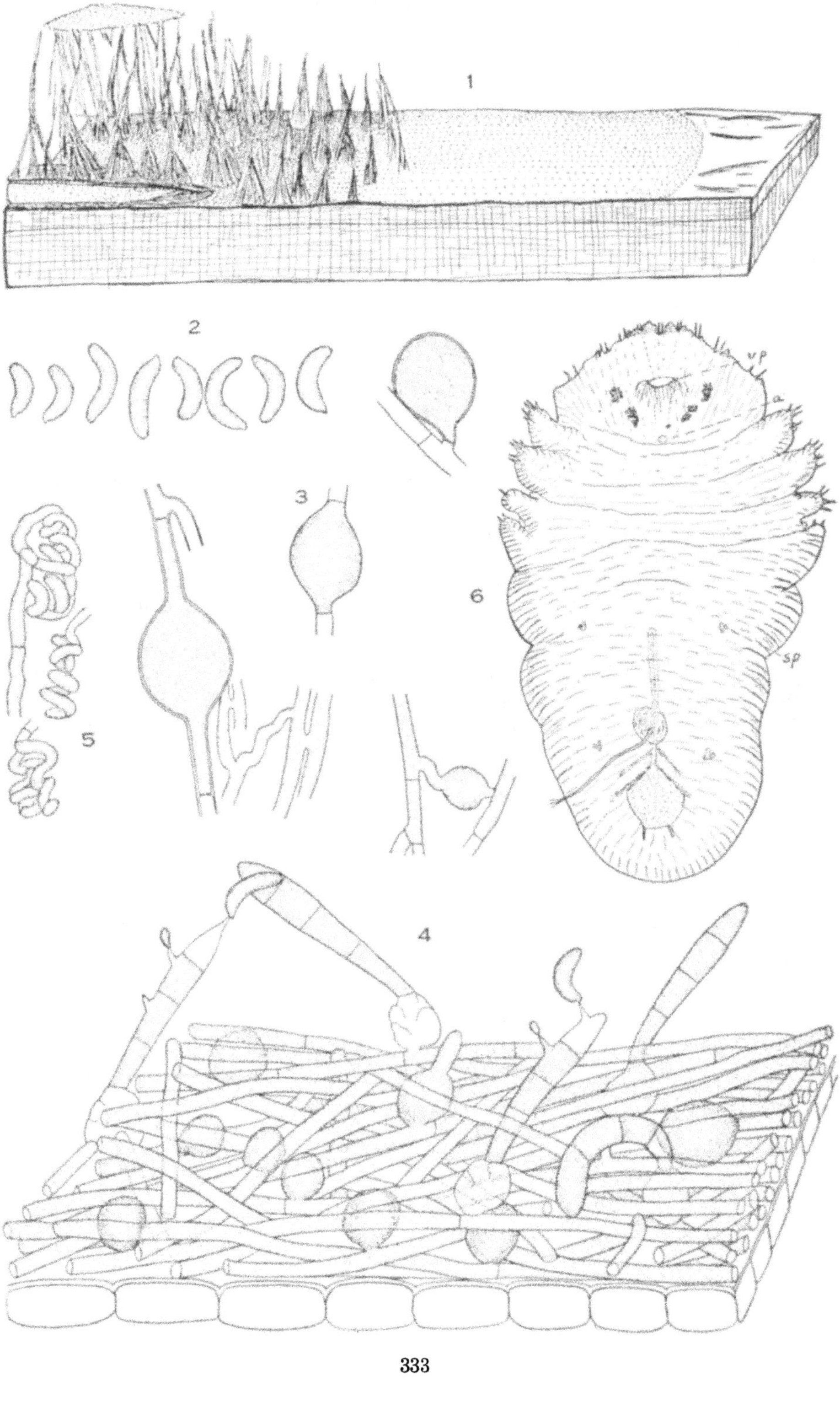

Plate 20

S. Carestianum

Fig. 1.—Sketch showing section of hymenium, pillars, parasitized insect, and wood. × 33.

Figs. 2–7.—Probasidia. Figure 5 also shows a thread from hymenial region. × 833.

Figs. 8 and 9.—Basidia. × 833.

Fig. 10.—Enlarged club-shaped bodies found in hymenial region in fall and winter. × 833.

Fig. 11.—*Chionaspis corni* associated with *S. Carestianum*. This female was giving birth to young when collected. × 60.

Fig. 12.—Section of hymenial region of young fertile plant in late spring. × 833.

Fig. 13.—Empty probasidia showing club-shaped bodies growing through two probasidial cells. × 833.

Fig. 14.—Haustoria. × 833.

PLATE 20

PLATE 21

S. alni

Fig. 1.—Section showing top layer with hymenium, pillars, one parasitized scale insect, and subiculum. No. 8325. × 33.

Fig. 2.—Section of hymenium showing probasidia, basidia, and spores. No. 8326. × 833.

Fig. 3.—Spores. No. 8326. × 833.

Figs. 4–7.—Empty probasidia, showing proliferation. Torrend, coll. In New York Bot. Gard. Herb. no. 373. × 833.

Figs. 8-10.—Empty and proliferating probasidia. Type from Sydow, no. 68. × 833.

Figs. 11 and 12.—Empty probasidia. From Georgia, Higgins, coll., no. 8498. × 733.

Fig. 13.—Proliferating probasidium (?). N. Y. Bot. Gard. Herb. no. 373. × 833.

Fig. 14.—Haustoria. × 833.

PLATE 21

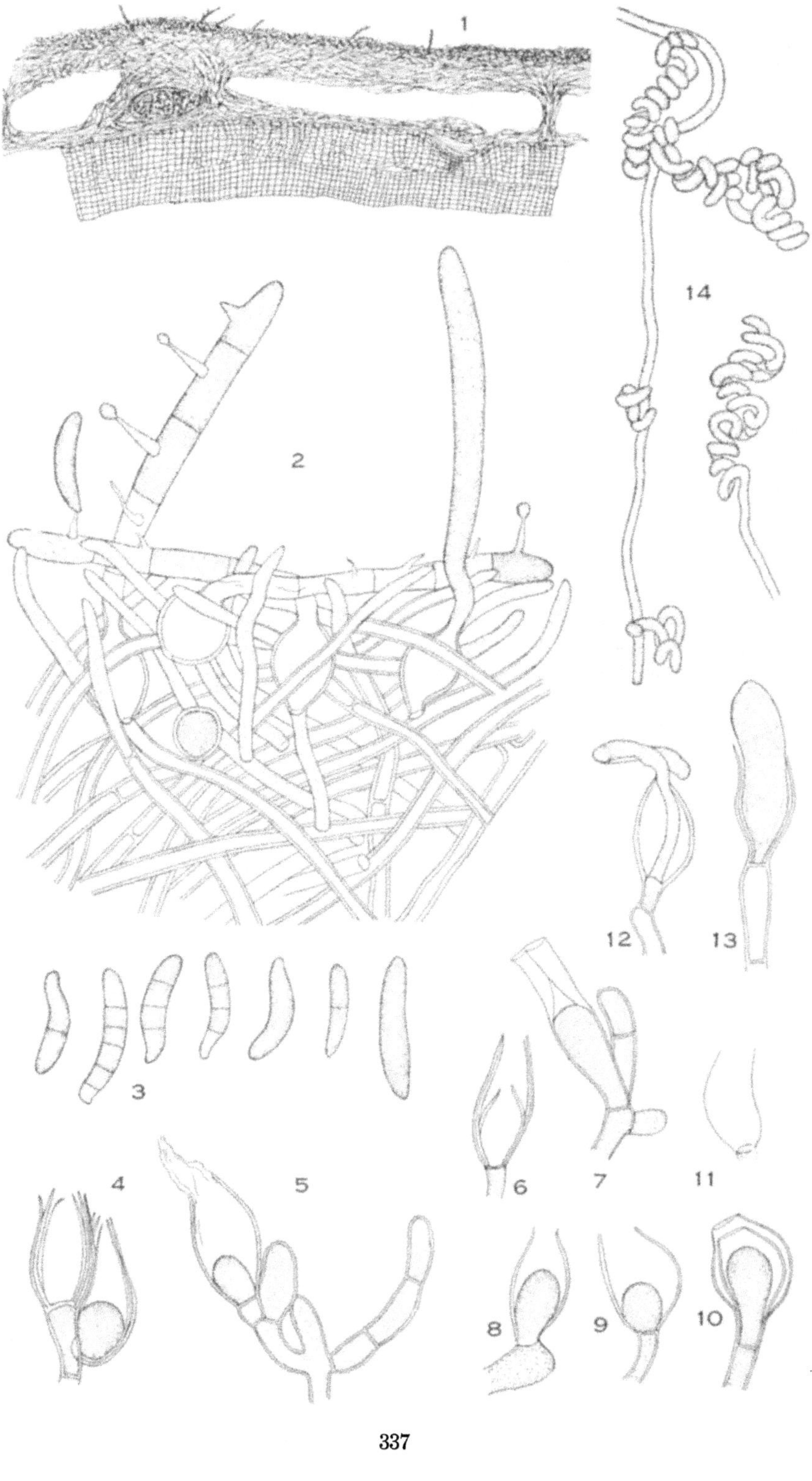

PLATE 22

S. sabal-minor

Fig. 1.—Sketch through outer part of fruit of *Sabal minor* showing fungus covering scale insect. × 38.
Fig. 2.—Hymenium showing probasidia and basidia. × 938.
Figs. 3–7.—Probasidia and basidia. × 938.
Fig. 8.—Spores. × 938.
Fig. 9.—Haustorium. × 938.

PLATE 22

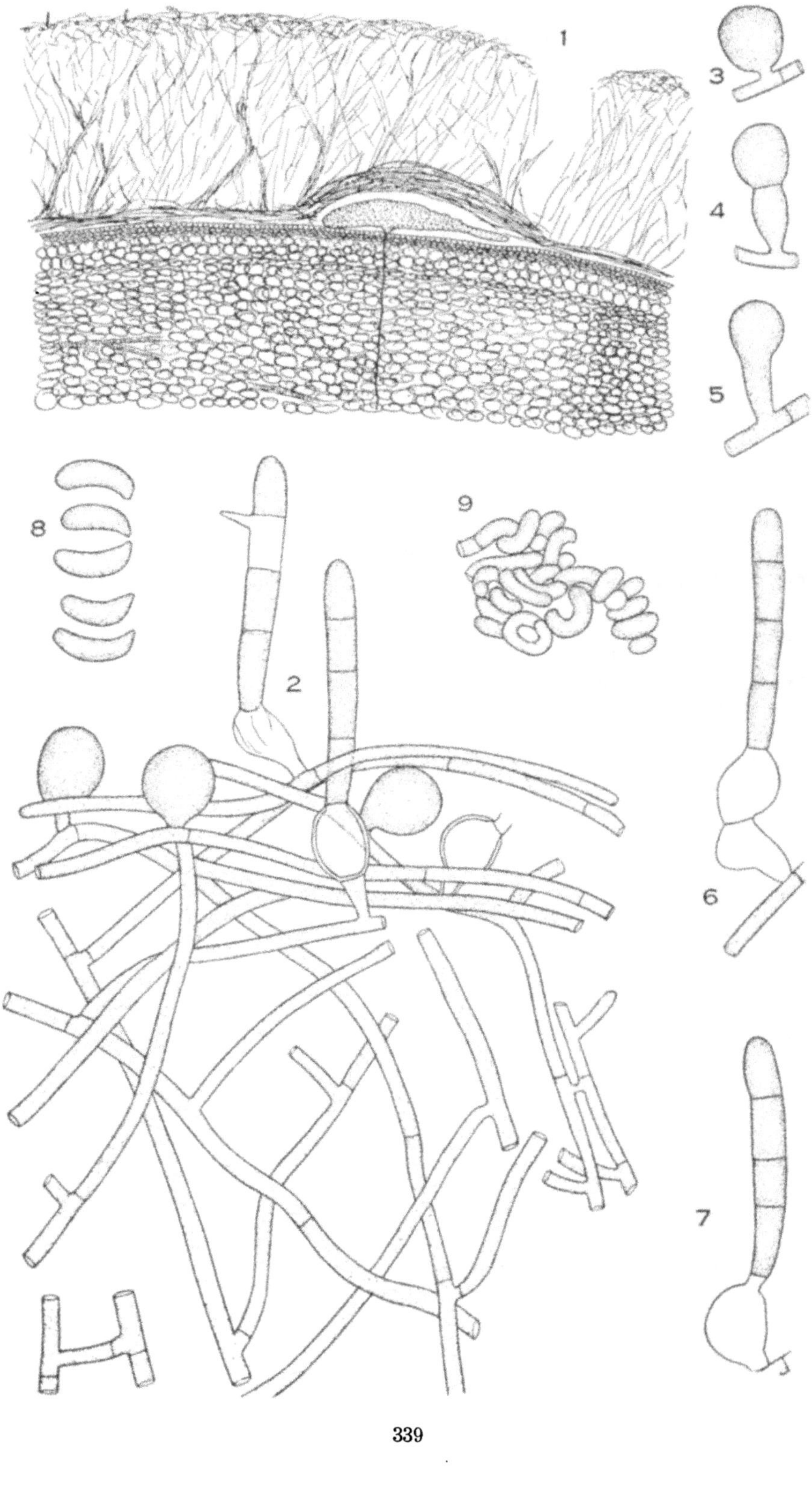

Plate 23

S. fumigatum

Fig. 1.—Section showing hymenium, context, and subiculum, also a scale insect. × 33.

Fig. 2.—Hymenium, with coiled paraphyses, probasidia, basidia, and spores. No. 8458. × 833.

Fig. 3.—Three basidia of type material. × 833.

Fig. 4.—Basidium of material from Alabama. Mo. Bot. Garden Herb. no. 20068. × 833.

Fig. 5.—Haustoria within *Aspidiotus tenebricosus* on *Cornus florida* from Florida. × 833.

PLATE 23

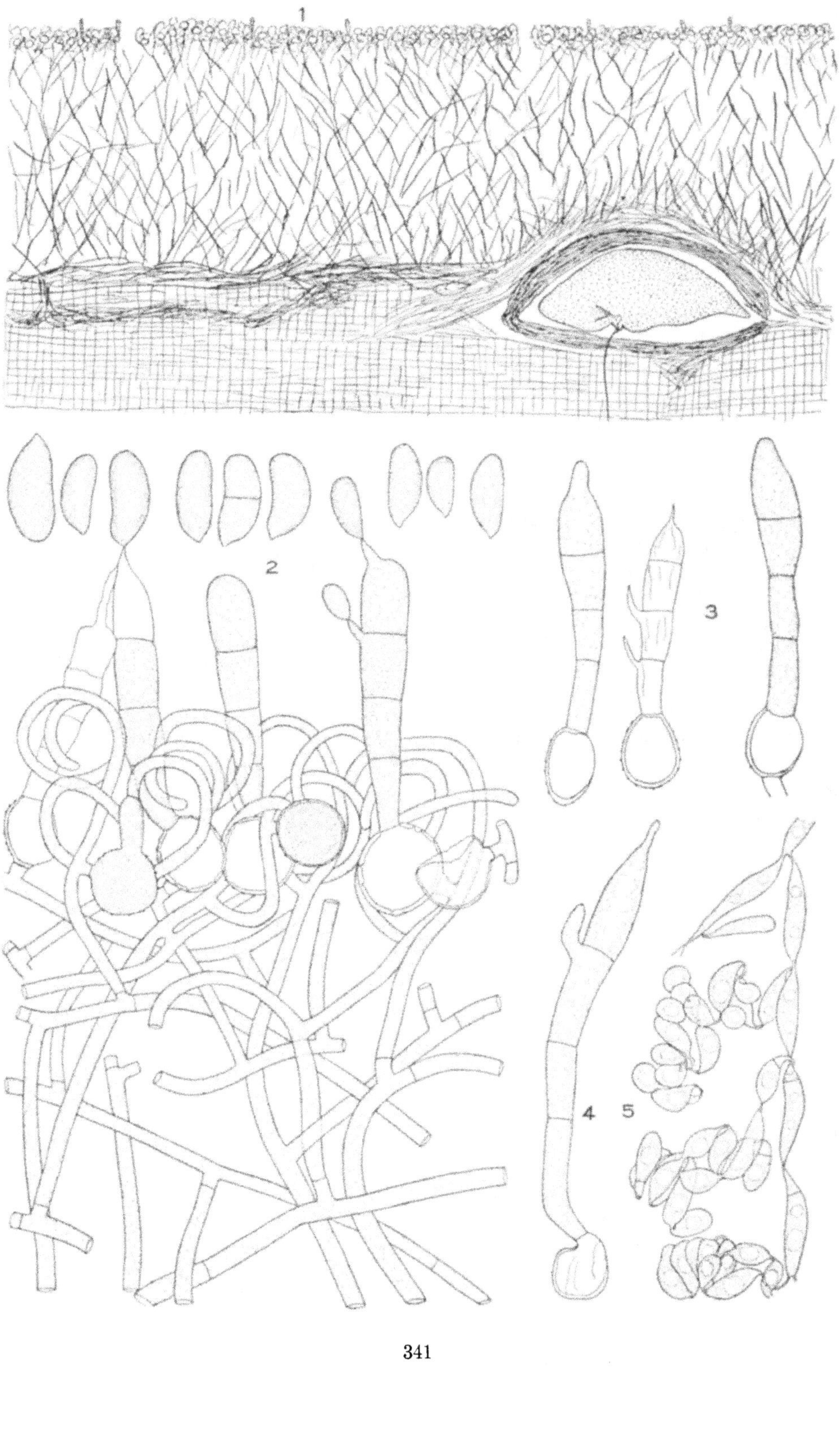

PLATE 24

S. fuscum

Fig. 1.—Section showing fertile region and thick tufts of upright hyphae over insects. × 33.

Fig. 2.—Enlarged section showing subiculum and hymenium with recurved paraphyses, probasidia, basidia, and spores. × 833.

Fig. 3.—Spores. × 833.

Fig. 4.—Haustoria. × 833.

Fig. 5.—View of parasitized insect in spiracular region showing a fungal coil surrounded by chitinous material. A connection is shown between coil and external hyphae passing through dermal pore. × 833.

Fig. 6.—Showing how fungal pad may overgrow part of insect. × 833.

PLATE 24

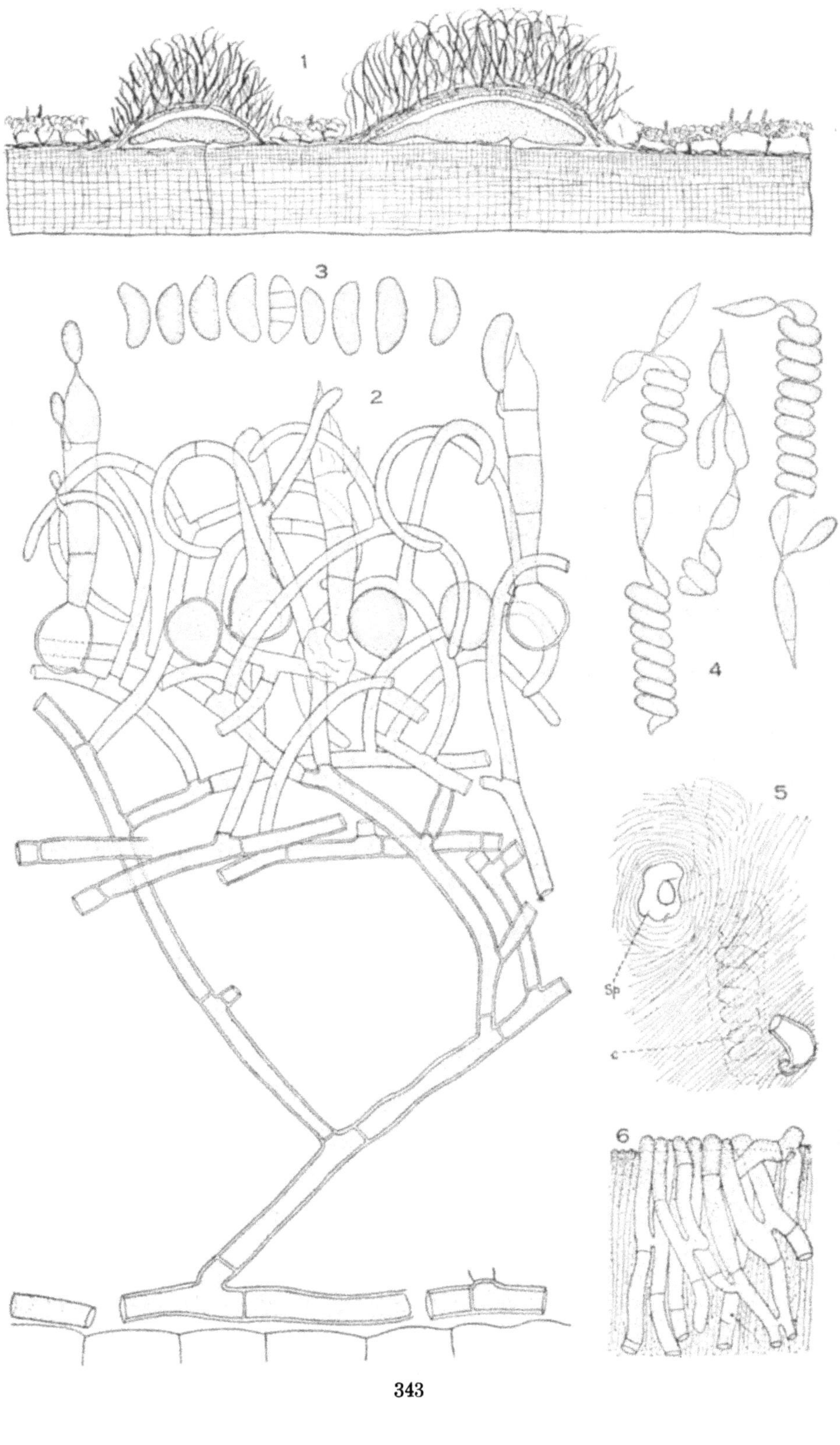

Plate 25

S. castaneum

Fig. 1.—Sketch of type showing very irregular marginal condition. Stubby pillars and three horizontal layers are shown in places, also note scale insects. × 9.

Fig. 2.—Hymenial region showing probasidia and basidia. × 833.

Fig. 3.—Spores. × 833.

S. alni var. *squamosum*

Fig. 4.—Sketch showing surface and sectional view. × 18.

S. Leprieurii

Fig. 5.—Section showing stratification of fungus, the top stratum being fertile with mature basidia. Note parasitized scale insect. × 33.

Fig. 6.—Section of hymenium showing probasidia, basidia, and spores. × 833.

Fig. 7.—Spores. × 833.

PLATE 25

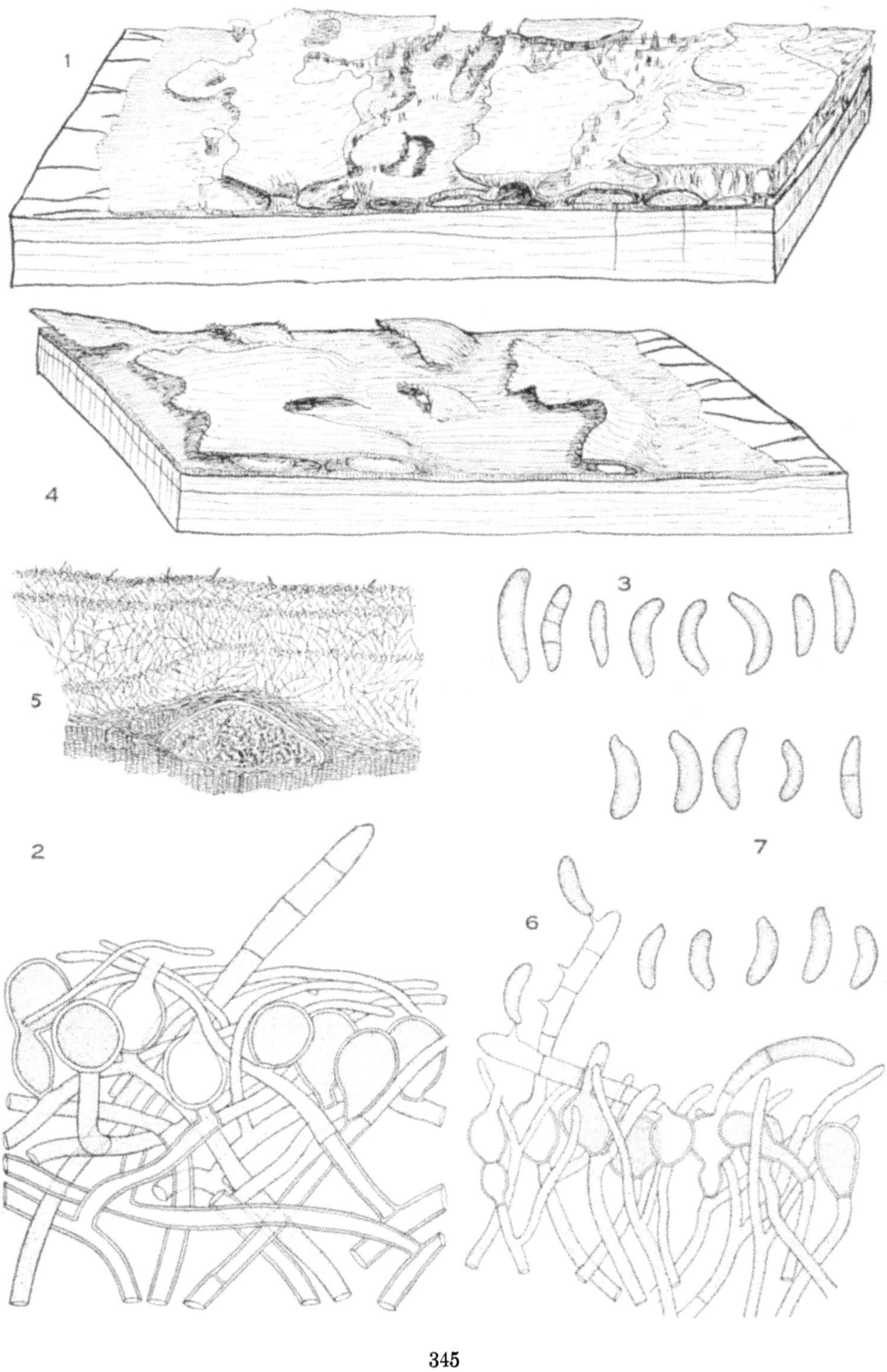

PLATE 26

S. Mariani

Fig. 1.—Section showing three horizontal layers of fungus, the hymenial surface showing probasidia and basidia. Note branched and anastomosing pillars, also parasitized scale insect beneath fungus. × 33.

Fig. 2—Anastomosing pillars. × 833.

Fig. 3.—Section through hymenial region showing probasidia and basidia. × 833.

Fig. 4.—Probasidia and basidium. × 833.

Fig. 5.—Spores. × 833.

PLATE 26

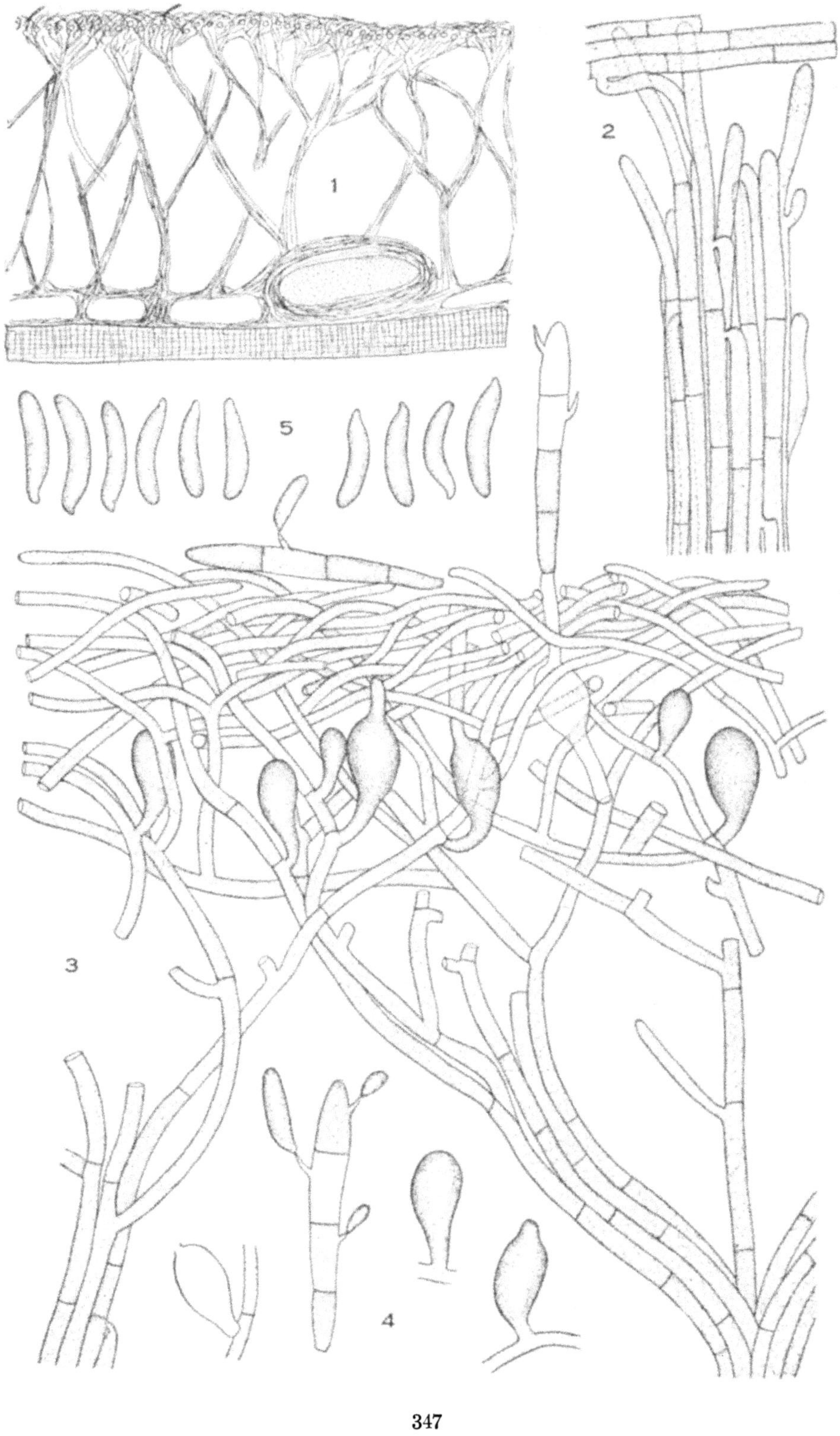

Plate 27

S. pseudopedicellatum

Fig. 1.—Sketch showing subiculum, pillars, and top layer; note tuft of short threads at base of pillars. × 18.

Fig. 2.—Section through hymenium showing probasidia, basidia, spores, and paraphyses. At left are two probasidial cells each of which has formed one basidium and into which new probasidia have proliferated. Type. × 700.

Fig. 3.—Spores. Above on *Carpinus*, no. 9389, from North Carolina; below on *Citrus* from Brazil, Müller, coll., no. 9722. × 700.

Fig. 4.—Spores. On *Fraxinus*, no. 8401. × 833.

Figs. 5 and 6.—Various shapes and sizes of basidia. × 833.

PLATE 27

PLATE 28

S. Burtii

(= *S. retiforme* sense of American authors but not Patouillard.)

Fig. 1.—Section of hymenium showing probasidia, basidia, and spores. × 938.
Fig. 2.—Spores showing two types of budding. The four to left show close type of budding, the other four show open type of budding. × 938.
Fig. 3.—Hyphae and bud cells on agar. × 938.
Fig. 4.—Germination on agar of coiled haustoria. × 825.

PLATE 28

PLATE 29

S. sinuosum

Fig. 1.—Hymenium showing probasidia, basidia, spores, and paraphyses. × 833.
Fig. 2.—Basidia of Mo. Bot. Gard. Herb. no. 44211. × 833.
Fig. 3.—Segments of haustoria from parasitized insects. × 833.

S. leprosum

Fig. 4.—Section through hymenium, context, and subiculum showing probasidia, basidia, and spores. × 533.
Fig. 5.—Spores. × 833.

PLATE 29

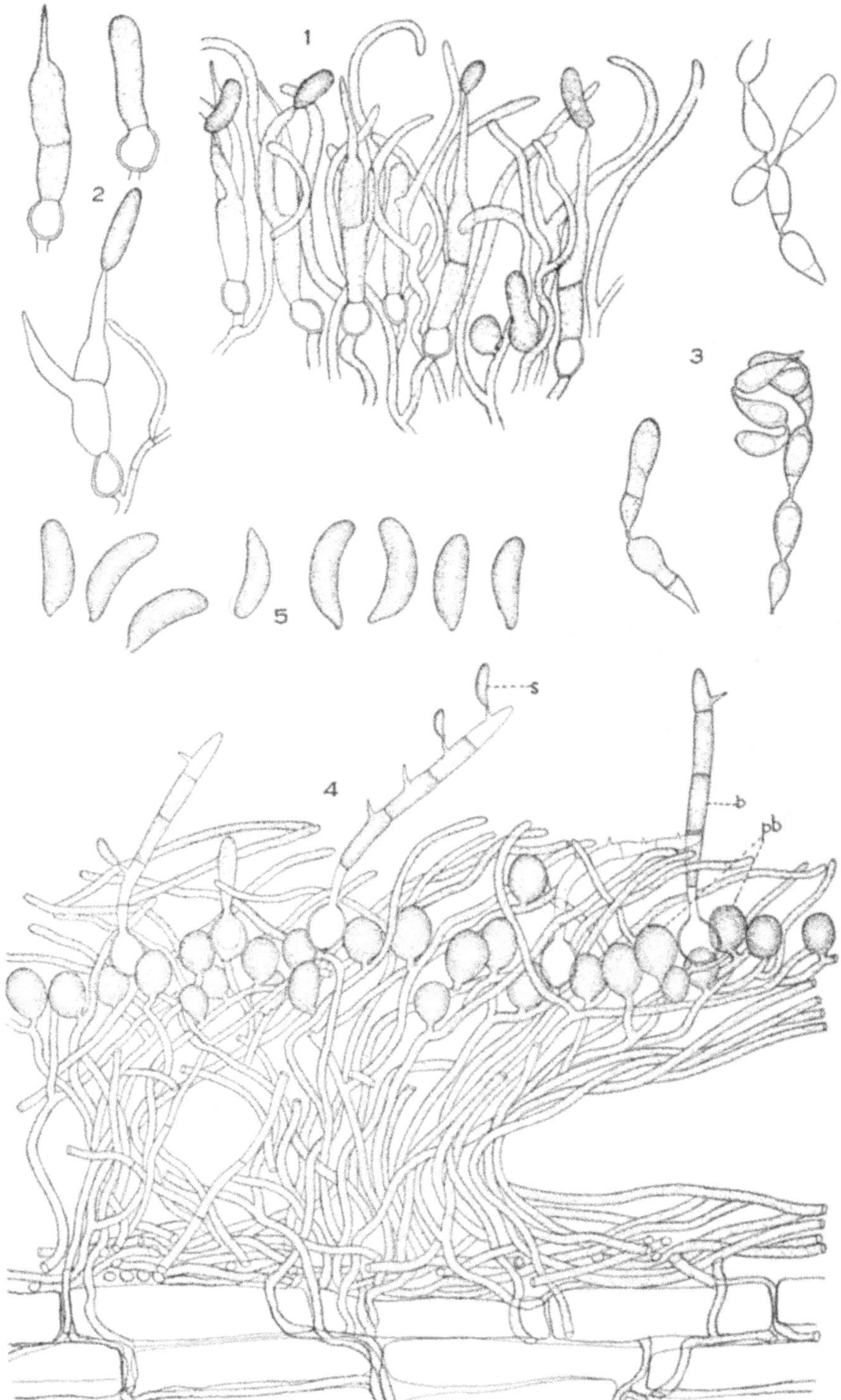

Plate 30

S. Sydowii

Fig. 1.—Section through fungus showing compact context and stratified hymenium. × 33.
Fig. 2.—Section of hymenium showing probasidia, coiled basidia and spores. Also note old collapsed probasidial cells at base of hymenium. × 833.
Figs. 3–5.—Basidia. × 833.
Fig. 6.—Spores. × 833.
Fig. 7.—Haustoria. × 833.
Fig. 8.—Anal segment of scale insect associated with *S. Sydowii*. × 233.

PLATE 30

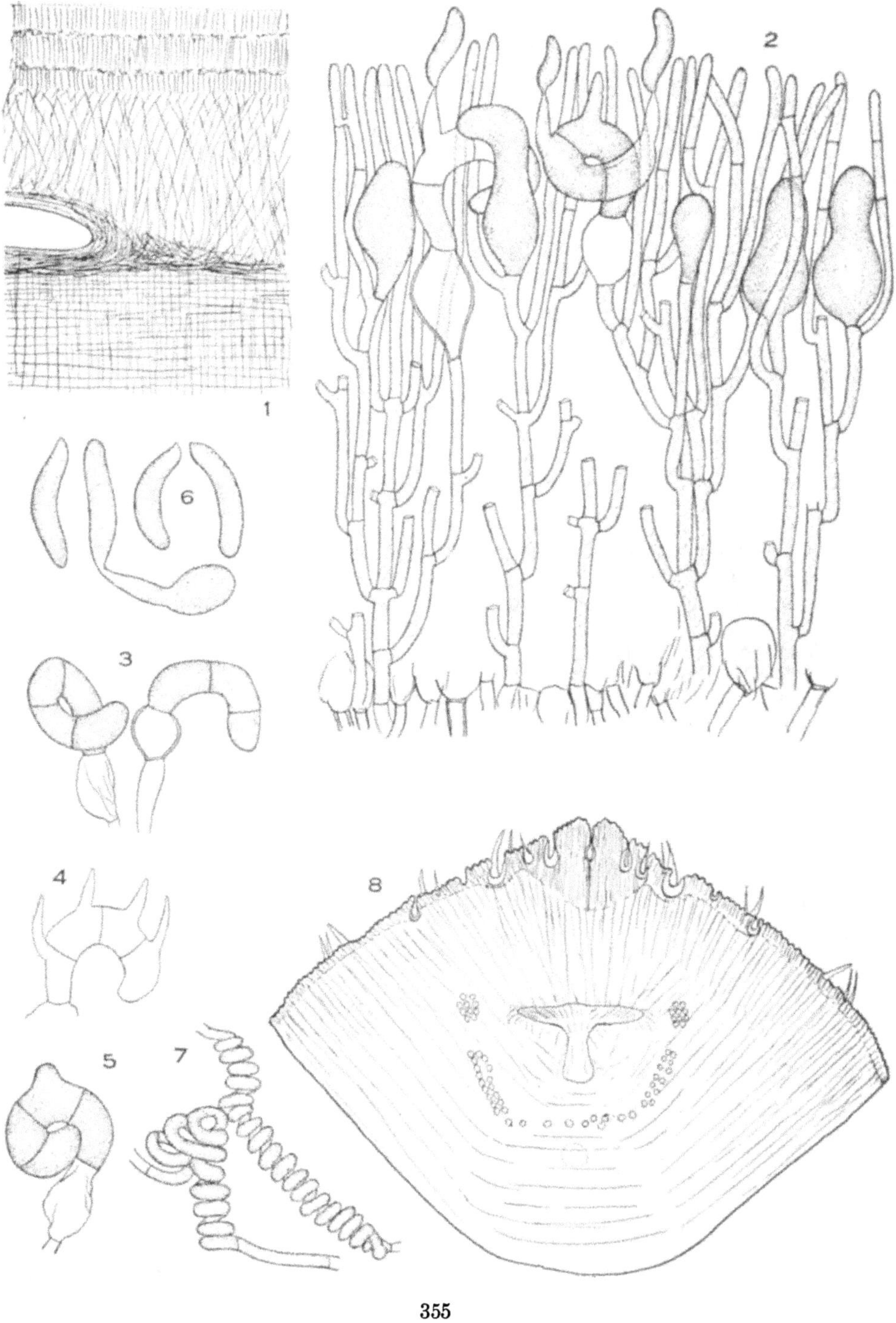

PLATE 31

S. cremeum

Fig. 1.—Section through marginal region of fungus showing hymenium with basidia, pillars, subiculum and one scale insect. × 33.

Fig. 2.—Hymenial region, showing empty probasidia, basidia, spores, and paraphyses.

Fig. 3.—Spores. × 833.

Fig. 4.—Haustorium from parasitized insect. × 833.

Fig. 5.—*Aspidiotus liquidambaris* Kot. × 60.

Fig. 6.—Anal segment of same scale insect, ventral surface. × 233.

PLATE 31

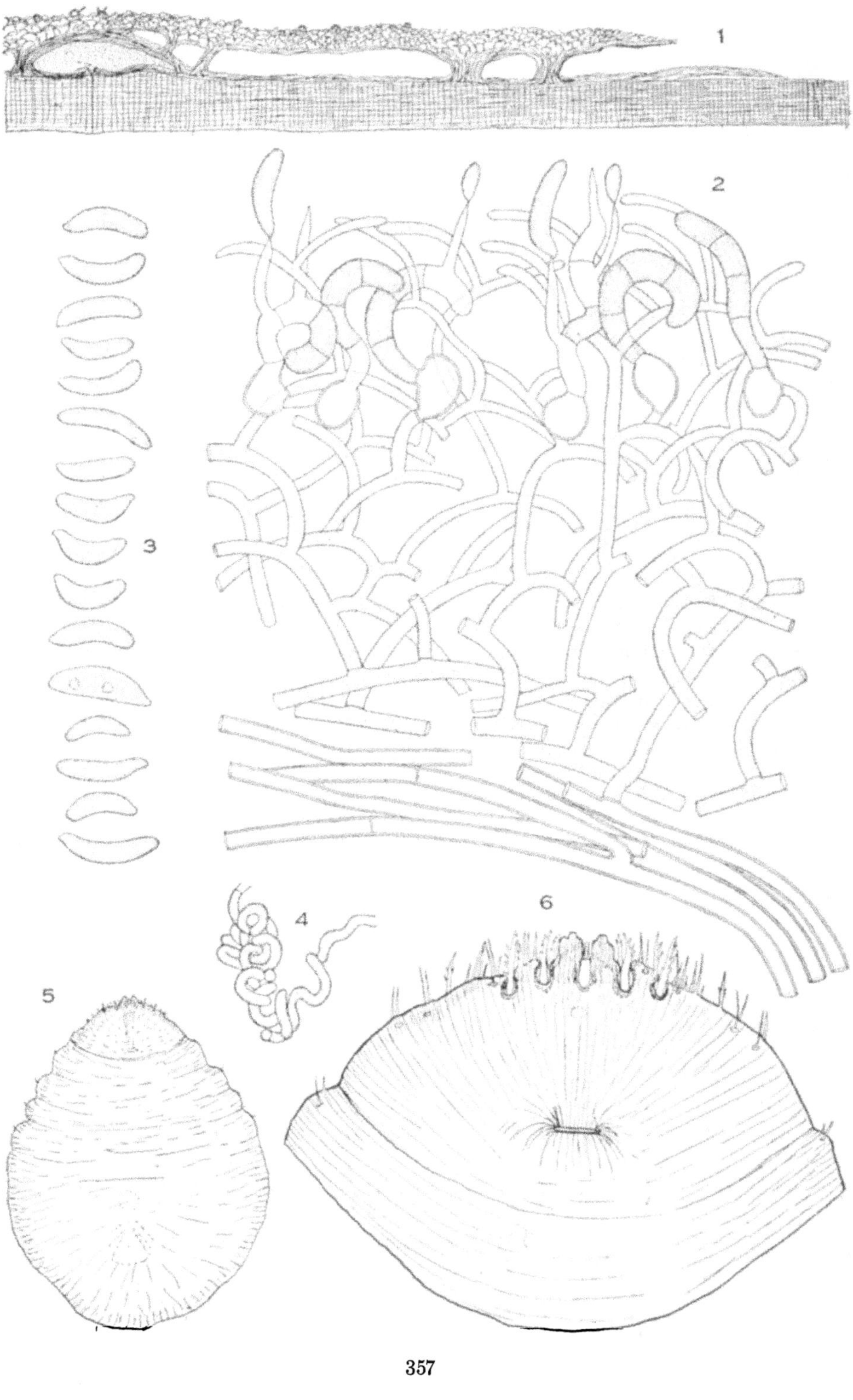

Plate 32

S. filiforme

Fig. 1.—Section of fungus showing hymenium, pillars, subiculum, and three insects. × 33.

Fig. 2.—Hymenium with probasidia, basidia, and very narrow threads which frequently break up into short bacilliform segments. × 847.

Fig. 3.—Young basidia.

Fig. 4.—Mature basidia and spores. × 833.

Fig. 5.—Haustoria. × 833.

PLATE 32

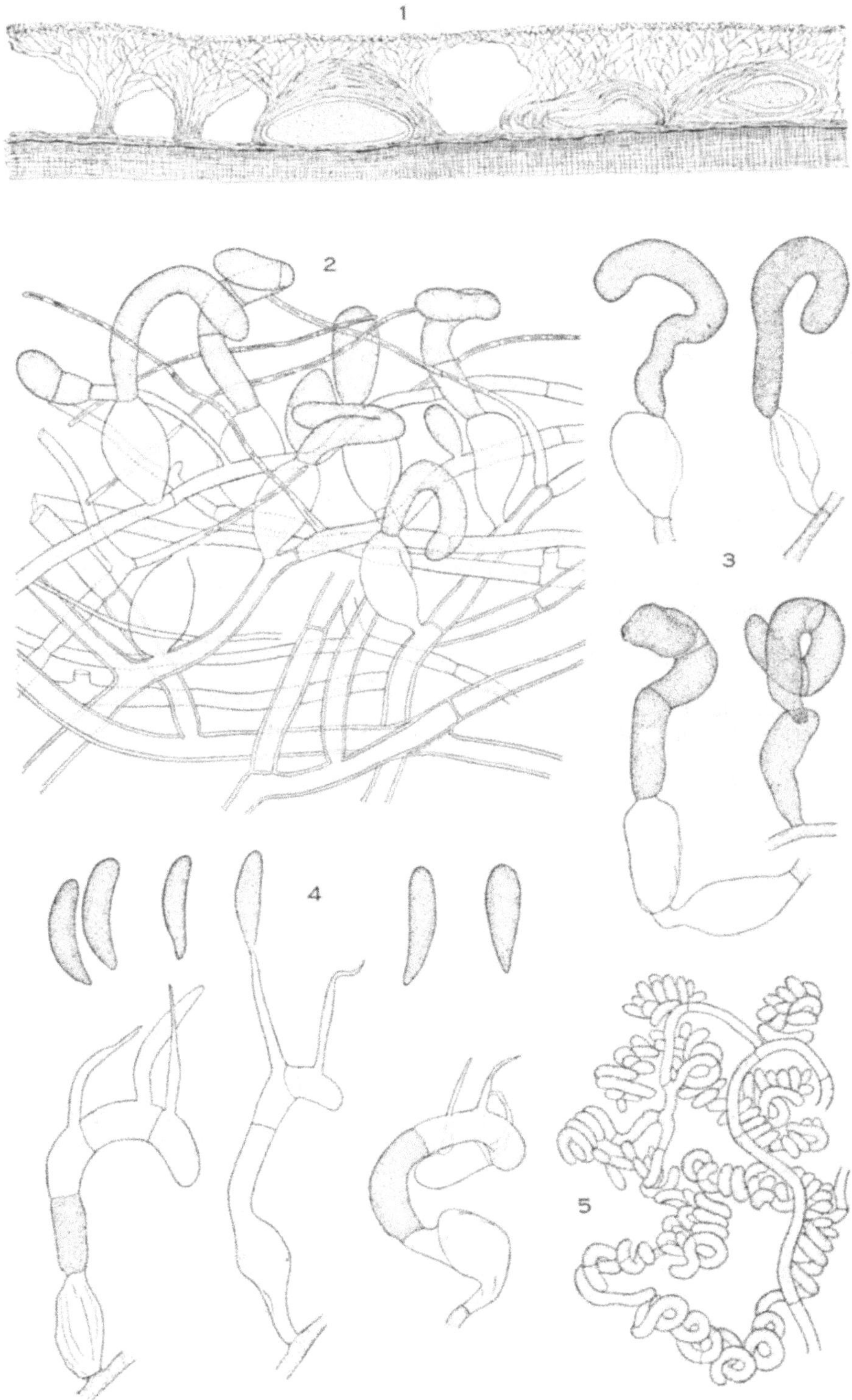

PLATE 33

S. Schweinitzii

Fig. 1.—Section of fungus showing hymenium, pillars, subiculum and parasitized insects. No. 5876. × 33.

Figs. 2 and 3.—Basidia bearing spores. × 847.

Figs. 4–8.—Probasidia (?) and resting basidia. Type material. × 833.

Fig. 9.—Hymenium showing upright thick-walled paraphyses, basidia and spores. Coll. March 14, 1929. No. 8397.

Fig. 10.—Hymenium showing resting, coiled, septate basidia. Coll. Dec. 28, 1921. × 833.

Fig. 11.—Spores. × 833.

PLATE 33

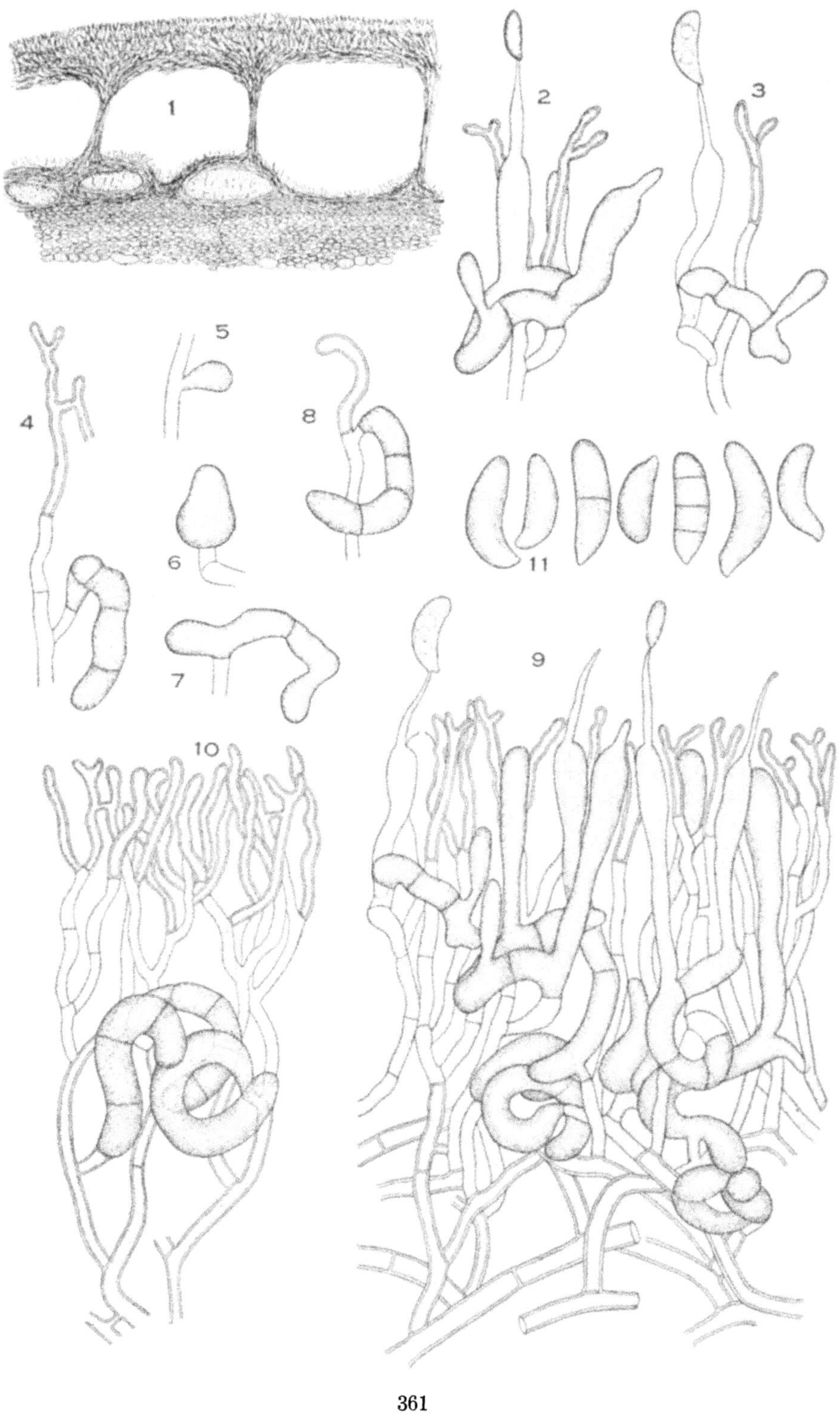

PLATE 34

S. lilacinoalbum

Fig. 1—Section of fungus showing hymenium, pillars, subiculum and two scale insects. × 33.

Fig. 2.—Hymenium with basidia, spores, and paraphyses. Note numerous anastomoses between paraphyses. × 833.

Fig. 3.—Probasidia (?) and basidia. × 833.

Fig. 4.—Spores. × 833.

Fig. 5.—Paraphyses. × 833.

Fig. 6.—Segment of hypha showing protoplasmic connection between cells. × 833.

Fig. 7.—Haustoria. × 833.

Figs. 8 and 9.—*Chionaspis gleditsiae* Sanders (?), the scale associated with this fungus. Fig. 8 × 60; fig. 9 × 233.

PLATE 34

PLATE 35

Hybrid between *S. pseudopedicellatum* Burt and *S. Schweinitzii* Burt.

Fig. 1.—Sketch of marginal region of hybrid showing "tents," which are characteristic of *S. Schweinitzii* and "pillars" characteristic of *S. pseudopedicellatum*. × 16.

Fig. 2.—Hymenium of *S. pseudopedicellatum*, showing typical probasidia, basidia and paraphyses. × 750.

Fig. 3.—Hymenium of hybrid, showing paraphyses characteristic of *S. Schweinitzii* and probasidia and basidia combining characters of both parents. × 696.

Figs. 4–10.—Stages in development of basidia of hybrid. × 750.

PLATE 35

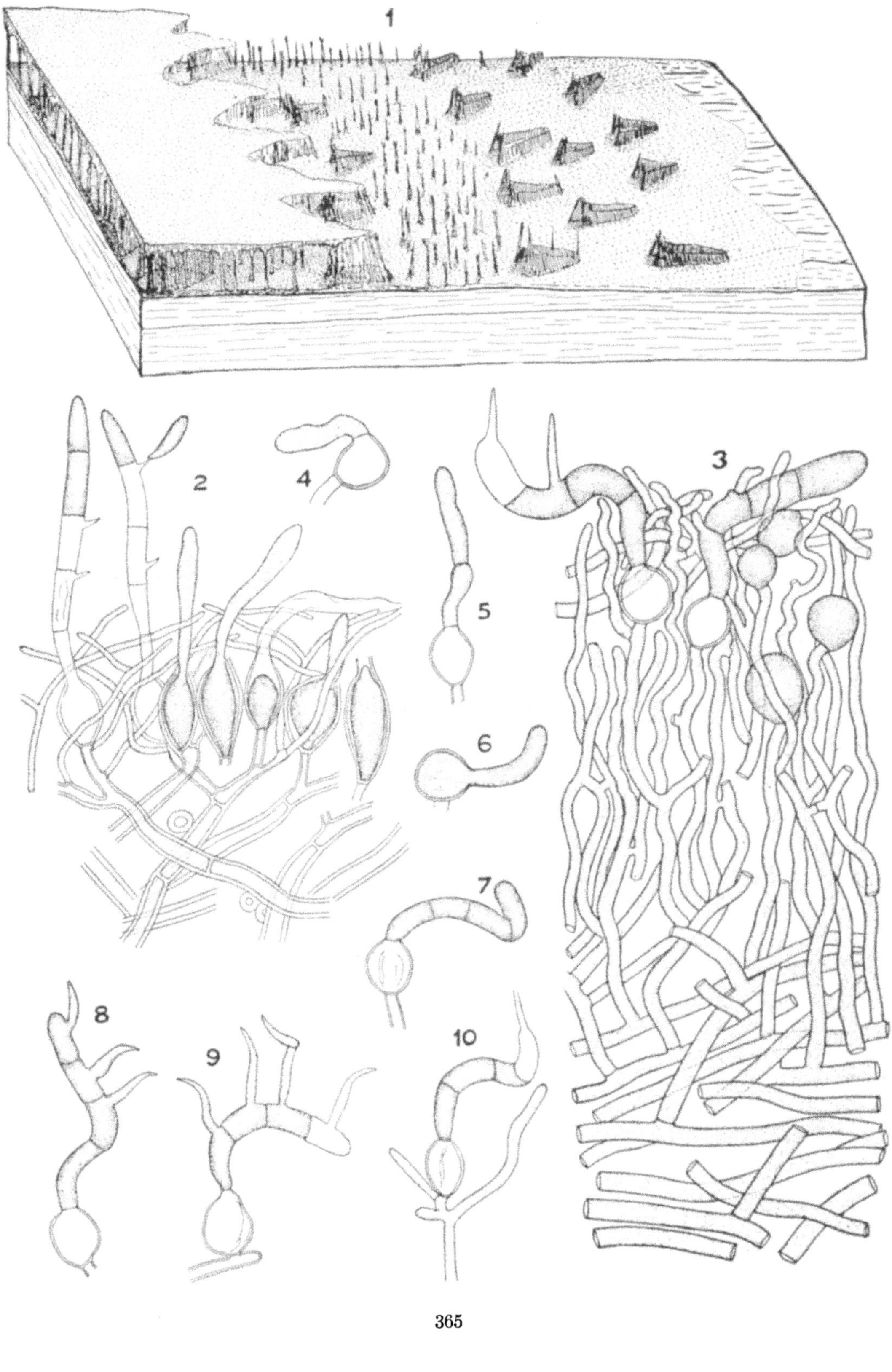

PLATE 36

S. rugulosum

Fig. 1.—Section of fungus, marginal region. × 50.

Fig. 2.—Section showing context and hymenial region with recurved, crystal encrusted paraphyses and basidia. No. 9251. × 938.

Fig. 3.—Spores. No. 8472. × 938.

Figs. 4–10.—Basidia, all of no. 8472 except fig. 9 (no. 9151). × 938.

Fig. 11.—Hyphal segments of context separated by crushing under a cover slip. × 938.

Fig. 12.—Associated scale insect, *Chionaspis gleditsiae* (?), from no. 9229. × 68.

PLATE 36

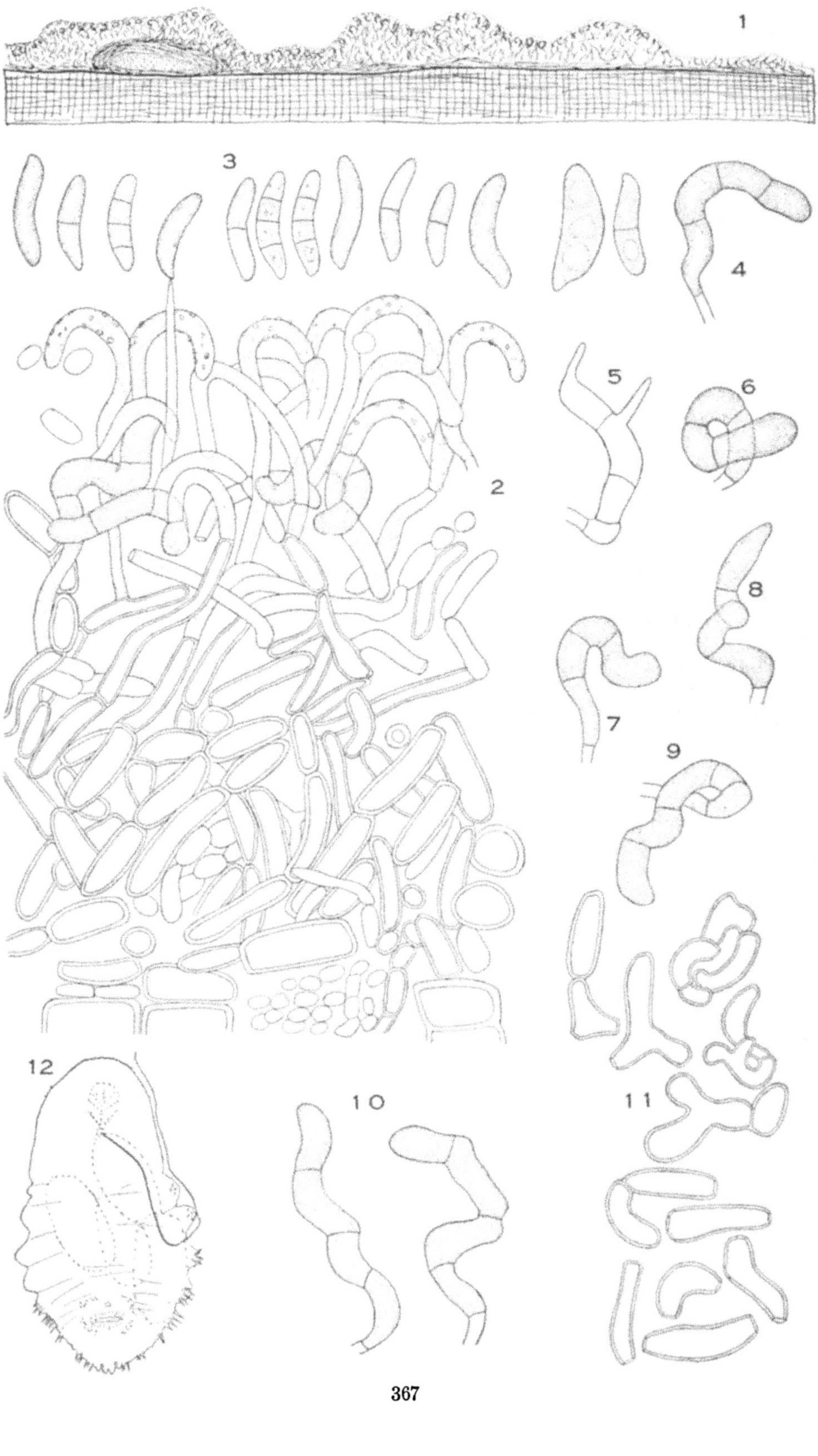

PLATE 37

S. tenue

Figs. 1 and 2.—Sections of fungus. Fig. 1, no. 9154; fig. 2, no. 9183. × 33.

Fig. 3.—Section of context and hymenial region showing basidia and spores, paraphyses with crystal encrusted tips. Note algal cells; also note that hyphae of context are not broken up into short segments. × 833.

Fig. 4.—Spores, below of no. 8456, above of no. 9154. × 833.

Figs. 5–9.—Basidia. × 833.

Fig. 10.—Haustoria from parasitized scale insect. No. 8456. × 833.

PLATE 37

PLATE 38

S. canescens

Fig. 1.—Sketch of section through marginal region of fungus, showing hymenium with probasidia and basidia, context, and one parasitized insect. × 33.

Fig. 2.—Section of hymenial region showing probasidia, basidia, paraphyses, and algal cells. × 833.

Figs. 3–6.—Basidia. × 553.

Fig. 7.—Spores. × 553.

Fig. 8.—Haustorium of "glomerulus" type. × 833. All figures from type material.

PLATE 38

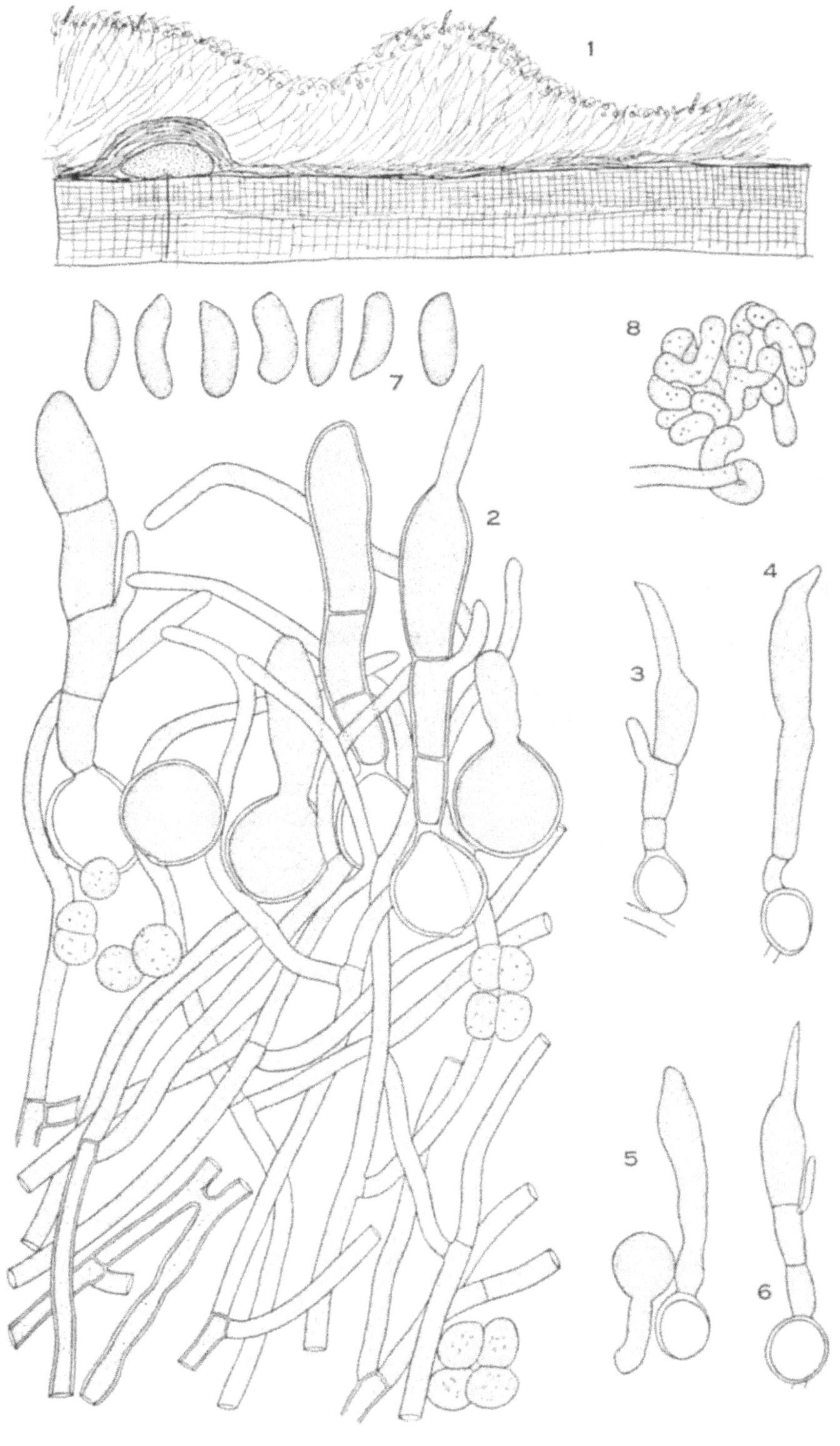

PLATE 39

S. apiculatum

Fig. 1.—Section of fungus showing stratified condition. No mound shown. × 33.

Fig. 2.—Hymenium showing probasidia (?), basidia and spore. × 833.

Fig. 3.—Spores. No. 9302. × 833.

Figs. 4-9.—Probasidia and basidia. × 833.

Fig. 10.—Haustoria. × 833.

S. Cokeri

Fig. 11.—Probasidia and basidia, the middle probasidium showing internal proliferation. × 700.

Fig. 12.—Spores. × 700.

PLATE 39

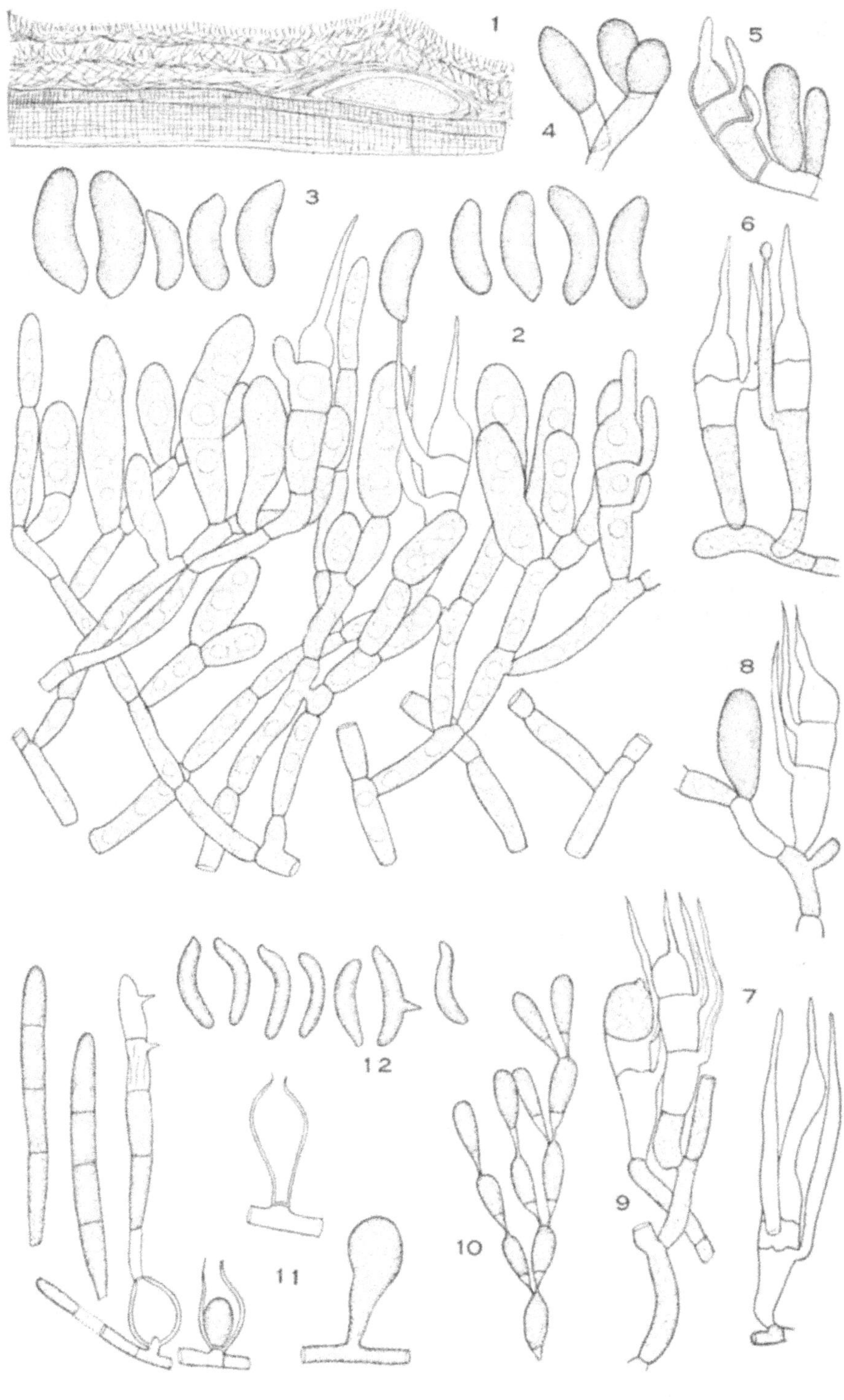

PLATE 40

S. sabalis

Fig. 1.—Section of fungus showing three horizontal layers, scale insects between subiculum and middle layer; note basidia at surface of top layer. × 23.
Fig. 2.—Hymenium showing probasidia, basidia, paraphyses, and spore. × 833.
Fig. 3.—Spores. × 833.
Figs. 4–6.—Basidia. × 833.
Fig. 7.—Haustorium. × 833.

PLATE 40

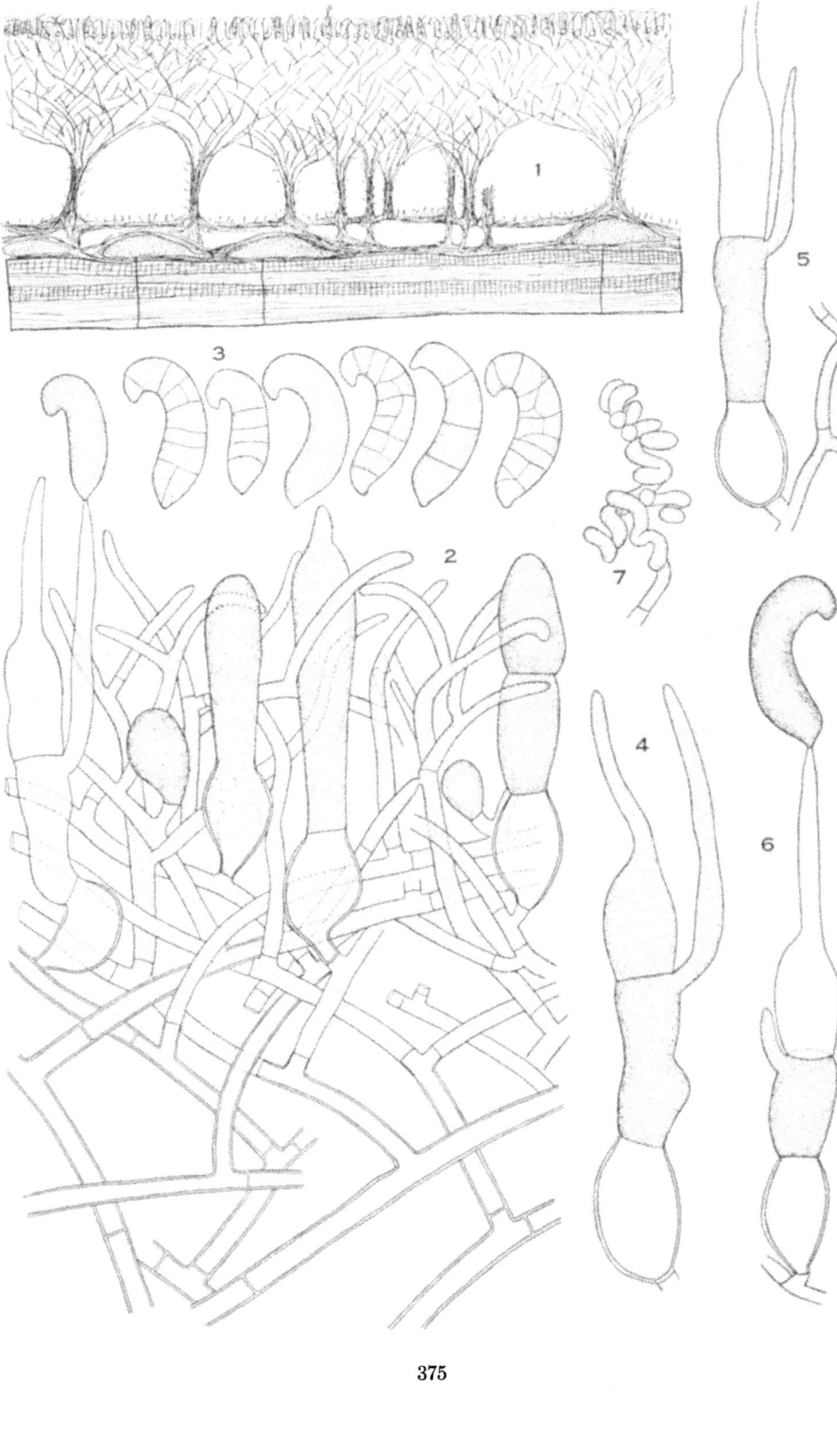

PLATE 41

S. Langloisii

Fig. 1.—Surface view showing marginal condition. No. 9196. × 9.
Fig. 2.—Section of context showing hymenium and scale insects. × 33.
Fig. 3.—Hymenium showing upright (hyaline) paraphyses, probasidia, basidia, and spores. Daytona Beach, Fla. Thaxter, coll. × 833.
Figs. 4–6.—Probasidia and basidia. No. 9196. × 833.
Fig. 7.—Basidium and three spores of type. No. 2995. × 833.
Fig. 8.—Haustoria. No. 9196. × 833.
Fig. 9.—Spores. No. 9196. × 833.

PLATE 41

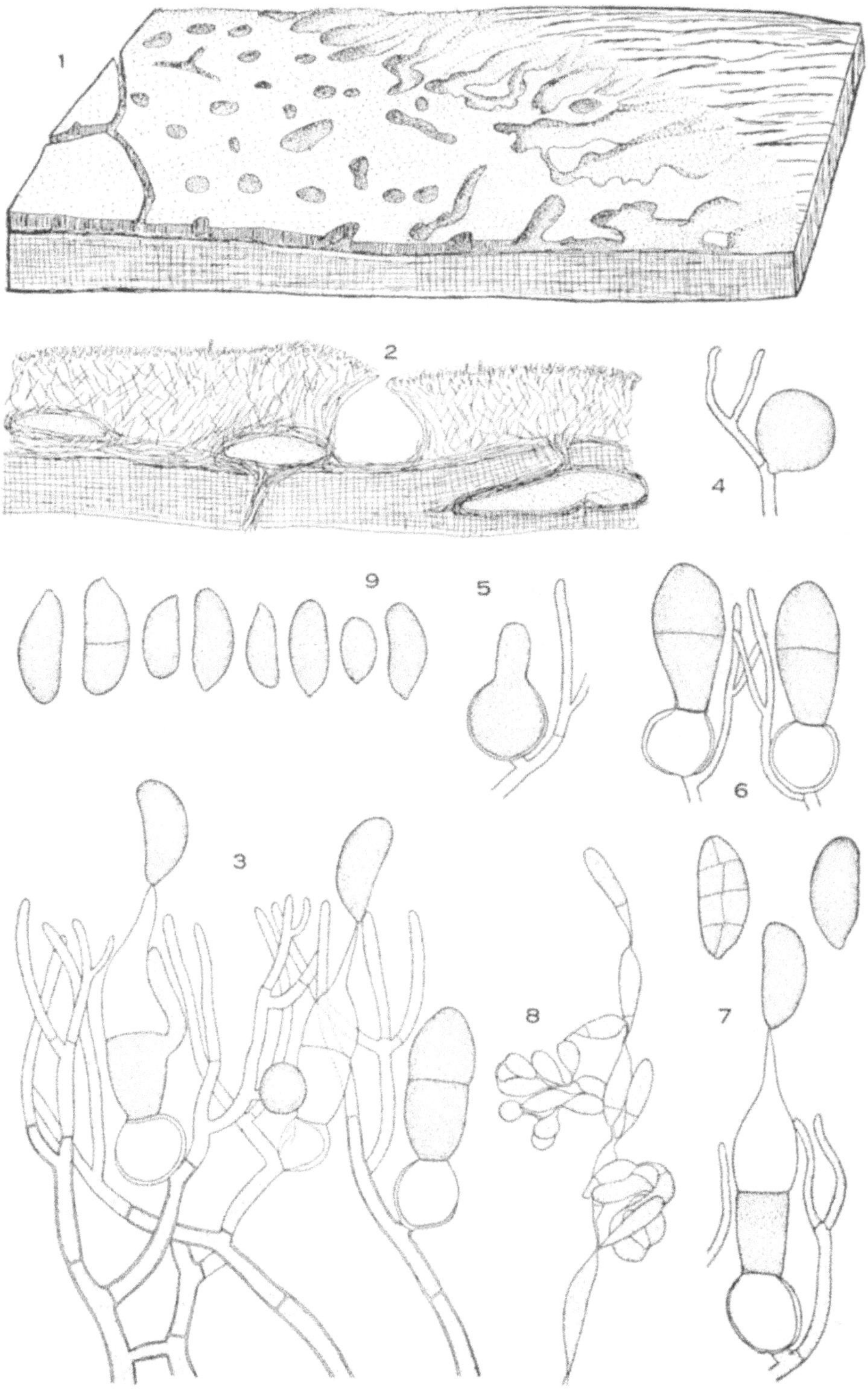

PLATE 42

S. Patouillardii

Fig. 1.—Section of fungus in marginal region, showing top layer, pillars, subiculum, and one parasitized scale insect. × 33.

Fig. 2.—Hymenium showing upright, dark colored, thick-walled paraphyses, probasidia (?), basidia, and spores. × 833.

Fig. 3.—Spores. × 833.

Fig. 4.—Basidium and paraphyses. × 833.

Fig. 5.—Haustoria of irregularly coiled type. × 833.

S. taxodii

Fig. 6.—Section through fungus showing insect covered by thick upright hyphae. × 33.

Fig. 7.—Paraphyses from hymenium. × 833.

Fig. 8–13.—Probasidia and basidia. × 833.

PLATE 42

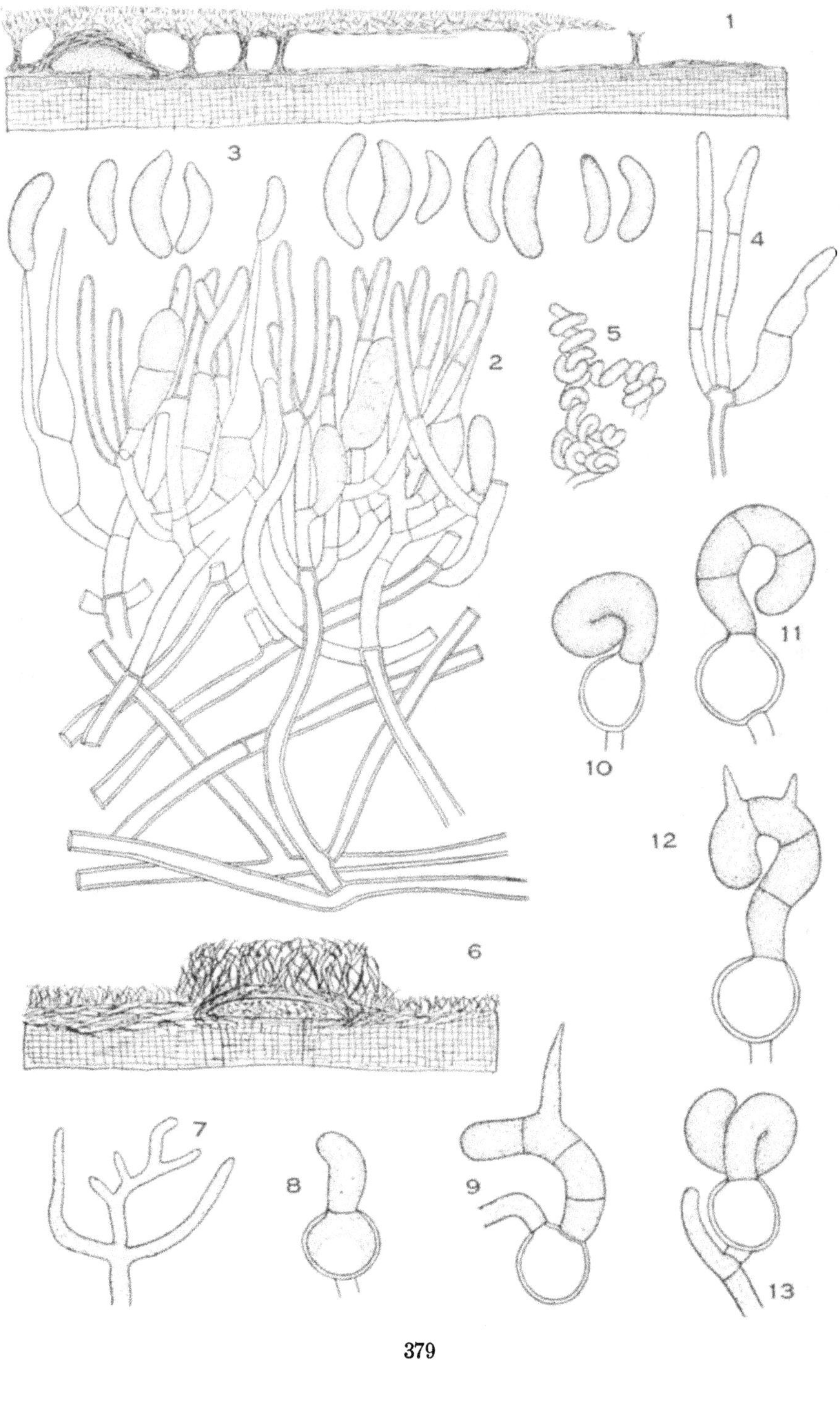

PLATE 43

S. grandisporum

Fig. 1.—Section of fungus showing hymenium, branching and anastomosing pillars, and scale insects. × 18.

Fig. 2.—Hymenium with upright paraphyses, probasidia, and basidia. × 833.

Fig. 3.—Basidia. × 833.

Fig. 4.—Spores, three of which have become divided into numerous small cells, the upper spore forming bud cells. × 833.

PLATE 43

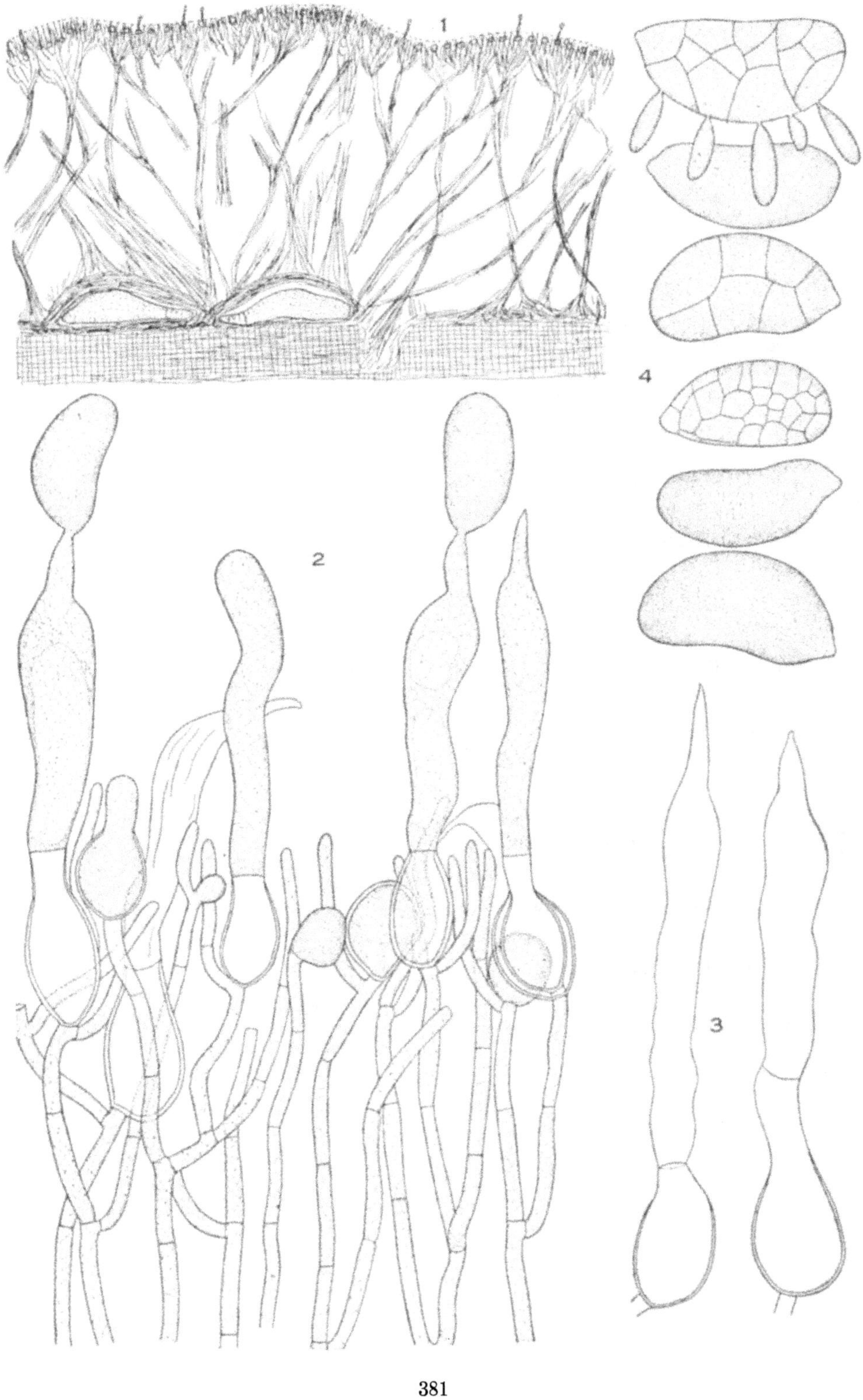

PLATE 44

S. Michelianum

Fig. 1.—Section view of context. × 25.
Fig. 2.—Section of hymenium. × 580.
Fig. 3, 4.—Basidia and spores. × 580.
Above material from Nimes, France, May 1922.

S. ficicolum

Fig. 5.—Section of context. × 25.
Fig. 6.—Basidia and spores. × 625.

S. alni var. *brasiliense*

Fig. 7.—Basidia and spores. × 525.

S. Carestianum var. *natalense*

Fig. 8.—Basidium. × 525.

S. Galzini

Fig. 9.—Probasidia, basidia, and spores. × 525.

S. mexicanum

Fig. 10.—Sections of context. × 18.
Figs. 11, 12.—Probasidia and basidia. × 580.

S. alveomarginatum

Fig. 13.—Probasidium, basidium, and thread from hymenium. × 625.
Fig. 14.—Section of hymenium. × 625.

S. suffultum

Fig. 15.—Probasidia, basidia, and spores. × 525.

S. pachydermum

Fig. 16.—Probasidia, basidia, and spores. × 580.

S. Mariani var. *japonicum*

Fig. 17.—Section of context. × 14.

PLATE 44

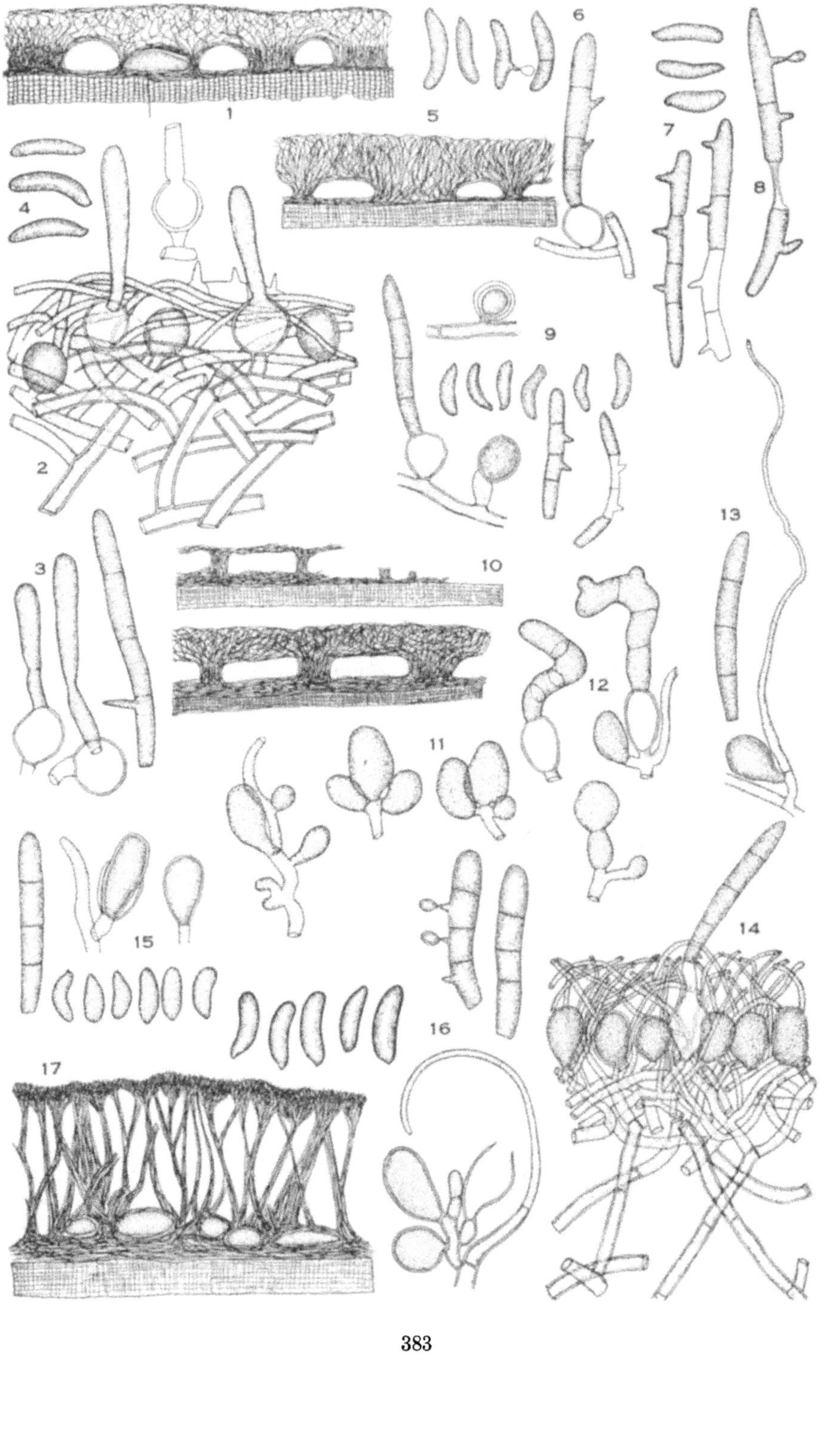

Plate 45

S. macadamiae

Fig. 1.—Section of context. × 25.
Fig. 2.—Hymenium showing old and new probasidia and basidia. × 525.
Fig. 3.—Spores. × 525.

S. fusco-violaceum

Fig. 4.—Section of context. × 25.
Fig. 5.—Probasidia and basidia. × 580.

S. piperis

Fig. 6.—Section of context. × 25.
Fig. 7.—Section of hymenium, probasidia, and basidia. × 525.

S. punctatum

Fig. 8.—Surface showing elevated bumps and margin showing openings into tunnels. × 10.
Fig. 9.—Section of context. × 25.
Fig. 10.—Section of hymenium. × 635.

S. Lagerheimii

Fig. 11.—Section of context. × 25.
Fig. 12.—Section of hymenium. × 525.

PLATE 45

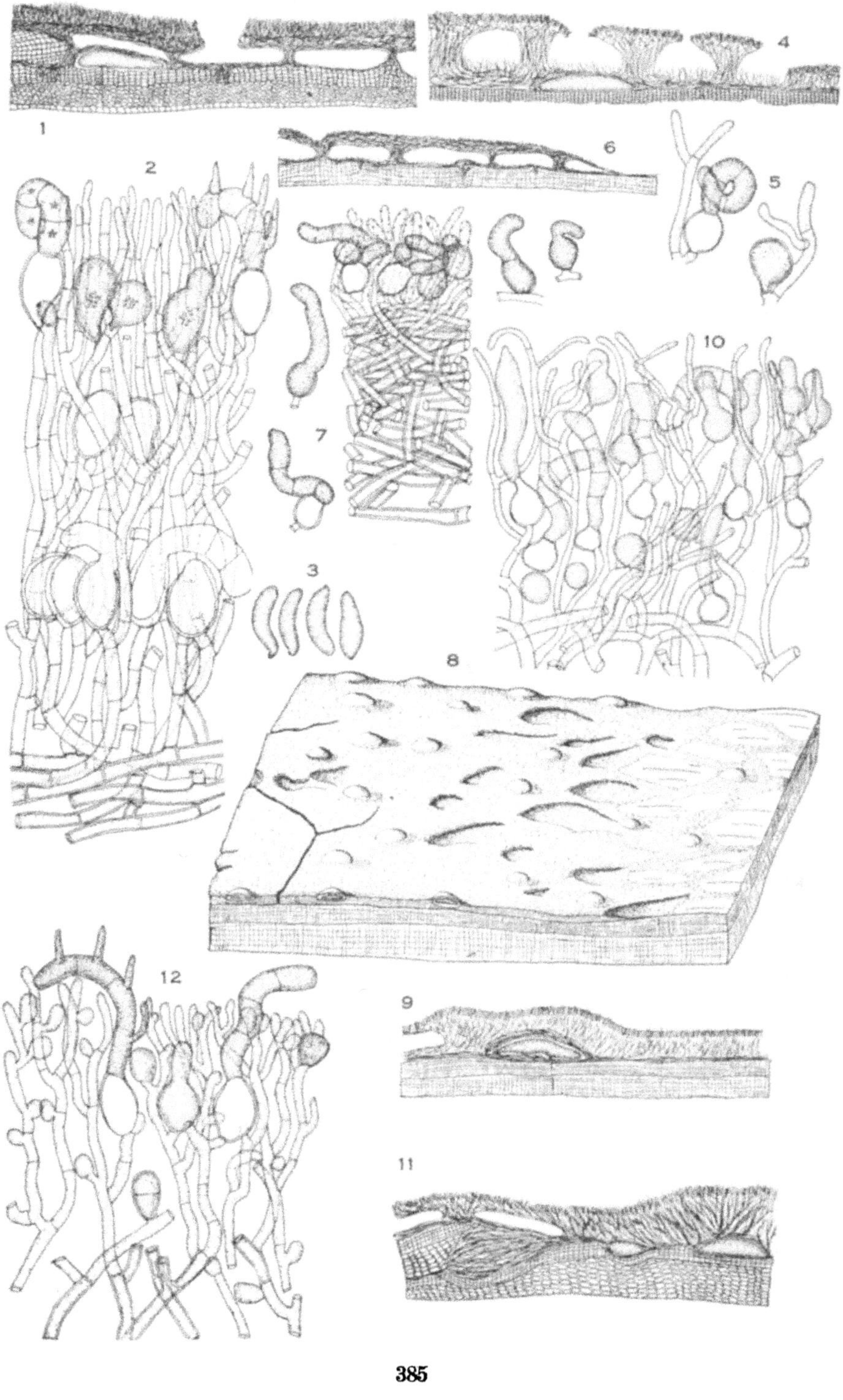

Plate 46

S. Thwaitesii

Fig. 1.—Section of context; h, subiculum, p, pillars, h_1, second horizontal layer, c, context, h_2 third horizontal layer, hy, hymenium; also note two insects. × 14.
Fig. 2.—Section of hymenium showing probasidia and basidia. × 525.
Figs. 3, 4.—Probasidia, basidia, and spores. × 525.

S. accumbens

Fig. 5.—Section of context. × 14.
Fig. 6.—Two conidial masses. × 525.

S. tigrinum

Fig. 7.—Section of context. × 25.

S. sinense

Fig. 8.—Surface of marginal region showing subiculum, the umbrella-like beginnings of top layer, and a large mound under which a female was giving birth to young, one of which was crawling out the opening. × 18.
Fig. 9.—Section of context showing thick compact subiculum, insects, second horizontal layer, and top layer. × 25.
Fig. 10.—Conidia and a piece of hypha broken by crushing. × 525.
Fig. 11.—Probasidia, basidia, and spore. × 525.

S. tuberculatum

Fig. 12.—Coiled hyphae from context. × 335.
Fig. 13.—Bladder cell from context. × 335.
Fig. 14.—Star-like arrangement of hyphae just after passing through derm pore of insect, haustoria of sausage-shaped type. × 525.

S. schizostachyi

Fig. 15.—Hymenial hyphae and spores. × 525.

PLATE 46

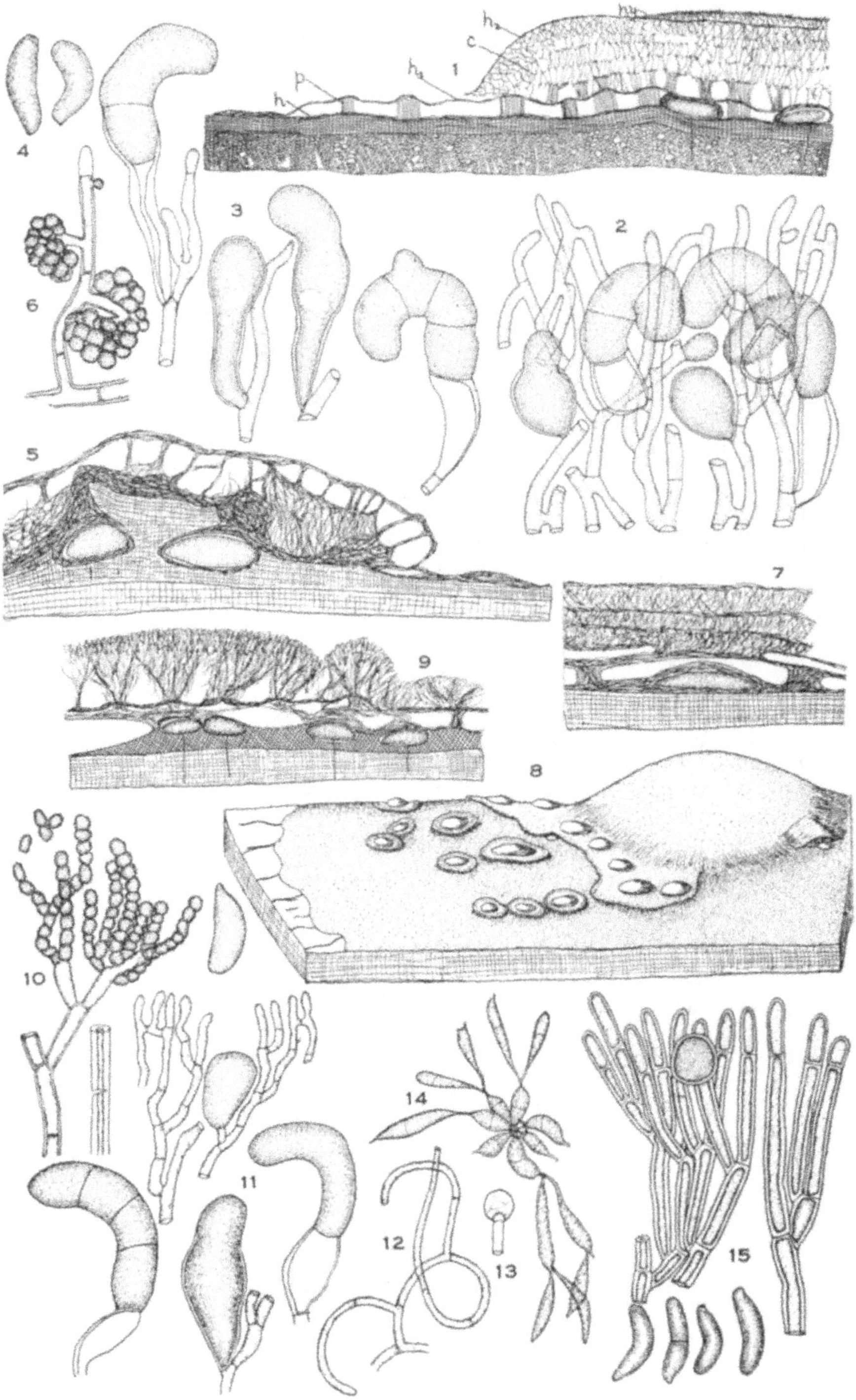

Plate 47

S. abnorme

Fig. 1.—Sketch of fungus. × 1.
Fig. 2.—Threads from context. × 580.
Above from type material.

S. lanosum

Fig. 3.—Section of context. × 19.
Fig. 4.—Bladder-like cells in context. × 525.
Fig. 5.—Jug-shaped cell in context. × 525.
Fig. 6.—Probasidia. × 525.
Figs. 7, 8.—Conidia. × 525.

S. paulense

Fig. 9.—Conidial hyphae near base and probasidial hyphae near top. × 19.
Fig. 10.—Jug-shaped cells in context. × 525.
Fig. 11.—Probasidia. × 525.
Fig. 12—Conidia. × 525.
Fig. 13.—Star-like arrangement of hyphae just after penetrating insect's derm. × 525.
Fig. 14.—Bladder-like cell from context. × 525.

S. makilingianum

Fig. 15.—Context. × 580.

S. myrsinae

Fig. 16.—Margin. × 18. Note pillars with hat-shaped tops, the brims of which grow out to form middle horizontal layer.

PLATE 47

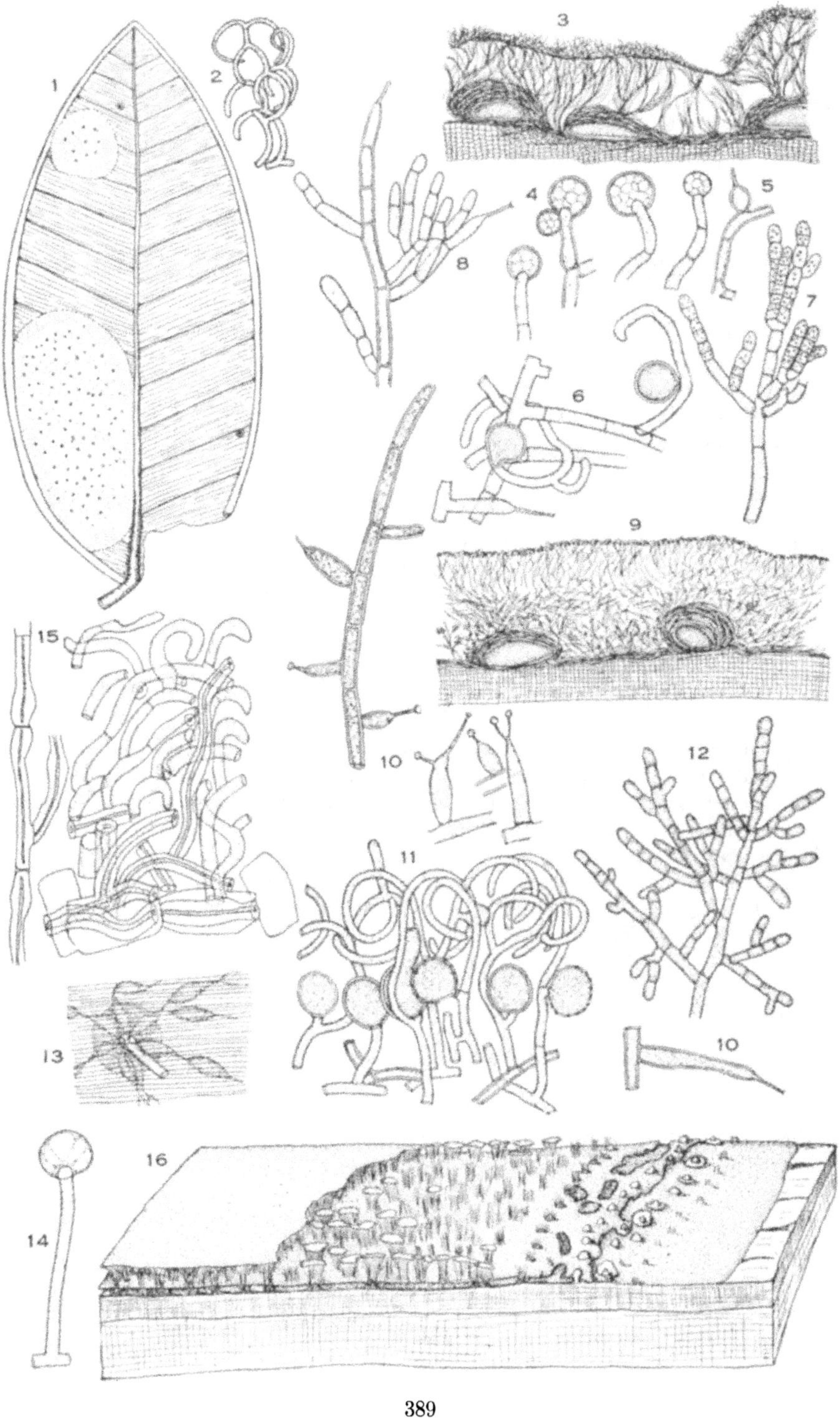

PLATE 48

Septobasidium sp. hybrid on *Liquidambar*

Fig. 1.—Section of context. × 25.

S. theae

Fig. 2.—Section of context showing compact (cross-hatched) tissue in which two insects are imbedded and loose tissue over which conidia are formed. × 19.

Fig. 3.—Bladder-like cells in context. × 525.

Fig. 4.—Conidia. × 525.

Fig. 5.—Jug-shaped cells in context. × 525.

S. conidiophorum

Fig. 6.—Section of context. × 25.

Fig. 7.—Chains of cells which give rise to small conidia. × 525.

Fig. 8.—Hyphae from near top layer showing dark thickenings at septa. × 525.

S. molle

Fig. 9.—Probasidia. × 525.

S. Rickii

Fig. 10.—Surface and section view, showing parasitized and healthy insects imbedded in thick spongy layer; insect parasitized by hymenopterous wasp; tunnels and chambers to far left; half-moon-shaped openings on surface through which young insects crawl; rounded opening formed by an emerging wasp. × 14.

Fig. 11.—Section of hymenium showing old gelatinized probasidia (?). × 525.

Fig. 12.—Hyphae from context, showing walls glued together by adhesive substance. × 525.

S. prunophilum

Fig. 13.—Section of context with compact tissue around parasitized insects. × 25.

Fig. 14.—Hymenium showing basidia and upright hyphae. × 580.

S. indigophorum

Fig. 15.—Context, showing pillars, two parasitized insects, and top layer. × 25.

PLATE 48

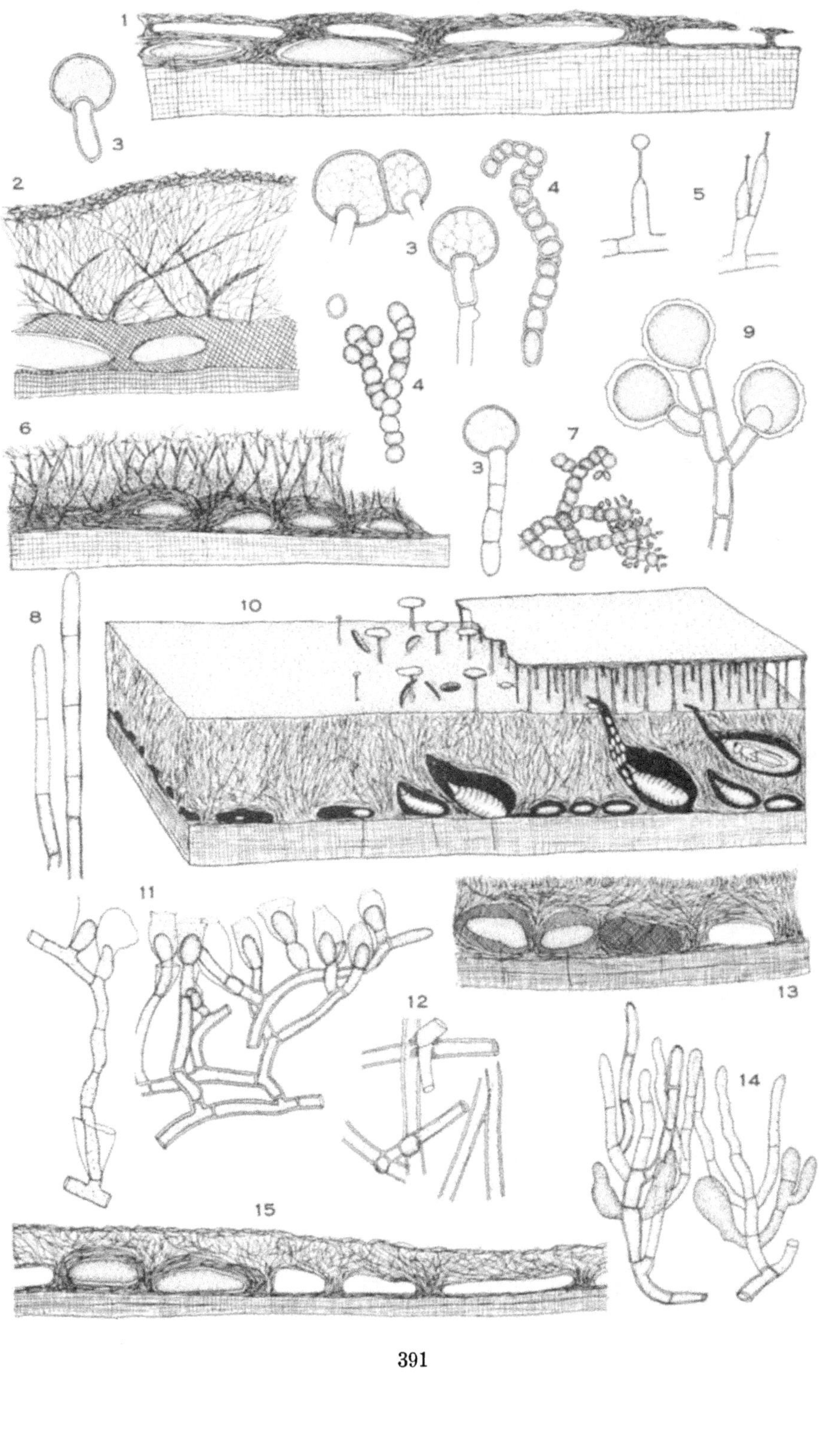

PLATE 49

S. formosense

Fig. 1.—Section and surface view of fungus on leaf showing very thin subiculum and mound over insect, also three dark spines on mound, perhaps multicellular conidia. × 19.
Fig. 2.—Part of one of the spine-like structures. × 335.
Fig. 3.—Probasidia and basidia. × 335.

S. Bagliettoanum

Fig. 4.—Surface and section view showing irregular patches of hymenium and rhizomorphs extending between cracks of bark. × 14.
Fig. 5.—Haustorium from parasitized insect. × 580.
Fig. 6.—Section of hymenium showing probasidia, basidia, and spores. × 625.

S. crustaceum

Figs. 7, 8.—Probasidia, basidia, and spores. × 930.

S. neglectum

Fig. 9.—Section of hymenium showing probasidia and basidia. × 335.
Fig. 10.—Three basidia. × 525.

S. proliferum

Fig. 11.—Basidium with one spore. × 525. (Note laminations in probasidial wall, indicating proliferation of new probasidium into empty one.)
Fig. 12.—Spores, three above and two below. × 525.

S. fisso-lobatum

Fig. 13.—Section of hymenium showing probasidia, basidia, and spores. × 625.

PLATE 49

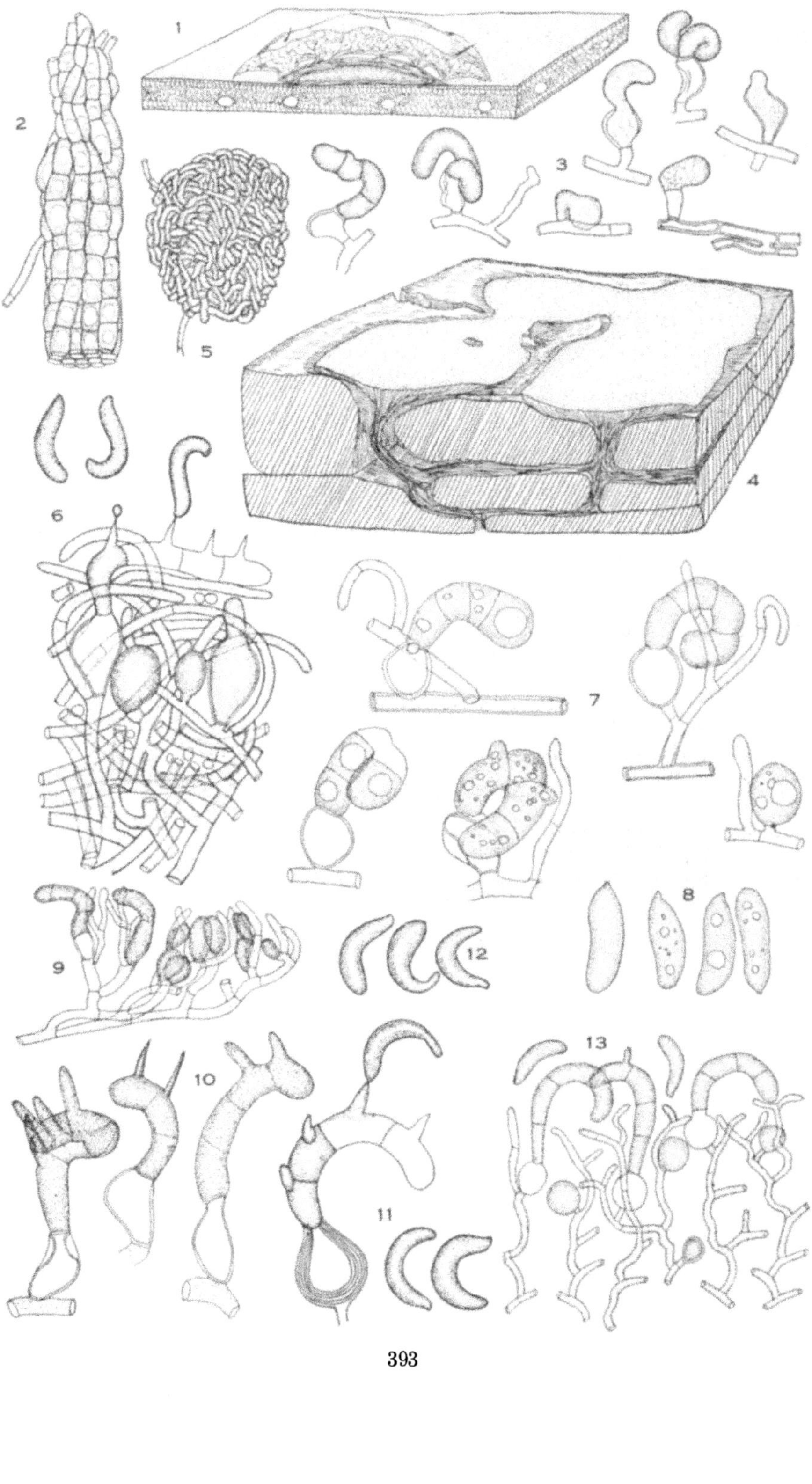

PLATE 50

S. Linderi

Figs. 1–11.—Probasidia, two-, three-, and four-celled basidia, branched sterigmata and spores. × 525.

S. Stevensoni

Fig. 12.—Section of context. × 25.
Fig. 13.—Probasidia, basidia, and spores. × 580.

S. cervicolor

Fig. 14.—Section of context. × 25.
Fig. 15.—Probasidia, basidia, and spore. × 625.

S. cinereum

Fig. 16.—Section of context × 25.
Figs. 17, 18.—Probasidia and basidia. × 625.
Fig. 19.—Spores. × 525.

PLATE 50

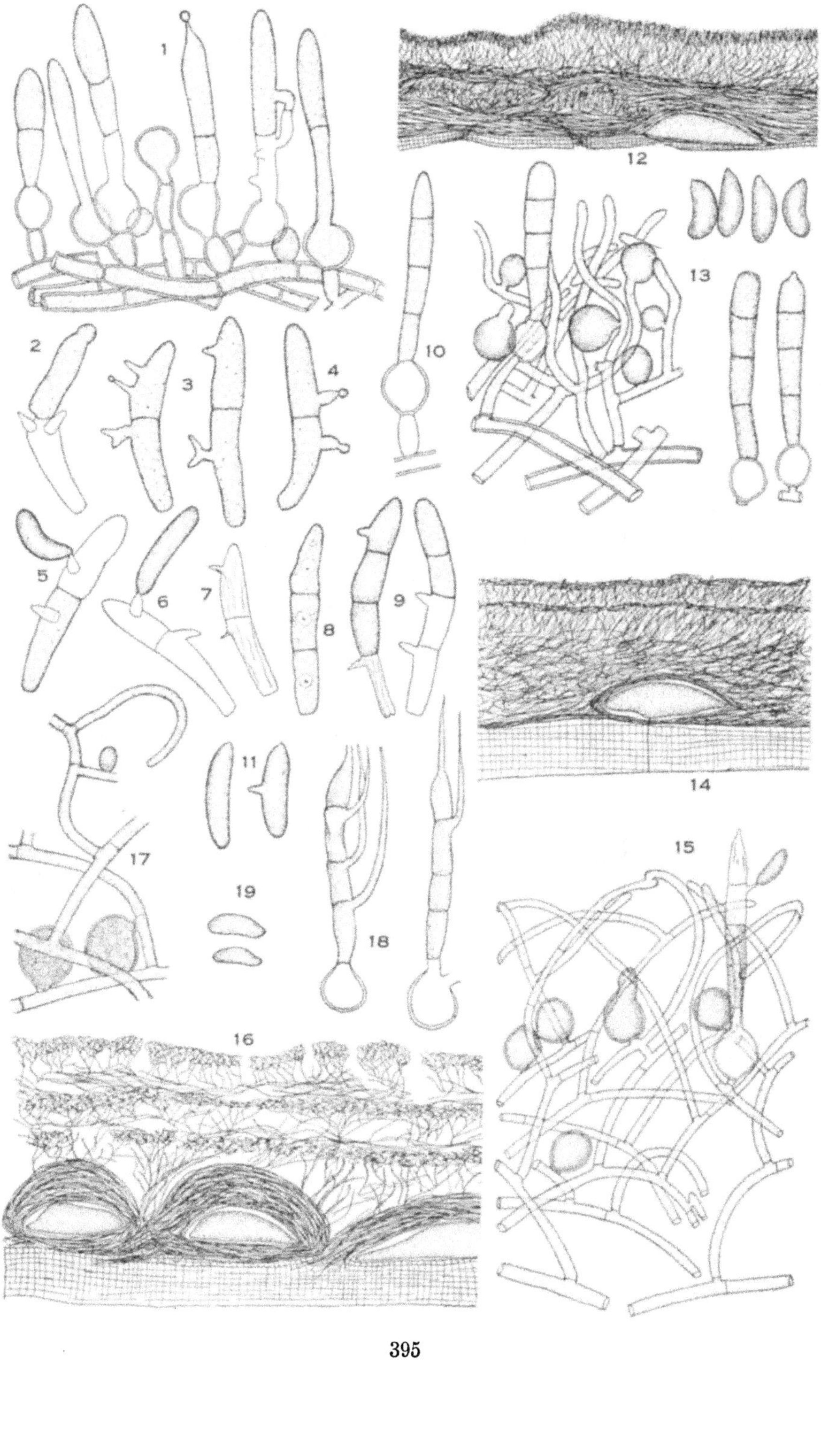

PLATE 51

S. Bresadolae

Fig. 1.—Section of context, showing scale insects and lack of pillars. × 25.
Figs. 2, 3.—Section of hymenium with probasidia, basidia, and spores. × 625.

S. rimulosum

Fig. 4.—Section of hymenium with probasidia, basidia, and spore. × 580.

S. stratosum

Fig. 5.—Section of context showing two scale insects and seven hymenial strata. × 18.
Fig. 6.—Section of hymenium showing probasidia, basidia, and spores above. × 525.

S. protractum

Fig. 7.—Section of context showing subiculum with two insects, entangled pillars, and top layer. × 14.
Figs. 8, 9.—Section of hymenium with probasidia, basidia, and spores. × 525.

PLATE 51

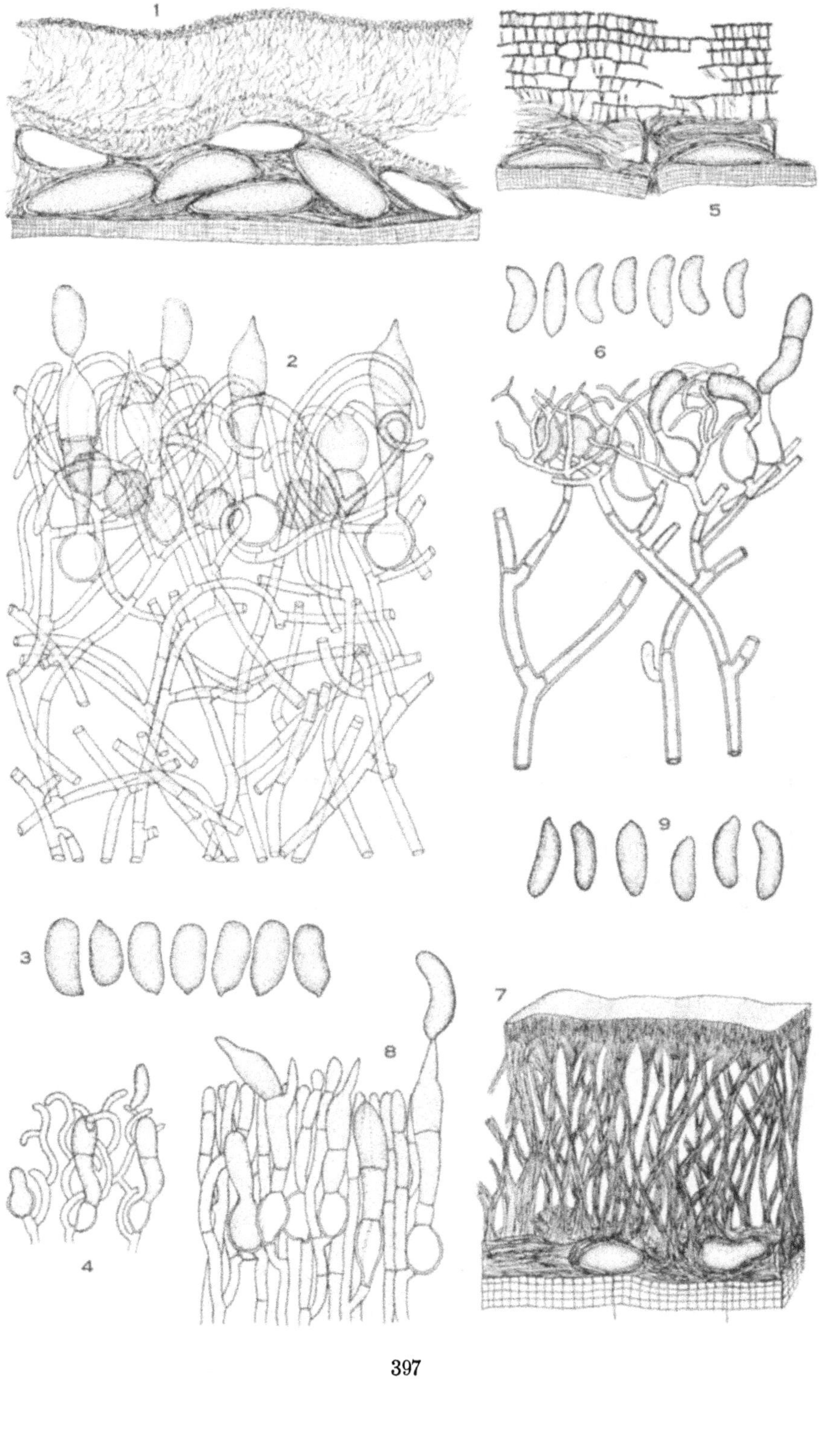

Plate 52

S. griseopurpureum

Fig. 1.—Section of context showing scale insect. × 25.

Fig. 2.—Enlarged section of context showing thin subiculum among threads of which are algal cells and (above) hymenium with probasidia and basidia. × 335.

Figs. 3, 4.—Basidium and spores. × 525.

S. dictyodes

Fig. 5.—Section of hymenial region showing probasidia, basidia, and four bladder-like cells. × 625.

S. velutinum

Fig. 6.—Section of context. × 25.

Fig. 7.—Section of hymenium showing basidia, a single basidium to left. × 625.

S. cinnamomeum

Fig. 8.—Sections of context: below, marginal region; above, older part where hymenium is well formed. × 14.

Fig. 9.—Section of hymenium with probasidia, basidia, and spores. × 525.

PLATE 52

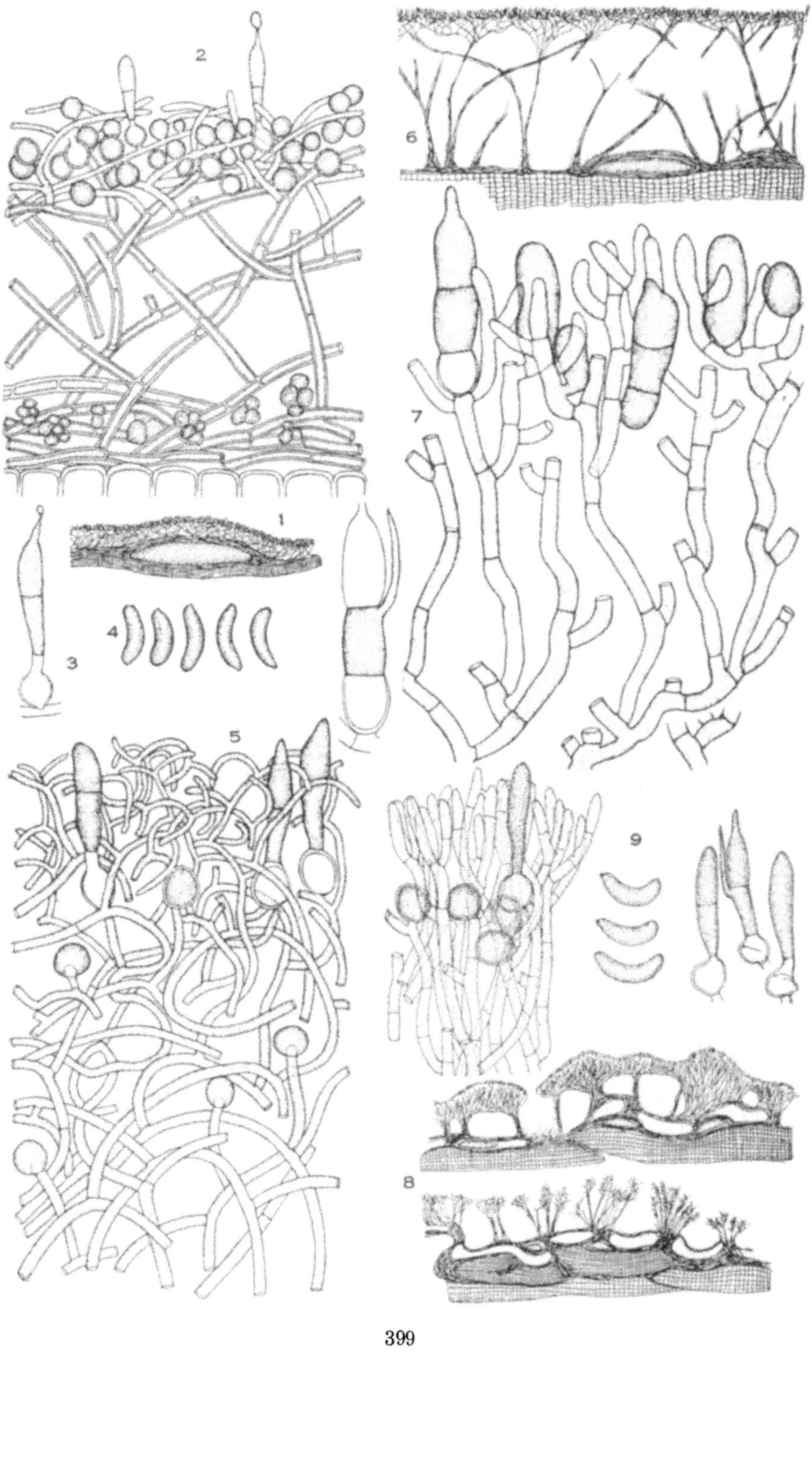

Plate 53

S. separans

Fig. 1.—Section of context. × 25.
Figs. 2, 3.—Hymenium, basidia and spores. × 625.

S. subcarbonaceum

Fig. 4.—Section of context. × 25.
Figs. 5, 6.—Hymenium with probasidia, basidia, spores, and one bladder-like cell beneath two spores. × 580.

S. fragile

Fig. 7.—Section of context. × 18.
Figs. 8, 9.—Probasidium and basidium. × 525.

S. lilacinum

Fig. 10.—Section of context. × 25.
Figs. 11, 12.—Hymenium with probasidia, basidia, and spores. × 580.

S. nodulosum

Fig. 13.—Section of context. × 14.
Fig. 14.—Probasidium and basidium. × 580.

S. Cavarae

Fig. 15.—Section of context. × 25. Type material.

PLATE 53

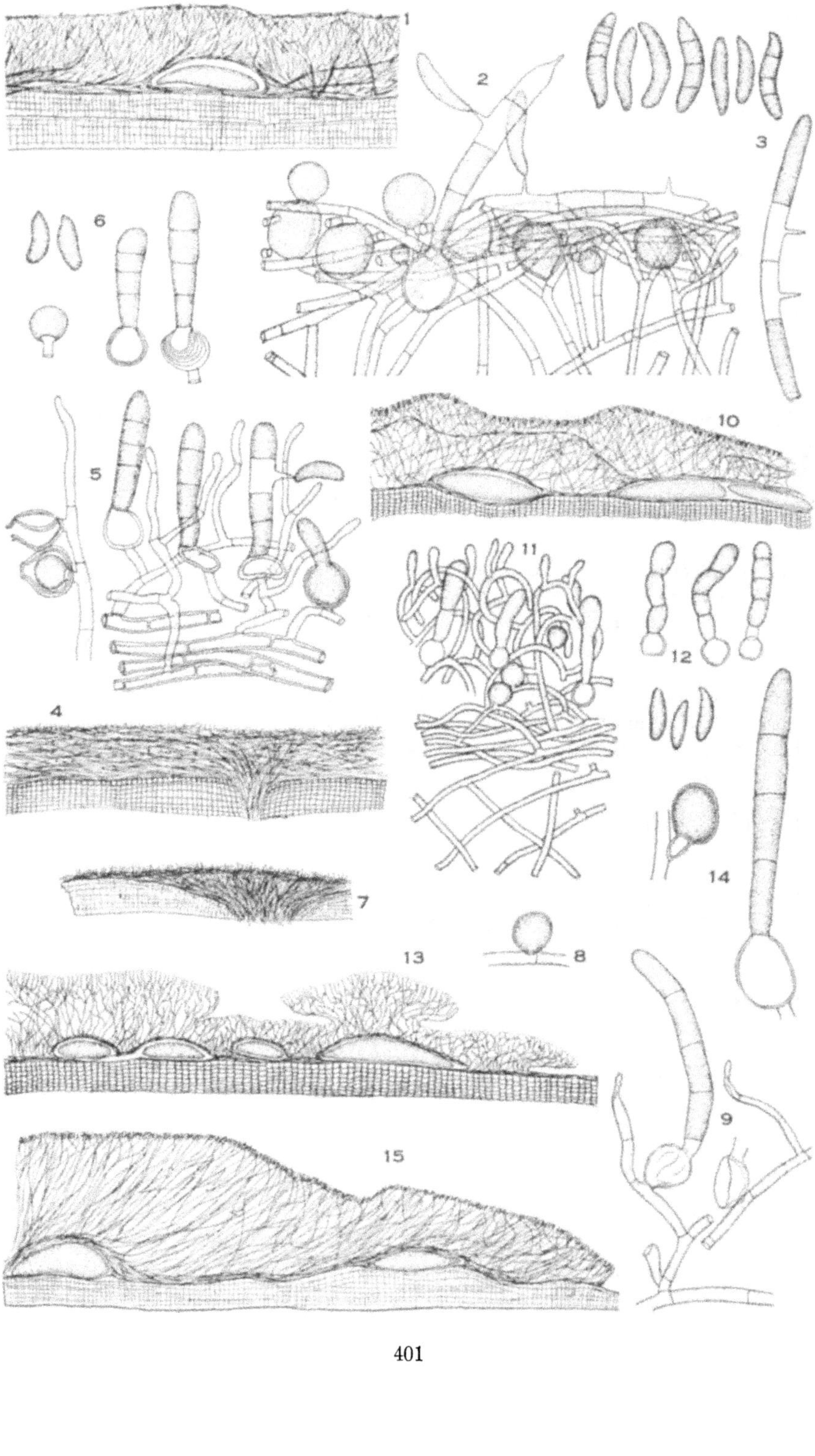

Plate 54

S. Clelandii

Fig. 1.—Section through wood and fungus showing parasitized larva of insect, over which is spongy bark tissue interwoven with fungal hyphae (cross-hatched) and over this is subiculum from which arise spines. × 4.

Figs. 2, 3.—Hymenium, basidia, and spores. × 580.

S. sclerotioides

Fig. 4.—Section of context showing sclerotic tissue at base (cross-hatched), hymenium above. × 14.

Fig. 5.—Hymenium, basidia and spores. × 525.

S. fusco-cinereum

Fig. 6.—Surface and section view of context. × 14. Type material.

Fig. 7.—Hymenium. × 525. Type material.

Figs. 8, 9.—Basidia and spores. × 525. Type locality.

S. jamaicaense

Fig. 10.—Basidia and spores (for comparison with former species). × 525.

S. retiforme

Fig. 11.—Basidia. × 525. Type material.

Fig. 12.—Threads from context. × 335. Type material.

S. areolatum

Fig. 13.—Basidia (for comparison with above). × 525.

PLATE 54

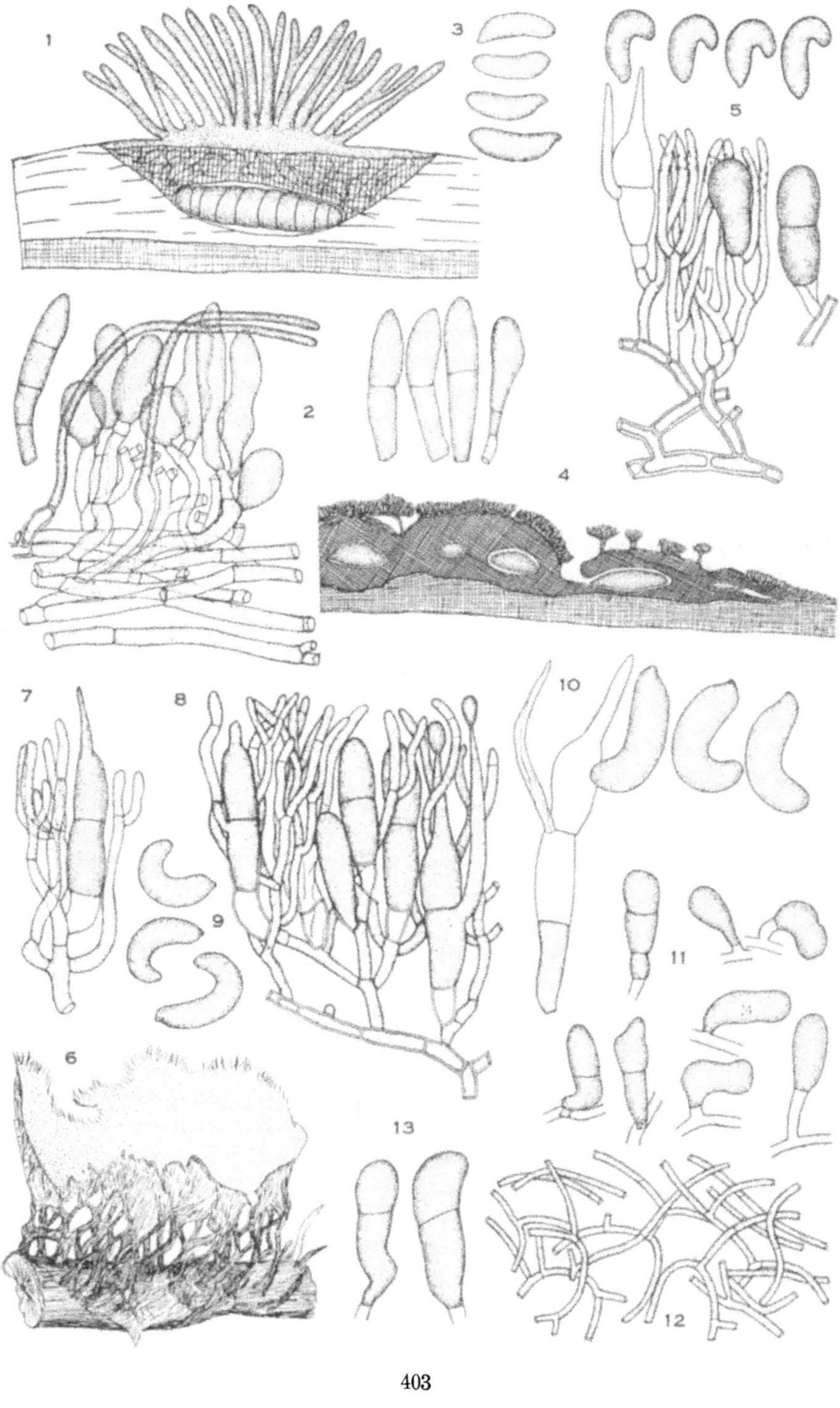

Plate 55

S. minutulum

Fig. 1.—Section showing leaf epidermis, subiculum, pillars, and top layer with probasidia and basidia. × 320.

S. irregulare

Fig. 2.—Section of context. × 19.
Fig. 3.—Sketch of insect house surrounded by bristles. × 25.
Fig. 4.—Hymenium, probasidia, basidia, and spore. × 525.

S. leucostemum

Fig. 5.—Section of context. × 25.
Fig. 6.—Section of hymenium. × 525.
Fig. 7.—Basidium and spore. × 525.

S. Merrillii

Fig. 8.—Section of context. × 19.
Fig. 9—Hymenium with probasidia and basidia. × 525.

PLATE 55

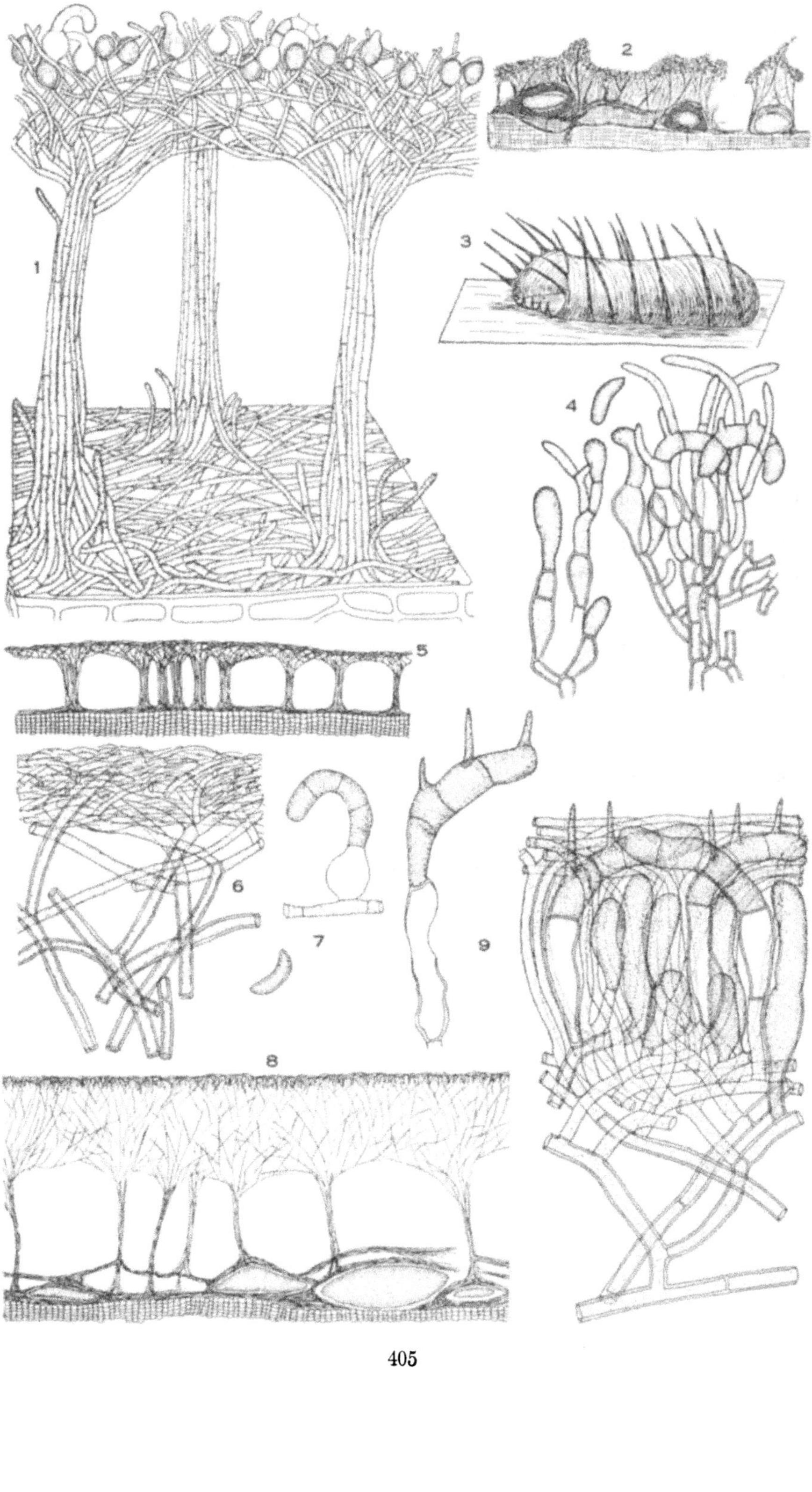

PLATE 56

S. Boedijnii

Fig. 1.—Section of context of fungus and leaf, showing insect house with hole at top. × 19.
Fig. 2.—Section of hymenium. × 325.
Fig. 3.—Probasidia and basidia. × 525.

S. laxum

Fig. 4.—Surface of margin showing mound-like insect houses. × 7.
Figs. 5, 6.—Probasidia and basidia. × 525.

S. phyllophilum (= *S. minutulum*)

Fig. 7.—Context. × 19.
Fig. 8.—Hymenium. × 525.

S. stratiferum

Fig. 9.—Section of context. × 19.
Fig. 10.—Section of hymenium showing young and old probasidia and basidia. × 525.
Fig. 11.—Spores. × 525.

PLATE 56

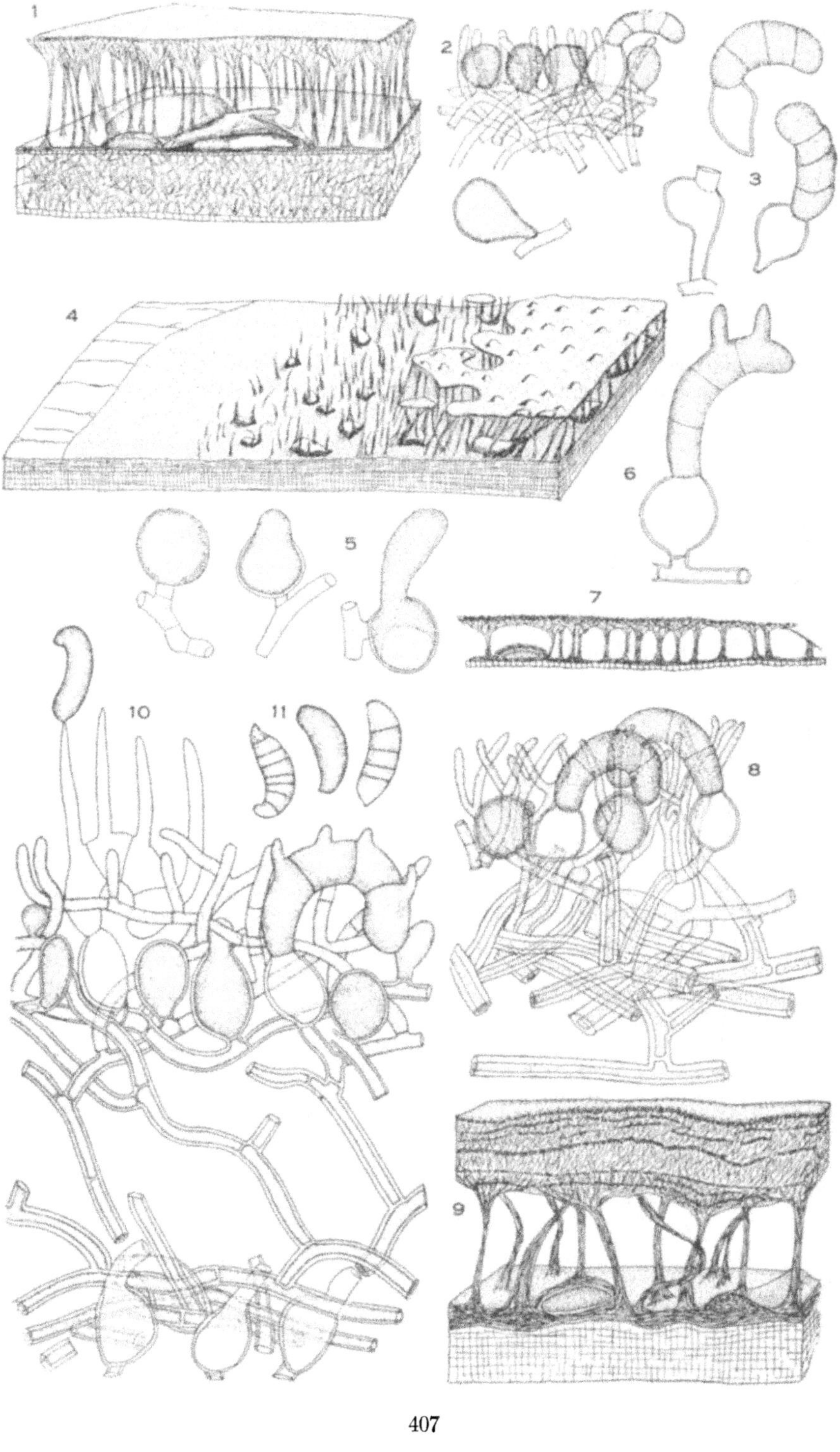

PLATE 57

S. atratum

Fig. 1.—Surface and section view showing insect houses on margin. × 14.
Fig. 2.—Hymenium, probasidia, basidia, and spores. × 625.

S. fumigatum var. *brasiliense*

Fig. 3.—Probasidia, basidium, two spores, and bladder-like cell in context. × 580.

S. siparium

Fig. 4.—Section of context. × 19.
Fig. 5.—Probasidia, basidia, and spores. × 525.

S. hakgalanum

Fig. 6.—Two basidia broken from probasidia and one spore. × 525.

S. spongium

Fig. 7.—Basidia and spore. × 525.

S. triviale

Fig. 8.—Section of context. × 25.
Fig. 9.—Basidium. × 580.

S. heveae

Fig. 10.—Hymenial threads, basidium, and two spores. × 625.

S. ugandae

Fig. 11.—Two sections of context. × 25.
Fig. 12.—Probasidia, basidia, and spores. × 525.

S. obscurum

Fig. 13.—Section of context showing region with and without pillars. × 14.
Fig. 14.—Hymenium, probasidia, basidium, and spore. × 525.

PLATE 57

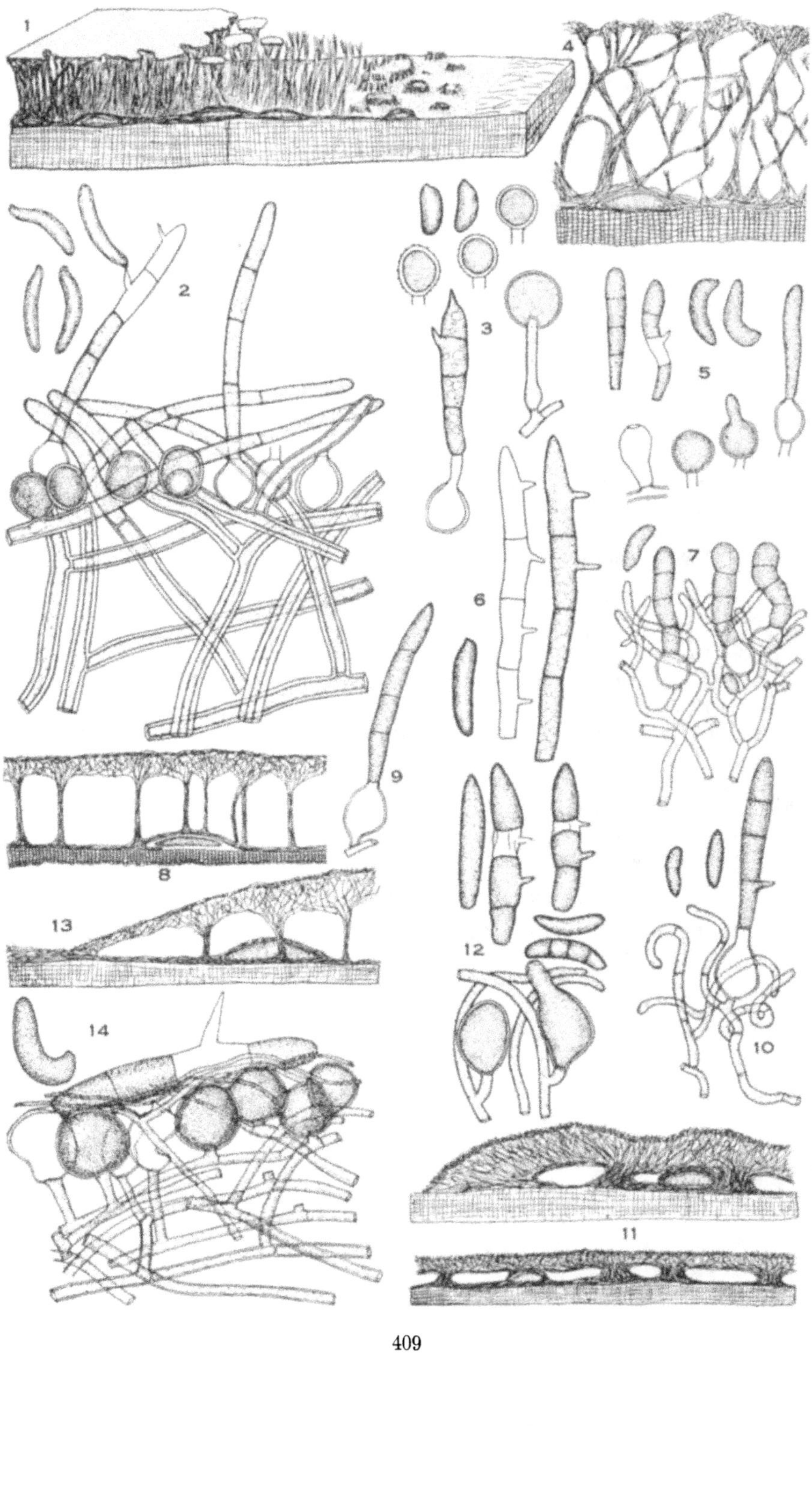

PLATE 58

S. Reinkingii

Fig. 1.—Surface and section view of older part of context showing stratose condition. × 25.
Fig. 2.—Hymenium. × 525.
Fig. 3.—Basidia and spores. × 525.

S. aligerum

Fig. 4.—Section of context. × 18.
Fig. 5.—Hyaline threads from subiculum. × 525.
Fig. 6.—Section of hymenium with old and new basidia. × 525.
Fig. 7.—Basidium and spores. × 625.
Figs. 4-6.—Drawn from material from Australia; 7, from Ceylon.

S. alatum

Fig. 8.—Sketch showing winged margin. × about 1.
Fig. 9.—Section of top layer with irregular mounds. × 25.
Fig. 10.—Basidia. × 625. Type material.

S. sulphurellum (= *S. rhabarbarinum*)

Figs. 11, 12.—Basidia and spores. × 525.

PLATE 58

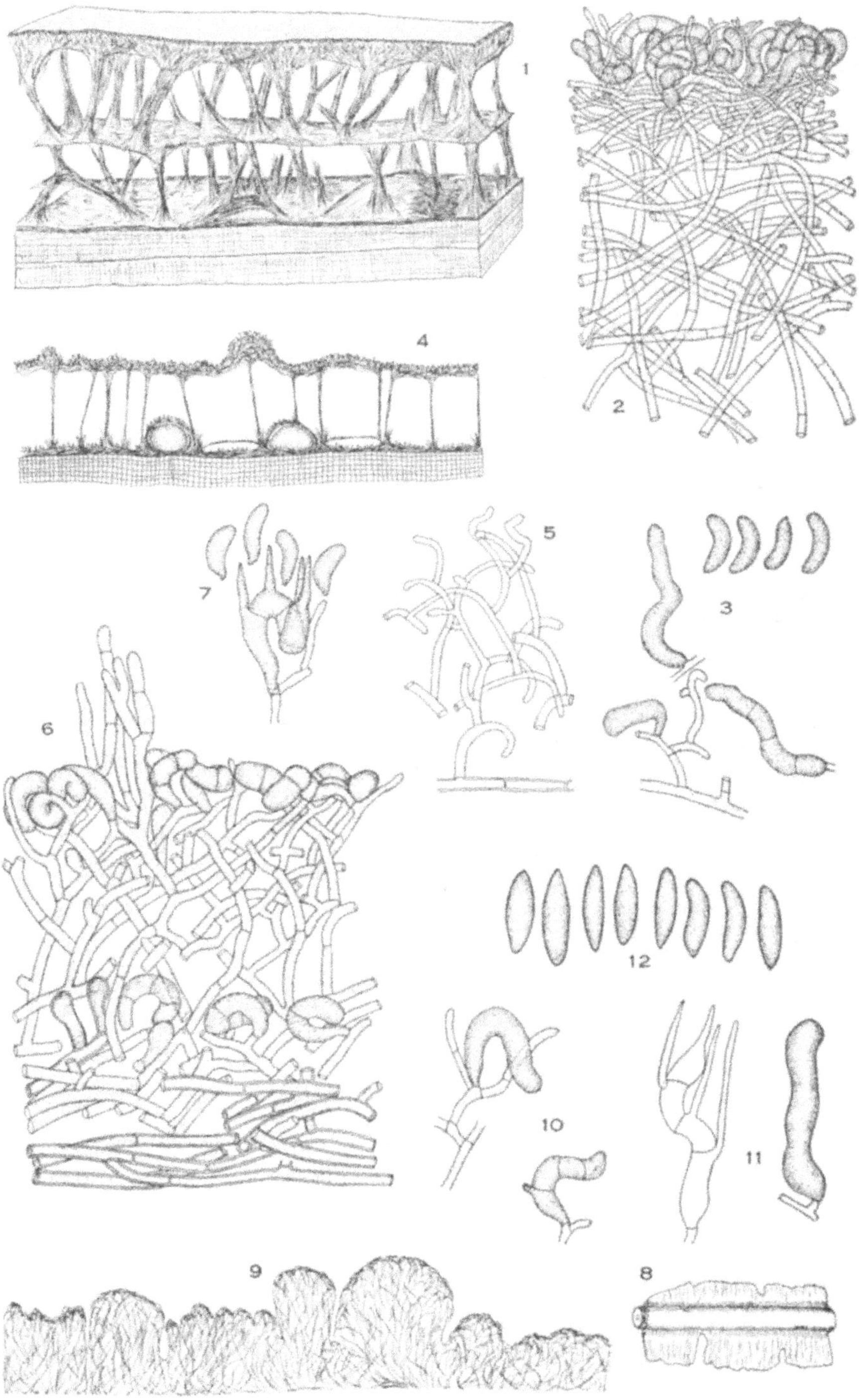

PLATE 59

S. Westoni

Fig. 1.—Section and surface views of context. Above × 25, below × about 1·
Fig. 2.—Basidia. × 625.
Fig. 3.—Conidia from surface of subiculum. × 625.

S. subolivaceum

Fig. 4.—Section of mound, showing old parasitized and young healthy insect, chamber of young opening to exterior. × 14.
Figs. 5, 6.—Context and basidia. × 525.

S. granulosum

Fig. 7.—Sketch showing wings. × 2.
Fig. 8.—Section of context showing bumps. × 25.
Fig. 9.—Section of hymenium. × 175.
Fig. 10.—Basidia and spores. × 625.

PLATE 59

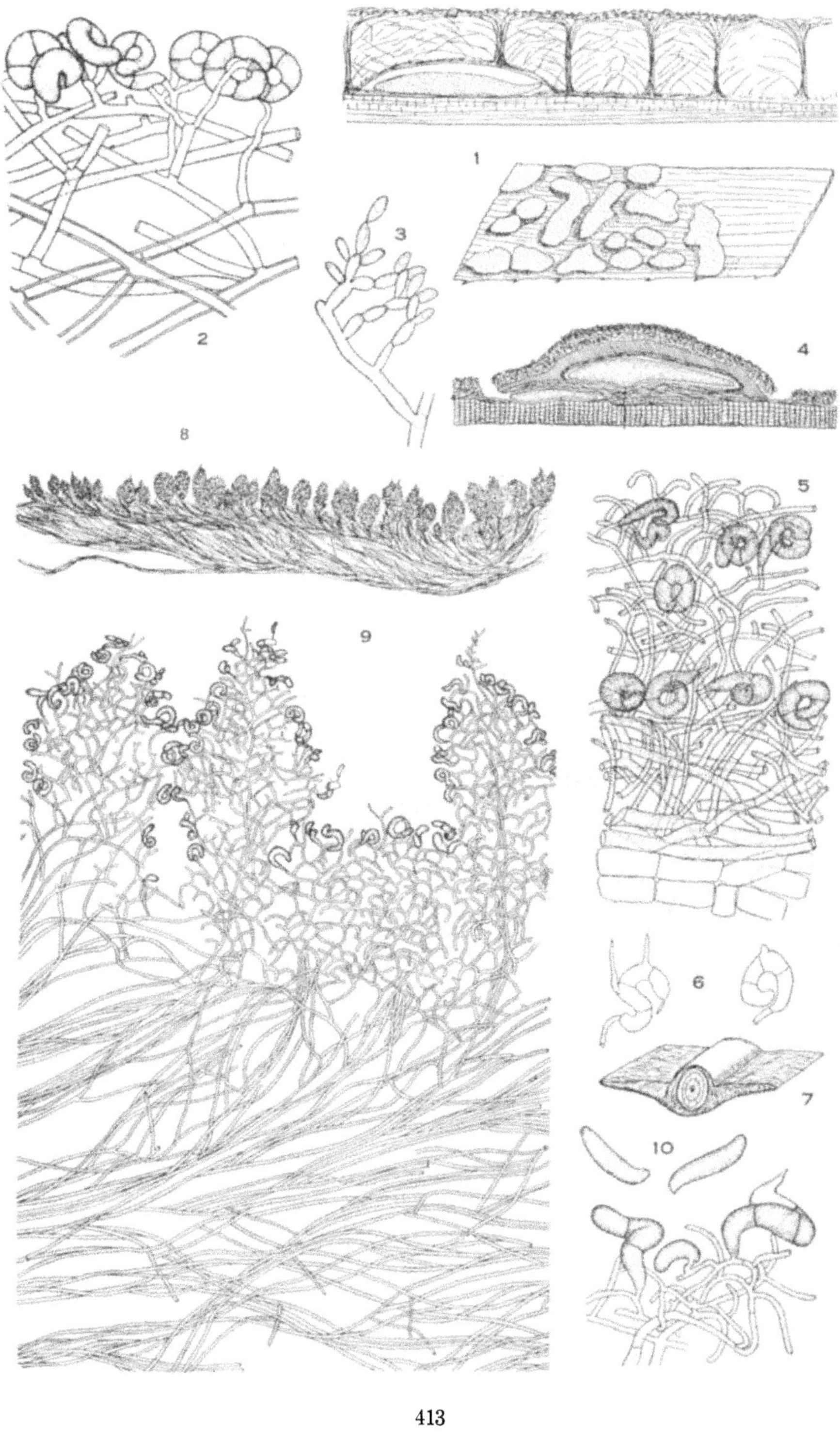

Plate 60

S. tropicale

Fig. 1.—Longitudinal section of context. × 14.
Fig. 2.—Cross section of context. × about 14.
Fig. 3.—Hymenium, probasidia, and basidia. × 580.

S. ussanguense

Fig. 4.—Section of context showing pillars and three horizontal layers. × 14.
Fig. 5.—Hymenium, probasidia, and basidia. × 525.

S. pedicellatum

Fig. 6.—Section and surface view of context. × 19.
Figs. 7, 8.—Probasidia, basidia, and spores. × 525.

S. rubiginosum

Fig. 9.—Section of older part of context. × 25.
Figs. 10, 11.—Hymenium, probasidia, basidia, and spores. × 625.

PLATE 60

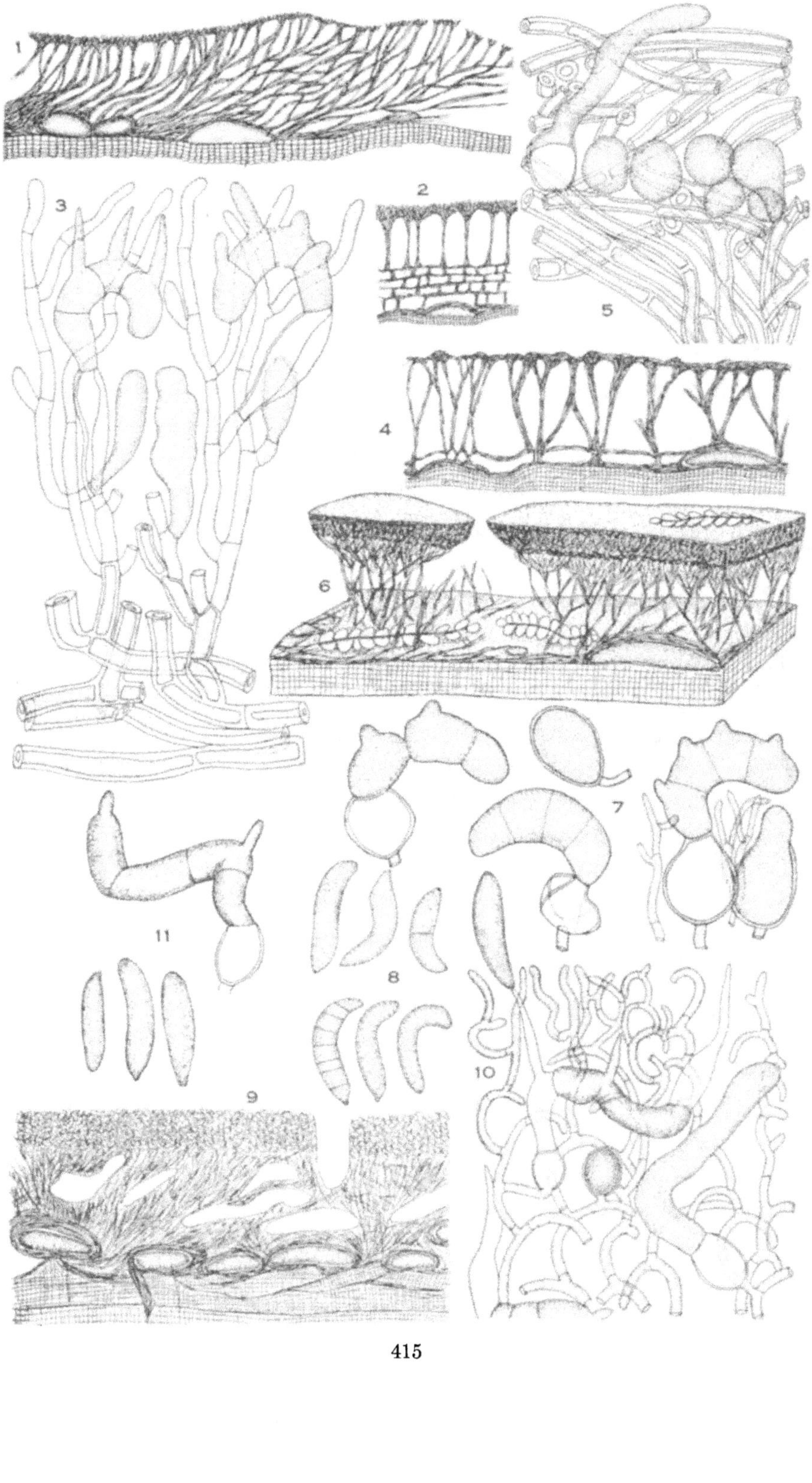

Plate 61

S. clethrae

Fig. 1.—Sketch showing subiculum, pillars, top layer, and two insect houses, one with roof turned back, showing female giving birth to young. × 19.
Fig. 2.—Basidia and spores. × 525.

S. bogoriense

Fig. 3.—Section of context. × 25. Material from Steinmann, Java.
Fig. 4.—Context. × 25. Duport, coll., from Tonkin.
Fig. 5.—Hymenium. × 625. Type material.
Figs. 6, 7.—Probasidia and basidia. × 580. Rands, coll., Java.
Fig. 8.—Basidia. × 580. From Steinmann, Java.
Fig. 9.—Basidia. × 580. Duport, coll., Tonkin.
Fig. 10.—Spores. × 580. Demange, coll., Tonkin.

S. murinum

Fig. 11.—Section of context. × 25.
Fig. 12.—Hymenium. × 525.

S. arachnoideum

Fig. 13.—Section of context. × 25.
Fig. 14.—Hymenium. × 580.

PLATE 61

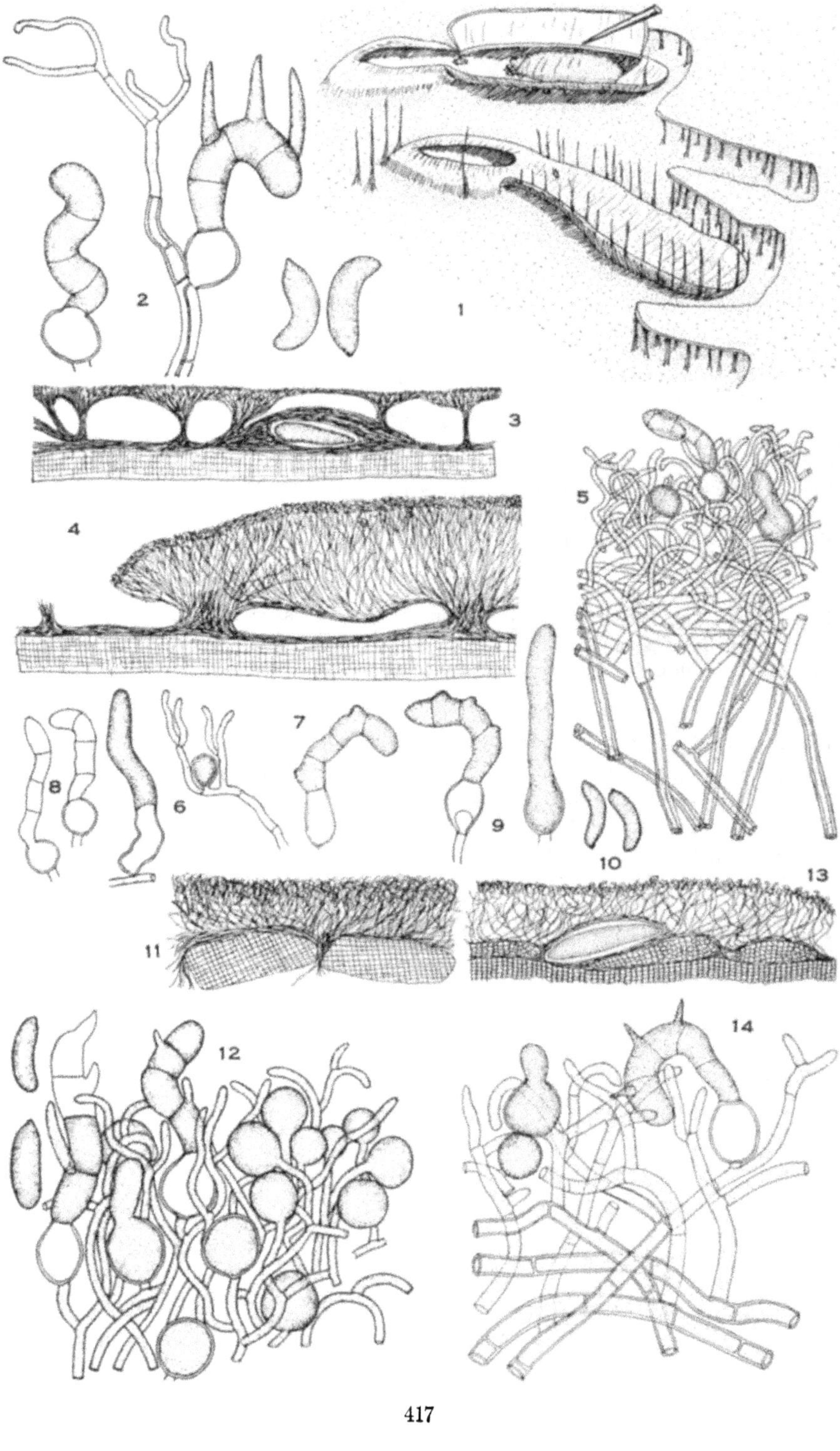

PLATE 62

S. humile

Fig. 1.—Surface view of fungus on leaf. × 25.
Fig. 2.—Section of subiculum, pillar, top layer with probasidia and basidia. × 335.
Fig. 3.—Spore. × 335. Type from Java.

S. citricolum

Fig. 4.—Context showing compact subiculum (cross-hatched), short pillars, thick top layer. × 19.
Fig. 5.—Hymenium, probasidia, basidia, spores. × 525. Type from Japan.

S. pannosum

Fig. 6.—Section of context. × 14. Type material from Brazil.
Fig. 7.—Section of context. × 18. Material from Guadeloupe.
Fig. 8.—Hymenium. × 525. Type.
Fig. 9.—Hymenium. × 525. Guadeloupe material.
Fig. 10.—Spores. × 525. Guadeloupe material.

S. gossypinum

Fig. 11.—Probasidia and basidia. × 580. Type from Tonkin, China.

S. cupressi

Figs. 12–14.—Probasidia, basidia, and spores. × 580. Type from Argentine.

PLATE 62

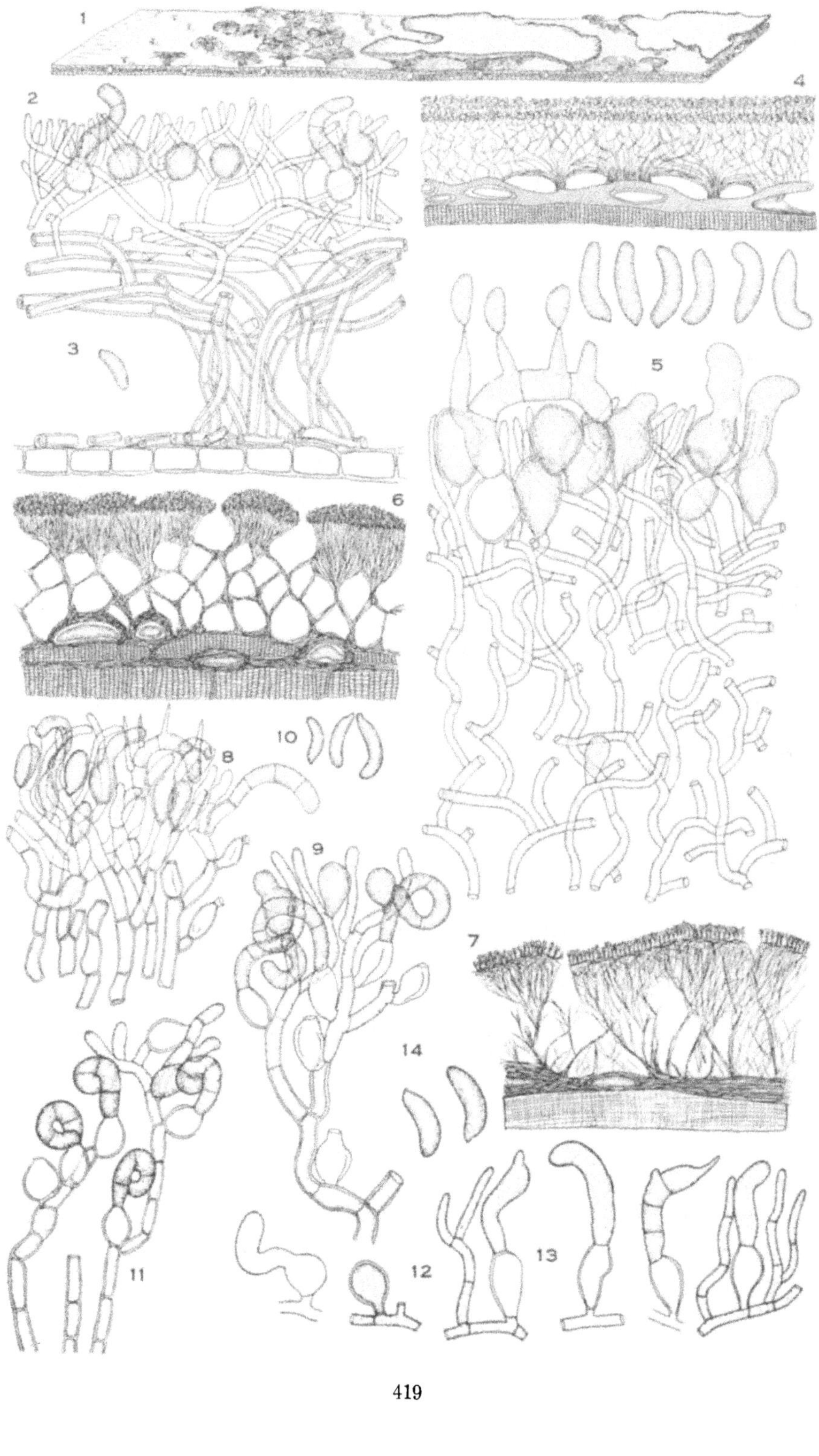

Plate 63

Septobasidium sp. (near *S. Henningsii*)

Fig. 1.—Context. × 18.
Fig. 2.—Section of hymenium. × 525.
Above from East African material, Volkens, coll., no. 1985 in Berlin Museum labelled *S. pedicellatum.*

S. Henningsii (= *tjibodense*)

Fig. 3—Context showing two hymenia. × 19. Type of *S. Henningsii.*
Figs. 4, 5.—Hymenium showing upright threads and basidia. 4 × 580, 5 × 525. Type of *S. Henningsii.*
Fig. 6.—Context. × 14. Type of *S. tjibodense.*
Fig. 7.—Spores. × 525. Type of *S. tjibodense.*

S. verrucosum

Fig. 8.—Section through subiculum and hard mounds, one insect giving birth to young, other three imbedded in sclerotic tissue and parasitized; two lighter colored strata are old hymenia. × 25.
Fig. 9.—Hymenial threads with young basidium. × 525.

S. flavo-brunneum

Fig. 10.—Section of context. × 19.
Fig. 11.—Hymenium and a few thick-walled cells from context. × 580.

S. Petchii

Fig. 12.—Section and surface view, parasitized insect beneath small mound, healthy insect giving birth to young beneath large mound; large mound composed entirely of sterile tissue, basidia formed only over context between small mounds. ×18.
Fig. 13.—Section of context showing peculiar threads and basidia. × 525.
Fig. 14.—Hyphal segments from context. × 525.
Fig. 15.—Spores. × 525.

PLATE 63

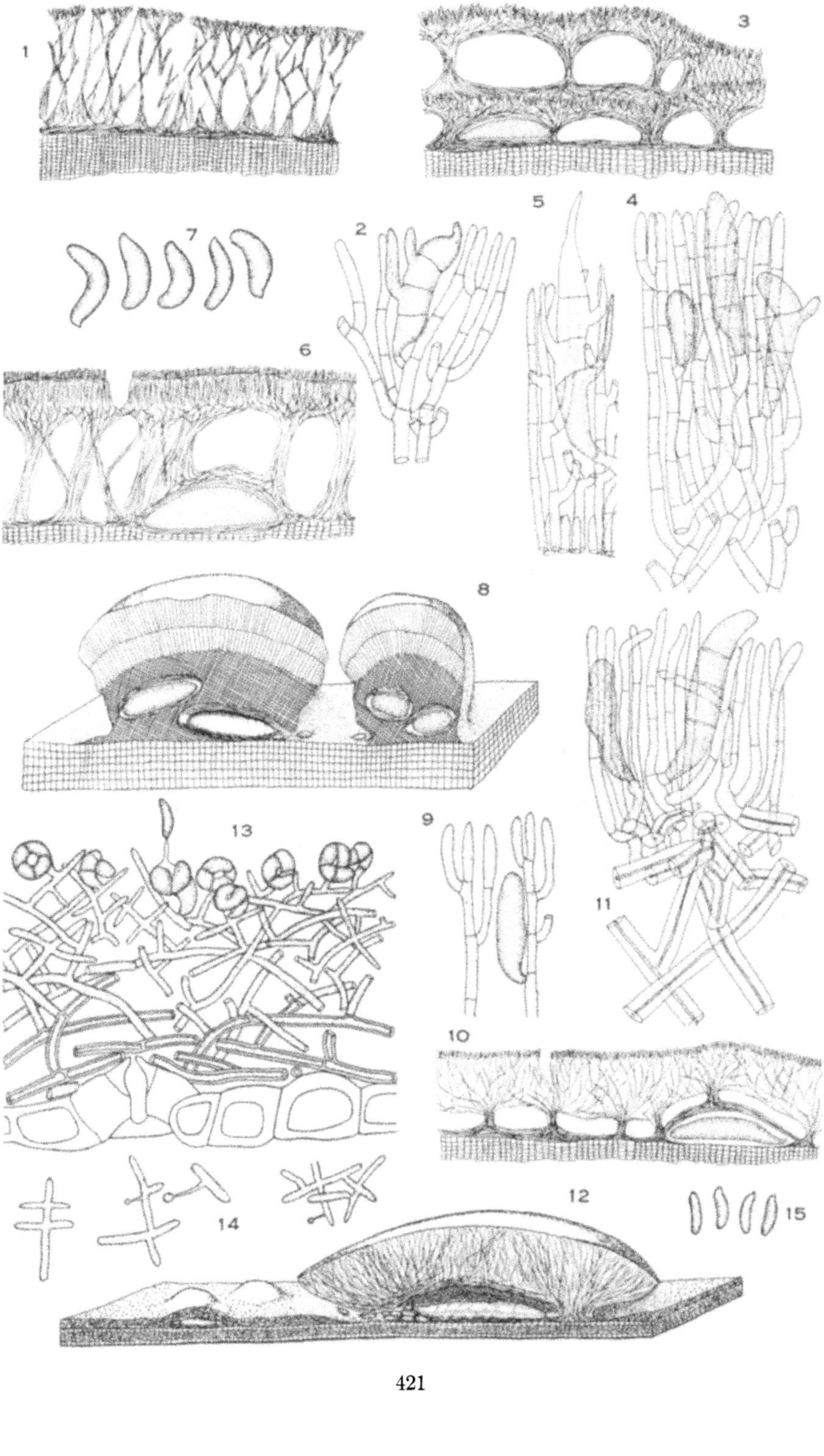

PLATE 64

S. albidum

Fig. 1.—Two sections of context, showing thick and thin regions. × 25. Type.
Figs. 2, 3.—Hymenium, basidia, and spores. × 625. Type.

S. pallidum

Fig. 4.—Section of context. × 25.
Fig. 5.—Hymenium, basidia, and spores. × 525.

S. philippinense

Fig. 6.—Sketch of surface and margin. × 7.
Fig. 7.—Section of context. × 25.
Figs. 8–10.—Hymenium, basidia, and spores. × 625.
Figs. 11, 12.—Basidia and spores. × 580.
Figs. 6–10.—From Philippine material; 11, 12, from South Africa.

PLATE 64

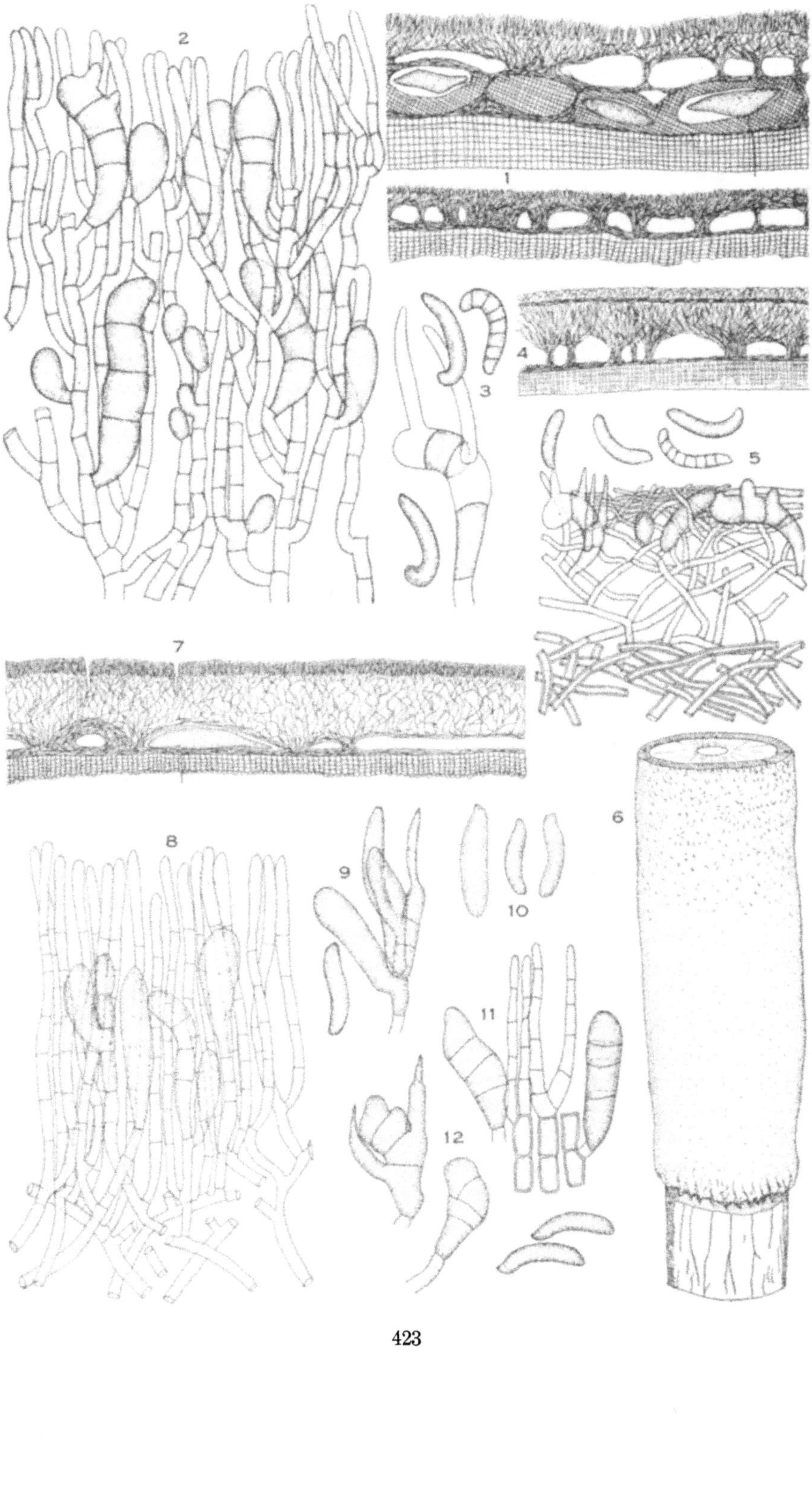

PLATE 65

S. cinchonae

Fig. 1.—Surface view showing subiculum, umbrella-like structures on margin, and nodulose top layer; also numerous young scale insects. ×19.
Fig. 2.—Sections of young and older regions of context. × 19.
Figs. 3–5.—Hymenium, probasidia, and basidia. × 580.
Fig. 6.—Thread from context broken by crushing. × 580.

S. lichenicolum (sense of Boedijn and Steinmann)

Fig. 7.—Surface and section view showing peculiar marginal condition; note crack in surface through which insects crawl. × 19. Boedijn, no. 11044.
Fig. 8.—Section of context showing stratose condition. × 19.
Figs. 9, 10.—Hymenium and basidia. × 580. Steinmann, coll.

S. lichenicolum (type)

Fig. 11.—Section of context. × 25.
Fig. 12.—Basidia. × 580.
Fig. 13.—Segment of threads from context broken by crushing. × 580.
Fig. 14.—Thick brownish hyphae from context. × 580.

PLATE 65

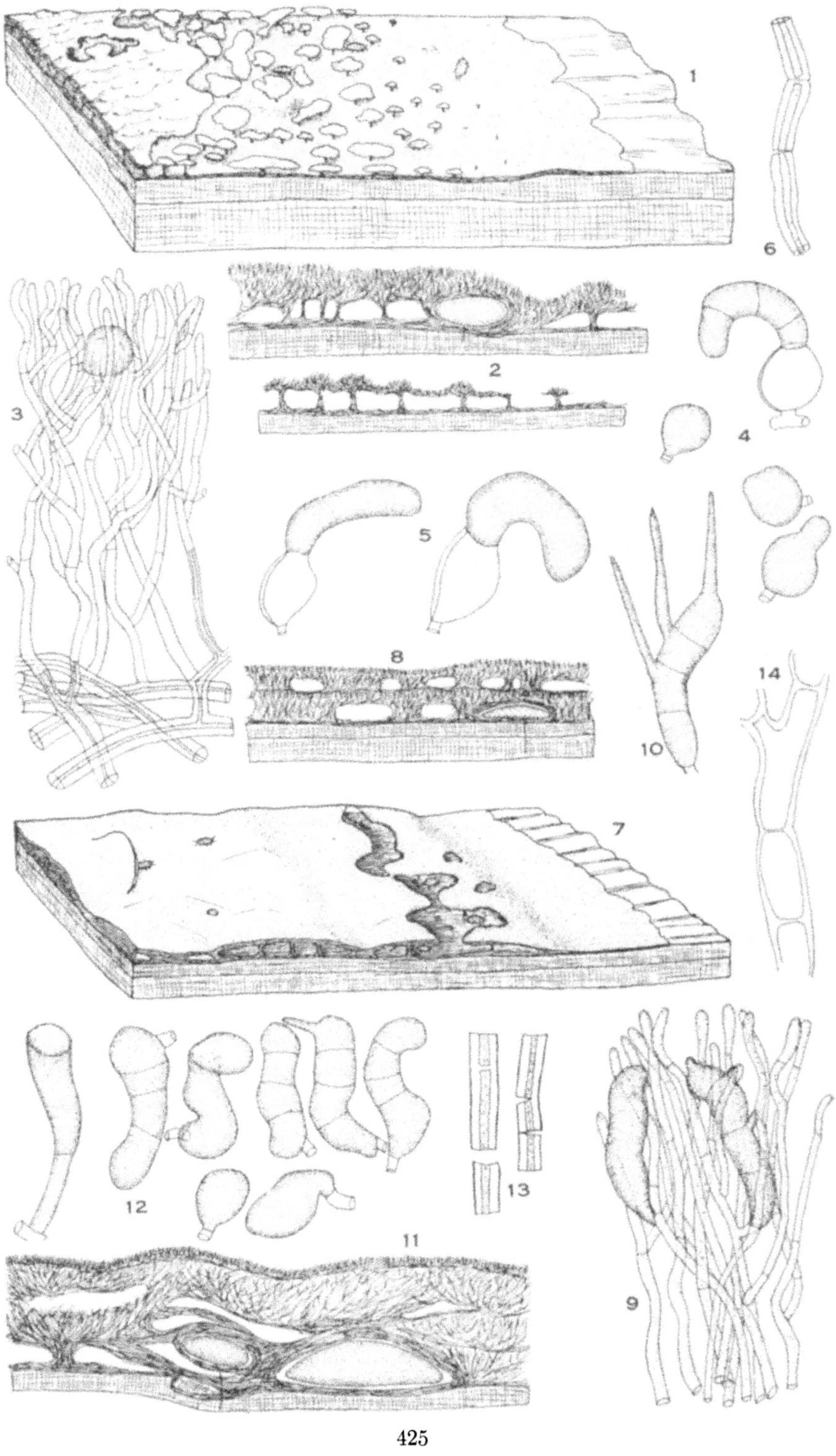

Plate 66

S. septobasidioides

Fig. 1.—Surface of subiculum in young part showing two insect houses, one with top layer partly removed exposing insect within. × 18.
Fig. 2.—Section of hymenium from mature part showing basidia. × 525.
Fig. 3.—Spores. × 525.
Above from type material.

S. cirratum

Fig. 4.—Section of context near margin. × 19.
Figs. 5-8.—Hymenium, basidia, and spores. × 525.
Fig. 9.—Bladder-like cells from context. × 525.
Above from type material.

S. crinitum

Fig. 10.—Surface of margin. × about 10.
Fig. 11.—Section of hymenium. × 625.
Above from type locality.

PLATE 66

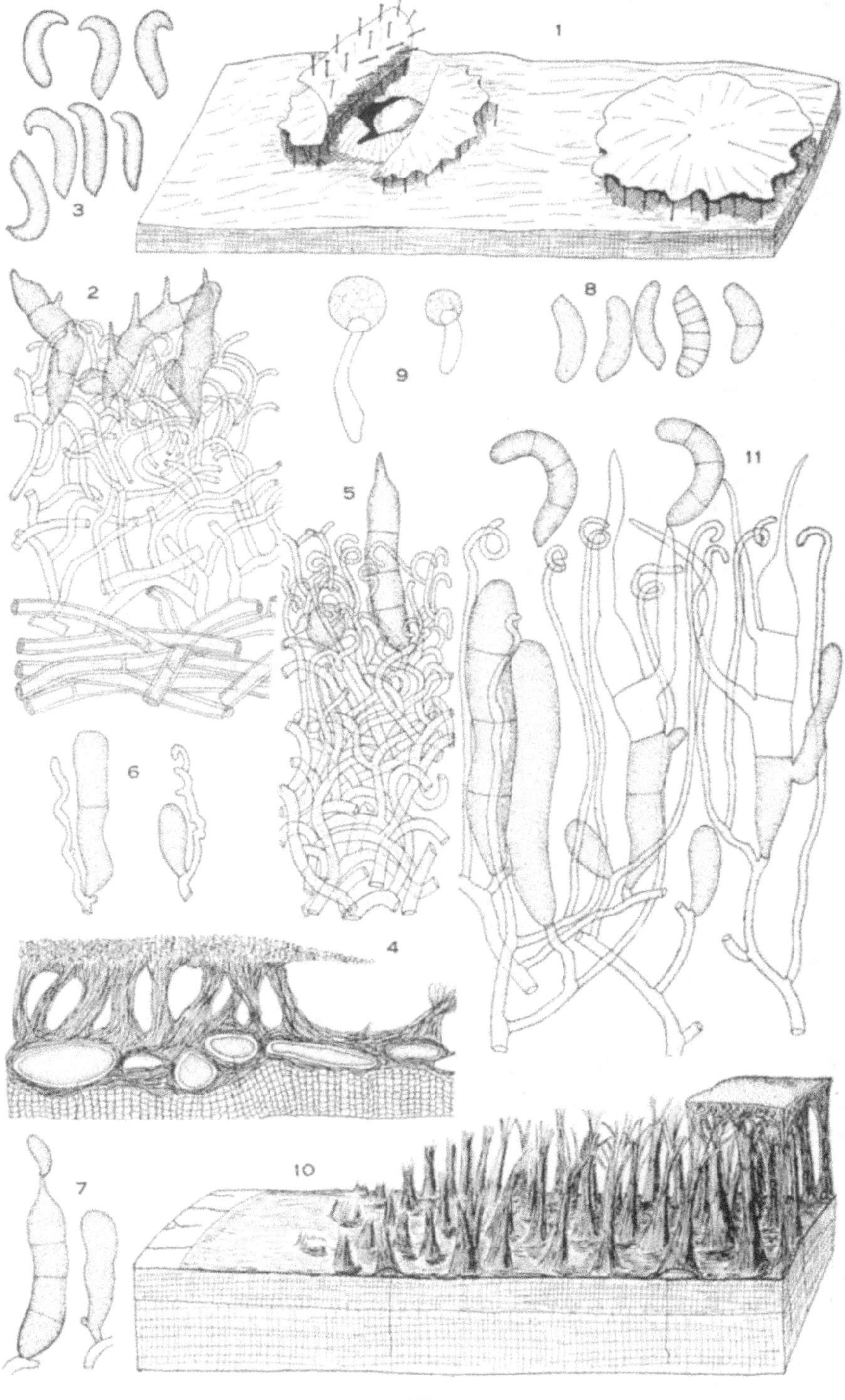

PLATE 67

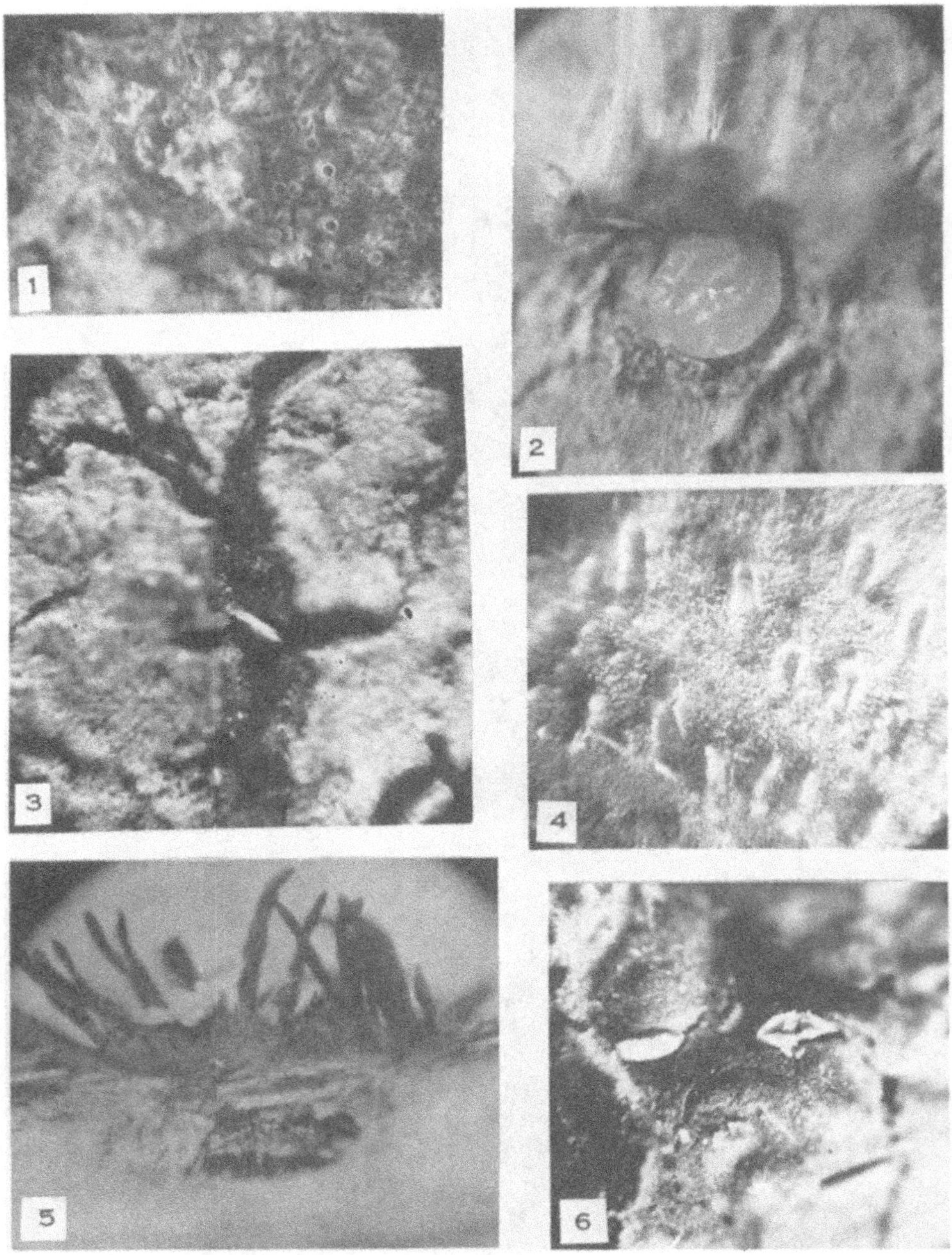

Figs. 1, 2, *S. apiculatum*; 1, young insects settled on surface; 2, adult healthy insect with fungal roof turned back. 3, *S. Sydowii*, surface showing adult female insects with scales projecting through cracks. 4, *S. Schweinitzii*, margin with insect traps; note young insect in middle upper trap. 5, *S. Clelandii*, section of fungus and wood showing insect in cavity. 6, *S. philippinense*, surface showing two white scales projecting through cracks; note young insect crawling out through opening near center.

Fig. 5 × 10; others × 22.

PLATE 68

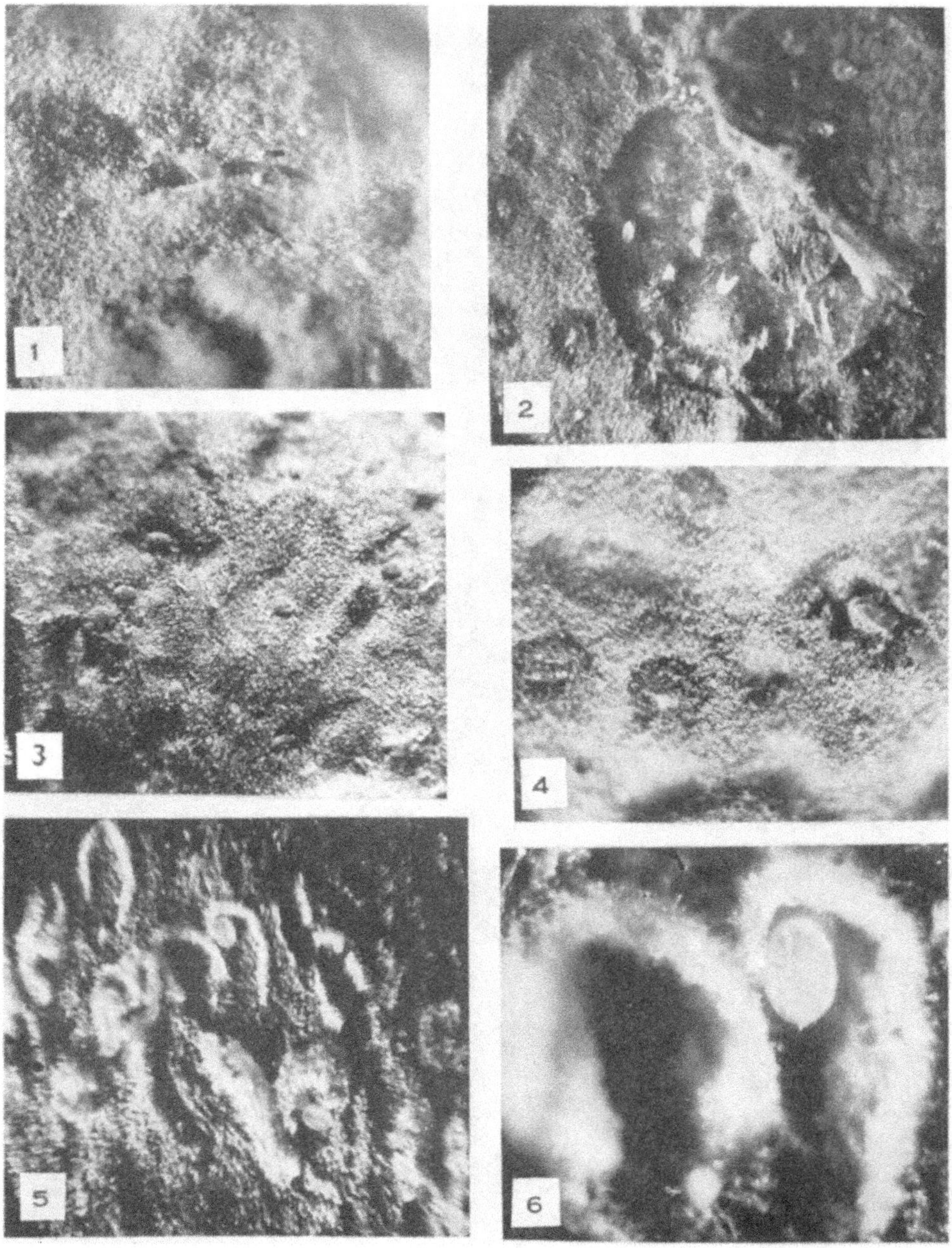

Figs. 1–4, *S. rugulosum*; 1, young insect emerging through new moon-shaped crack, × 33; 2, top of fungal house shown in fig. 1 turned back showing adult female on right (now dried) and white empty "egg shells" of young, × 22; 3, four young insects settled on surface, × 22; 4, surface several days after young have settled, showing how they sink into fungal tissue, × 33. 5, 6, *S. Curtisii*; 5, insect traps in which are two insects, × 22; 6, one insect in trap, × 75.

PLATE 69

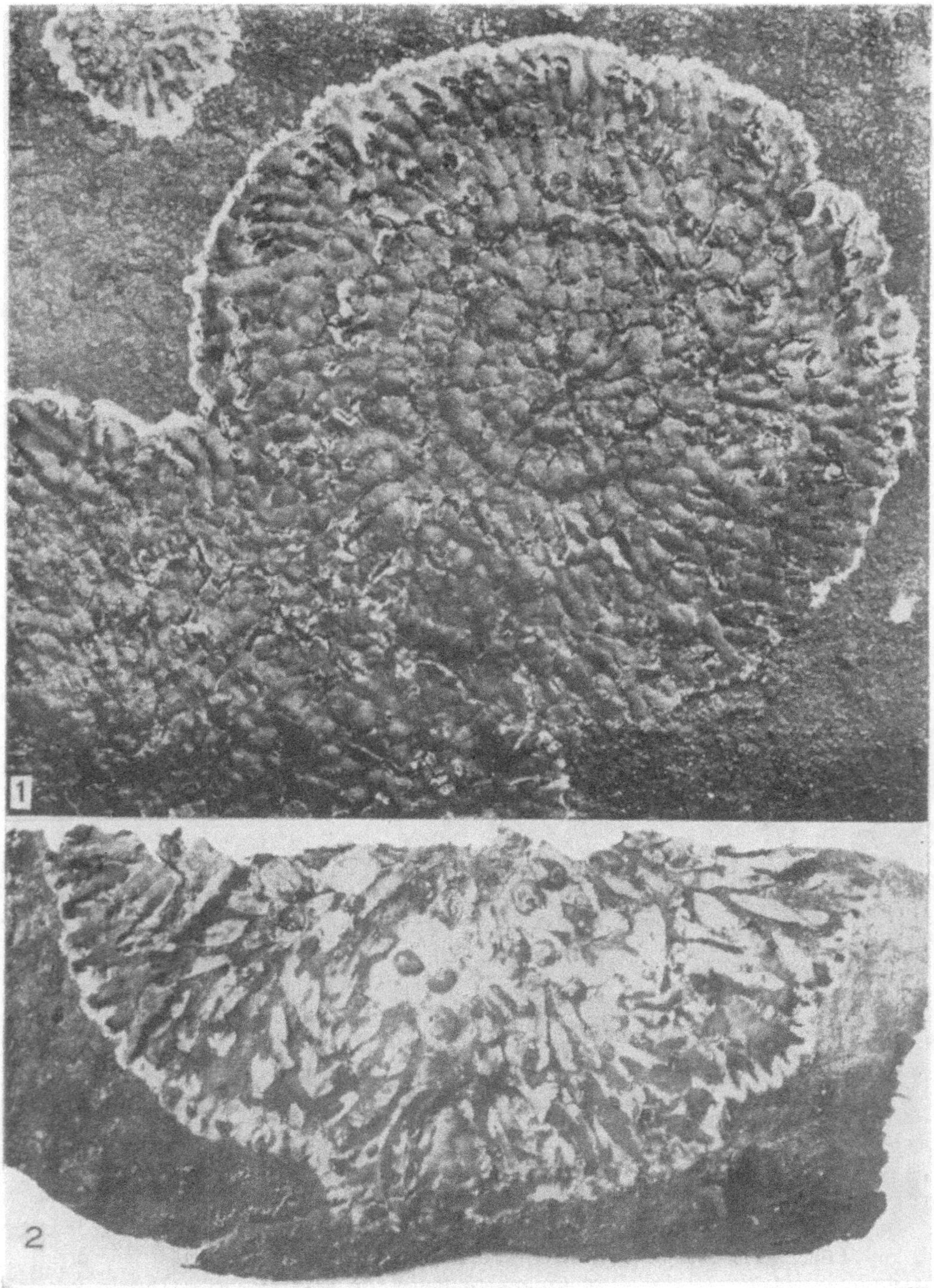

S. Burtii

Fig. 1, surface view showing openings to tunnels, radiating ridges and depressions, × 4. 2, Top layer removed exposing numerous healthy, non-parasitized scale insects, × about 11.

PLATE 70

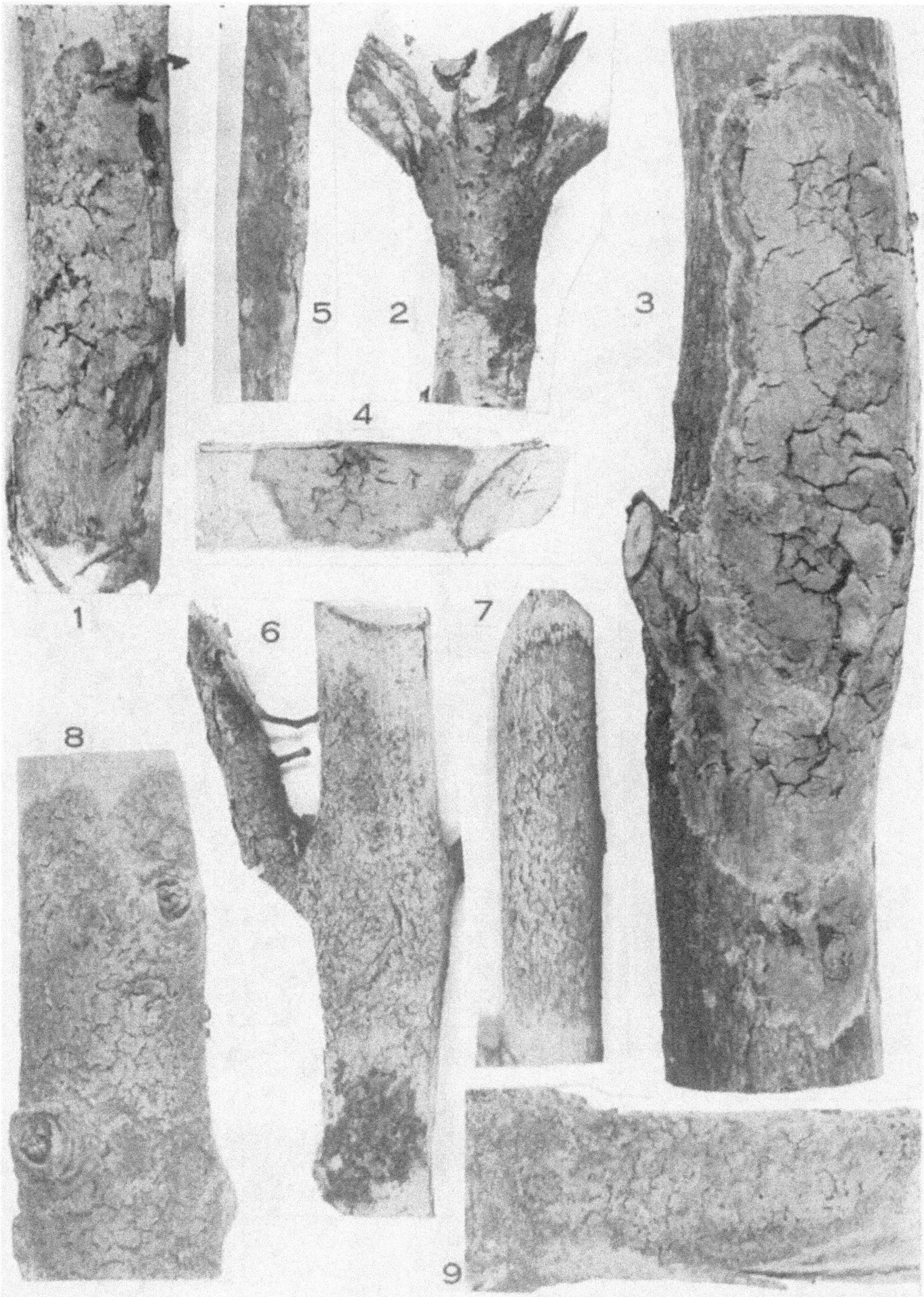

Figs. 1–5, *S. alni*; 1, 2 on oak; 3, on *Hicoria pecan* enlarged about one-third; 4, on oak; 5, type. 6, 7, *S. alni* var. *squamosum* on oak; in 6 note *S. castaneum* on lower end of stick. 8, 9, *S. Leprieurii* on *Magnolia virginiana*.

Figs. 1–5, *S. castaneum*; 1, type, slightly enlarged; 2, × about ½. 6, 7, *S. Leprieurii* on *Magnolia virginiana*; in fig. 7 note whitish lichen.

PLATE 72

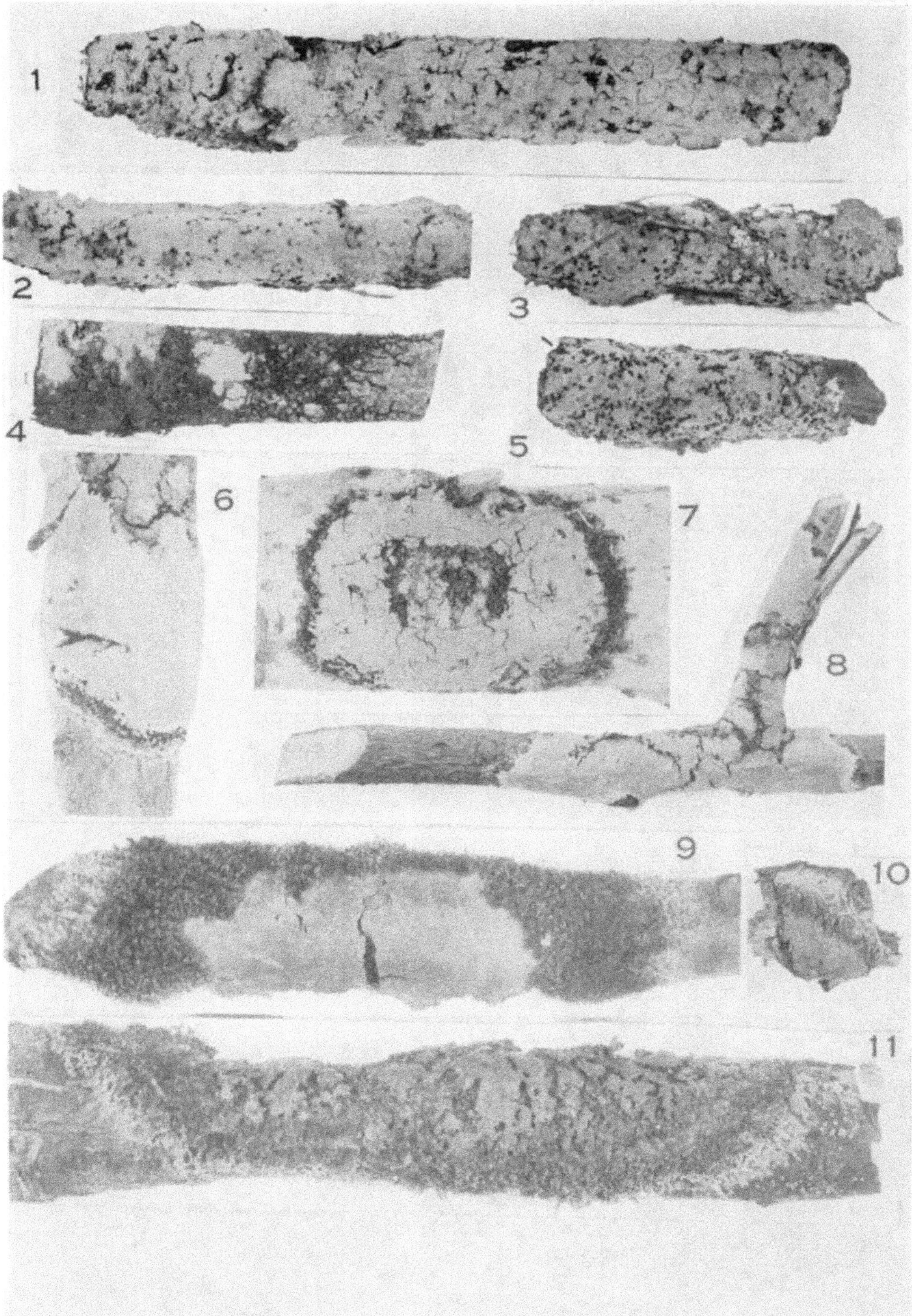

Figs. 1–5, *S. fumigatum.* 6–11, *S. Mariani*; 10, type; 6, 8, on *Pyrus japonica*, spring condition; 9, 11, fall condition. In fig. 11 note young scale insects covered by round white pellicle, these hatched beneath fungus, crawling out to settle down on bark. 7, on *Liquidambar*. 9, 11, × about $2\frac{1}{2}$.

PLATE 73

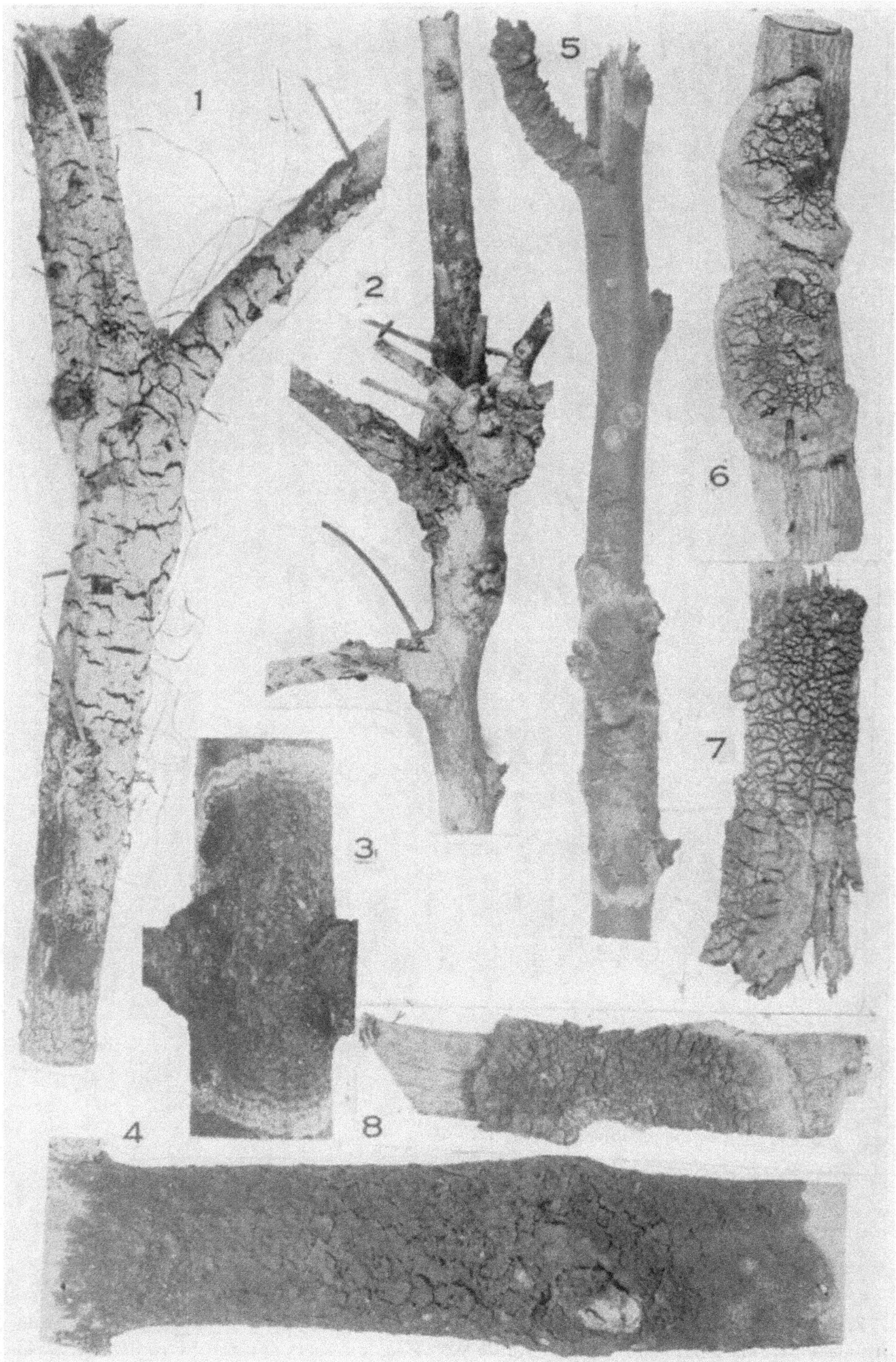

Fig. 1, *S. grandisporum*, × ½. 2–4, *S. Curtisii*. 5, *S. Burtii*, variety with nearly smooth surface, × 2. 6–8, *S. Sydowii*, type.

PLATE 74

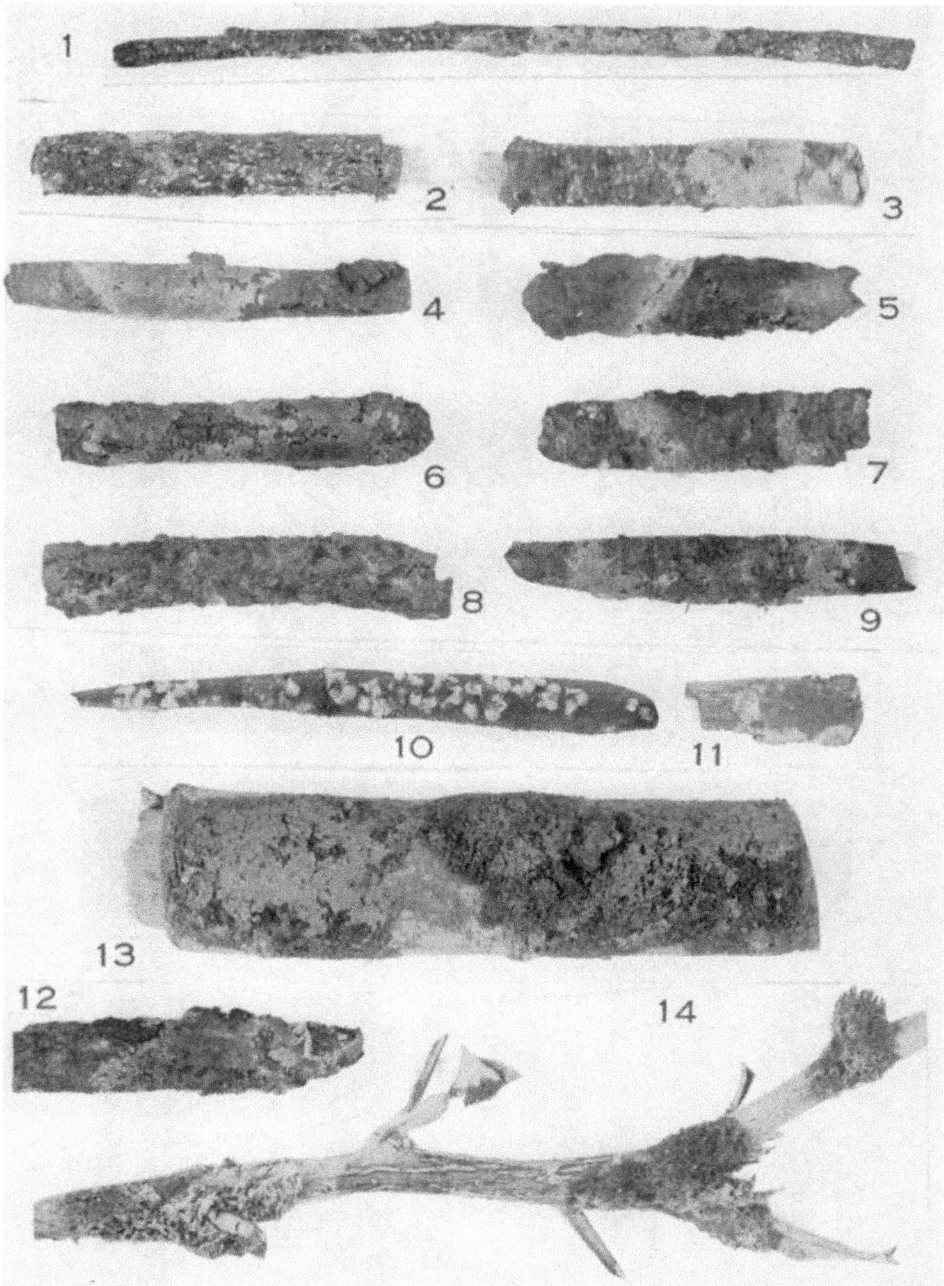

Figs. 1–12, *S. Carestianum*; 1, on *Cornus sanguinea*, France; 2, 3, on *Salix incana*, leg. Otto Jaap; 11, type, others from Canada. 13, 14, *S. lepidosaphis*; 13, material from Cuba; 14, on *Citrus* from Florida.

PLATE 75

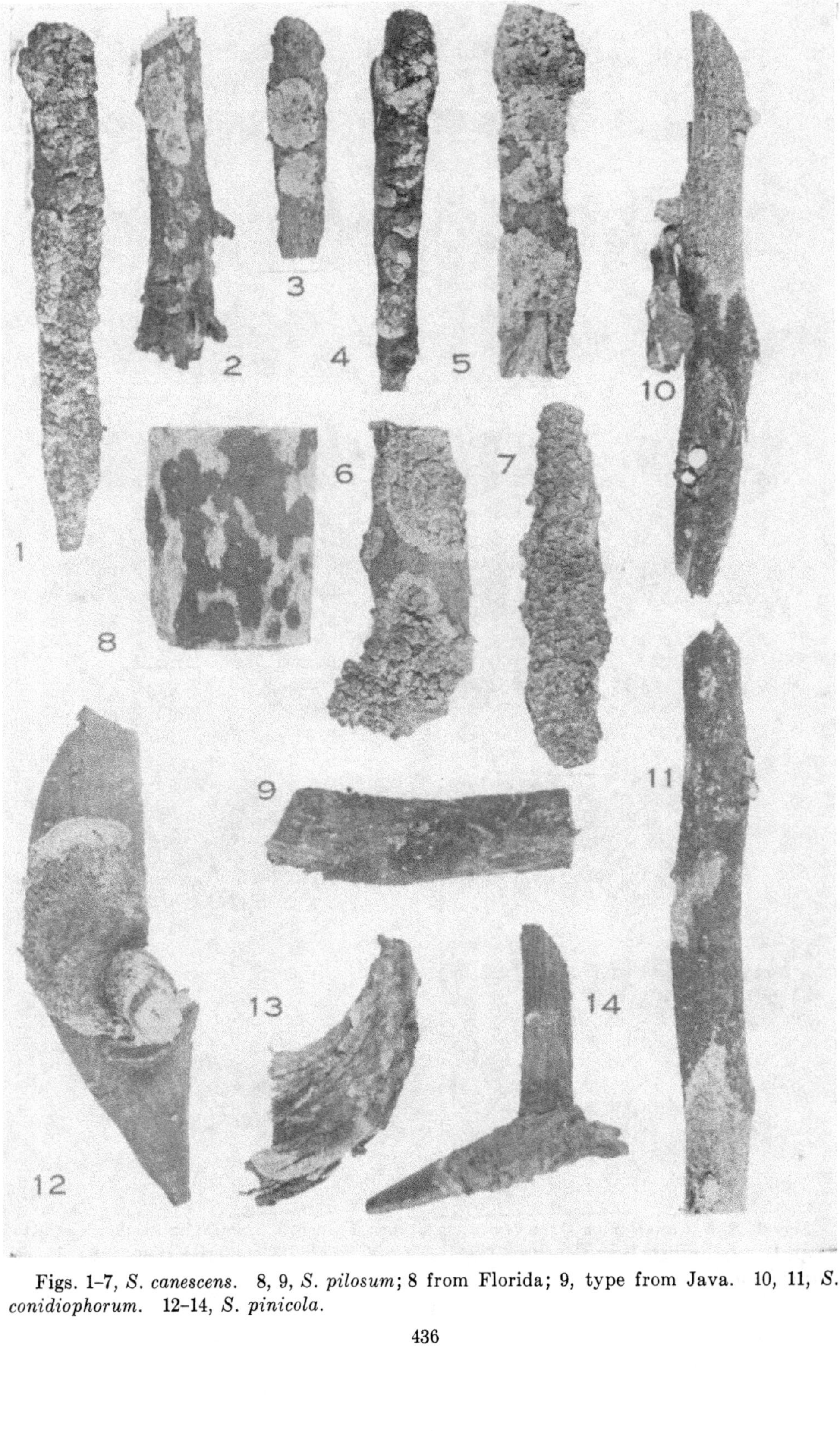

Figs. 1–7, *S. canescens*. 8, 9, *S. pilosum*; 8 from Florida; 9, type from Java. 10, 11, *S. conidiophorum*. 12–14, *S. pinicola*.

PLATE 76

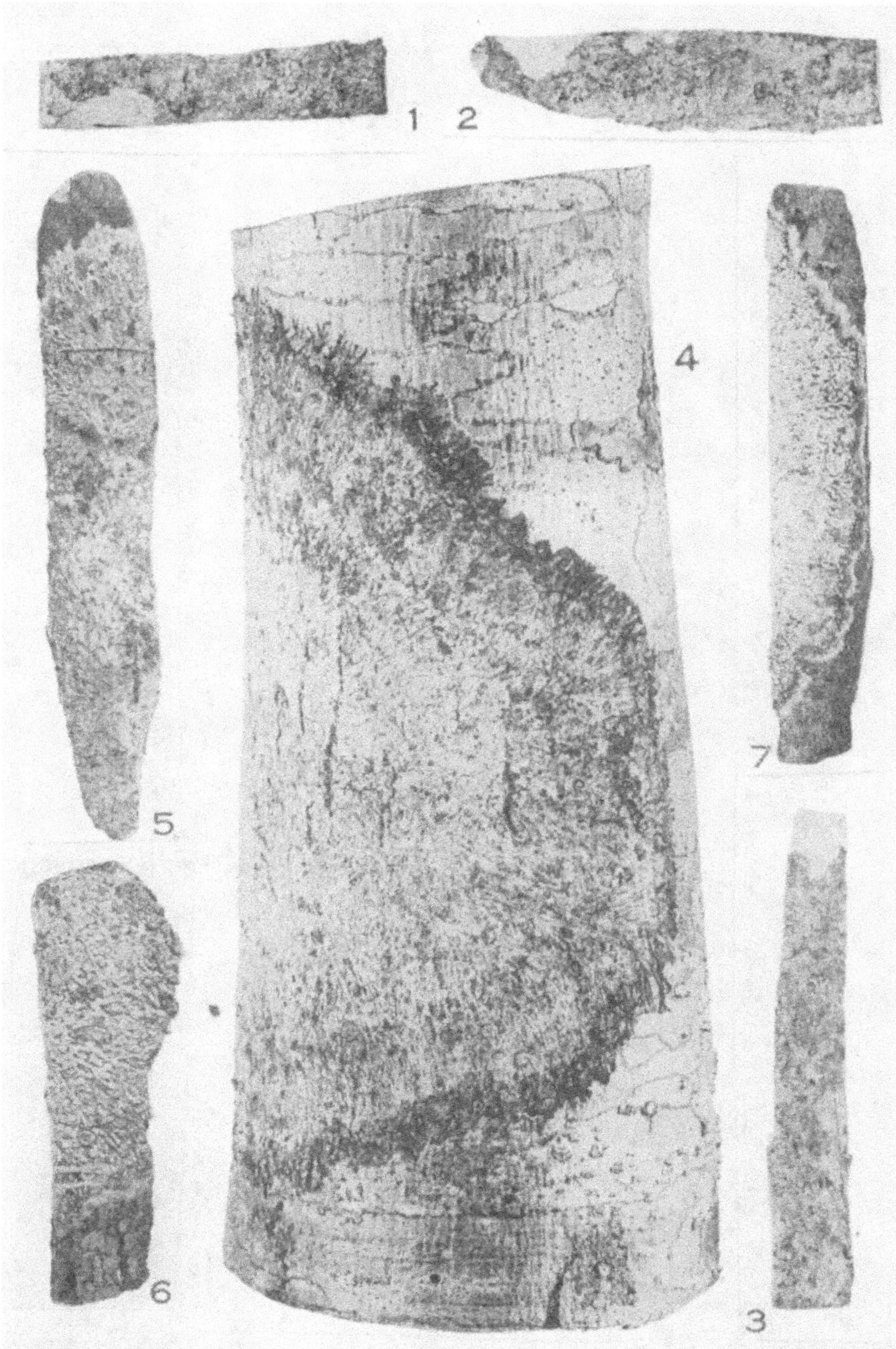

Figs. 1–3, *Septobasidium* sp., close to *S. Langloisii*. 4, *S. Langloisii* on *Magnolia virginiana* from Florida, × about $\frac{3}{4}$. 5–7, *S. sinuosum*.

PLATE 77

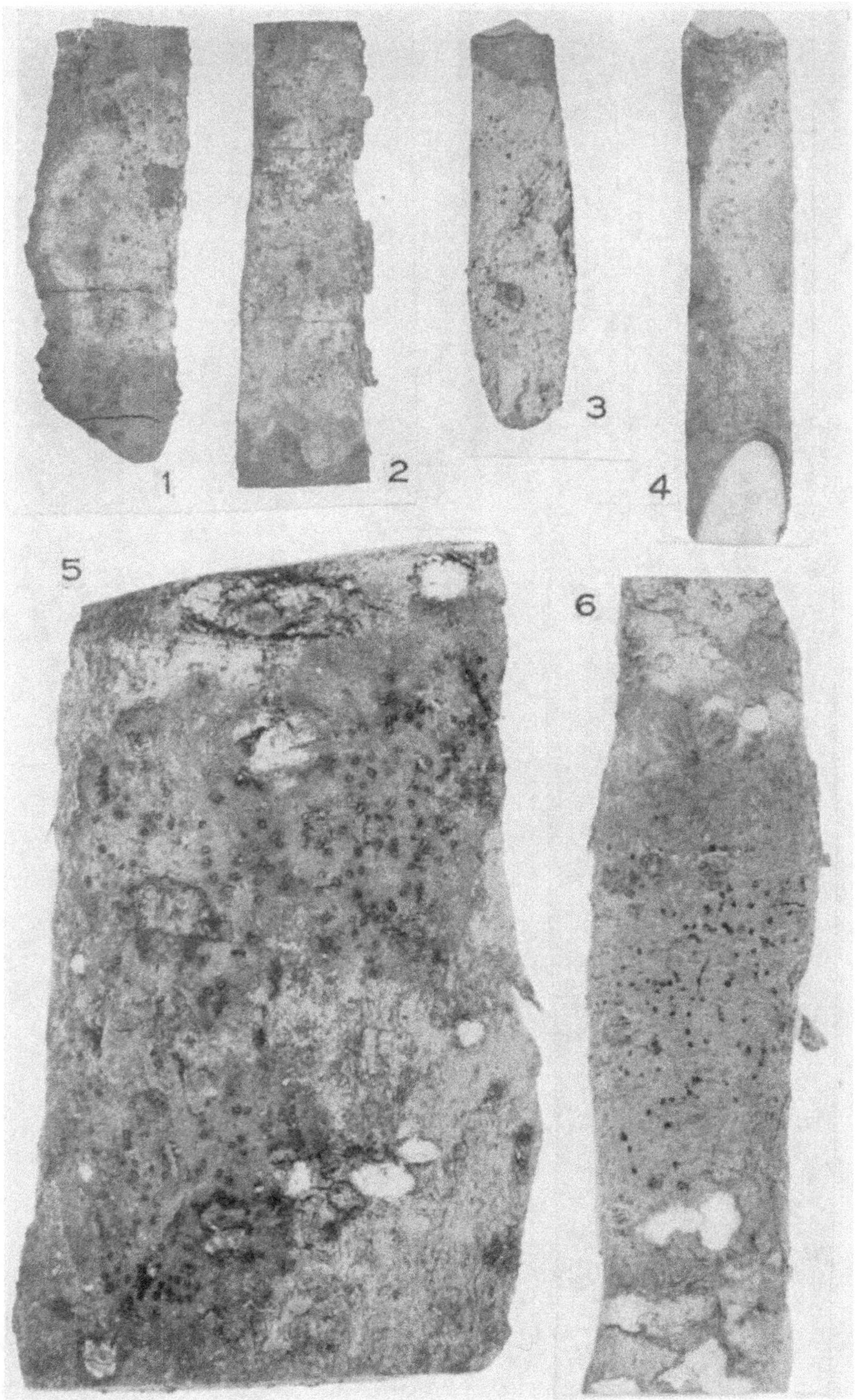

Figs. 1–4, *S. tenue*, black dots are elevated mounds each one covering a scale insect. 5, 6, *S. rugulosum*; 5, × about 4; 6, × about 2; in both the black dots are depressed areas in each of which is a scale insect.

PLATE 78

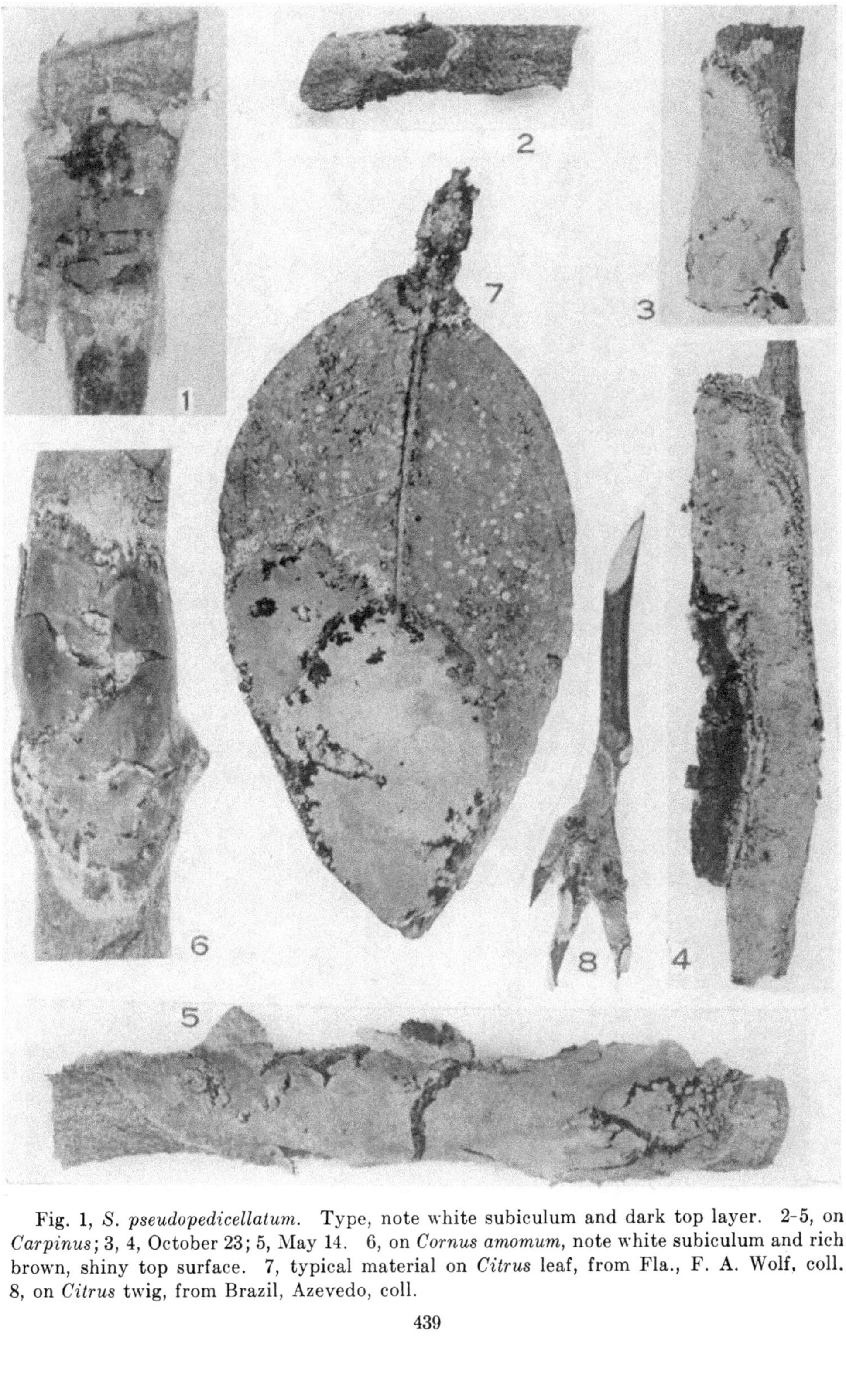

Fig. 1, *S. pseudopedicellatum.* Type, note white subiculum and dark top layer. 2–5, on *Carpinus*; 3, 4, October 23; 5, May 14. 6, on *Cornus amomum*, note white subiculum and rich brown, shiny top surface. 7, typical material on *Citrus* leaf, from Fla., F. A. Wolf, coll. 8, on *Citrus* twig, from Brazil, Azevedo, coll.

PLATE 79

Figs. 1–3, *S. pseudopedicellatum.* Showing wood injury, note *S. Schweinitzii* on lower right side of piece to left.

PLATE 80

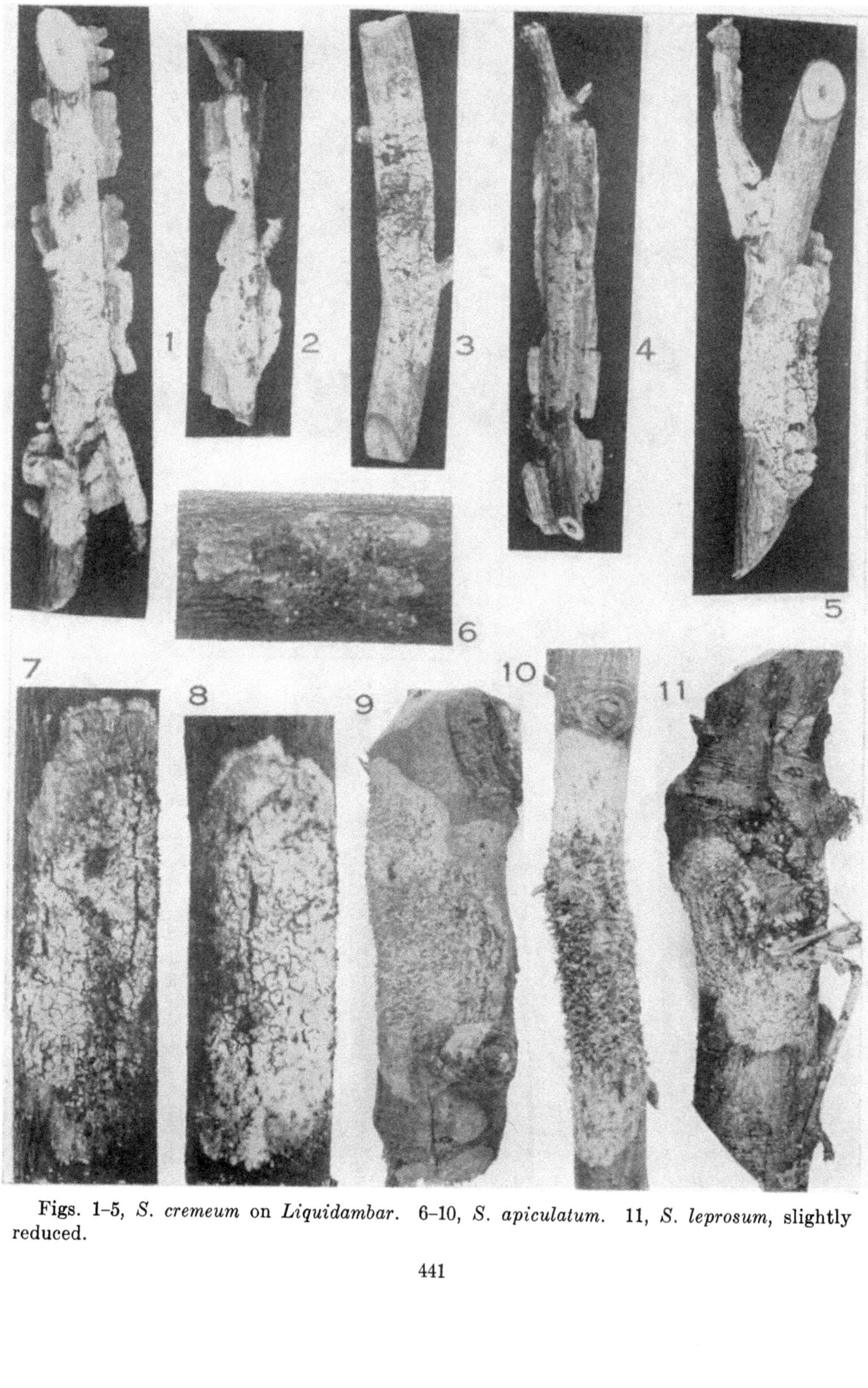

Figs. 1–5, *S. cremeum* on *Liquidambar*. 6–10, *S. apiculatum*. 11, *S. leprosum*, slightly reduced.

PLATE 81

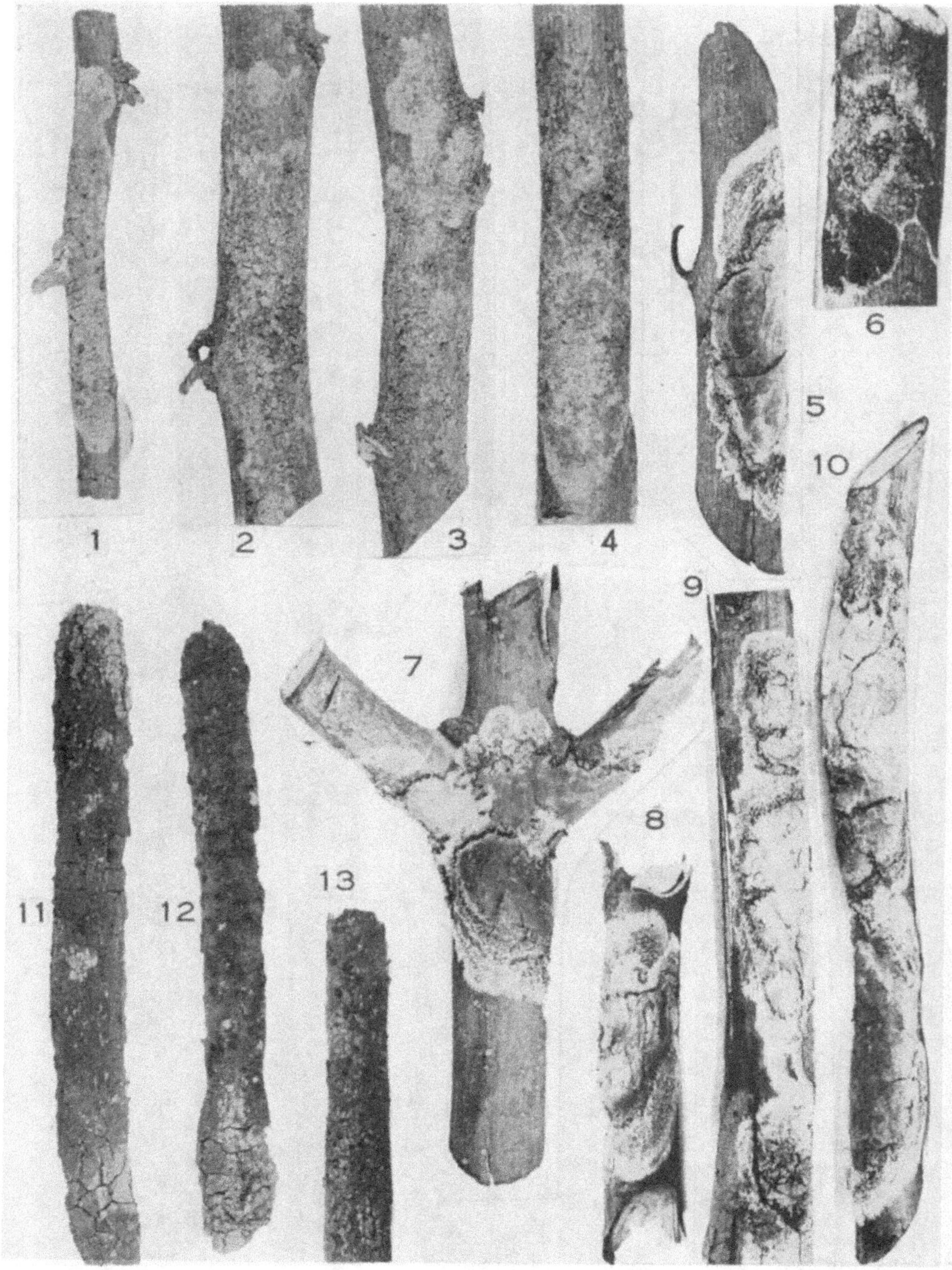

Figs. 1–4, *S. Hesleri.* 5–10, *Septobasidium* hybrid species between *S. pseudopedicellatum* and *S. Schweinitzii*; fig. 6 also shows some *S. Patouillardii.* 11–13, *S. fuscum*; note dark elevated mounds in 12 and 13.

PLATE 82

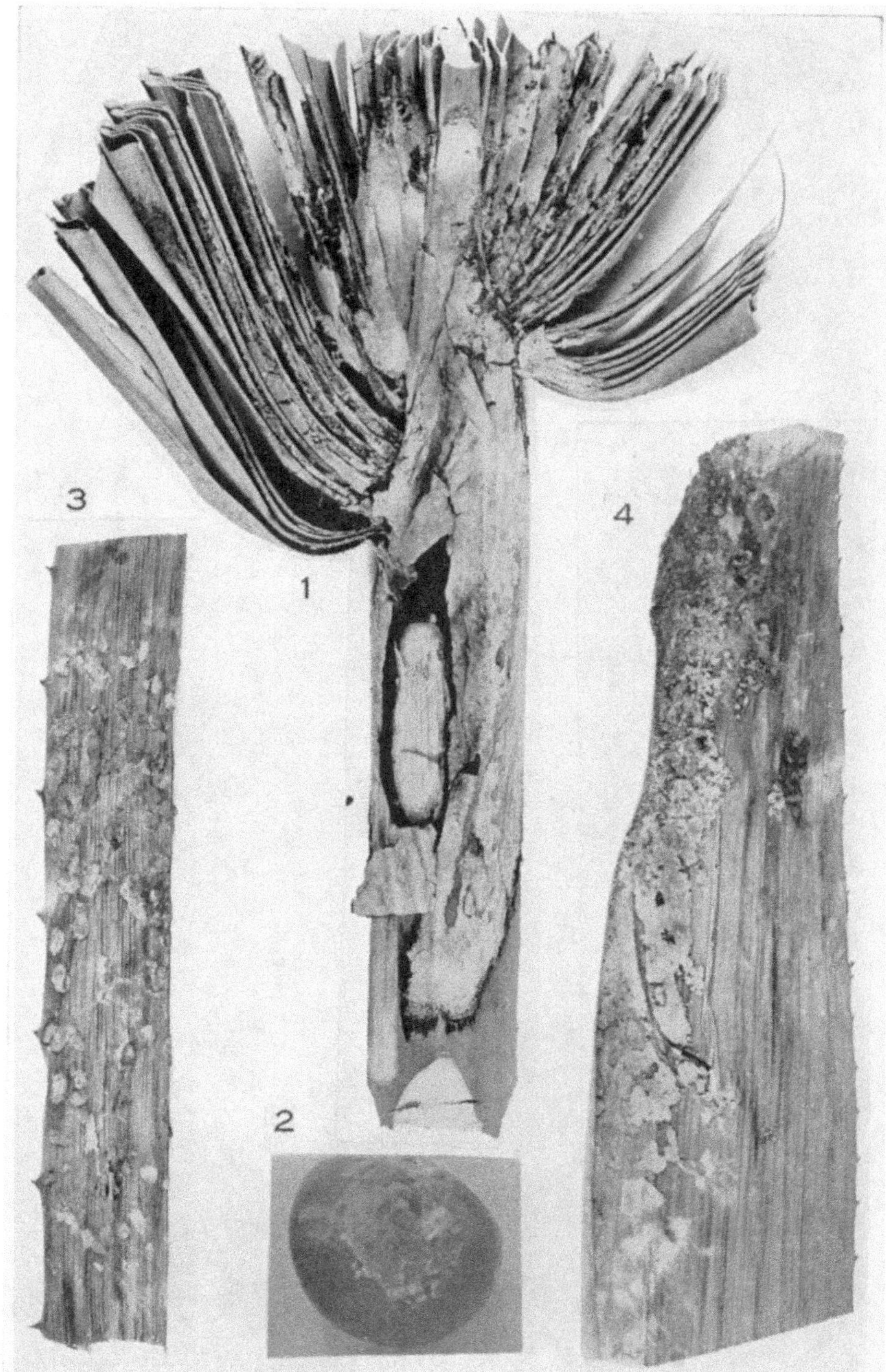

Fig. 1, *S. sabalis*, from Louisiana, Brown, coll., type. 2, *S. sabal-minor*, from Florida, Bomhard, coll., type. × about 3. 3, 4, *S. Westoni*, on pineapple leaf from Canal Zone, Weston, coll., type.

PLATE 83

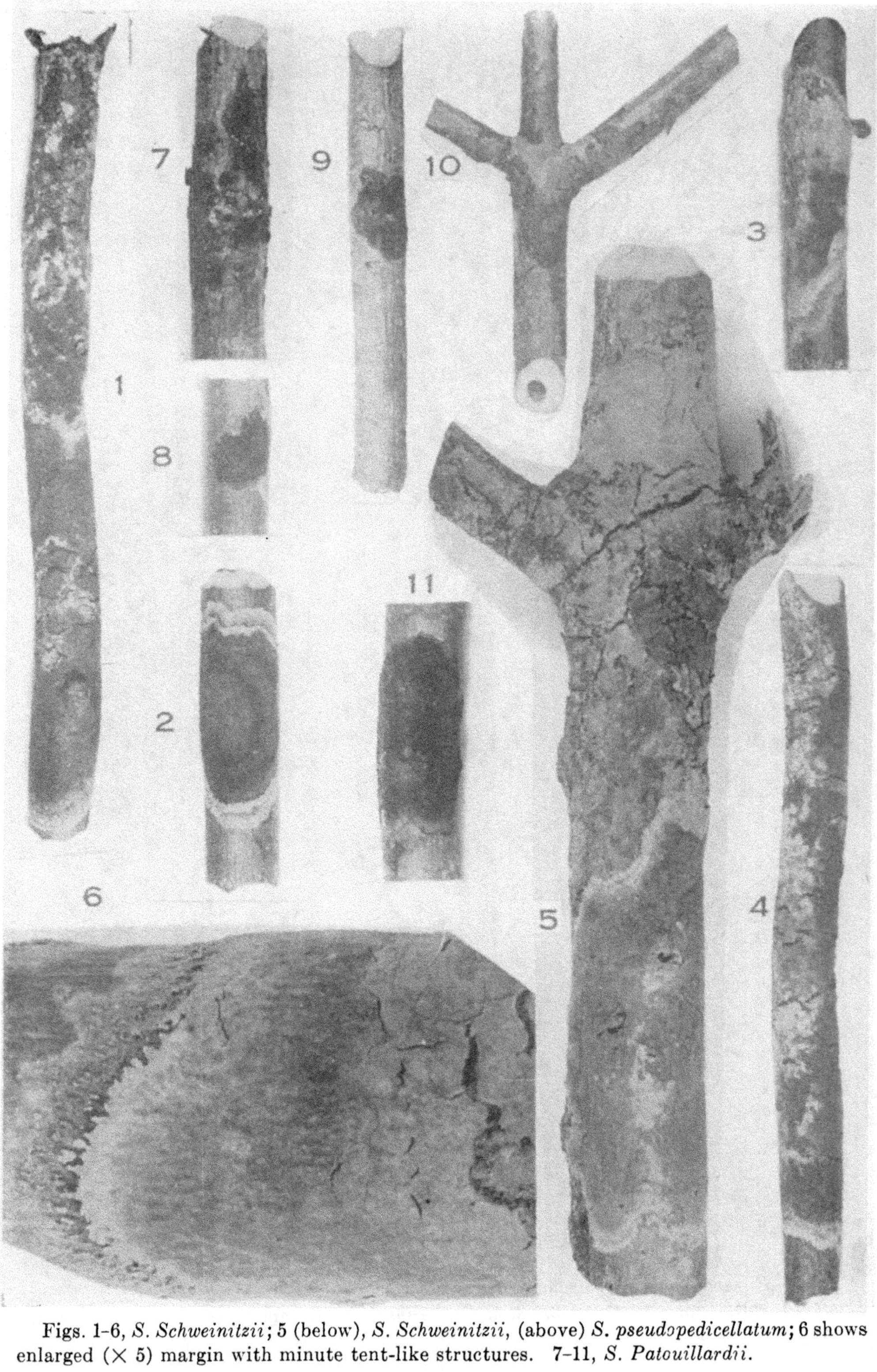

Figs. 1-6, *S. Schweinitzii*; 5 (below), *S. Schweinitzii*, (above) *S. pseudopedicellatum*; 6 shows enlarged (× 5) margin with minute tent-like structures. 7-11, *S. Patouillardii*.

PLATE 84

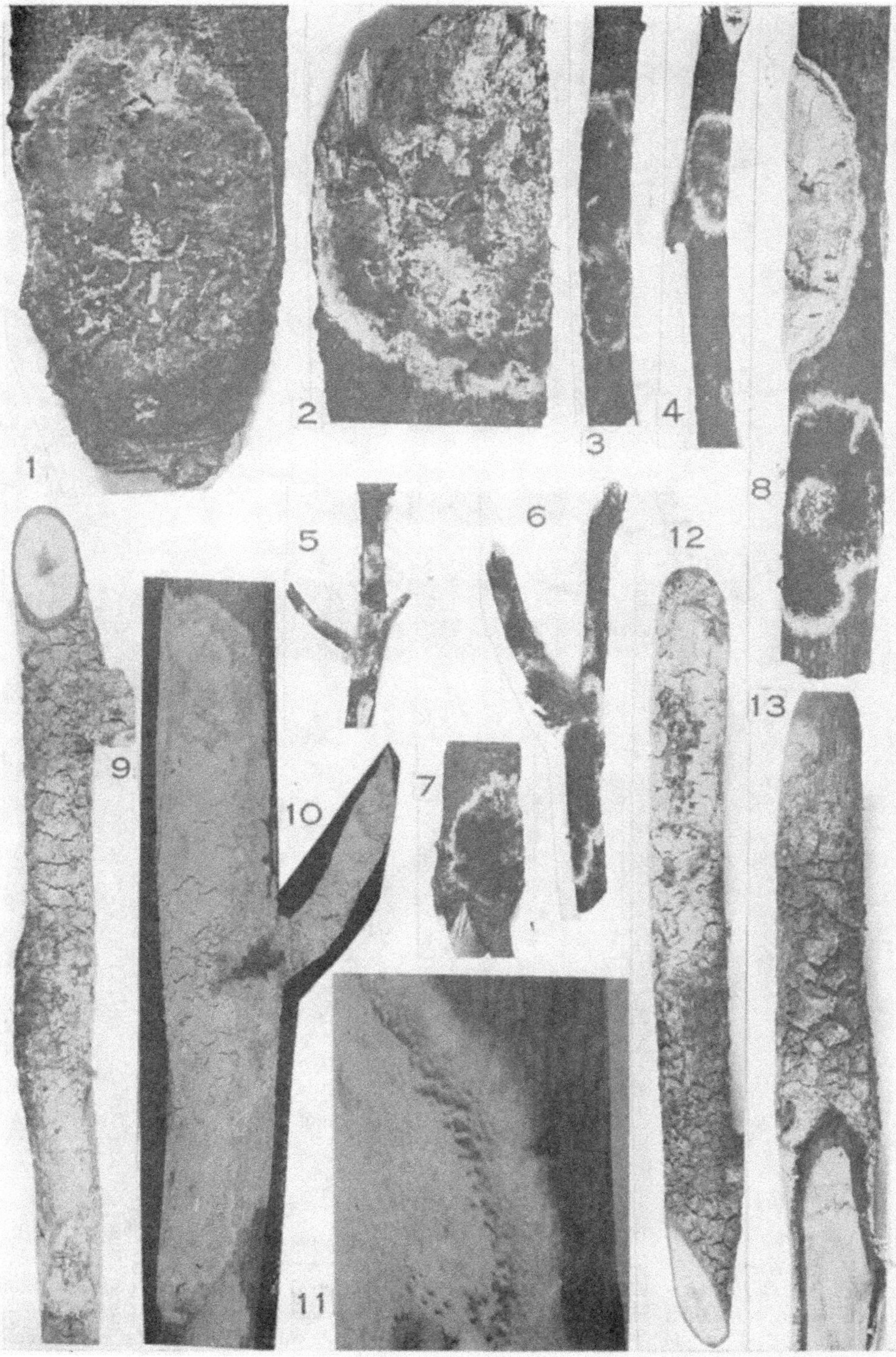

Figs. 1–8, *S. filiforme*; 8 (above) *S. Cokeri*, (below) *S. filiforme*. 9–13, *S. lilacinoalbum*; 11 shows margin; × about 4.

PLATE 85

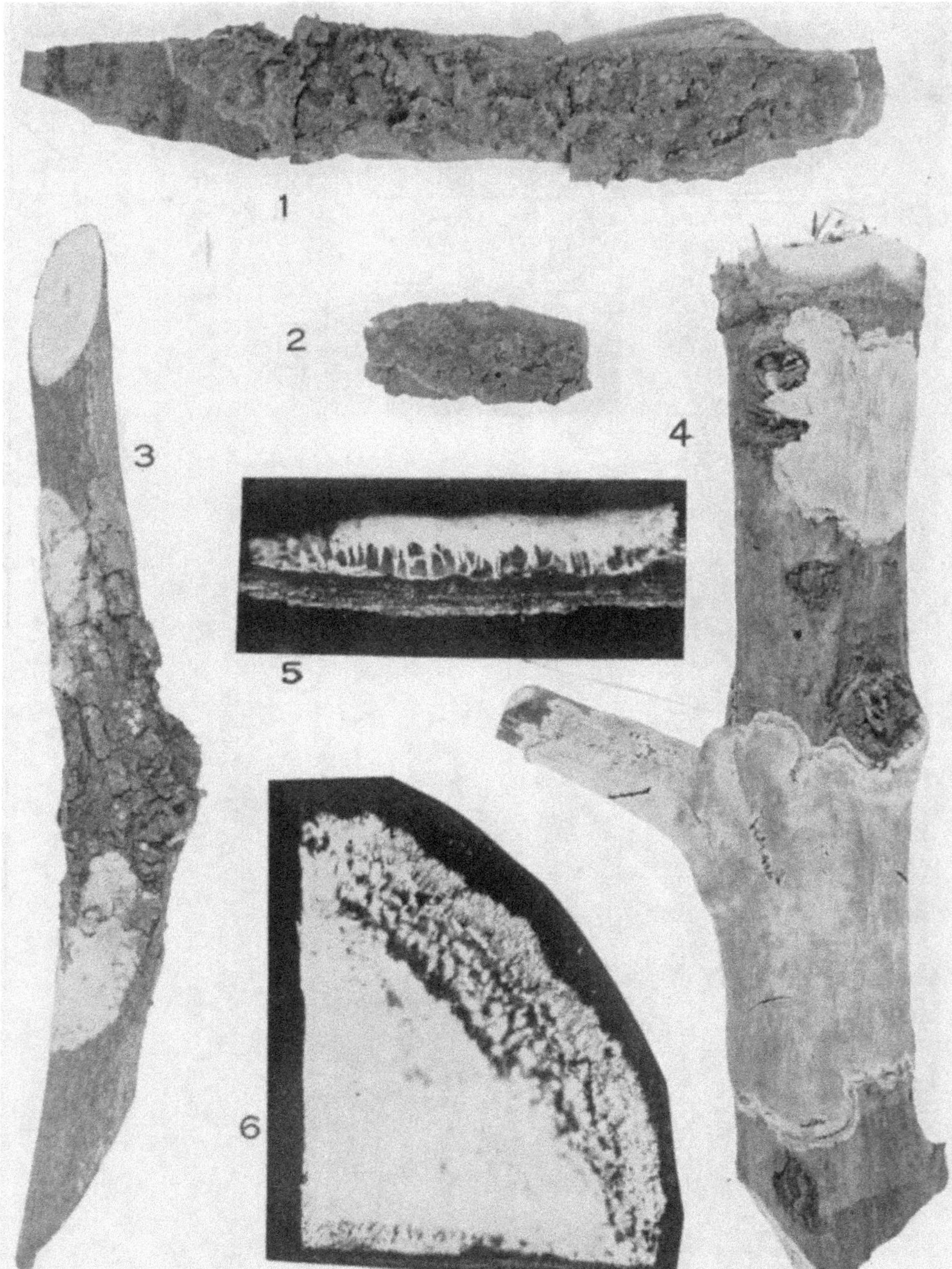

Figs. 1–2, *S. Peckii*, × 1.5. 3–6, *S. Cokeri*; fig. 3 shows wood injury on dogwood; 5, section view showing subiculum, pillars, and top layer; 6, surface view of part of same specimen as shown in 5; 5, 6, × about 10, both photographed in fresh condition, note pure whiteness throughout.

PLATE 86

Septobasidium Curtisii. Two pieces on left on *Fraxinus.* Photographed July 18, 1928, by A. B. Couch, × 1. The half-moon shape of the growth is due to the fact that one-half of it was cut away the preceding year. Note whitish margin, new growth zone for season 1928. Two pieces on right on *Nyssa*, photographed Feb. 1933, slightly reduced, showing large cracks in bark. Margin of fungus concolorous with other parts during dormant season.

PLATE 87

Septobasidium septobasidioides

Note dark clumps of pillars on piece to left and second from right; under each is a scale insect. Whitish hymenium is turned up in places. Photograph by P. F. Shope. × 1.

PLATE 88

Septobasidium jamaicaense (above)
Note the distinct rhizomorphs.
Septobasidium areolatum (below)
Note areolations. The grayish areas are hymenium. Photograph by Paul F. Shope, × 1.

PLATE 89

Septobasidium purpureum (above and two pieces on left below)
Septobasidium rhabarbarinum (three pieces on right below)
Photograph by Paul F. Shope, × 1

PLATE 90

Septobasidium atropunctum

Note the minute dark dots on the piece to right near bottom and next piece near top. Photograph by Paul F. Shope, × 1.

PLATE 91

Septobasidium alveolatum (top, middle, and right)

Note the irregularly honeycomb appearance of the piece to the right. The grayish areas are hymenium.

Septobasidium arboreum (four pieces on left below)

Photograph by Paul F. Shope, × 1

PLATE 92

Figs. 1, 2, *S. fusco-cinereum*; 1, type locality, n. s.; 2, type, × 3. 3–5, *S. Clelandii*; 3, n. s.; 4, 5, × 2. 6, 7, *S. retiforme*, type; 6, × 3; 7, × 2. Photograph, fig. 7, from D. H. Linder.

PLATE 93

Fig. 1, *S. rimulosum*, n. s. 2, *S. protractum*, type, n. s. 3, *S. dictyodes*, type, × 1.9. 4, *S. griseopurpureum*, × 1.5. 5, 6, *S. cinnamomeum*, type, × 3. 7, *S. stratosum*, × 1.6.

PLATE 94

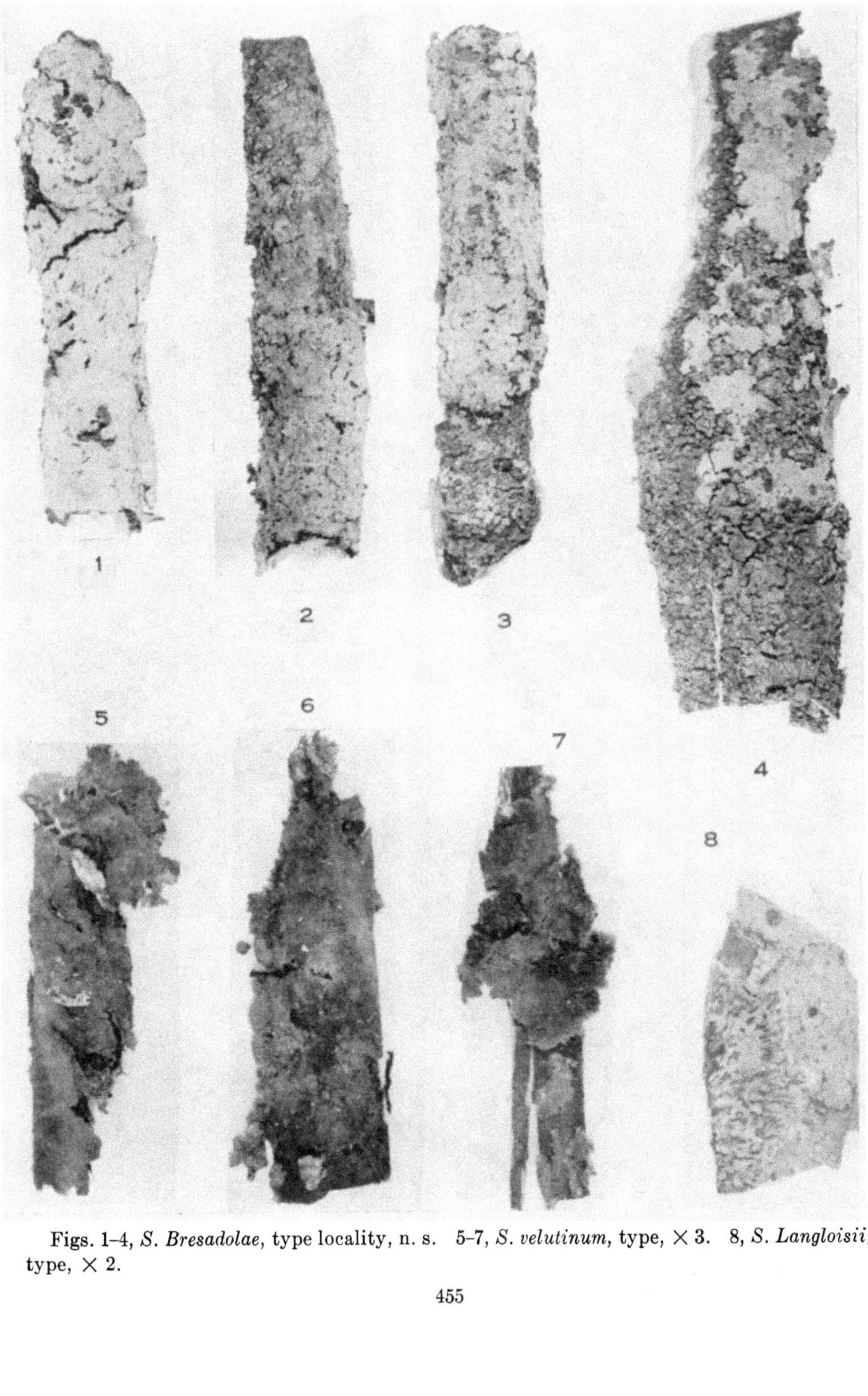

Figs. 1–4, *S. Bresadolae*, type locality, n. s. 5–7, *S. velutinum*, type, × 3. 8, *S. Langloisii*, type, × 2.

PLATE 95

Fig. 1, *S. subolivaceum*, type, n. s. 2–4, *S. Petchii*; 2, type; 3, 4, atypical forms; all × 2. 5, *S. natalense*, n. s.

PLATE 96

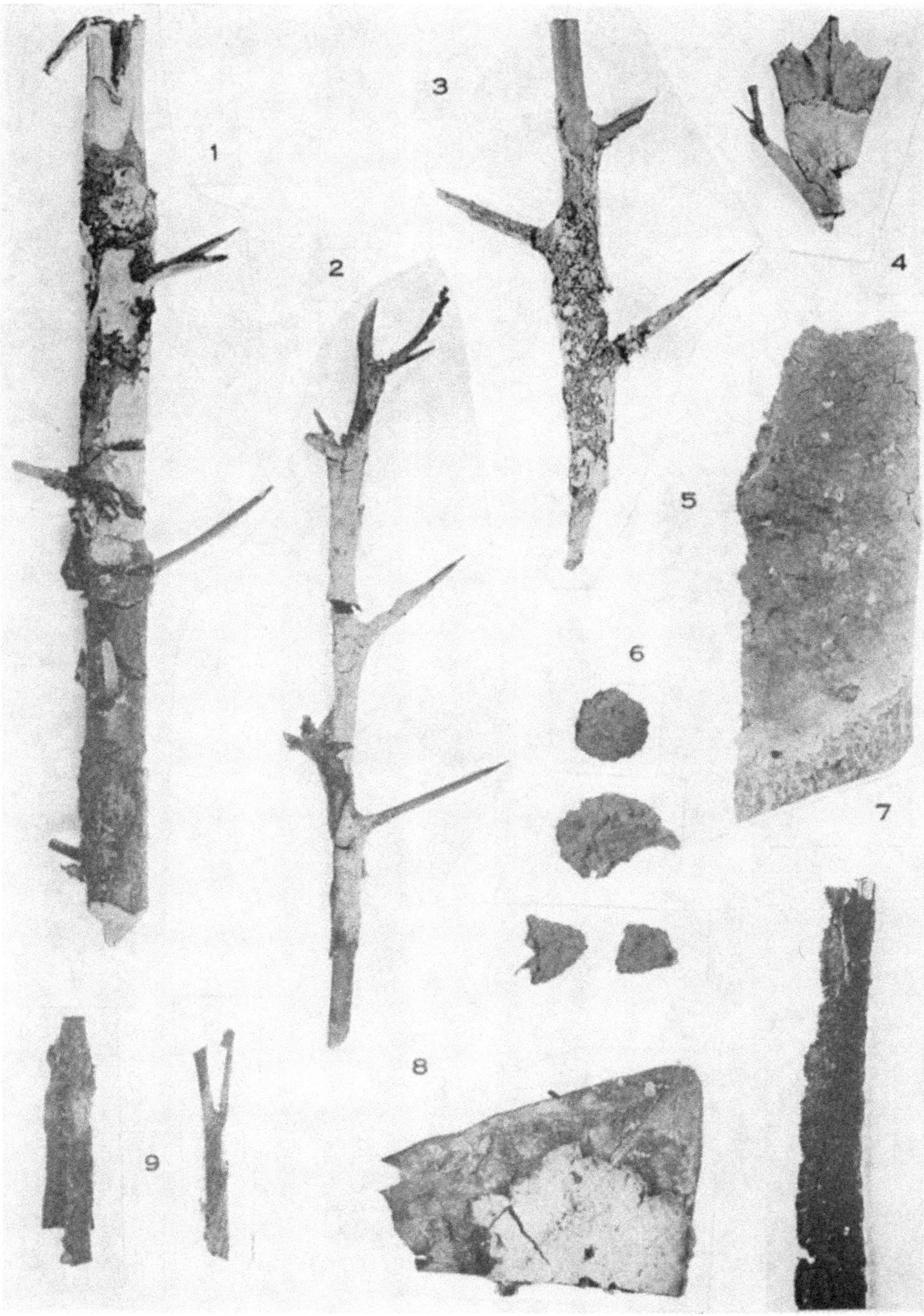

Figs. 1–4, *S. Reinkingii*; 4, type, others type locality. 5, *S. rhabarbarinum* from Petch, Ceylon, n. s. 6, *S. rhabarbarinum*. 7, *S. granulosum*, type, n. s. 8, 9, *S. aligerum*, from Petch, n. s.

PLATE 97

Fig. 1, *S. siparium*, type, × 1.25. 2, *S. cervicolor*, n. s. 3, *S. acaciae*, type locality, n. s. 4, *S. lilacinum*, type, n. s. 5, *S. Stevensoni*, n. s. 6, 7, *S. hakgalanum*; 6, × 2; 7, margin of same, × 8.

PLATE 98

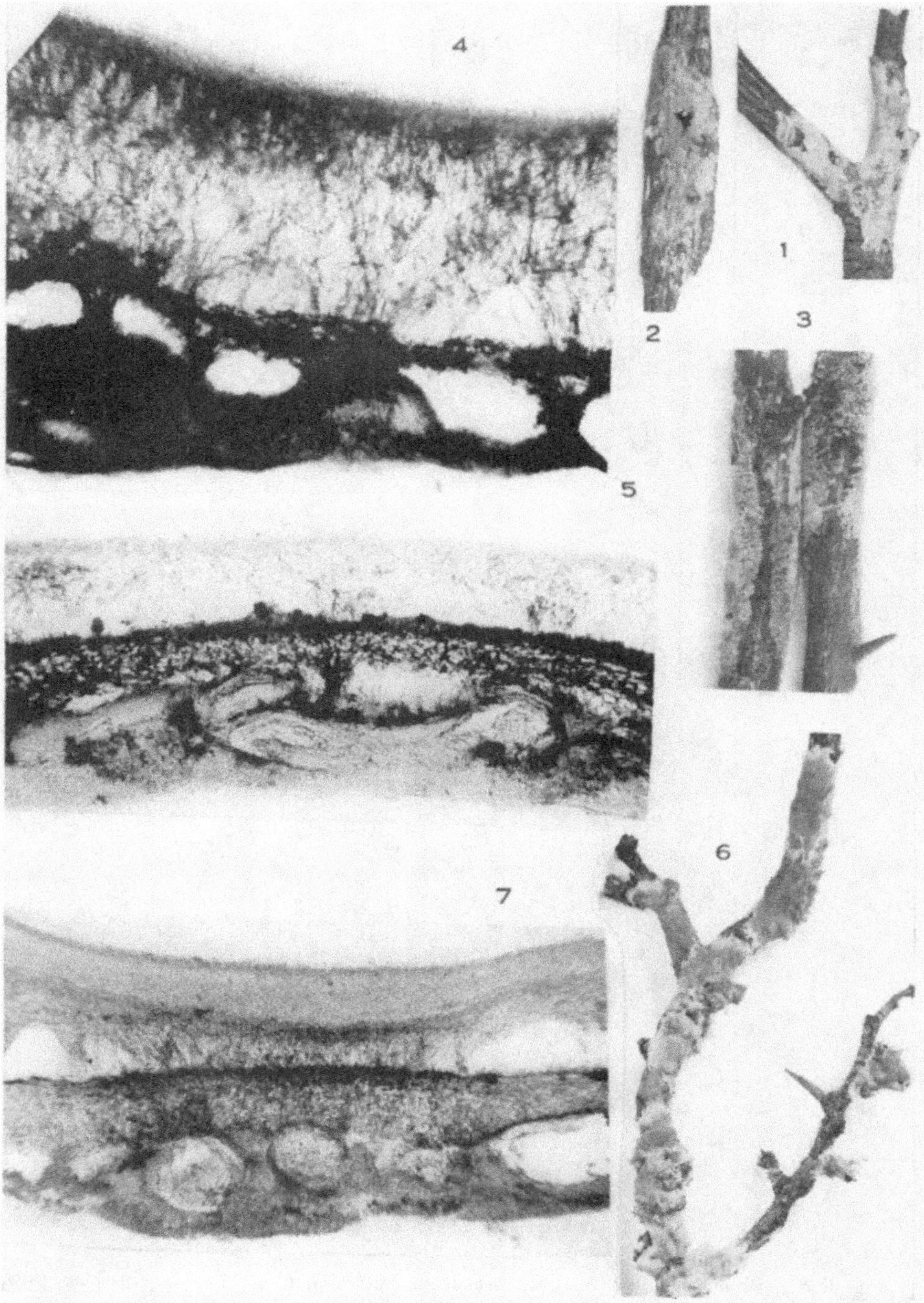

Figs. 1–5, *S. alveomarginatum*; 4, 5, sections of context, × 50; others n. s. 6, 7, *S. grandispinosum*; 6 n. s.; 7, section of context, × 50.

PLATE 99

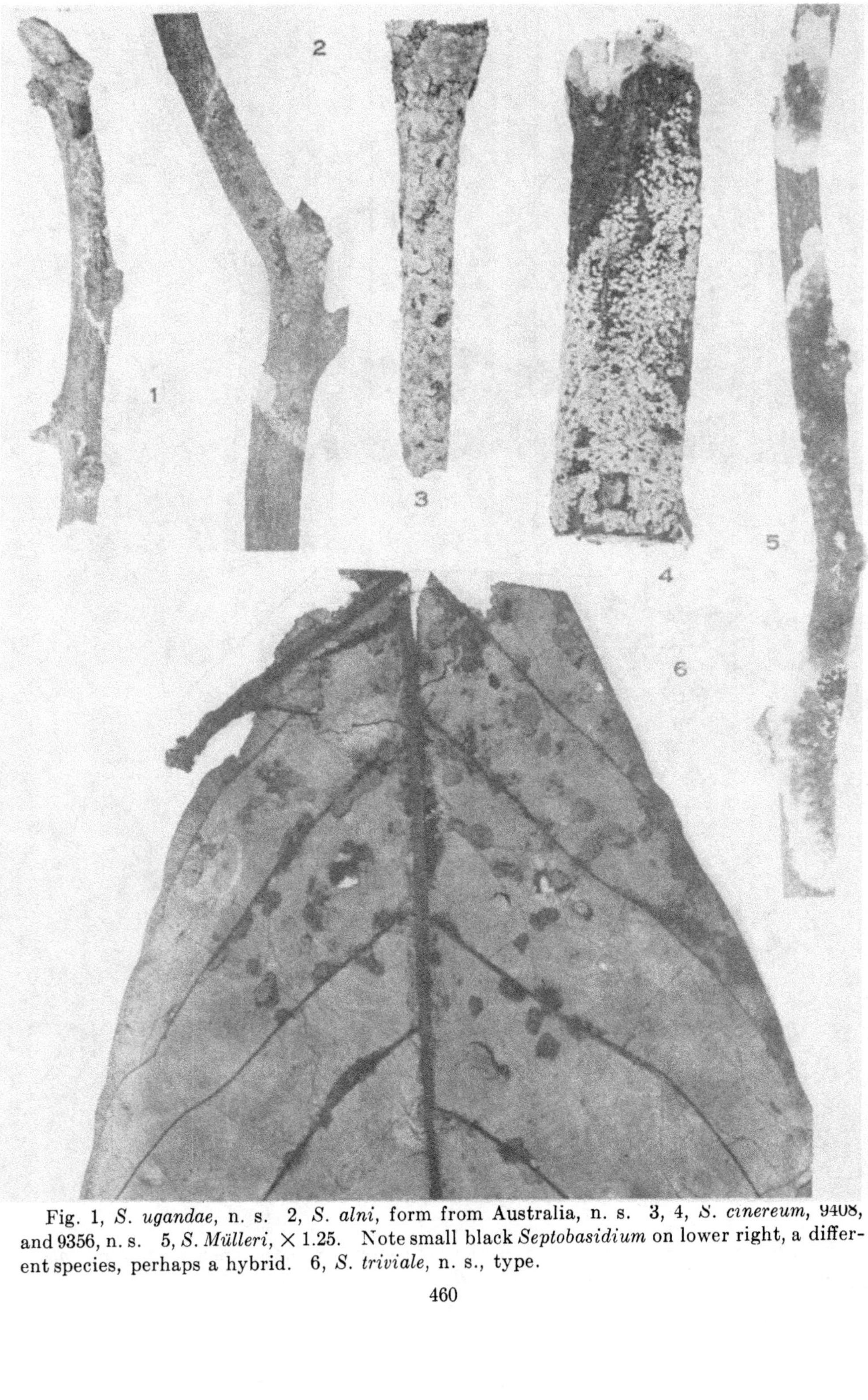

Fig. 1, *S. ugandae*, n. s. 2, *S. alni*, form from Australia, n. s. 3, 4, *S. cinereum*, 9408, and 9356, n. s. 5, *S. Mülleri*, × 1.25. Note small black *Septobasidium* on lower right, a different species, perhaps a hybrid. 6, *S. triviale*, n. s., type.

PLATE 100

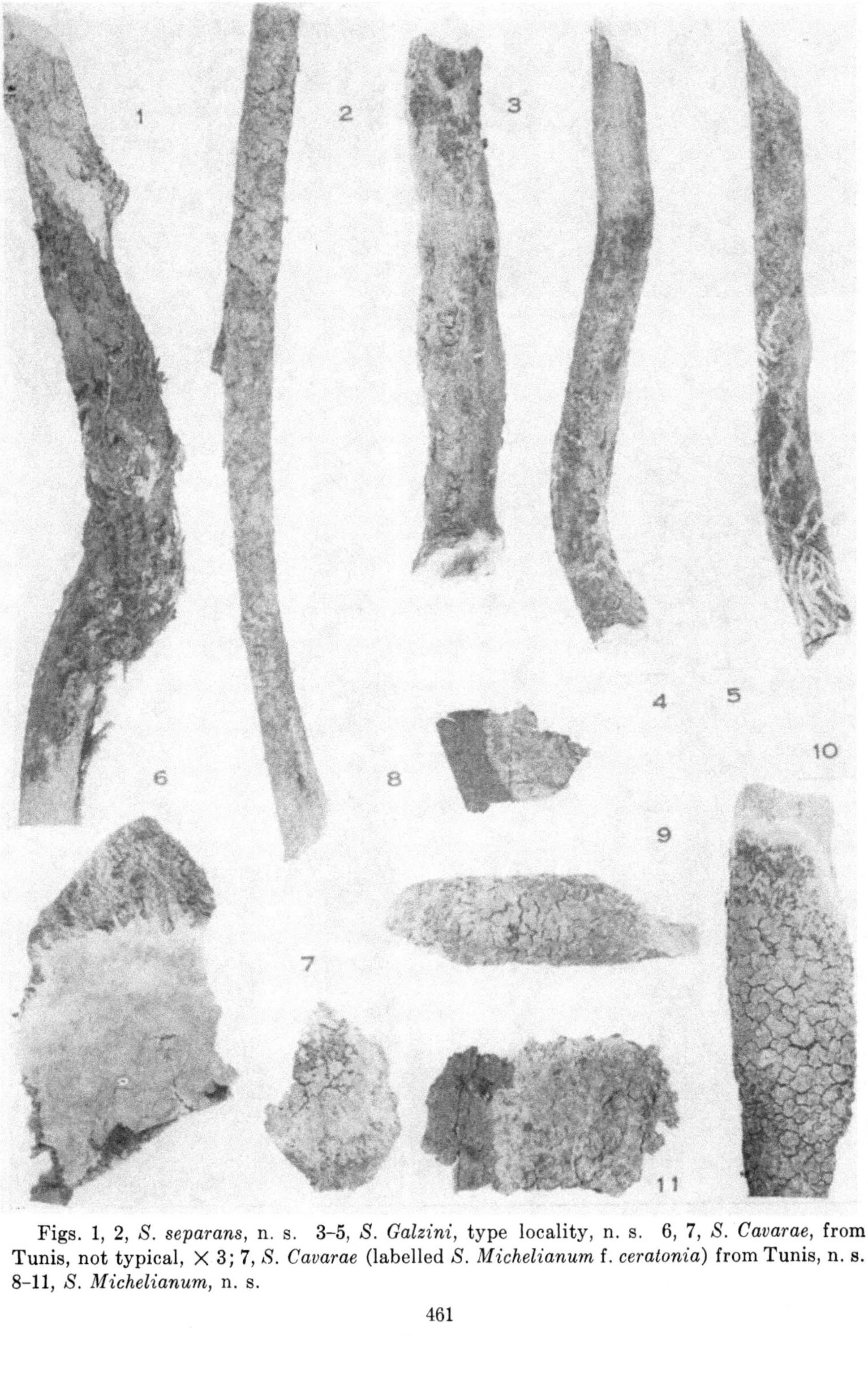

Figs. 1, 2, *S. separans*, n. s. 3–5, *S. Galzini*, type locality, n. s. 6, 7, *S. Cavarae*, from Tunis, not typical, × 3; 7, *S. Cavarae* (labelled *S. Michelianum* f. *ceratonia*) from Tunis, n. s. 8–11, *S. Michelianum*, n. s.

PLATE 101

Figs. 1, 2, *S. suffultum*, type, 1, note margin, tangled pillars and top layer, × 1.5; 2, margin, × 10, showing rounded insect traps. 3–5, *S. suffultum* from Petch, note absence of top layer and elongate insect houses each protected by a dense tangle of hyphae; 3, × 1.5; 4, × 4; 5, × 10. 6, *S. atratum*, type, no. 89, × 2.5. 7, *S. obscurum*, type, n. s.

PLATE 102

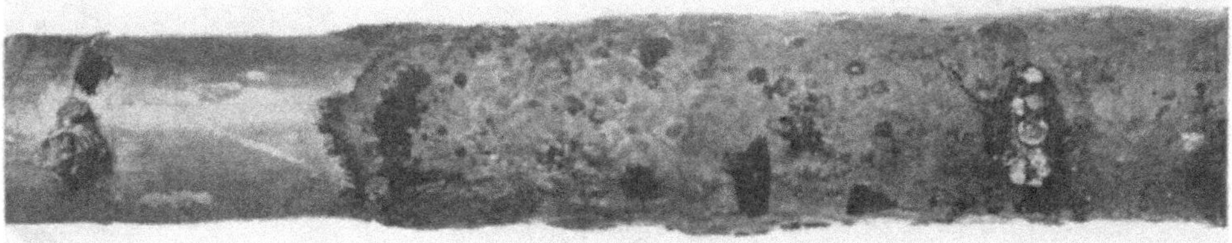

Figs. 1, 2, *S. pachydermum*, type, both photographs from Royal Botanic Gardens, Kew; above, half original specimen, × about ⅓; below, margin, × about 2.

PLATE 103

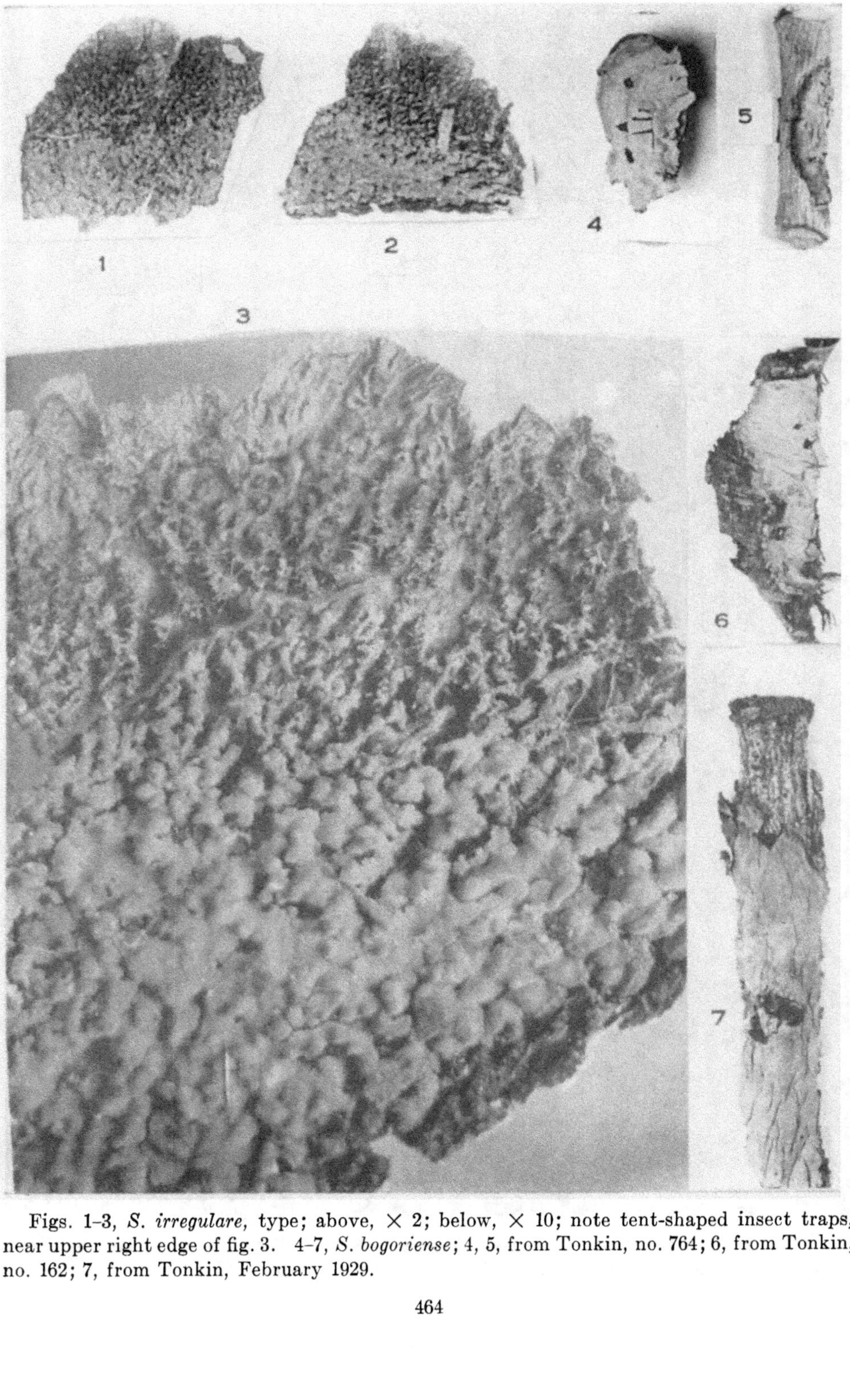

Figs. 1–3, *S. irregulare*, type; above, × 2; below, × 10; note tent-shaped insect traps, near upper right edge of fig. 3. 4–7, *S. bogoriense*; 4, 5, from Tonkin, no. 764; 6, from Tonkin, no. 162; 7, from Tonkin, February 1929.

PLATE 104

Figs. 1, 2, *S. sinense*, × 2. 3, *S. humile*, type, × about 2. 4, *S. Thwaitesii*, from Petch, × 1.5. 5, *S. neglectum*, type, n. s. 6, *S. proliferum*, type, note small tufts of pillars, n. s. 7–9, *S. rubiginosum*; 9, type; 7, 8, ×2.5; 9, ×3.5. 10–12, *S. bogoriense*; 10, Rands, coll., Java, n. s.; 11, F. W. Smith, coll., West Indies, as *T. pedicellata*, × 1.5. 12, Steinmann, coll., Java, n. s.

PLATE 105

Fig. 1, *S. citricolum*, type locality, × 1.2. 2, *S. Bagliettoanum*, type locality, × 2. 3, *S. cupressi*, type, photograph by Grodsinsky, n. s. 4, 5, *S. minutulum*; 4, type; 5, *S. minutulum* (= *S. phyllophilum* Sydow); all n. s.

PLATE 106

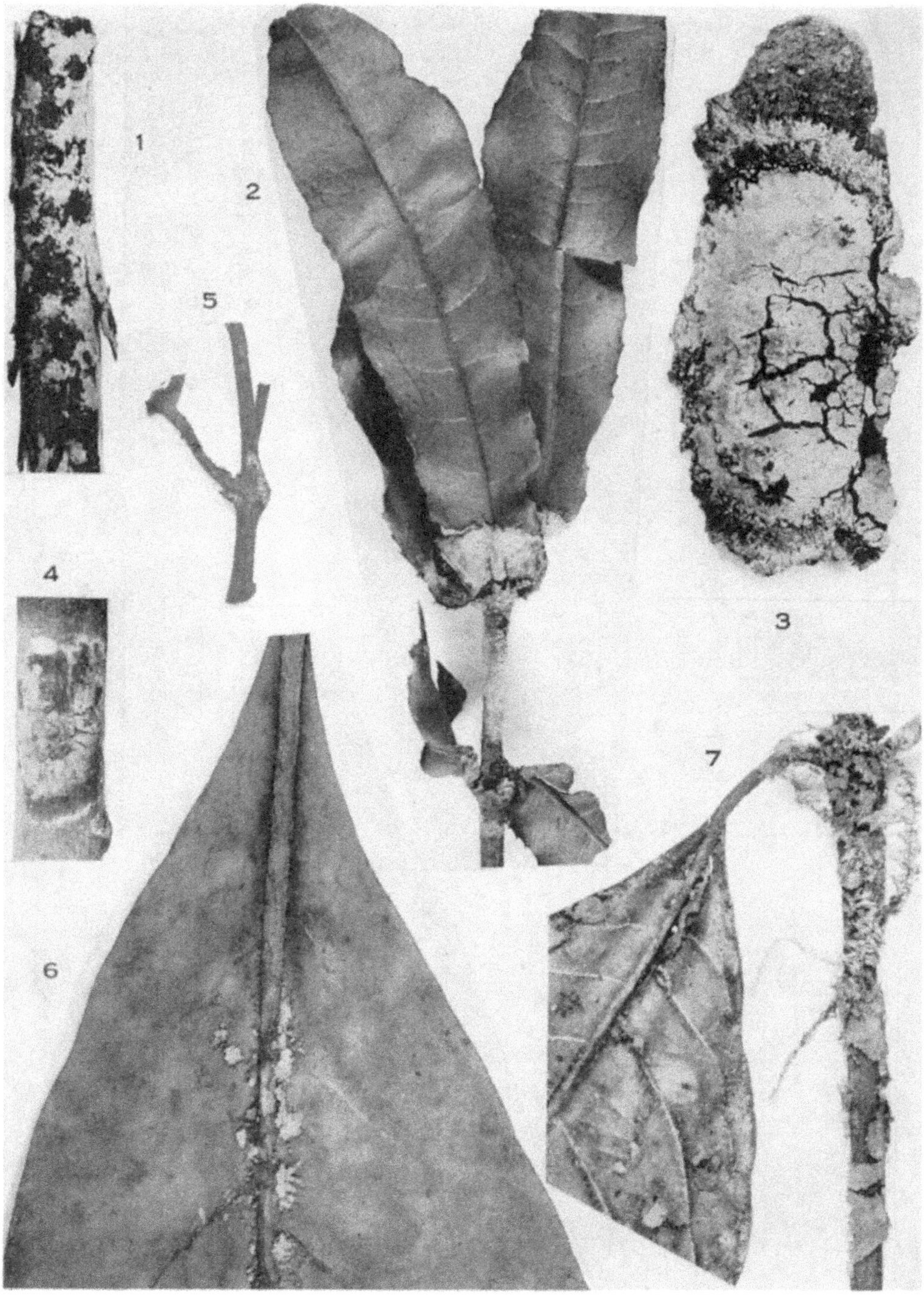

Fig. 1, *S. clethrae*, type, n. s. 2, *S. macadamiae*, type, n. s. 3, *S. tropicale*, type, × 1.5. 4, *S. mexicanum*, type, n. s. 5, *S. piperis*, type, n. s. 6, *S. Boedijnii*, type, n. s. 7, *S. Boedijnii* from Petch as *S. molliusculum*. × 1.5.

PLATE 107

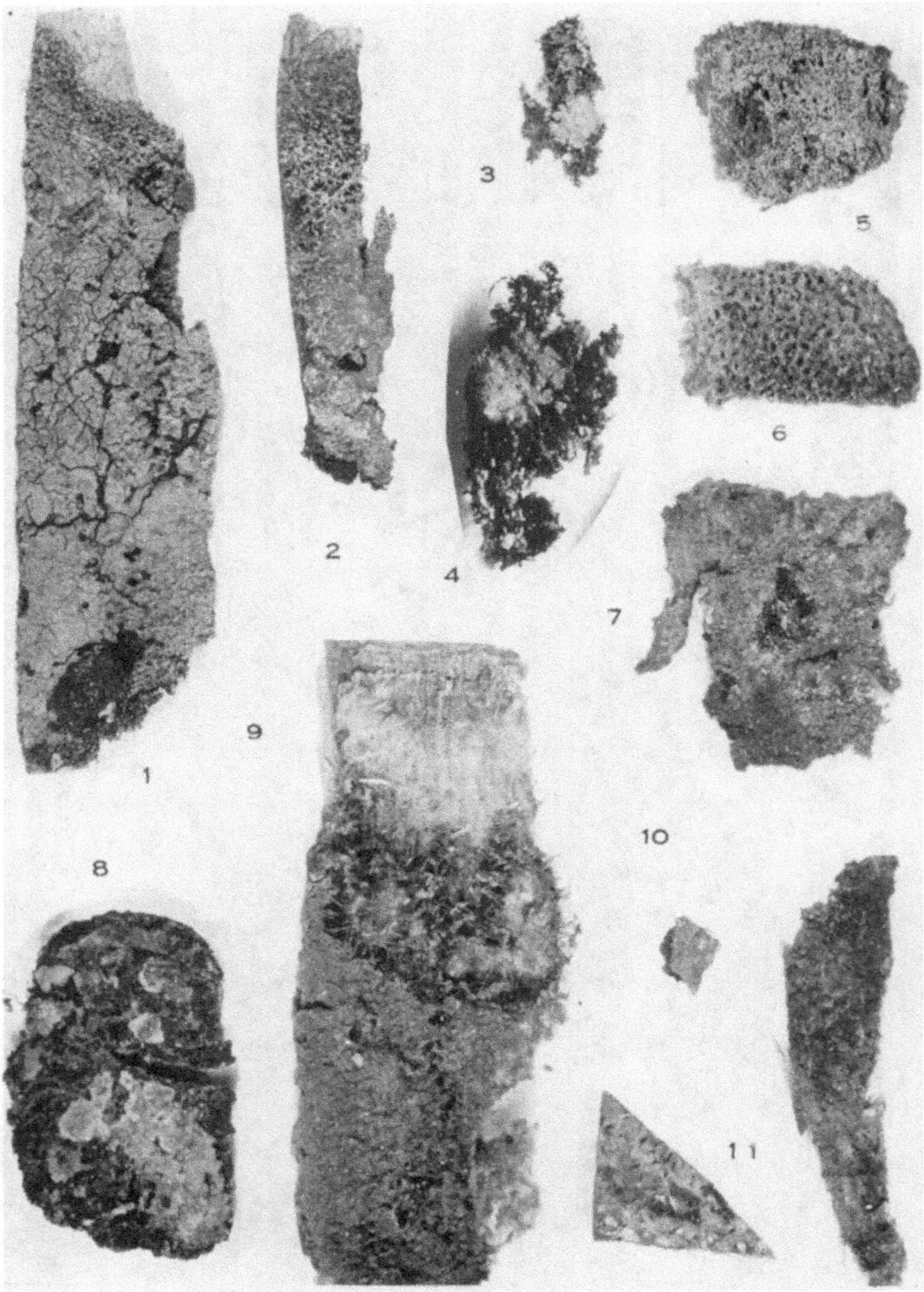

Figs. 1, 2, *S. pannosum*, 1, type, n.s.; 2, as *S. spongia* in Pat. Herb., × 2. 3, 4, *S. pedicellatum*, type; above, × 3; below, × 6. 5–7, *S. gossypinum*, type. 6 shows marginal region, × 3. 8, *S. stratiferum*, type, n. s. 9, *S. laxum*, type, × 3. 10, *S. Merrillii*, type, n. s. 11, *S. fusco-violaceum*, type, × 3.

PLATE 108

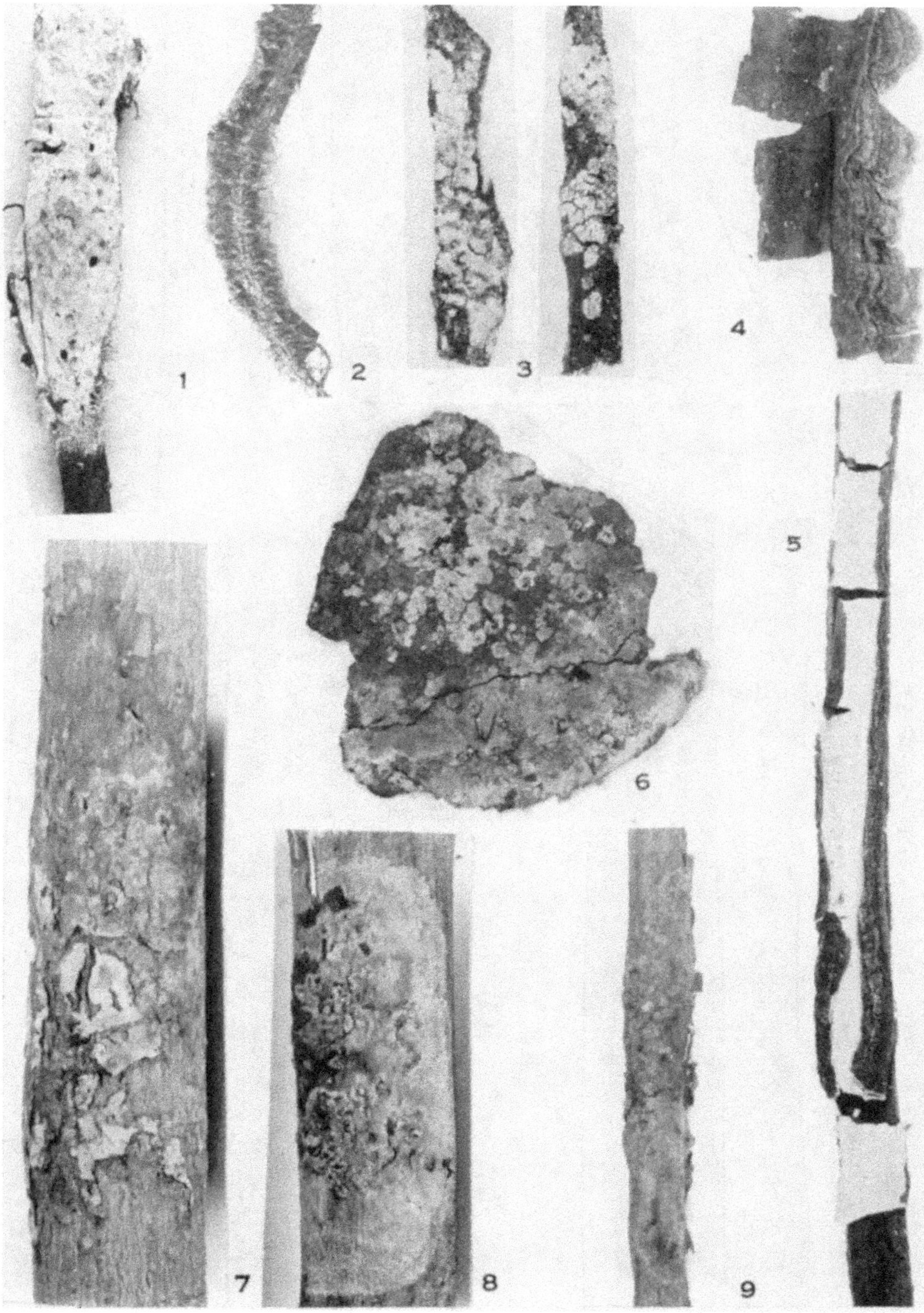

Figs. 1, 2, *S. crinitum*, type locality, n. s. 3, *S. cirratum*, type, n. s. 4, 5, *S. septobasidioides*, type, × 2. 6–9, *S. lichenicolum*; 6, type; 7, 8, from Petch on mango; 9, from Boedijn and Steinmann as *S. cinchonae*; all n. s.

PLATE 109

Figs. 1, 2, *S. albidum*, type; 1, × 1.8; 2, part of same, × 4.5. 3, 4, *S. albidum*, type locality, × 1.8. 5, 6, *S. philippinense*; 5, on *Derris* sp., × 2; 6, on *Citrus*, × 2, type. 7, *S. verrucosum*, type, × 1.25. 8, *S. Henningsii* (as *S. tjibodense* B. & S.) 9, *S.* sp. near *albidum* from Philippines, n. s. 10, *S.* sp. near *albidum* from China, n. s.

PLATE 110

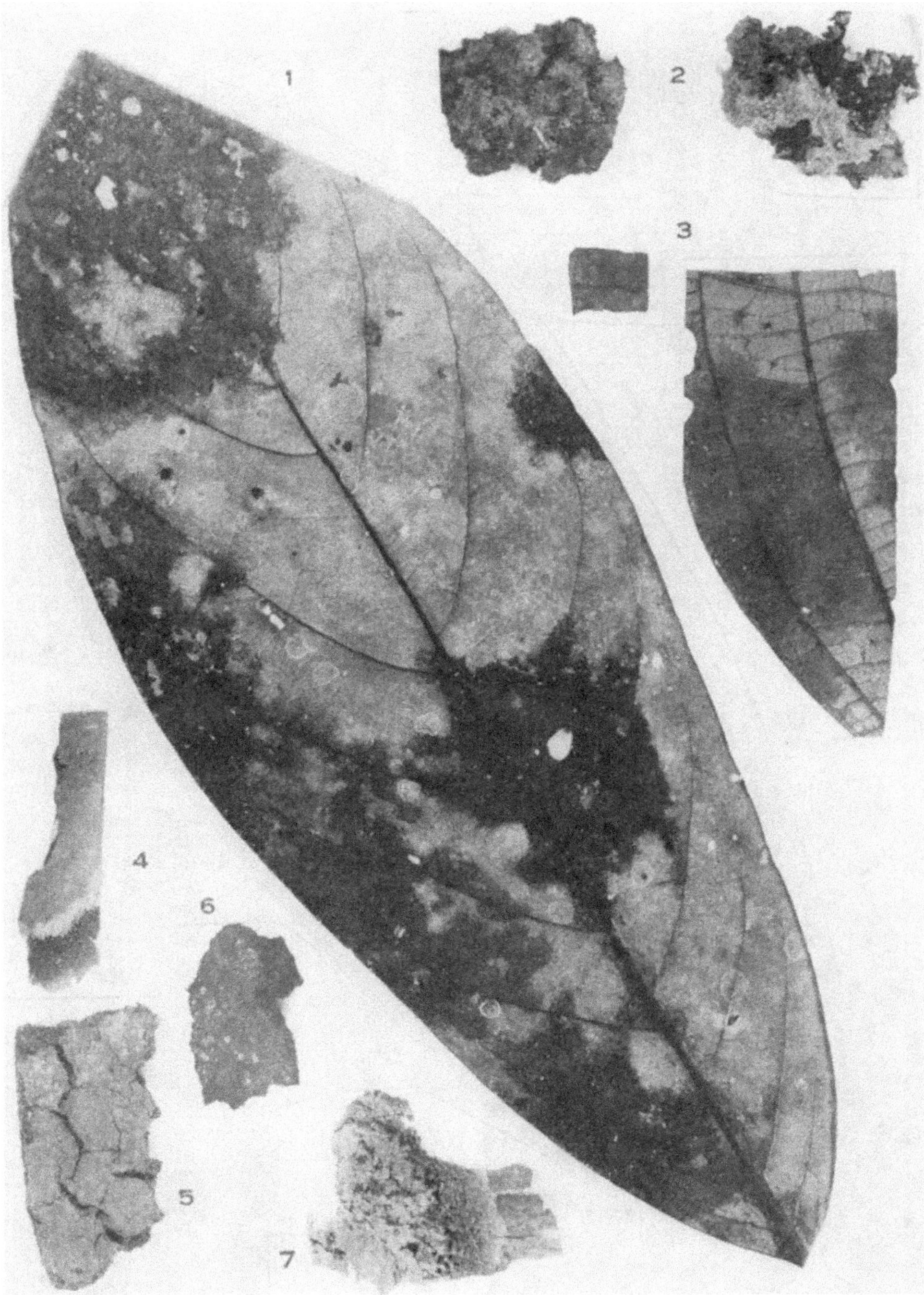

Fig. 1, *S. molliusculum*, type, n. s. 2, *S. carbonaceum*, type, n. s. 3, *S. makilingianum*, type, n. s. 4-6, *S. Bakeri*, note pillars on margin, cracked surface of older part and white scales of insects on lower surface of fungus. 7, *S. Mariani* var. *japonicum*, type, slightly enlarged.

PLATE 111

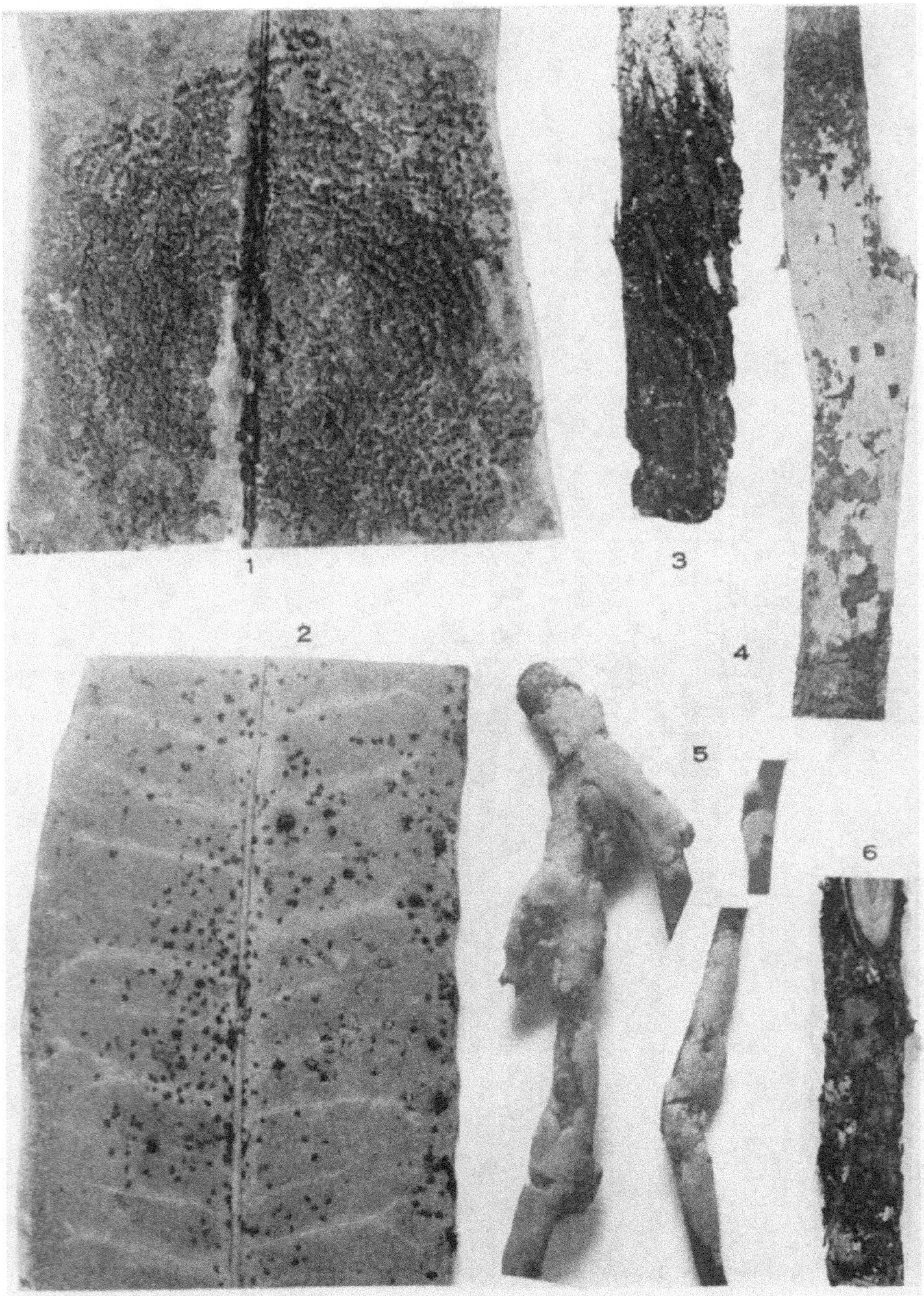

Figs. 1, 2, *S. scabiosum*; 1, type, × 1.3. 3, *S. spiniferum*, type, × 3.2. 4, *S. myrsinae*, type, × 2. 5, *S. Simmondsii*, type, × 1.5. 6, *S. robustum*, type, n. s.

PLATE 112

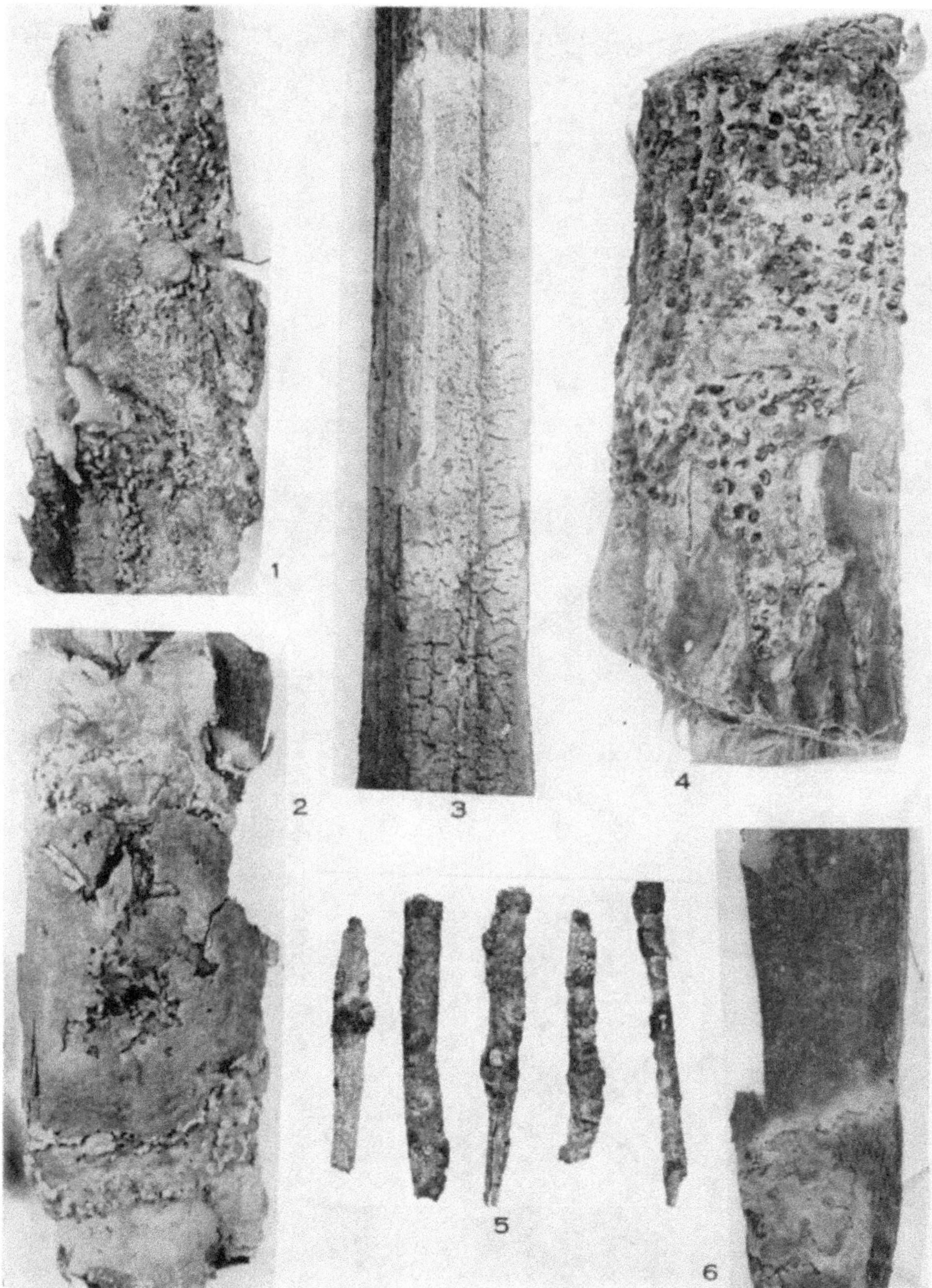

Figs. 1, 2, *S. Rickii*, type, × 1.6. 3, *S. perforatum*, type, × 1.6. 4, *S. lacunosum*, type, note holes each partly covered by flap of fungal tissue, × 3.3. 5, *S. sublilacinum*, type, n. s, 6, *Septobasidium* sp. (labelled *S. Thwaitesii* in Pat. Herb.), × 2.

PLATE 113

Figs. 1–3, *S. pteruloides*, Montagne's figures. 4, *S. pteruloides*, coll. by Petch, Ceylon, × 2.2. 5, Margin of above showing minute ∧-shaped insect traps, × about 9. 6, *S. scopiforme*, type, × about 3. 7, *S. elatostemae*, type, n. s. 8, *S. paulense*, type, n. s. 9, 10, *S. accumbens*; 9, type; 10, from Tonkin; both n. s.

PLATE 114

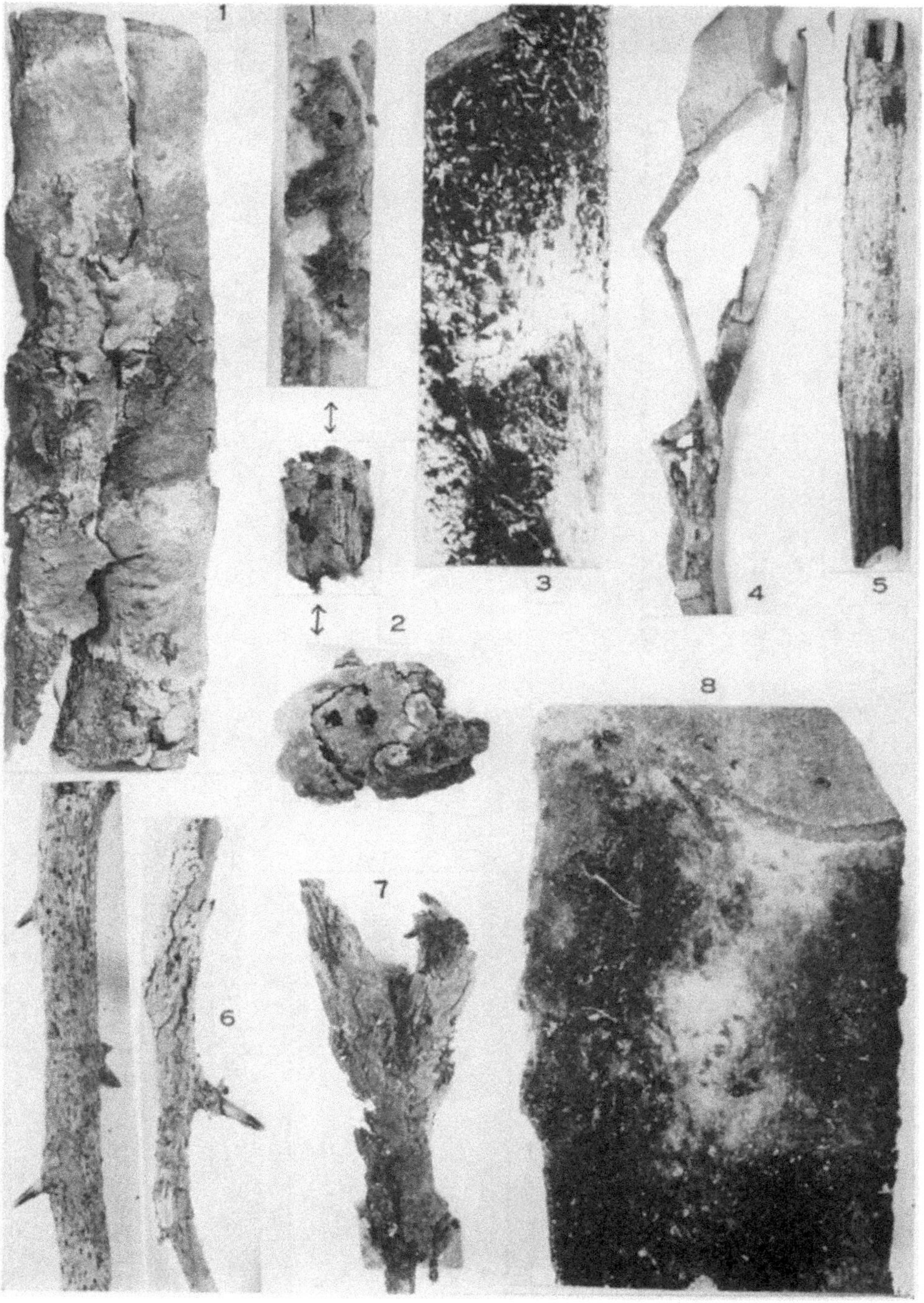

Fig. 1, *S. alni* var. *brasiliense*, type, n. s. 2, *S. nodulosum*, larger piece from Philippines; two smaller, type, × 2. 3, *S. alatum*, type, marginal region enlarged to show pillars, × 6. 4, *S. Carestianum* var. *natalense*, type, n. s. 5, *S. cinchonae*, type, × 2. 6, *S. punctatum*, type, × 2. 7, *S. ussanguense*, type, × 2. 8, *S. theae*, type, n. s.

INDEX

www.ingramcontent.com/pod-product-compliance
Lightning Source LLC
LaVergne TN
LVHW082026200926
848481LV00010B/2931